SECOND EDITION

CARTONS, CRATES and CORRUGATED BOARD

Handbook of Paper and Wood Packaging Technology

Diana Twede, Ph.D.,
School of Packaging, Michigan State University

Susan E. M. Selke, Ph.D.,
School of Packaging, Michigan State University

Donatien-Pascal Kamdem, Ph.D.,
School of Packaging, Michigan State University

David Shires, BA (Hons)
Smithers Pira

DEStech Publications, Inc.

Cartons, Crates and Corrugated Board, Second Edition

DEStech Publications, Inc.
439 North Duke Street
Lancaster, Pennsylvania 17602 U.S.A.

Printed in the United States of America
10 9 8 7 6 5 4 3 2 1

Main entry under title:
  Cartons, Crates and Corrugated Board: Handbook of Paper and Wood Packaging Technology, Second Edition

A DEStech Publications book
Bibliography: p.
Includes index p. 561

Library of Congress Control Number: 2014958357
ISBN No. 978-1-60595-135-5

HOW TO ORDER THIS BOOK

BY PHONE: 877-500-4337 or 717-290-1660, 9AM–5PM Eastern Time

BY FAX: 717-509-6100

BY MAIL: Order Department
DEStech Publications, Inc.
439 North Duke Street
Lancaster, PA 17602, U.S.A.

BY CREDIT CARD: American Express, VISA, MasterCard, Discover

BY WWW SITE: http://www.destechpub.com

This book is intended to be a textbook for students in graduate and under-graduate university packaging courses. It describes the properties, manufacturing processes, function, specification, and design of wood, paper, paperboard, and corrugated fiberboard packages, beginning with the trees. It reviews the history of wood and paper-based packaging, and it explores the opportunities and threats of the future. The book's description and discussion of commonly-used tests and terminology make it a useful reference for practicing packaging professionals as well.

The book began in the lecture notes of some legendary MSU professors: Gary Burgess, Don Abbott, Dennis Young, Ruben Hernandez, Bruce Harte, Tee Downes, Hugh Lockhart, Gunilla Jonson, Pete Rafael, and Jim Goff. It reflects the School of Packaging's creation in 1952 in the MSU Department of Forest Products. It draws from the current literature of our field and from the assistance of MSU packaging alumni and others in the packaging industry. All chapters in the first edition (2005) were thoroughly reviewed by leading practitioners who are acknowledged there.

Ten years later, this second edition has updated chapters, more science, and two new co-authors. Dr. Pascal Kamdem, a scholar of Forest Products, adds depth to our understanding of the science behind the applications of wood in packaging. David Shires adds an international perspective from his experience (consulting and testing) in the British packaging industry and as Editor-in-Chief of the international research journal, *Packaging Technology and Science*. Recent research is cited to provide a route for the reader who wants to learn more.

We have also incorporated feedback from previous students in the two MSU courses for which the textbook is used, *Packaging with Paper and Paperboard and Packaging Materials*, and from instructors in other packaging education programs who have used the first edition. We have added metric units where they seem appropriate. Thanks to everybody for the great recommendations. Likewise, we welcome your recommendations for continual improvement.

DIANA TWEDE, Ph.D.
SUSAN E.M. SELKE, Ph.D.
Michigan State University School of Packaging

CHAPTER 1

Introduction and Historical Background

The brown paper bag is the only thing civilized man has produced that does not seem out of place in nature. Crumpled into a wad of wrinkles, like the fossilized brain of a dryad; looking weathered; seeming slow and rough enough to be a product of natural evolution; its brownness the low-key brown of potato skin and peanut shell—dirty but pure; its kinship to tree (to knot and nest) unobscured by the cruel crush of industry; absorbing the elements like any other organic entity; blending with rock and vegetation as if it were a burrowing owl's doormat or a jack rabbit's underwear, a No. 8 Kraft paper bag lay discarded . . . and appeared to live where it lay.—Tom Robbins, *Even Cowgirls Get the Blues* (1976)

Packages made from paper, paperboard, corrugated fiberboard, and wood have a special relationship to life. Paper and wood hold the memory of life and offer the potential for recycling and resource renewal. They are transitory, like us, and their nature determines the unique properties of packages made from them.

But they are not all so plain as the brown paper bag. Paper and paperboard can also play well-dressed supporting roles in our economy. A distinctive feature is their ability to carry brilliant print. They contain, protect, and advertise all kinds of products, from cereal to electronic games. They have been responsible for the dazzling graphic display in retail stores for the past century, providing substrates for colorful brand and advertising communication on bags, cartons, and labels. The food pictured on a prepared food package can stimulate appetites and sales. The information can instruct and save lives.

The roles played by wood and corrugated fiberboard are less glamorous, but no less significant. The wooden pallet is a standard shipping platform, and corrugated fiberboard boxes are the shipping containers of choice for most

1

products in many supply chains in the world today. Efficient transport, material handling, storage, and inventory control depend on their strength, dimensions, and printed symbols.

Paper and wood are the basis for the most widely used packaging materials in the world. They are particularly significant in North America, with its abundant forests and well-developed paper industry, where they represent about half of the weight (and one-third of the value) of all packaging materials used.

This book describes the properties of wood, paper, paperboard, and corrugated fiberboard, and explores their use as packaging materials. It describes the manufacturing process for materials and packages. It discusses commonly used tests, terminology, and design principles.

This introductory chapter provides some historical background and current statistics. Paper and wood are significant materials in the history of packaging. They have played a key role in the development of the United States and its commerce, and commercial growth, in turn, has stimulated more packaging developments.

The historical background that follows is useful for two reasons. First, it helps us to understand the industry's vocabulary. Since the use of wood and paper is so old, many processes and specifications have a historical basis. For example, the designations for corrugated board flute grades A, B, and C have no relationship to their order of size, but are related to the order in which they were introduced. On the other hand, names like "pinch bottom open mouth" have a very specific meaning to the bag industry, and names for common measures like "ton" derive from historical names, and it is useful to know why. Understanding the source of our wood and paper vocabulary helps to remember what everything is named.

But the most important reason to study our packaging history is to remember that we are heirs to a craft and industry that continues to change. Package materials and forms come and go based on available technology, resources, and demand.

For most of humankind's history, some packages have been made from wood and fiber. There have been marvelous and useful innovations. Baskets were invented over 12,000 years ago. Paper was invented only 2,000 years ago (and for half of that time the process was a secret known only in Asia). The use of wood fibers is even newer, only 150 years. We've used corrugated fiberboard for about 140 years, and paper/plastic laminations for only the past 60 years.

Paper and wood are used in packaging in higher volume today than ever before, and will continue to play a role in the future, because trees are a plentiful, natural, renewable, biodegradable resource.

What will the future hold for wood- and paper-based packaging? You, the packagers of the future, will be the ones to decide. This textbook shows how paper and wood will continue to be part of that future.

1.1 WOOD PACKAGING HISTORY

Wood and woody fibers are among the oldest packaging materials. Early humans used hollowed-out pieces of wood and gourds, as well as reeds, leaves, and clay as packaging materials. Soon the use of wood and fibers extended to the weaving of baskets.

1.1.1 Baskets

Older than the weaving of cloth and more ancient than the early ceramics, the interlacing of twigs into baskets first occurred during the Neolithic period, contemporary with the first arrowheads (Bobart 1936). Early men and women used baskets for gathering food, carrying goods, and for ceremonial and decorative purposes.

Evidence of early basket use can be found in all parts of the world. However, very little ancient basketry has survived because of its tendency to biodegrade. Baskets as artifacts have been found primarily in arid places like Egypt (dated 10,000–8,000 B.C.) and the North American Southwest (about 9,250 B.C.). But there is plenty of evidence from secondary sources. The first Sumerian priest-king of Babylon in 3,000 B.C. was depicted carrying a basket on his head. Early "corn mother" harvest festivals from Egypt to Peru to Borneo employed the basket symbol.

Baskets have much in common with other packages throughout history. They were made from the least costly materials available. Whatever suitable fibrous material was at hand could be used, ranging from parts of a tree— twigs, split wood strips, leaves, palms, bark, roots, and willows (the Roman osier)—to grasses, reeds, straw, and bamboo. Furthermore, the techniques of *twining, coiling,* and *plaiting* were well-developed; basket-work was used for other household and functional items.

Like other packages, form followed function. Basketry's flexible construction methods invited variations in shapes and styles. Creative basket makers developed many designs that were cleverly adapted to each use.

Most of the packaging functions served by early baskets were related to food. Hunters and gatherers used baskets for collecting the food and for carrying leftovers to the next nomadic location. The transition to crop cultivation was accompanied by the development of special baskets for planting seeds, harvesting, winnowing, sifting, and storage, with different shapes for each activity. For example, harvest baskets must be large; winnowing baskets are shallow; and storage baskets are tightly woven with lids.

Baskets played an important role in distribution from the very earliest trading and markets (Sentance 2001). Traders carried goods to market in baskets designed to be lightweight and ergonomic. Some were designed to be bal-

FIGURE 1.1. *Basket designed for back packing, South Vietnam circa 1950. From the collections of the Michigan State University Museum.*

anced atop a person's head, shallow with a low center of gravity. Others were designed to be carried on a person's back; in the interest of stability they were longer than they were wide, and they had straps that were supported by the shoulders, chest, or forehead (see Figure 1.1). Pannier baskets carried by a beast of burden usually came in a pair, one slung over each side. The first "crates" used for shipping were wicker hampers (the word, like "cradle," derived from the Roman *cratis*). Glass and ceramic vessels were sometimes protected by basketry.

Baskets used for display in markets were (and still are) shallow to allow a clear view of the produce. Baskets used by shoppers to carry home the goods had/have handles that can be carried with a hand or by an arm, as shown in Figure 1.2.

Asian packagers added other woven forms to the functional theme. There were many clever shapes of intertwined wrappings, specific to the product and the materials at hand. Traditional Japanese packaging even had the reputation for endowing its contents with symbolic meaning. For example, a holder for five eggs made from intertwined wisps of rice straw ingeniously uses a farmer's natural resource to form a strong and functional package that also enhances the feeling of the freshness and warmth of newly laid eggs (Oka 1975).

A notable basket-like open crate was traditionally used in England for shipping pottery. Made from *withies*, a name used for willows, the crates were fashioned with *round-stick joinery*, twisted into an interlocking open framework with upright hazelwood rods and horizontal branches ranging from one

FIGURE 1.2. *Early Southeastern American wicker basket designed for shopping. From the collections of the Michigan State University Museum.*

to two inches thick. The pottery was protected by straw that had been stuffed in and around it. The crates were strong and resilient, and the structure gave them the ability to absorb shocks. Making them was a craft that required 5–7 years of training. At one time there were 40 firms, employing about 350 journeymen and 150 apprentices in Staffordshire (Castle 1962). The withie crate is just one example of the myriad creative basket-like package forms that have been crafted throughout history, and an example of a package whose time has come and gone.

Baskets are still used in some supply chains that are short and local, where the baskets are re-used. The familiar *bushel basket* made from splints of wood is still widely used to harvest and bring fresh produce to farmers' markets in the United States, and has long been used as a unit of volumetric measure for marketing purposes. In many parts of the world, baskets are still used for distribution of fresh food ranging from melons to live chickens.

1.1.2 Barrels (Casks)

Wooden barrels made from *staves* were invented during the time of the Romans and have been used for over 2,000 years. Barrels were arguably the most significant shipping container form in history.

In the United Kingdom, *cask* is the general term and *barrel* refers to a specific size and shape. In the United States, the word *barrel* is used to refer to all cylindrical wooden containers with a bulging bilge. Both terms are used interchangeably in this book. (*Keg* refers to a specific size of barrel. The word *drum* is used for straight-sided containers made from steel, plastic, or fiberboard.)

Barrel-making may have developed concurrently with ship building technology. Materials, modes of construction, and tools are similar: thin *staves* of wood, bent to curved shapes in the presence of heat and water, are precisely fitted and bound together (Elkington 1933).

The shape was ideal for the logistical systems in which they were shipped. The bilge shape makes it easy to roll with little friction and easy to pivot into position; a single person can handle much more than his or her own weight. They can be transported on a simple wagon chassis, and they are easy to load into a ship using simple fixtures for rolling or lifting. They also fit nicely into ships' holds in the *bilge* and *cantline* arrangement shown in Figure 1.3. The shape is very strong, and uses the properties of the wood to their best advantage.

For the Romans, casks served as substitutes for an older type of shipping container, clay transport amphoras (Twede 2002). They were used to ship liquid products like wine and olive oil, and other goods from the northern parts of the Empire where wood was a more abundant natural resource than clay. Casks were more convenient, lightweight, easy to transport, and not subject to breakage like pottery.

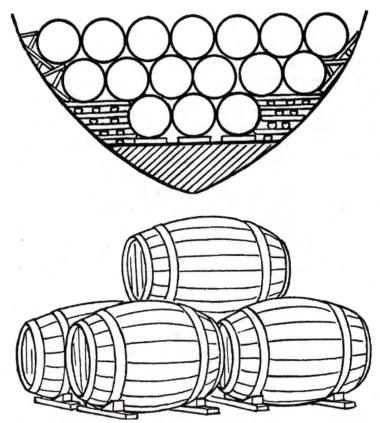

FIGURE 1.3. *Bilge and cantline stacking pattern; stowage aboard ship. Reprinted with permission of Cornell Maritime Press, from Garoche,* Stowage, Handling and Transport of Ship Cargos *(1944).*

By the time of the Crusades, the cask was the primary European shipping container form. Casks carried supplies for the knights seeking the Holy Grail. In the Middle Ages and the Renaissance, the most common cargo, especially in the Mediterranean where most sea trade was conducted, was wine in casks called *tuns*, which became a standardized unit of volume measure for the purpose of import tax collection. (The derivative *ton* measure has, over time and various circumstances, indicated either weight or volume. A ship's *tonnage* was originally the number of tun-sized barrels that it could carry.)

A barrel-maker is called a *cooper*, and in the middle ages, coopers' *gilds* (also spelled *guilds*) formed all over Europe. Such medieval gilds were the first trade unions, and oversaw the training and apprenticeship of new tradesmen for all types of production. The Worshipful Coopers Company of the City of

London was a typical minor livery "company." It and its Gildhall still exist as a social club, but today has little connection to casks.

The Worshipful Coopers Company of the City of London worked with the government to standardize casks and influence the market (not unlike today's packaging trade associations). A 1420 city ordinance required every cooper to brand his barrels with a distinctive mark registered at the Gildhall, an early form of packaging quality assurance. The Gild instituted an inspection and certification regime to combat fraud and imports. In 1501 a Royal Charter granted by Henry VII incorporated the Coopers as a fraternity-in-Gild forever. It was most active in the 1500s and 1600s, when it lobbied to raise the government's fixed price on casks, fought vertical integration whereby brewers tried to employ more than two of their own coopers (the legal limit), and supported the rebellion against the King by the Lord Mayor of London and Parliament. They standardized the hogshead at 52.5 gallons and wine cask at 31.5 gallons under the authority of Richard III, and under Henry VIII they set standards for beer barrels. They also set standards for so-called *slack* barrels that contained dry products like gunpowder and soap, based on weight (Elkington 1933; Foster 1944).

Some gild traditions, like commemorating the completion of an apprenticeship, continued into the twentieth century. The education and apprenticeship of these coopers was as comprehensive in its way as our modern packaging education. The completion of an apprenticeship was celebrated in a ceremony called the *Trusso*, which bears a vague similarity to some of our own zany graduation rituals. As a final exam, the apprentice was required to make a cask; the process was graded by a Master Cooper, a member of the coopers' union and a gild member. If the apprentice "passed" the test, he was placed inside the still-warm barrel, doused with water, shavings and ashes poured on his head, and then the barrel was rolled down the street with the cooper still inside. The procession ends at a party where he "dispenses hospitality" and receives his certificate of service (Elkington 1993; Hankerson 1947; Gilding 1971), as shown in Figure 1.4.

The history of the London coopers gild illustrates the widespread nature of the trade. There were coopers dispersed throughout the city and suburbs, usually making barrels for a single product like beer, wine, flour, or fish. There were coopers working throughout supply chains, repairing (*recoopering*) casks whenever they leaked. Ship's coopers kept the cargo in sound condition; dock coopers supervised the quality of the cargo as it was unloaded, and bond coopers inspected cargo stored in customs warehouses and vaults. "Sound condition" was an apt description, because a cooper could tell, just by the sound of a mallet tap, whether a cask was full (Gilding 1971).

By the 1600s, princes along the Rhine were competing for who had the biggest, most ornately carved barrel (filled with wine "tax" from his peasants).

FIGURE 1.4. *Master Cooper initiation ceremony and Trusso or David J. French in 1955. Photographs courtesy of Mrs. Janice Freeman.*

The largest and most famous was in Heidelberg, filled for the first time in 1752 with 50,000 gallons of wine (Hankerson 1947). The *Heidelberg Tun* (which still exists) was so famous that over 100 years later writers like Herman Melville and Mark Twain could still refer to it without needing to explain.

In America, coopers were likewise hard at work, although they were less well organized. The colonies were supplied with goods packed almost entirely in barrels. The Nina, Pinta, and Mayflower carried casks of water, biscuits, meal, meat, vegetables, and gin. The Mayflower hired America's first cooper (John Alden) to come with them and brought cooperage tools so that the pilgrims could make the essential containers themselves.

Colonization in America stimulated demand. The colonists needed fish, meat, sugar, tobacco, shoes, and hardware. More colonists came, and made products like soap, butter, candles, cider, and syrups. Each new product increased the demand for barrels and coopers.

Barrels were especially necessary for American exports. The first ships returning to England carried barrels of salted codfish. American barrels carried two-thirds of the high volume triangle trade, bringing molasses from the Caribbean islands to Boston and Newport where it was converted into rum that, in turn, was shipped in barrels to England and other destinations. Exports of rice and Virginia tobacco stimulated the earliest demand for *slack barrels*. The heavily forested American colonies even exported staves to be fabricated into barrels elsewhere, in England, Ireland, Spain, and Portugal (Coyne 1940).

Whale oil, used for candles and lamps, was another early American commodity. The whales were "processed" and barrels were filled on board. Whaling ships had at least one cooper to oversee the operation (Howard 1996). Melville's description makes vivid the dramatic role (roll) of the casks:

> *While still warm, the oil, like hot punch, is received into the six-barrel casks; and while, perhaps, the ship is pitching and rolling this way and that in the midnight sea, the enormous casks are slewed round and headed over, end for end, and sometimes perilously scoot across the slippery deck, like so many land slides, till at last manhandled and stayed in their course; and all around the hoops, rap, rap, go as many hammers as can play on them, for now* ex officio, *every sailor is a cooper.*—Melville (1851)

Barrels supplied the American military, from George Washington's band of rebels through the War of 1812, the Civil War, and the two World Wars. They appeared in pirate booty and Niagara Falls stunts. They filled the westward wagon trains with water, nails, china, food, oil, gunpowder, and beer.

Coopering remained a hand operation until 1837 when machinery was first developed for cutting and dressing staves. Economies of scale improved even more when the machinery for assembling them was developed over the next 30 years. The discovery of oil in Pennsylvania and New York (and later in Ohio, West Virginia, California, and finally, Texas) created a huge demand for oil barrels and mechanization. Large cooper shops sprang up in the oil regions to get the black gold to refineries and then to market. New and rapidly growing industries like meat packing, brewing, distilling, flour milling, and corn products stimulated more growth (Coyne 1940).

Mechanization, however, produced barrels of inferior quality. Unlike the coopers who matched their staves from the cylindrical shape of the tree, the power saws cut straight lines that could not be so effectively shaped. The ability to make barrels that were leak-proof was somewhat improved with glue and other lining materials like silicate of soda (for oil-based products) and paraffin (for water-based products).

Furthermore, the barrels' geometry, and the fact that they had to be shipped by rail standing up, did not fit into or onto railroad cars as well as they had fit into ships where they lay on side, packed tight like fruits. About 1870, the first tank cars (wooden tanks on flat cars) for oil were developed to replace leaky barrels. Other substitutes like steel drums followed, but not right away.

Even flour was still packed in 196-pound barrels in Minneapolis mills up until the early 1900s, at a time when wood and woodworkers were more plentiful than textiles for bags. Henry Chase and John Batchelder developed a machine to sew bags in 1849 (based on Elias Howe's first sewing machine in

1846). But in 1880, barrels still outnumbered bags by ten to one. Bags were most successful in smaller sizes like 24-1/2 pounds, which is one-eighth barrel, cross-referencing standard barrel quantities even when there was none (Steen 1963).

By the early 1900s, barrels began to lose market share to consumer packages (Twede 2005). The cracker barrel was soon seen as old-fashioned, a symbol of a bygone era. The famous water-resistant Uneeda biscuit package symbolized the birth of the consumer packaging industry, and most retail merchandise quickly moved into consumer packages, as described in Section 1.2.3.2.

Oak barrels are still used for aging wine and whisky, and as such are valued for the nuances of flavor they impart. The principles of their construction are described in Chapter 4. But in the United States, most liquid products have long ago moved to alternative semi-bulk shipping container forms, such as steel and plastic drums and tanks, as technology became available. Most dry granular products moved to bags. And for shipping everything else, the better alternative was boxes.

1.1.3 Wooden Boxes to Pallets

There have probably been wooden boxes for as long as there have been carpenters. The shape is simple and easy to form. *Chests* were used for personal belongings, as people traveled, and later as furniture at home (like the so-called chest of drawers). But the use of wooden boxes as routine shipping containers is relatively recent, with a notable exception: tea chests.

The first widespread use of wooden boxes was for tea, shipped from China by the East India trading companies beginning in the late 1600s. Tea chests were made from thin planed wood, and later were the earliest use of plywood. They were lined with lead (later aluminum) to keep the tea free from damp and foreign odors, with a paper barrier between the lead and the tea. For the better teas, painters "adorned the exterior with grotesque flowers and fanciful devices." The chests were filled and coolies (Chinese workers) stamped down the tea with their feet to compress it and maximize the cube utilization. Finally the chests were sewn up in rough matting and secured with rattan. Every chest had two paper labels, one inside the burlap and the other outside, which named the ship it was to travel on (Goodwin 1991).

Weights varied, but by the mid-1800s, the standard tea chest was reduced to 100 pounds, a weight that was capable of being manually handled. They were dimensioned to perfectly fit the hold of clipper ships, as shown in Figure 1.5. A pair of workers loaded the ship tightly, chests chock-a-block, from the outer hatch wings inwards, hammering in the final center chest in each layer as if it were a keystone of an arch, with a wooden mallet. This maximized the amount of cargo and minimized the amount of shifting in transit as the ship swooped and shuddered under pressure of the sea and sail. Although tea is mostly now shipped in bags, standard specifications for tea chests are still used; now they are sized to fit pallets and 40-foot ocean containers (Forrest 1973).

"The ubiquitous tea chests" evolved into the symbol of Britain's wealth and expanded trade influence in the East. Millions of the original "Useful Box" were sent to tea-crazy England over 300 years, where they have a reputation for being reused for moving households and for storage. "The attics of England creak beneath their collective weight," (Goodwin 1991).

The chests were also sent to the colonists in America, who found their symbolism to be more useful than their function. For the Boston Tea Party crowd who dumped them overboard in 1773, the chests symbolized taxation without representation, and the collective act of destroying them with tomahawks and dumping them overboard was the catalyst for the American Revolution.

For other products, widespread use of wooden boxes did not become common until the Industrial Revolution, coinciding with the development of saw-mills, railroads, and an increase in trade (Marquis 1943). Earlier, wooden boxes had not been widely used as shipping containers because of the high cost to transport and handle them. They were heavy (especially when loaded with something more dense than tea) and awkward to handle. Trans-loading boxes

FIGURE 1.5. Tea chests in the hold of a clipper ship. Reprinted with permission of Conway Maritime Press, from MacGregor, The Tea Clippers (1972), 17.

from one transport conveyance to another is difficult, especially compared to cylindrical casks which could be rolled.

This is a classic case of how geometry relates to the history of transport and packaging technology. The advent of railroad transport created concentrated hubs of material handling and therefore the opportunity to mechanize it with the use of hoists and wheeled conveyances. This made it possible for economical wooden boxes to displace barrels for many dry commodities and manufactured goods shipped by rail. At the same time, *boxcars* were developed to ship the boxes, creating the squared-off world of distribution that now dominates in all transport modes. Plaskett (1930) calls wooden boxes "the first modern shipping container."

Box making in the United States was at first a completely local industry, utilizing readily available local timber made for local users. A mutually beneficial relationship developed between the box and railroad industries, since the railroads had been granted much timber land that could be made into boxes. Sawmills, first established for making railroad ties, later specialized in making boxes or planing wood for other box factories.

Boxes are much more economical to make than cylindrical barrels, and the process was more successfully mechanized. The introduction of the *band re-'saw*, which slices thin boards, made it possible to produce box parts known as *shooks*, to be assembled in the user's *box room*.

Productivity was improved by using power planers, matching machines, squeezing machines for tongue and groove work, nailing machines, and stapling machines. The size and number of box factories increased steadily between 1860 and 1910. At the same time, demand rapidly increased for rail shipments, creating a tremendous stimulus to the box industry during the closing years of the 1800s (Marquis 1943).

In the United States, the box and railroad industry trade associations (National Association of Wooden Box Manufacturers and Association of American Railroads) worked together with the USDA's Forest Products Laboratory (which was developed by the industry in 1910) to establish standards for wooden box construction, and later for corrugated boxes. The standards were intended to minimize damage in transit.

In the history of shipping containers, compared to amphoras, baskets, and barrels, the American wooden box had a relatively short life cycle. They were most widely used as shipping containers from the mid-1800s to the 1920s when they were displaced by the newly invented corrugated and solid fiberboard shipping containers.

Corrugated fiberboard boxes quickly gained markets because they used less material, were lighter weight, and were more economical to ship in a knocked-down fashion to the packer-filler. The decision of the U.S. Interstate Commerce Commission to permit their use in 1914, coupled with their low cost and the growth of motor carrier transport, signaled the end of the predominance of the hardy wooden box (Howell 1940).

Wirebound boxes made from thin veneer plywood were an intermediate solution for some products such as plumbing fixtures and iced produce. They offered some of the same advantages as corrugated fiberboard boxes since they could be mass-produced and shipped knocked down, but they were sturdier. The industry developed other hybrid packages made from plywood with cleats and hinged corners. Some of these are still used, and are described in Chapter 4.

The wooden box and barrel industries in America slowly faded for all but military and other specialized extreme uses. The industrialization that had so successfully mechanized and increased the production of wooden boxes, in the end created their strongest competition: the totally mechanized production of corrugated fiberboard shipping containers. Some of the uses that persisted the longest were for regional distribution of farm commodities such as eggs and bottles of milk, soft drinks, and beer, where the boxes were returned and reused.

Fresh fruits and vegetables were still shipped in wooden boxes into the 1950s, and produce boxes had regional standards: for example apple boxes were 1-1/5 bushel for soft varieties in New England and 1-1/8 bushel in other parts of the East.

Packing for export has been another persistent application. Wooden boxes predominated in ocean shipments until the 1970s when the use of containerization (e.g., 40-foot ocean containers) became commonplace. For non-containerized shipments, wooden boxes are often still required in order to withstand the rigors of break-bulk ship loading and stowage. Wooden boxes, especially cleated plywood and wooden crates, are still used for heavy and large products and in "extreme" applications such as military shipments of equipment and supplies. Crates (which differ from boxes because they are only frames with open sides) were used until the 1950s for appliances, and are still used for some large equipment. Their construction is described in Chapter 4.

But even as World War II signaled the demise of the wooden box and barrel industries, wooden packaging was simultaneously reborn with the concept of unit loads. Wooden pallets were key for the logistics needed to provision the two war fronts (Europe and the Pacific) simultaneously.

Prior to the 1930s when forklifts were invented, shipping containers were all handled manually or with the use of small wheeled hand trucks. The invention of forklifts and the creation of pallets and skids for unit loads revolutionized material handling. During WWII, the U.S. military bought more than 50 million wooden pallets and skids to move the materials of war (LeBlanc and Richardson 2003).

In the decades after WWII, forklifts mechanized material handling throughout the commercial world, creating a large demand for wood pallets. Most of the material handling in the world today takes place with the use of pallets. Once again the geometry of transport and packaging changed, this time to focus on optimizing the cube of palletloads.

Because of the demand for pallets, there is more wood used for packaging today than ever before. Even with increasing competition from plastic, pallets are still the highest-volume category of wood packages in use because of their low cost, durability, and low-tech ease of repair. The principles of their construction are described in Chapter 5.

Wood packaging was also responsible for the creation of the first university degree-granting program in packaging, which in turn is responsible for the writing of this book. When Korean War boxes were found to be deficient, Michigan State College introduced a program in Packaging in the Department of Forest Products in 1952, to provide education and research in the performance of wood and corrugated fiberboard boxes. That idea grew into the School of Packaging at Michigan State University, the leading Packaging degree-granting program in the United States (and this story explains why our program is in MSU's College of Agriculture and Natural Resources).

But even as the types and uses of wood packaging have declined, its valuable properties and favorable economics have not. Wood's strength and renewable nature are also responsible for the success of the paper-based packaging forms discussed in Chapters 11–15.

1.2 PAPER-BASED PACKAGING HISTORY

The most significant role of wood in packaging today is the use of its fibers for the raw material for paper and paperboard. But wood fibers have been used for only a short part of the 2000-year history of papermaking.[1]

A persistent theme throughout history is that packaging is made from the lowest cost materials. But most paper was, for a long time, too expensive to use for packaging. It was hand-made from rags for 17 centuries after its invention. Paper use was hindered by the laborious process of papermaking and the limited supply of raw materials until the invention of processes for pulping straw, wood, and waste paper in the mid-1800s. The ability to make paper from abundant natural resources has made it the inexpensive commodity with the wide range of packaging uses that it is today.

The earliest paper-like material used for packaging was *papyrus*, used as a

---

[1]The history of papermaking is well documented, especially with respect to printers' paper. Unless otherwise noted, most of the facts in the papermaking sections are found in Hunter 1947, Hills 1988, Weeks 1916, and Bettendorf 1946, although these are not referred to throughout the text. It should also be noted that in some cases reported dates vary by a couple of years from each other since inventions and new equipment installations take place over time, and only patent issue dates are documented.

writing material by the Egyptians, Romans, and Greeks from 3000 B.C. Papyrus is made by laying parallel thin strips from the heart of the tall highly fibrous papyrus plant, with a second ply laid perpendicular to the first. The sheet is pressed flat and/or pounded, removing the water and crushing the plant cells, liberating a natural gum adhesive that glues the strips together.

The lowest grades and recycled papyrus were used for packaging. The Romans recycled papyri for wrapping incense, a clever marketing move since the papyrus itself smells nice, especially when burned. They used a cheap grade of papyrus, called *charta emporetica*, for wrapping merchandise. Pliny, writing in the first century A.D., tells how it was made from the lower quality outer layer of the plant, useless for writing, and was sold in sheets 4.3 inches wide, half as wide as finer writing paper (Lewis 1974).

Later, the Mayans and Aztecs developed a similar non-pulped sheet called *amatl*, from the bast (inner bark) layer of ficus and mulberry trees. The bast was peeled away in strips, soaked to soften, beaten with a rough stone that macerates the fibers against a smooth flat surface that stabilizes the sheet, and then left to dry in the sun. There is no indication that amatl was used for packaging as we would recognize it, but it was used in ceremonies, including wrapping sacrificial victims (Von Hagen and Wolfgang 1944), an unusual manifestation of "disposable packaging."

Although our word paper comes from the Greek word for papyrus, neither it nor amatl are true paper in the sense of pulping and blending fibers. But the development of all three was motivated by the same need for a writing substrate in cultures where the need for documentation was growing.

Paper as we know it originated in China in 105 A.D.. The earliest process known is attributed to Ts'ai Lun. It was invented as a calligraphy substrate, as a substitute for fabric and bamboo strips. Calligraphy was one of the highest forms of Asian art and scholarship, and it led the craft of papermaking to grow throughout China, Japan, and Korea over the next several hundred years.

Fundamentally, Ts'ai Lun's method is the same as that used for making paper today. He shredded the inner bark of mulberry trees, softened it with lime, mixed it with scraps of linen and hemp, and then beat it into a pulp. Mulberry was plentiful in China, cultivated as the food source for silkworms, and the inner bark was porous and easy to pulp. Water was added to the pulp, and a sheet mold with a woven floor was dipped into the mixture. The water was allowed to drain through the porous cloth, leaving a mat of fibers. The molded paper was dried in the sun, and then removed from the cloth.

It took several hundred years for the ancient Chinese art of papermaking to reach the West. The reputation of Chinese paper was widespread, but the technique was not. In 750, Chinese knowledge about papermaking reached North Africa from prisoners taken at the battle of Samarkand. The first recorded use of paper for packaging was in 1035, when a Persian traveler visiting markets in Cairo noted that vegetables, spices, and hardware were wrapped for the customer after they were sold (Hunter 1947).

The Moors introduced papermaking to Spain and Italy, and the first European paper mill was built in Xativia (on the Mediterranean coast near Valencia) in the 1100s. Papermaking reached Germany and France in the 1300s and 1400s. It came late to England where John Tate's mill in Hertfordshire was the first, built in the early 1490s.

Before that, documents in the West were parchment and vellum, made from the skins of sheep, goats, or calves that were treated with lime, scraped and stretched thin. When Gutenberg first printed his Bibles in 1450, parchment and vellum were still used for most manuscripts, and only a third of the Gutenberg Bibles were printed on rag paper. The skins of 300 sheep were needed to print one Bible.

Once paper and printing came to Western Europe, literacy followed. Paper dramatically reduced the cost of printed materials. In the 1500s, England developed an extensive paper industry, mostly using old linen (and later cotton) rags as raw material. Without paper (and printing), the works of authors like Chaucer and Shakespeare might not have existed and certainly would not have survived.

The first paper mill in the United States was built by William Rittenhouse near Philadelphia in 1690. In addition to paper for printing, the Rittenhouse mill later made brown and blue paper for wrapping (Rickards 2000). Benjamin Franklin, as a printer, publisher, and forward-thinking patriot, patronized and encouraged the building of mills, helping to start 18 of them in Pennsylvania and Virginia. Papermaking provided fuel for the American Revolution against Britain and its Stamp Act, a tax on all paper made in the Colonies.

Before the 1800s, the paper that was used for wrapping was low quality or had already been used for other purposes. For example, books that failed to sell in London in the 1600s were sold as wrapping paper to grocers and apothecaries. This change of use was facilitated by the fact that books were stocked in sheet form, folded but uncut and unbound until they were purchased (Davis 1967).

Paper was costly because of its raw material. All paper in Western Europe and the Americas was made from rags, which grew increasingly scarce compared to the demand generated by the growth of printing. The process was also costly; each sheet was made by hand. Figure 1.6 shows the *vatman* scooping up the pulp on the mold, the *coucher* depositing the wet sheets between sheets of wool felt for pressing, and the *layman* removing the paper from the felt.

The rags were pulped by first wetting them and allowing them to rot and partially disintegrate. Then they were beaten with a large mortar and pestle in a water-powered *stamping mill*, seen in the background of Figure 1.6. The invention in the 1700s (in Holland) of the *Hollander beater*, which revolves and cuts

FIGURE 1.6. *The handmade paper process showing the vatman, coucher, and layman. Reprinted with permission of Alfred A. Knopf (Hunter 1947).*

the fibers against a stone bed plate, improved productivity and eliminated the need (and risk) associated with partially rotting the fabric first.

There was an extensive collection system of *rag pickers*. A poem dating from the 1700s by an unknown author sums up the supply chain (Hunter 1943):

> *Rags make paper,*
> *Paper makes money,*
> *Money makes banks,*
> *Banks make loans,*
> *Loans make beggars,*
> *Beggars make*
> *Rags.*

The rag supply was never very stable. There were many other uses for recycled clothing, and until the process of weaving and sewing were mechanized, used clothing was relatively scarce. The supply was interrupted during the Black Plague by the order to burn rags in the belief that they spread the disease.

In America the situation was even worse. People wore their clothes until they wore out, and settlers had plenty of homespun uses for the scraps. Cotton was more prevalent than linen in America, and its rags made weaker pulp. Some coarse inferior papers, made from printed calicos, sack fabric, and worn-out sails and tarred rope were used for wrapping.

1.2.1 Papermaking Inventions in the 1800s

Three key technological inventions in the1800s set the stage for the mass production of paper-based packaging by the end of the century. The papermaking machine and the process for pulping straw, wastepaper, and wood provided the cheap materials. And lithographic printing dressed them up to match the advertising in the publications that inexpensive paper also made possible.

These inventions reduced the cost of printed paper enough to make it useful for a wide range of disposable packages. They also made possible a burst of creative and journalistic writing during the second half of the 1800s, enabling the widespread distribution of books, newspapers, and magazines (serials). These inventions came at the right time, just as demand for packaging began to grow due to the growth of manufacturing and railroad distribution of consumer goods. Paper made it possible for packaging to join the Industrial Revolution!

1.2.1.1 Papermaking Machines

The papermaking machine was invented first. In 1799, the earliest machine to make roll-stock paper in a continuous process was invented by Nicholas-Louis Robert in France. The first practical machine was based on Robert's plans and built in England, financed by Henry and Sealy Fourdrinier, two London stationers. An engineer, Bryan Donkin refined the design, added several patented features, and installed the first production machine at Frogmore Mill in Hertfordshire in 1804.

The machine deposited watery pulp onto a moving wire mesh belt mold, and a series of felts removed most of the water. It is the basis for the most common paper forming process used today, named after its investors: fourdrinier. The first American fourdrinier machine was built in England by Donkin, and was put into operation in 1827 at Saugerties, New York, in the mill of Henry Barclay (Donkin played several important roles in packaging history; he also built the world's first canning factory in the United Kingdom.)

The *cylinder* or *vat machine* was invented shortly thereafter, in 1809 by

John Dickinson. It operated on the principle of rolling a mesh cylinder mold through a pulp slurry, to scoop it up and deposit it onto a moving felt belt. The first cylinder machine was used in Dickinson's Hertfordshire, England, mill in 1809. In 1824, he also invented the first machine for pasting sheets of paper together to make "cardboard."

The first papermaking machine in America was a cylinder machine, installed in 1817 in the Thomas Gilpin mill on Brandywine Creek, near Wilmington, Delaware. But the great advantage of the cylinder former was not realized until the 1830s when George Shryock's mill near Chambersburg, Pennsylvania, found the first way to make thick straw-based paperboard on a machine, described in the following section. In 1870, Shryock was the first to successfully combine a number of the cylinder formers in series to produce multi-ply strawboard.

It took longer to invent a way for fourdrinier machines to make paperboard because of insufficient drainage capacity on the wire mesh belt for a thick sheet. Furthermore, the short broken straw and wastepaper fibers that were used for board do not drain as freely as do longer rag and (later) kraft fibers. The cylinder machine, with its many molds permitting drainage of each ply of the board through the wire of each cylinder, was more adaptable to the manufacture of board.

By 1850, the U.S. Commissioner of Patents reported, "The cylinder machine, more simple and less costly than the other, is in more general use; but the paper made on it is not equal in quality. Notwithstanding it does very well for news, and for the various purposes which a coarse article will answer for," (Weeks 1916). One of these various purposes was wrapping paper.

By the mid-1800s, most paper mills were equipped with fourdrinier or cylinder paper machines. Today, cylinder machines are used only to make paperboard. The fourdrinier and cylinder processes are still the most common ways to make paper and paperboard, and they are described in Chapter 7.

The continuous web produced by a papermaking machine (a reeling operation was added by Bryan Donkin in 1850) is key to making all modern paper-based packages. From roll stock, paper is converted into bags and labels, containerboard is converted into corrugated board, paperboard is converted into cartons, and all of these are printed in a continuous process—as the paper runs continuously through the converting machines on a roll, in the same way that it came off of the papermaking machine.

Mechanization made paper more plentiful and reduced its cost. An 1850 letter notes that the cost of machine-made paper was one-eighth the cost of hand-made. As the price fell, the demand for paper, including wrapping grades, began to rise.

As paper demand increased, so did the price of rags. They were valuable enough to import to America from abroad, especially considering that the foreign rags had a higher linen content. Before the Civil War, supply was so scarce that the Stanwood mill in Maine imported Egyptian mummies to America in 1855 in order to pulp the cloth wrappings (30–40 pounds each) and papyrus fillers. Stanwood had hoped to make high quality paper, but found that the linen was deeply stained with resin that could not be bleached out. Instead, he made a coarse brown paper that was, in turn, used by grocers and butchers for wrapping food. Recycling of mummy wrapping was not new; in 1140, a Baghdad writer had identified mummies as the source of fiber for paper destined for food markets.

As the 1800s progressed, the need for an alternative source of fiber intensified. Patents were granted for producing paper with fibers from cornstalks, hemp, jute, raw cotton, sugar cane, bamboo, peat, straw, and wastepaper. These produced coarse lower grades of paper, some of which did not have very good printing properties, but were nevertheless suitable for wrapping.

1.2.1.2 Straw, Jute, and Wastepaper Pulp

The ability to commercially pulp straw, jute, and wastepaper was developed earlier than wood pulping. The paper made from all three was used for packaging. This left rags and later wood pulp to predominate in the publication market.

In a 1797 pamphlet by Thomas Greaves addressed to the East India Company, importers, grocers, drapers, gunpowder makers and paper makers, jute was first proposed as a source of fibers for "useful paper for use of Grocers, Chemists, &c," (Hunter 1943). Jute is a highly fibrous plant used in India, twined to make rope and woven into burlap bags.

Jute paper, also known as *rope paper*, was made from recycled wastepaper reinforced by strong jute fibers that had been recycled from used rope and burlap bags.

Used rope and burlap were cheap in port cities. The tough, flexible paper came to be used for heavy wrapping paper and bags for seeds and cement (Smith 1935).

Even in the early 1900s, when jute fibers were replaced with long, strong kraft wood fibers, the names "juteboard" and "rope paper" persisted, coming to indicate it was made from reinforced wastepaper. For the next 50 years (up until the 1950s), corrugated board was made with kraft-reinforced juteboard facings.

But the most commercially successful pulp alternative was straw, the stalky by-product of rye, wheat, and oats. Straw had long been a useful material for purposes ranging from thatched roofing to bedding, and it was a plentiful natural resource. Its pulp made a coarse, inexpensive paper and board that was widely used in packaging for over 100 years. It is a good example of a packaging material that came—and went.

Strawpaper was first produced by Matthias Koops in 1800 (patented in 1802), based on experiments in 1765 by Dr. Jacob Christian Schäffer. Koops established the first straw paper mill in the United Kingdom. He also patented the first processes for pulping (and de-inking) wastepaper and wood, and published a book that was printed on papers made from straw, wood, and other fibers. Koops was a visionary who saw beyond the world of printing and fine papers to predict the use of straw and wood pulp paperboard as a strong building material. He also envisioned the recycling of paper into a useful material (Koops 1800).

The first commercial straw paper in the United States was made in 1831 by George Shryock's Chambersburg, Pennsylvania, Hollywell mill on a cylinder machine. The pulping method involved cooking the straw with potash, and later lime. The process was developed by Col. William Magaw, a potash manufacturer who noticed that it reduced straw to a pulpy mass.

Although it was first used for printing, primarily newspapers and decorative paper like wallpaper, straw paper was coarse, yellow in color, and was not worth bleaching to match the whiteness of rag paper for publications. The fibers were short, and so longer fibers were usually added to give strength. Some of the earliest straw paper was only 20% straw, mixed with rope and jute/burlap bagging, and later it was made from 100% straw or straw mixed with waste paper. Paper made from straw had a hard stiffness and "rattle," but low tensile strength and poor tearing resistance.

Strawpaper became a widely used packaging material. It was used for wrapping paper, butcher paper, and "packing paper." Twisted cones of straw paper used by grocers preceded the paper bag, and were common up to 1890.

But straw's greatest use was to be for box board and binder board (book covers). Shryock describes how he made the first strawboard by stacking up wet sheets and pressing them together:

> *I soon discovered that when the paper broke between the press roll and layboy it accumulated in (sometimes) six or eight lamina round the press roll, and formed a solid and beautiful binders' board.*

Shryock invented a special grooved roll that built up windings of wet paper until a board achieved the desired thickness. A sharp knife was used to cut through the groove, freeing the thick sheet of pulp. The sheets were then pressed and dried. Later, in 1870, the Shryock mill was the first to combine cylinder machines in series to make a multi-layer board (Quoted by Bettendorf 1946).

The natural stiffness of strawboard made it well suited for packaging, a use that persisted long after wood pulp was in common use for paper. Strawboard was used for common items such as setup boxes, egg-carton partitions, and stiffener backing for paper pads. It was used in the first folding cartons and corrugated board, and ultimately was used only for corrugating medium, a use that persisted until the 1950s.

For fifty years, four mills in Columbia County in New York were the leading producers, using rye straw that was abundant there, and cheap. As demand and the nation expanded, mills were built in the midwestern grain belt: Ohio, Indiana, and Illinois (Bettendorf 1946).

An on-site description of the Lafayette Box Board mill in Indiana illustrates the scale of the strawboard business in the early 1900s. In 1909 the plant was known as the largest single machine strawboard factory in the world with the largest output, selling strawboard as far east as Maine, south to Mexico, and north to Canada:

> *One machine is almost 300' long with 87 huge rollers each weighing 3,000 pounds, around which pulp is run to shape into a board to dry. This machine makes a sheet of strawboard 120 inches wide.*

The factory contained 10 hollow spheres or globes known as rotaries. Each measured 14 feet in diameter and held seven tons of pulp cooking in steam and lime water while the spheres revolved.

> *Here is daily produced marketable strawboard in all weights and sizes; mill-lined in all colors; sheet-lined in white and special colors; booklined for high grade candy boxes; black-lined for picture-making; manila-lined for folding suit boxes; double-lined for other requirements. Also heavy strawboard for trunk, case and file manufacturers.*

Indiana grain fields supplied the mill with 30,000 tons of straw per year. Every November at harvest time, baled straw arrived and was stored in sheds. The plant survives and is now a paper recycling plant, apparently a common fate for many straw mills (Kriebel 2003).

In England, straw was not much used for pulp. Rags were more readily available there and were competitive with the cost of the poorer quality paper made from straw. But during roughly the same period, 1860–1950, the British made paper from esparto grass, imported from North Africa and Spain. Esparto made a high quality printing paper, but was only economically feasible because it was backhauled in ships sent south with fuel for Britain's fleet. It does not appear to have been used in packaging (Hills 1998).

Most of the strawboard used in the United Kingdom for machine-made set-up boxes was made in Holland. *Dutch strawboard* was valued for its stiffness, but it had drawbacks: it had a propensity to absorb moisture that made it change dimensions and give off a musty smell that transferred to food wrapped in it. The waste-based *chipboard* made in British mills and used for folding cartons was not as stiff, but was more dimensionally reliable, ensuring that lids consistently fit their boxes. It led the British mills (about the time of WWII) to develop a substitute material, called *rigid board*, which combined the benefits of both Dutch strawboard and the home-produced chipboards (Paine 1991).

Straw was rarely used for pulp. It was more costly to collect and store than wood. But with the worldwide demand for paper rising, papermakers in countries like Denmark, Britain, and Japan with few forest resources were interested in reviving the practice (O'Brien 1990, Bower 1996).

1.2.1.3 Wood Pulp

It is surprising for most people to learn that there was no commercial wood pulp until the 1850s. Investigations had begun in 1719 when a French naturalist, René Antoine Ferchault de Réaumur, observed that the nest of the American wasp is made from fine white paper made from masticated wood. In the following hundred years, there were many experiments, including Koops' patent in 1802, but there was no commercial process until a grinding machine, which literally grinds the wood to a pulp, was patented by Friedrich Gottlob Keller in 1840 in Germany.

The first U.S. mill producing wood pulp for paper by the mechanical *groundwood* process was built in 1867. The quality was decidedly inferior to rag paper, and it was not accepted right away. Groundwood fibers are shorter, weaker, and darker than those from rags, and some proportion of longer rag fibers had to be added to the wood pulp to strengthen it (*rag content* is still used in the finest writing and book papers). The prejudice was largely overcome when the paper made from wood pulp was found to have good printing qualities and was particularly acceptable as newsprint.

The biggest advantage of groundwood pulp was its dramatically lower cost, especially in heavily forested North America where wood is a plentiful natural resource. In the early 1860s, rag-pulp newsprint sold for about 25¢/lb. As groundwood was increasingly added, the price came down to 14¢/lb in 1869, and fell to as low as 2¢/lb by 1897 (Hunter 1943). A high percentage of newsprint pulp is still made by a mechanical process, and much paperboard used in packaging is now made from such recycled newsprint.

The invention of chemical pulping processes for wood quickly followed. These processes dissolve the lignin and separate the wood's cellulosic fibers by cooking it under pressure in hot chemical solutions, like the process that was used for pulping straw. Chemical pulping results in a more durable paper with longer pure cellulosic fibers, but with lower yields.

The *soda process*, the first chemical pulping method, was invented in 1851 by Hugh Burgess and Charles Watt in England; they also established the first plant in America, near Philadelphia.

The *sulfite process*, which dissolves the wood with sulphurous acid, was developed in 1867 by Benjamin Chew Tilghman in Pennsylvania and commercialized in Sweden in 1875. The first commercial production of sulfite pulp on the American continent was in Ontario, Canada, in 1887. International Paper and Fibre Company bought the rights to the patent and the first sulfite mill in the United States was built in Alpena, Michigan, developed there to take advantage of lumbermill waste. The sulfite process became the most common chemical pulping method until the 1930s. Its sulphur burner, which makes the chemical liquor, produces the sulphurous smell that is associated with paper mills.

The plentiful new wood-pulp paper supply fueled a creative burst of literature and journalistic publishing after the Civil War (1865). American authors from the second half of the 1800s, such as Herman Melville and Mark Twain, owe some measure of their enduring popularity to wood-based paper. This is when newspapers, magazines, and books became available to common readers, fueling a growth in literacy.

Paper soon assumed more roles. The 1860s and 1870s were known as the "Age of Paper," the title of a popular song in London. Paper was used to make collars, cuffs, aprons, curtains, cups, and carpets. Durable items like furniture, luggage, tea trays, and even coffins were made from printed sheets that had been glued together and varnished. By 1900, *compressed paper* had become a standard material used in railway carriages, steamer trunks, and building construction.

This is when packaging uses for paper accelerated. As the 1800s drew to a close, the developments in printing and papermaking gave birth to the package converting and filling industries, discussed below. Paper-based packaging was getting ready to move onto the center stage of grocery shelves.

At the same time, the grocery shelves (especially in America) were being transformed by the development of chain stores. The relationship to the consumer was passing from shopkeepers to brand manufacturers. Mass production required mass-produced packages. Mass marketing required a print media to carry the sales message. And mass distribution required an inexpensive disposable shipping container. The new paper supply was well positioned to meet the new demand (Twede 2012).

A third chemical pulping process was to have great significance for packaging: the *kraft* (also known as *sulfate*) *process*. It was invented by C. F. Dahl in Germany in 1884 by adding sodium sulfide to the cooking liquor in the soda pulping process. First made in Sweden, *kraft* means "strong" in German and Swedish. The first kraft paper in the United States was produced in 1909, and it was soon preferred for uses where strength is valued.

Kraft pulping was able to utilize the almost weed-like growth of pinewood in the Southern United States. When they are young (9–10 years old), Southern pines are sufficiently free from pitch to make good pulp. This was the key to unlocking the resinous pinewood forests of the South for making paper and board. Kraft wrapping paper was first made from Southern pine pulp in 1911 in Orange, Texas, and other Southern mills soon followed in Louisiana and Mississippi.

The kraft process has come to predominate for high-strength paper-based packaging. The coarse brown paper is not very good for printing, but it has long tough fibers that can be made into a thinner, stronger paper than other pulp, yielding a greater area of paper from a ton of pulp. Furthermore, recycled kraft paper pulp retains much of its strength, increasing the amount of recycled paper and board that can be successfully used in packaging.

Kraft paper quickly became popular for wrapping and shopping bags. It was used to make multiwall bags and corrugated linerboard, and for replacing jute/rope paper and juteboard. *Solid kraft* board was used to make beverage carriers and bleached food-grade board. Kraft pulp is still used in most of these applications.

The mechanical, chemical, and semi-chemical wood pulping processes used today are described in more detail in Chapter 6. The use of wood pulp changed the process, the paper's properties, and its supply chain. As wood pulp increased the paper supply, packaging uses for it were also increasing.

The following three sections show how paper-based packaging technology and industries developed in the late 1800s and early 1900s to take advantage of the new marketplace. Although the timeline for each package form is presented sequentially, the dates make it clear that many of these developments were in parallel, as they collectively formed the modern paper-based packaging industries.

1.2.2 Paper Packages

Prior to the 1800s, most of the references to paper-based packaging are to the unprinted wrapping or bags used by merchants. For example, the first English patent pertaining to papermaking was granted to Charles Hildeyerd in 1665 for, "the way and art of making blew paper used by sugar-bakers and others," (Hunter 1943). *Sugar blues* were a common type of wrapping paper made from waste paper (blue from logwood dye, as is litmus paper) and used to wrap cones of sugar.

Retailers would purchase products in barrels and weigh out and wrap up a quantity for each consumer. Wrapping required skill and time, and the customers would wait while their orders were made up. There were different patterns of wrapping that can still be seen today in many parts of the world. Paper cones were twisted up to hold loose items, a practice common in British sweet shops up until the mid-1900s: "the grocer was most dextrous at conjuring up his own bags or twists," (Opie 1989). Cone-shapes were also mass produced using prison labor (Drasner 2004).

Wrappers were usually the lowest quality paper or reused paper. For example, the first mention of grocers' paper bags was in 1630, describing European paper that was such poor quality that it was "not even fit for grocers' paper-bags," (Long 1964). A 1719 History of Kent (England) describes, "some Paper-Mills . . . which make a great deal of ordinary Wrapping Paper for Tobacco, Grocery Ware, Gloves, and Milleners Goods, etc.," (Hills 1988). If a wrapper was printed, the print was probably from a previous use. This is a practice that continues in some places today, where, for example, fresh fish is wrapped in yesterday's newspaper.

There were various grades of wrappers, and these increased after papermaking was mechanized. Shorter (1999) notes an 1876 reference to British mills making "Browns, Middles, Shops, Wrappers and Skips." Labarre (1952) defines them: *browns* were coarse wrapping paper, *middles* were used in the center of pasteboard and covered with *pasters*, *shops* were white and machine glazed, and *skips* were thin packing papers used to line crates. *Bottle wrapping* was a special tissue paper. Wrapping paper was used as sheets or purchased in *counter reels*.

Many specialty papers for wrapping were developed. By the 1900s, properties were tailored to the product intended to be wrapped, and most had names that reflected their use. Greaseproof and glassine paper for baked goods, bleached water-resistant butcher paper for meat, pouch and envelope papers, and other specialties like cartridge paper and tissue paper were common (Labarre 1952). Those that are still used in packaging are described in Chapter 9. Many others have been replaced by plastic films and laminations.

Molded pulp was developed in 1903 by Martin L. Keyes. The process differs from paper-making inasmuch as the mold is three-dimensional. It has the advantage of being able to use recycled fibers, and recycled newspaper has predominated. The highest volume molded pulp packaging has been for eggs, a purpose for which it is still used (Wever and Twede 2007). The manufacturing process and newer uses, like cushioning and dunnage, are described in Chapter 13.

1.2.2.1 Lithographically Printed Wrappers and Labels

Germans were leaders in printing from the time of Gutenberg, and the earliest printed labels were also German. The first recorded use of printed labels was by the Fugger family of German traders in 1500. They traded throughout Europe and the East in silks, spices, and wool that were wrapped and identified for shipment (Bruno 1995).

Some of the earliest printed wrappers were used by the paper industry itself. German papermakers in the 1550s used printed wrappers for their reams of paper. The wrappers were practically a by-product of the papermaking process, using the dregs at the bottom of the vat to make a coarse paper that was then printed in the center with the papermaker's design or crest. Likewise, some of the earliest printed wrappers in America were for paper reams. An early ream label is shown in Figure 1.7 (Hunter 1947, 1951; Davis 1967).

The French word for label is *etiquette*, which began as the designation for manners at court in the 1600s, and was the source of the word "sticker." The court used the word to indicate a paper list of permitted modes of behavior, and these were the predecessors of the modern label (Bükle and Leykamm 2001).

FIGURE 1.7. *Paper ream label of German papermaker, Andreas Bernhart, Papiermacher auff der Ocker, circa 1550. Reprinted with permission of Alfred A. Knopf, from Hunter,* Papermaking *(1947).*

In the 1700s, printed trade cards and wrappers for tobacco, powdered medicines, health aids, and pins began to be used throughout Europe. Tobacco papers identified the retailer who blended and measured out the tobacco into the wrapper (Davis 1967).

Growing literacy in the 1700s contributed to the growth of labels. The earliest known wine label dates from 1756, used for Port from Real Companhia (Rickards 2000). By the late 1700s, printed paper labels were used on glass bottles for patent medicine, wine, and condiments like sauces. The first labels were merely used to identify the maker. As products began to be sold further from their place of origin, labels gave buyers a means of identifying their manufacturer and tracking their quality (Long 1964).

Early wrappers and labels were printed in one color by letterpress, with the illustrations printed using woodcuts or engraved copper or steel plates. All-over colored and patterned papers, like the marbled paper used for bookbinding, were sometimes used to give color, and some labels for perfume and pomade in the early 1800s were hand-colored. But most packaging was black-and-white (and usually a greyish-white at that) (Davis 1967).

The third significant invention in the early 1800s set the stage for modern packaging: color printing of illustrations. Lithographic printing was to become particularly well suited to mass reproducing multi-colored designs on packages at a relatively low cost. Combined with the papermaking machine and the straw and wood pulping processes, it brought the cost of illustrated paper-based packaging low enough so that its marketing value outweighed its cost.

Lithography, from the Greek and Latin words for "stone writing" was invented by Alois Senfelder in Munich in 1796, who was experimenting with printing. He wrote his laundry list on a stone with greasy wax, and then wetted the surface with a mixture of gum arabic and water, so that the design area repelled the mixture and the blank areas absorbed it. An oily ink rolled onto the stone then coated the greasy design but not the wet blank area. It made a clean impression of the design when a sheet of paper was pressed against the surface of the stone. The quality became sharper yet when the offset process was added, first for printing tinplate cans and boxes in the 1880s (Meggs 1998). Offset lithography and other printing methods used today in packaging are described in Chapter 10.

Lithography's clean impressions were ideal for reproducing illustrations, and this added pictures to packages. Black-and-white illustrations on labels, probably for cigars and tea, were the earliest lithographed packaging materials. By the 1830s, one-color lithographed or letterpress labels were used on a wide range of products and packages including glass bottles, metal boxes, and early paperboard boxes. Paper printing was widespread—most towns had printers, and labels were cheap (Meggs 1998). Early labels were affixed by glue or water-activated gum, or fastened mechanically with string or rivets.

Color printing began with letterpress and engraved plates about the same time, and color labels made their first appearance on matchboxes in the 1830s (Davis 1967). This was quickly followed in 1837 by the invention of chromolithography by the French printer Godefroy Englemann, who invented the means for analyzing colors in a subject and separating them into component colors to be printed in sequence. Chromolithography was first known as a fine art medium, and got its biggest boost toward recognition when Currier and Ives popularized it for art reproductions in the 1850s (Meggs 1998).

Early American printers were some of the earliest paper-based packaging makers. For example, Robert Gair, who played a central role in the development of paper bags, folding cartons, and corrugated fiberboard boxes, was also a lithographer.

From the 1840s to the early 1900s, chromolithography was the predominant way to print award-winning labels. At the Great Exhibition in England in 1862, printed packaging was on display: labels, wrappings, and box tops reportedly "in gold and colors." There were also set-up boxes covered with fancy papers from France (Davis 1962). In1863 Ferdinand Revoul, Valréas' lithographer and carton-maker, received a bronze medal at the exposition in Nimes (Locci and Baussan-Wilczynski 1994). Figure 1.8 shows a cigar box with lithographed label.

Some of the earliest chromolithographed labels were for holiday gifts like handkerchiefs and chocolates. Fry and Cadbury, in 1868, were the first to print special labels for Christmas and Easter candy. Christmas wrapping paper also originated in Europe during this period (Opie 1987).

Some of the best commercial printing was on cigar bands and box labels in the 1870s to 1890s; these are highly collectable because of their beauty and the fact that people have prized them for over a century. But ordinary products were also dressing up:

> *Victorian design was often ebullient, frivolous, detailed and colorful. Such delightful jewels' on the grocer's shelf must have added greatly to the appeal of more mundane products like washing soap or health salts, (Opie 1989).*

Some of the lithographic label masterpieces from the late 1800s were produced in up to 14 different color printing operations (Bükle and Leykamm 2001).

When the invention in the 1870s of photoengraving and half-tone engraving was also applied to labels and cartons, illustrated labels became even more commonplace. By then, decorative labels could be mass-produced in a variety of sizes.

Cans for heat processed food had some of the most elaborately printed labels by the 1890s. Most metal boxes (also called canisters, which are not heat processed) were being directly printed by transfer decal or offset lithography,

FIGURE 1.8. This mahogany cigar box, made in a factory in New York in 1875 for K. C. Barker and Company of Detroit, held fifty cigars. The color lithographed label was created by F. Heppenheimer of New York City. From the collections of the Michigan State University Museum.

an effect that printed labels and cartons sought to emulate. The brilliant lithography was essential for selling canned food that was not visible during a purchase based on faith. Butter, cigarettes, and soap were packaged in lithographed wrappers.

Fruit crate producers in California began another colorful labeling trend, creating a way to brand fruit. For smaller producers, stock labels were printed for products like fruit crates and bottles containing wine, vinegar, and jams that could be customized with the retailer's or grower's name (Opie 1987).

The second half of the 1800s marked the beginning of the power shift from retailers to manufacturers as an increasing number of products were packed by manufacturers. Packing in a factory gave the incentive to develop mechanized methods for high volume packing, wrapping, and labeling. For example, one of the first packaging machines to be invented was the gravity labeler introduced in 1893 by Standard-Knapp, a company that still exists today.

Although many labels are still made from paper, alternatives have been developed, including printing directly on containers (especially paperboard and metal). Stan Avery's introduction in 1935 of "Kum Kleen" pressure sensitive price stickers led, by the 1940s, to pressure sensitive labels on a release backing and by the 1950s, to the labeling machines to apply them (Klein 1994). Paper labels are discussed in Chapter 11.

Wrapping groceries in printed paper declined in the early 1900s as the use of printed bags and folding cartons grew, especially once bagging and cartoning equipment were developed. But there was an increase in specialty wrapping papers like waxed paper, first made in 1854, and glassine, produced in 1910 (Cornucopia of Progress 1977). Such specialty papers added protection like being greaseproof or being a water barrier, and lengthened the shelf life of tobacco and dry food products. Ainsworth, in 1959, provided a listing of over 100 paper markets (ranging from banks to upholsters) and on every list "wrapping paper" appears.

Cellophane was the ultimate wrapping "paper;" it was the first see-through flexible packaging material. It was used as a glossy transparent wrap for luxury goods. Cellophane was invented by Dr. Jacques Edwin Brandenberger in Switzerland, and in 1911 he designed a machine to produce it by dissolving cellulose and casting it as a film. The first factory to produce cellophane was built in France, and the DuPont Company acquired the American rights to produce it in 1923. DuPont scientists added a moisture resistant coating because of its poor barrier properties.

DuPont sold cellophane as a crisp, shiny wrap for high end candies and cosmetics, and also as a moisture barrier wrap for food products and tobacco. Cellophane launched the flexible packaging revolution. For over 40 years it enjoyed a steady growth in packaging applications. It was expensive, available in beautiful sheer colors, used for the packaging and over-wrapping of expensive items in cartons. It could be reverse-printed so that the surface retained its gloss. It was made into "cello" tape by Richard Drew at 3M in the 1930s (Personal Touch 1977).

Of course, the flexible packaging revolution did not end with cellophane, and it was ultimately replaced by various less-costly plastics with better physical properties. The wrapping machines and flexographic printing methods that were developed for cellophane evolved into the equipment that is used today for plastic film sealing and printing.

1.2.2.2 Paper Bags and Flexography

Some of the first consumer packages made from paper were printed and pasted seed envelopes made by members of Shaker communities in New York, Connecticut, and Ohio beginning in the 1790s. They were packed in a multi-sectioned wooden box with a colorful label, and were displayed in country stores throughout the United States. The Shakers were also the first to make colored picture labels for canned food (Raycraft 1980).

Paper bags developed from grocers' wraps. Printed paper grocery bags that advertised a specific department store or small shop became popular in the mid-1800s as paper availability grew. At first the paper bags were hand-made and printed with the retailer's name or an illustration of the shop.

For example, two firms in Bristol, England, competed for the bag business throughout the United Kingdom and as far away as New Zealand. In 1848, Elisha Robinson's brother went out as a salesman "all through the West country as far as Penzance, breaking new ground and introducing in face of much prejudice the novel idea of ready-made bags." The son of Robinson's competitor, James Mardon, noted that many customers in the grocery and drapery trades "desired to have views of their premises on their billheads, bags and tea papers," (Davis 1967).

Retail bags were also used in the United States. After the Civil War, Robert Gair sold bags, at first hand-made by girls and printed with a treadle press, to New York City's growing number of department stores and small shops. The largest bags were used for immense billowy crinolines. He also sold seed and hardware bags made from strong jute/rope paper (Smith 1939).

The mid 1800s were a creative period for paper converting inventions, and many were related to machinery for making or filling packages. For example, John Horniman, the British tea merchant, in 1826 was one of the first packers to invent an elementary device to accurately fill tea packets (Davis 1967).

According to an 1874 U.S. index, by that time there had been 79 patents issued for making paper bags and 66 issued for making paper boxes (Weeks 1916). For example, Francis Wolle, of Bethlehem, Pennsylvania, patented the first bag-making machine in 1852, and in 1869 formed the Union Paper Bag Machine company, which at one time owned or controlled 90% of the paper bag business in the United States.

James Arkell and his partners Benjamin and Adam Smith (Arkell & Smiths) patented another machine in 1859 and established the first U.S. bag-making firm in Canajoharie, New York. Gair also invented his own machines and advertised "Patented machine-made bags for grocers," (Davis 1967). Margaret Knight invented the first square bottom bag-making machine in 1870 (Kleiger 2001; McCully 2006). Mechanization reduced the price and increased supply.

The new mass-produced paper grocery bags speeded up retail transactions:

> *Nothing has had a greater influence in making possible the rapidity with which certain branches of retail business are conducted, as compared with 10 years ago—more especially in the sale of groceries—than the cheap and rapid production of paper bags. . . . With machinery have also come many improvements: square bags that stand up of themselves and*

need only when filled from a measure to have the top edges turned down
to make the package at once ready for delivery, (Wells 1899).

After the 1850s, manufacturers began to use paper bags too. They were first used for flour and sugar when the Civil War disrupted the supply of cotton for textile bags from the South (PSSMA 1991). In 1862, George West (later Gair's business partner) developed a paper sturdy enough to carry 50 pounds of flour. It had crisscrossing manila rope fibers and was claimed to withstand more splitting pressure than cotton (Smith 1939). The early heavy bags were made from one ply of jute paper (Labarre 1952; Steen 1963).

Paper flour bags, shown in Figure 1.9, gained popularity fastest in the smaller sizes. Millers' associations, beginning in 1881 with the sanction of the U.S. government, established pricing differentials for flour in smaller bags to cover the additional cost. By the 1940s, almost all flour packed in quantities of less than one-half barrel (98 pounds) was in paper bags (Steen 1963).

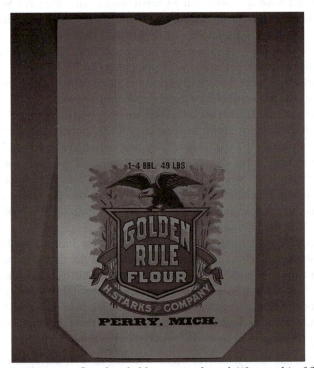

FIGURE 1.9. *This paper flour bag held a quarter-barrel (49 pounds) of flour. It was made in the first quarter of the 1900s by the Adams Bag Company of Cleveland, Ohio, under their "Neverburst" trade mark. From the collections of the Michigan State University Museum.*

One of the most successful paper bag styles was the valve bag, invented in 1898 by Adelmar Bates and John Cornell and commercialized in 1924 by the Bates Valve Sack Company, which had developed the valve style for cotton and jute textile bags 35 years earlier. The style, a closed bag with a filling opening that would close automatically when the bag was filled, had been "revealed" to Bates during a feverish dream. The valve style became successful because it is clean, easy to fill, and seals itself. Because the filling head is enclosed in the bag, valve bag packing dramatically increased worker safety in industries like cement, grain, flour, and carbon black where dust caused debilitating respiratory illnesses (PSSMA 1991).

Bates' paper valve bag packer worked best with cement, replacing cloth-lined barrels. Portland Cement had been patented in 1824, and in the following year construction of the Erie Canal created the first large scale demand. By 1880, about 42,000 barrels were produced annually in the United States, and a decade later the number had increased to 335,000, as the production process was improved and demand grew for large-scale construction projects (like the Statue of Liberty's foundation). This incredible growth was an incentive to find a less costly and less wasteful form of packaging. Valve bags made from jute paper were invented at the right moment to serve this very dusty, growing industry. By 1920, over 1,800 valve bag filling machines had been installed by the cement industry (Twede and Drasner 2013).

Bates made valve bags first from manila rope paper, and after WWI, when rope became scarce, he began to also include kraft fibers. This made the bags too stiff, so a manila rope/kraft paper two ply construction was developed. The early valve bags were sewn; in 1926 the first pasted valve bag was introduced (Drasner 2004).

Because of their strength and reusability, cotton and burlap were often preferred for larger quantities until a tubing machine to make multiwall paper sacks was invented in 1917 by Fischer and Krecke. Judson Moss Bemis was one of the first to make multiwall bags, although he also was a leading manufacturer of cotton and burlap bags (Personal Touch 1977).

During the late 1940s, multiwall shipping sacks replaced most refillable cotton and burlap bags that had been used for shipping flour from mills to bakers. The multiwall bags cost more than the reused cloth bags, but they were cleaner. After prohibitions were issued against reused flour bags by state boards of health, bags ceased to be refilled, giving paper bags a cost advantage (Steen 1963).

Multiwall bags were initially flush cut and sewn closed, which caused sifting problems. The St. Regis invention in the 1950s of a machine to make pasted stepped-ends solved the sifting problem. The *pinch* (or *pasted*) *bottom open mouth* bag is so named to differentiate it from valve or sewn bags. Other innovations that accelerated bag popularity were improved filling and closing equipment and material improvements like the use of stronger *extensible* kraft to add strength, plastic liners to add a water vapor barrier, and hot melt adhesive for closing the bags. Most of the innovation came from the technical labs of large integrated paper producing companies like St. Regis and International Paper, as well as some non-integrated ones like Bemis and Bancroft. Chapter 11 describes the styles and uses of multiwall paper bags today.

The fact that letterpress printing methods could not keep up with bag-making machines provided the impetus for the invention of the most significant printing process used in packaging today: *flexography*. The bag makers Bibby and Baron in Lancashire, England, designed, in 1890, the first press to print a continuous web from rubber blocks or plates, using fast drying aniline inks and a central impression cylinder (Davis 1967).

Flexographic bag printing was introduced in the United States by H. H. Heinrich, who sold his first flexographic press to the Packaging Paper Co. in 1925 in Holyoke, Massachusetts. It could print a 24-inch wide web in two colors. Heinrich pioneered improvements in rubber platemaking and processes that contributed to the growth of flexography, and he helped to create new markets for it. The inks were developed by Interchemical Corporation to print on paper, cellophane, linerboard, stock boxes, and cartons. ICI's Douglas E. Tuttle, in 1939, invented the anilox roll inking system that has made modern flexographic printing possible (Personal Touch 1977).

The term "flexography" was adopted in 1952 as the result of an industry poll to replace the earlier term, "aniline" printing, which referred to the coal tar-based inks originally used. Aniline inks were banned by the U.S. Food and Drug Administration in the 1940s as unsuitable for food packaging, and so the name was changed when safer inks were approved. Flexography is now used for corrugated board, milk cartons, paper bags, folding cartons, and plastics of all kinds.

Paper bag-making technology revolutionized package converting. It was for bags that the practice of straight line automatic folding and gluing was first developed. The straight line tubing concept has been used ever since for flexible package converting, including the form-fill-seal process that is most common for small bags. Furthermore, straight line folding, tubing, and gluing is used for almost all folding carton and shipping container converting to this day.

But the paper bag form is in the decline stage of its lifecycle. Lower-cost plastic bags have replaced them for most uses. Small, consumer-sized paper bags were used in the mid-1900s for many kinds of food, especially bakery products, as a substitute for wrapping. For example, the "greatest thing since sliced bread" was probably the waxed paper bag or cellophane used to contain it and lengthen its shelf life—now replaced by a polyethylene bag. Plastic is made into shipping sacks, grocery bags, and landscape product bags. As the first flexible package, paper bags led the way for these plastic film alternatives.

1.2.3 Paperboard Converting

The earliest paper-based boxes were made from *pasteboard*. It was made by pasting together pages of old books or other printing scrap; some of it was more like a molded paper maché material than a sheet, and it was widely used for book binders. In 1580, the first commercial pasteboard was manufactured in Europe, although board of this kind had been made in China and Persia centuries earlier (Hunter 1947).

Many of the significant developments in paperboard happened in the United States. The earliest recorded U.S. paperboard was made by hand in 1728 at the Milton Massachusetts mill, produced on the mold by either multiple dips in the fiber slurry or by pouring a heavy single layer into the mold. Some of the earliest paperboard was called "bonnetboard" and was used as a stiffener for hats in Massachusetts. Milled board was a thick handmade sheet made from rope or jute that was "milled" between heavy iron rolls. Shryock's Pennsylvania mill made the first straw cylinderboard, discussed previously (Bettendorf 1946; Davis 1967[2]).

1.2.3.1 Set-Up Boxes

Rigid paperboard boxes are believed to have been first used in China, as a package for fine teas. It is also known that the Persians made coffins from laminated paper (Hunter 1947). The first paperboard set-up boxes in the West were made in the 1700s when Germany was a leading producer, and many were imported to America. Pulpboard cartons were made in France before 1751, and our word *carton* is from the French word for pasteboard box.

Decorated *band boxes* (cylindrical set-up boxes, so named because they were originally made to hold gentlemen's collar ruffs, called bands) were popular during the early 1800s, with their greatest popularity during 1820–1945. They were used as hat boxes and for other accessories, and were used by travelers to carry almost anything (Lynn 1981). A later hat box is shown in Figure 1.10.

The earliest bandboxes were made from strips of pasteboard or handmade chipboard made from slivers of wood and waste paper pulp. Later they were made from strawboard produced on cylinder machines. They were shaped on wooden forms that are similar to those used for making hats and jewelry, and the joints were stitched and/or glued. Since the coarse board was unsuitable for

---

[2]*Reference note:* Most of the information in this paperboard section, unless otherwise noted, derives from Bettendorf (1946) and Davis (1967), although they are not cited throughout the text. Bettendorf focuses on the United States where most of the carton innovations occurred, and Davis on Britain where paperboard is called *cardboard*, but both outline the same basic facts. Bettendorf includes appendices that list the early US box makers and paperboard mills and presents Shryock's account of how he first made strawboard.

FIGURE 1.10. *Stetson Hat Box, circa 1930s, with artwork by Conrad Dickel who did similar Early American illustrations for Christmas gift boxes for the Interwoven Stocking Company of New Jersey. From the collections of the Michigan State University Museum.*

printing, the boxes were covered with decorative paper, much of it printed by wallpaper makers, and glued in place. The paperboard band boxes were similar to earlier boxes made of wood veneer strips, leather, or metal, covered with printed paper. Most were cylindrical (round or oval) because of the difficulty of making clean creases and sharp corners.

The earliest American paperboard set-up boxes appeared in the early 1800s, made in Philadelphia and Boston. A Philadelphia directory shows six set-up box makers during the period of 1785–1800. Box making was a labor-intensive craft until machines were developed. Much of the assembly work was performed by women, sometimes working from their homes. Before strawboard was made on cylinder machines, the paperboard was made by hand too.

The commercialization of set-up boxes is generally credited to the Dennison family in Boston, still a leading name in paper labels. Although he did not invent the package form, Aaron L Dennison, a jeweler, in 1839 began making set-up boxes in his shop using a jack-knife, rule, paste, and wooden forms. He made them for himself and other jewelers, competing with imported boxes from Germany. His father, Col. Andrew Dennison, who was a cobbler, took over the production and invented tools to make the job easier, like the first corner cutter and scorer, a hand shear that was a lever-type paper cutter equipped with gauges for cutting at predetermined angles, and a clamping device.

By 1850, Dennison's business had expanded to include boxes for a wide assortment of small products including cakes, combs, spectacles, pencils, hairpins, plaster, soda, needles, and botany samples, as well as mailing boxes. For jewelers, they developed display cards and price tags, which led to shipping tags and to Dennison's eventual leadership in the gummed label industry (Hayes 1929). Further strengthened by the merger with Avery, Avery-Dennison is still a leading label manufacturer today.

By 1860, others in Philadelphia, Chicago, and New England had established box-making plants and began developing their own specialized equipment to speed production and reduce costs.

Fancy set-up boxes began to be used for drugs, candy, jewelry, cosmetics, and larger items like hats and wedding cakes. The boxes were rectangular, cylindrical, or oval. Even coffins were made from laminated sheets in 1869. By the 1870s, set-up boxes were being used by some manufacturers to differentiate their goods. Matchboxes and candy boxes were some of the first to use colored labels. By 1900, fantastically elaborate chocolate boxes were being hand-made for the Christmas season with pleated silk and satin, metal clasps, and hand painting. A simple Christmas candy box is shown in Figure 1.11.

Shoes were one of the early products to be sold in set-up boxes. In the early 1870s, George Kieth, a shoe manufacturer in Brockton, Massachusetts, was the first to have the idea. He established a box plant and other shoemakers followed, as did a number of shoe box plants, supplied with strawboard and cover paper by Spaulding and Tewksbury, a leading mill in New England. The shoebox has been an essential element in shoe retailing ever since (although now it is likely to be a folding carton or even a corrugated box, rather than a set-up box).

During the 1870s and 1880s, the newly commercial multi-cylinder strawboard manufacturing process was joined by machinery for making the boxes. Equipment was developed for rotary knife scoring, corner cutting, setting up the box, taping the corners, and applying the covering (using paper from a roll).

At first, white news and manila paper were used for covering set-up boxes. Mechanization and increased demand for the boxes encouraged lithographers to develop more fancy covering papers, especially for candy and for Christmas gifts like men's socks and handkerchiefs. *Box enamel* was the name for cheap paper coated and highly glazed on one side.

As folding cartons, described next, became more common for lower priced goods like cereal, set-up boxes dominated the high end of the market where presentation and reuse was desired. They are still used for premium products like greeting cards, gifts, jewelry, and games, as discussed in Chapter 12.

FIGURE 1.11. *Fanny Farmer Christmas set-up box from 1949 held 42 ounces of candy. From the collections of the Michigan State University Museum.*

1.2.3.2 Folding Cartons

The difference between a folding carton and set-up box is clear from their names. The set-up box is set up by the box maker and cannot be shipped to the packer/filler in a flattened form like folding cartons. As set-up boxes became popular, American inventors looked for a more efficient way to pack products in boxes, which led to the folding carton. A folding carton is more efficient than a set-up box; it uses less paperboard, has a simpler manufacturing process and is shipped knocked-down, saving transport and storage costs.

The predecessor of the folding carton appeared in about 1850. Carpet tacks, which are obviously hard to handle, were packed by the store clerk in folded paperboard that was shaped into a tube around a wooden form. One end was folded in and held in place with a glued label. The shape was removed from the form, filled with tacks, and the other end was secured. The package, which was then tied up with string, was known as a *paper of tacks*. Although it had no preliminary creasing, this package influenced the eventual development of the folding carton.

The *folding carton* developed, primarily in the United States, over the next two decades. The first folding cartons were scored on a platen printing press from which the typeset had been removed and replaced with metal strips (brass printing rules). Then the cartons were cut in a separate operation with a guillotine knife.

But since most folding cartons are made in the form of a tube, similar to paper bags, it is no surprise that it was the paper bag maker, Robert Gair, who made some of the first folding cartons. The early versions were expensive and could not be produced in large quantities since they were manually cut and scored, but they offered great advantages:

> Being made of one piece of paper the cover is always in place ready for closing, they pack flat, thereby saving bulky freight, store room, and leave less surface exposed to danger of being soiled (Gair catalog, 1878, quoted by Smith 1935).

Gair's greatest role in packaging history was inventing a mechanized method for cutting and scoring folding paperboard cartons in a single stroke. The idea resulted from a mistake in 1879. A careless pressman in Gair's paper bag factory in Brooklyn failed to notice that a type rule had been set too high and had cut through and ruined 20,000 seed bags. Rather than igniting Gair's thrifty Scottish wrath, it fired his imagination:

> The clean incisions across the seed bags struck the eye of Robert Gair at a moment when his mind was ready and receptive for the sight. It came to him, in a flash, that there was a way of constructing a multiple die that would cut and crease box board in a single operation (Smith 1939).

The ruined bags gave Gair the simple idea to set sharp cutting blades a little higher than blunt creasing rules in a press, thus cutting and creasing in one operation. Gair could now mass produce folding carton blanks. His first $30 press could cut 750 sheets/hour, each with 10 dies. This one press produced as many cartons in 2-1/2 hours as his whole factory had previously made in one day (Smith 1939).

At first, unprinted folding cartons were used for containing small items that were formerly packed in bags. The first cartons used by manufacturers were labeled or were covered with a printed wrapper.

The decade of the 1890s was a period of folding carton market penetration. Gair at first made cartons for retailers, like Bloomingdales and the Great Atlantic & Pacific Tea Company, for cosmetics companies like Colgate and Ponds, and for the tobacco manufacturer P. Lorillard. The cigarette carton was developed in the 1890s and replaced the printed stiffener cards that had been added to cigarette wrappers since 1879. At first, cartons were used for local products, frequently the kind of products with a snake oil reputation, such as "Hungarian Cough Balsam" and "Plantation Chill Cure" (Smith 1939).

The folding carton earned prestige in 1896, when the National Biscuit Company began to sell Uneeda Biscuits in cartons. The crackers were wrapped inside a waxed paper liner inside a tray-style carton, and the colorful printed wrapper featured a boy in a raincoat to emphasize the moisture barrier, as shown in Figure 1.12. The package had been developed and tested by Frank Peters in the first documented water vapor transmission test (Twede 1997). Robert Gair's factory produced the initial order of two million cartons, and his son, George, takes credit for inspiring the name for the new cartoned crackers when he counseled, "You need a name," (Smith 1939).

FIGURE 1.12. *Uneeda Biscuit Advertisement. Reprinted by permission of Kraft.*

For the first time, the consumer could buy crackers in a clean unit-size package, rather than having the retailer measure out a quantity from the large cracker-barrel where they were exposed to moisture, odors, vermin, and breakage. The Uneeda Biscuit package is often cited as the birth of "consumer packaging" (which it is not—there were plenty of bottles, cans, set-up boxes, and bags before this) because of its widespread distribution and the dramatic effect that folding cartons were to have on the retailing business in the century to come.

The Uneeda Biscuit carton does represent the birth of modern packaging; brand advertising that relies on the package as a sales tool. It symbolizes the shift in power from retailers to manufacturers. By packaging at the factory instead of in the store, advertising directly to consumers in magazines and on billboards, and by making their packages easy to recognize, manufacturers were able to take control of the market (Twede 2012).

With the increasing demand for graphics to identify food and drug manufacturers and stimulate sales, Gair foresaw the value of printing cartons directly:

> *On the theory of Polonius that "the apparel oft proclaims the man," I added designs by printing, lithographing and embossing to the exterior of the folding box, thus establishing a standard whereby the merits of the contents could be judged (Bettendorf 1946).*

Printed cartons assured consumers of the consistency and quality of the product. However, most early cartons were only printed in one or two colors, using wood engravings and a letterpress. Print was limited to names and decorative borders. In the 1890s, Gair's catalog listed cartons printed for candy, Smith Brothers' cough drops, and Sweetheart Soap.

Lithographically-printed illustrated cartons were introduced in about 1900. Cereal companies were among the earliest adopters; clean cartons contributed to the healthful image that they cultivated, and the branded graphics promised consistency and quality, like the one shown in Figure 1.13. In 1903, a machine that converted a roll of printed board to tray-style cartons was invented for Quaker Oats and used later for other cereals (Quaker Oats, in 1877, was the first cereal marketed in a carton; the familiar cylindrical Quaker Oats fiber canister was introduced later.)

The first cartons were folded, glued, and labeled by hand, but mechanization soon ramped the carton industry into mass production. There were many inventions to automatically fold and glue cartons from 1895 to 1910. All of the machinery inventions were based on the idea of straight line folding and gluing similar to that used in tube style bag making machines.

The folding carton innovation was slow to move to Europe where many products were already packed in decorated metal and set-up boxes, but once American products in cartons began to be sold in English shops and two American box-makers established factories in England, the British and continental

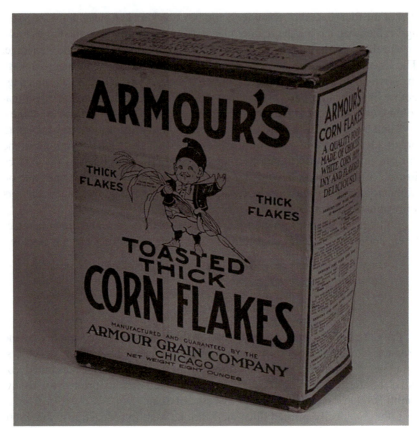

FIGURE 1.13. *Corn Flakes carton from the Armour Grain Company of Chicago. The carton was made by the Michigan Carton Company of Battle Creek, Michigan. It dates to the first quarter of the 1900s. From the collections of the Michigan State University Museum.*

manufacturers began to follow the trend, buying mostly American-made machinery. Robinsons of Bristol and Tillotsons of Bolton were some of the first, although some British boxmakers had previously installed German-made machines that manufactured cartons by "grooving and stamping" rather than the simpler creasing (Davis 1967).

Paperboard was a good use for wastepaper, and in 1918 systems for more efficiently pulping wastepaper began to be developed. The first successful breaker-beater was installed that year in the Haverhill Boxboard division of the Robert Gair Co. In 1939 the hydrapulper, developed by the Dilts Machine Works, Fulton, New York, permitted the pulping of wastepaper in higher quantities in one batch. The design was based on an early washing machine. This started a whole new segment of the paperboard industry by establishing paperboard factories in cities near consumers, rather than near forests or farms.

Other significant innovations were in automatic filling and sealing equipment. The first companies that developed the machines were primarily scale manufacturers such as Pneumatic Scale Corp. and Automatic Weighing Machine Co., as well as midwestern specialists like the Automatic Carton Sealing Machine Co. in Chicago and the Johnson Automatic Sealer Co. in Battle Creek, Michigan. This was the beginning of packaging systems that integrated machinery and material, like Klik Lok, Mead, Olin, and Ex-Cell-O.

Plenty of products that are packed in plastic today got their start in folding cartons. In 1950–1960, potato chips were packed in a waxed paper bag-in-box, similar to breakfast cereal packaging, formed on a Pneumatic Scale double package maker. When form-fill-seal systems were developed by Woodman, Mira Pak, and DuPont for cellophane and later for oriented polypropylene, potato chip manufacturers quickly dropped the carton, and the lower priced package's sales increased exponentially (Brody 2004).

Folding cartons of all package forms played the most central role in the branded consumer product revolution. Consumers came to trust the reliability of brands, and found that shopping could be done more quickly with packaged products. Retailers sold more when the merchandise was clean and attractive, and packages reduced their handling cost. Instead of bulk bins, products could be displayed on shelves. Manufacturers could mechanize the packing operation at the factory, replacing the use of costly manual packing in the retail store. Folding cartons are discussed in Chapter 12.

Bettendorf sings in praise of the humble paperboard carton:

> *Man could not have achieved his present high level of mental, spiritual and physical welfare without paperboard, for the paperboard box today is a vital key to orderly and sanitary transportation and distribution of goods. . . . Without the mass production box, there would be no mass production . . . we'd move and count things in piles instead of units; and in general we'd 'coolie' our way through life.*
>
> *And thus, out of the piles, confusion and dirt of the earlier period came the cleanliness, order, precision and efficiency of mass production goods through the employment of mass production packages of paperboard (Bettendorf 1946).*

The artistic and creative side of the folding carton business was developed by companies like Container Corporation of America (CCA), which had a magnificent design center and laboratory. Its 1960s advertising featuring "Great Ideas of Western Man" was illustrated by commissioned artists, and it sponsored a Bahaus art festival. This strategy was abandoned when CCA was acquired by Mobil in the 1970s (Brody 2004).

For all of their great graphics, folding cartons have hidden as much as they revealed. They hide the product and fill level, which led to the Fair Packaging and Labeling Act. They hide the fact that they are made from wastepaper with a brilliant clay coating and colorful print. They might also hide a toy or prize, and there is a long history of premiums and boxtop redemption schemes that enticed purchase as much as the product itself. For much of their history, cartons have implied something special that goes beyond the product.

Probably the biggest surprise is the value that an antique carton now adds to antique products. Rather than hiding the prize, the carton now has become a valuable part of it. Appraisers such as those on Antiques Roadshow always find that the value of an old item increases if it is in the original box.

1.2.3.3 Milk Cartons, Drink Boxes, Fiber Cans, and Beverage Carriers

Once set-up boxes, folding cartons, and paper bags proved themselves for dry goods, paper-based packaging inventors took on the challenge of liquid products and food. Water-resistant paper inventions started in the 1880s with paper cups (some were produced by Robert Gair). Some of the earliest water and grease resistant paper-based packages were described in 1889:

> *A purchaser can now take his butter or lard in paper trays that are brine and grease proof; his vinegar in paper jars that are warranted not to soak for one hour . . . and his ice cream and oysters in paper pails that will hold water overnight (Wells 1899).*

Efforts to package liquid in a paper-based container posed obvious problems: paper is absorbent and porous to liquids. Even when this was addressed with wax or other coatings, there remained the problem of how to make watertight seams. Three package forms succeeded: the gable top milk carton, the aseptic drink box, and (eventually) fiber cans. Paperboard also served as a carrier for the growing volume of cans and bottles filled with beverages. Milk cartons, drink boxes, fiber cans, and beverage carriers as they are used today are described in Chapter 13.

Gable top milk cartons were first patented in the United States in 1915 by John Van Wormer. The first report of milk being sold in a paper container on a commercial scale was in 1908 in San Francisco and Los Angeles by G. W. Maxwell. But it was not until the 1930s that the first commercial installation by Ex-Cell-O began to operate at a Borden Company plant. During the 1930s, there were many competing paper beverage cartons, but Ex-Cell-O's Pure-Pak was the most successful. This was largely because of its promotional efforts to change the operations of dairies and the minds of consumers, and its political efforts to change legislation regulating fill levels that had been based on glass milk bottles (Robertson 2002).

Compared to glass bottles, milk cartons reduced the cost of package filling, but required investment in new machinery. Their lighter weight reduced the cost of delivery, and they protected the milk better since they were more sanitary and safe. They eliminated the need for the dairy to collect bottles and wash them, contributing to the end of home milk delivery. Sold in a disposable carton, milk became like every other commodity sold in stores.

The gabletop folding format is one of the oldest and most basic end closures possible for a paperboard carton. Early gabletop milk cartons had a glued side seam and bottom. Before filling, they were dipped in hot wax for sanitation and water resistance, and were stapled closed on top. The earliest ones did not use the integral pouring spout, but had a perforated patch opening on the upper part of one side panel. In 1961 there were two more improvements: precoated blanks that simplified the filling process, and a modified sealing material to facilitate easy opening of the spout (Figure 1.14) (Lisiecki 1997).

FIGURE 1.14. *This version of the wax-coated Pure-Pak milk carton was patented in 1956 by Ex-Cell-O of Detroit, Michigan. The paperboard was made by International Paper Company and converted by Potlach Forests, Inc., Dairy Service Division. From the collections of the Michigan State University Museum.*

In Sweden, Ruben Rausing's company, Åkerlund & Rausing, began development of a paper-based milk carton after investigating U.S. gable-top cartons during the glass and metal shortage during WWII. By 1950 he had invented the Tetra Pak process that formed and filled a tetrahedron shape from a tube of waterproof plastic-coated paperboard, and sealed it at alternate right angles. By 1960 the packaging system was used by dairies throughout Europe for cream and milk (Robertson 2002).

The shelf-stable drink box evolved from the tetrahedron system. The aseptic packaging process combines the ultra-high temperature sterilization of milk with a process for sterilizing the packaging material before forming and filling. The multi-layer material combined paper for printing, foil as a barrier, and polyethylene for sealing, waterproofing, and adhesive. The Tetra Brik aseptic carton was commercialized in 1969, and markets have been growing ever since, especially in areas where refrigerated distribution of milk is problematic.

There were other patented liquid-tight paper-based packaging systems that succeeded in the 1950s and 1960s. The Perga carton system was a 1931 German invention, using a ribbed wax-coated sleeve, a circular bottom, and square top. It would run on glass bottling equipment. Its market was interrupted by WWII, but it returned in the 1950s. American Can made cartons with a flat top. Its Combibloc system, patented in 1956, was like the Tetra Brik, but formed from knocked-down blanks rather than rollstock (Robertson 2002).

The set-up box industry gave birth to *fiber cans*, which at first were very similar to the cylindrical set-up boxes with side walls made by winding paper several times around a form or mandrel. Paper tubes with paper bodies and ends were first used during the Civil War for ammunition (Hanlon 1998).

The first fiber cans were made in about 1875 by the Chicago boxmaking firm, Ritchie and Duck. The ends were round die-cut discs that had been pressed into a cap-like form and could be slipped over the end (Bettendorf 1946).

Ten years later, in 1886, the W. C. Ritchie Company developed a lap-tube winder that made a tube with a lap seam from a sheet, in a manner similar to the way that convolute composite cans are made today. The early cans and tubes were made from straw paper and board, and most of the producers were in the Midwest. By 1900, fiber cans were used for powders (like baking powder), drugs, yeast, candy, and malted milk powders. Paper tubes were used for textile roll cores and for mailing, and for dry cell batteries, shell casings, fireworks, and dynamite (Bettendorf 1946; Henderson 2013).

A number of methods for spiral winding were developed during 1900–1910. Spiral winding offers a production advantage because it feeds from a continuous roll. During the same time, metal can ends were adapted to the fiber tubes. The Gem Fibre Package Company in 1905 patented a spiral-wound metal-ended sifter can used for Old Dutch Cleanser. Oatmeal was packed in fiber cans by 1909. Spiral winding proved to be a better way to make cans. It was quicker and less expensive than convolute winding, even after American Can Co. got into the fiber can business and developed automated convolute winding equipment (Bettendorf 1946).

Frozen orange juice was commercialized by Florida Foods (which became Minute Maid) in composite cans in the 1950s, making "fresh" orange juice available year-round. Foil-lined fiber cans that were able to pop open along the spiral seam were developed for ready-to-bake biscuit dough in the early 1950s by Bill Fienup of R. C. Can Co. in St. Louis, which became part of Sonoco Products (which had its start in 1900 making spools for yarn). Motor oil moved from steel to composite cans in the 1960s (Kelsey 1989).

As molded plastic alternatives grew, fiber can share fell steadily. Container Corporation of America sold its previously large composite can operation when one of its largest markets, motor oil producers, moved on to high density polyethylene bottles in the 1980s.

The convolute process proved successful for making large fiber drums. In about 1904, cylindrical fiberboard shipping containers were first made for cheese in Sturgis, Michigan, using fiberboard made by the Monroe Binder Board Company. In the 1920s, Herbert Carpenter developed the process for mechanically forming a drum tube with a series of windings. He used metal rims to hold the wooden disc heads in place. Fiber drums competed with wooden slack barrels and steel drums, and won the market for granular products like coffee, dry products like detergent, and semi-solid products like lard (Bettendorf 1946).

Fiber drums share a history with corrugated and solid fiberboard shipping containers, having developed at the same time. The early ones were made from juteboard, using sodium silicate as an adhesive, and were later replaced by kraft paper and starch adhesive. Fiber drums played an important role in WWII, and vice versa. The war revealed the fiberboard containers' flaws and led to the structural improvements that ensured their success.

Another kraft paperboard success was in multipack beverage carriers for cans and bottles. Bottles of beer and soft drinks were at first packed in wooden basket-style carriers. The introduction of paperboard multipacks exponentially accelerated the marketing of carbonated beverages, which further accelerated with the use of cans for beverages. But it required strong kraft (also known as solid unbleached sulfate) board to give the carriers their strength. The model for the business was developed by Atlanta Paper (now part of Mead/Westvaco), which pioneered systems of multipacking and formed strategic alliances with Coca-Cola and Anheuser-Busch (Brody 2004).

Paperboard beverage carriers have come and gone and come back again in the intervening years, especially for soft drinks. In the 1960s, HiCone (a trademark of ITW) plastic ring carriers captured the 6-pack can market. More recently, refrigerator dispenser packs have once again turned back to paperboard.

Ever since about 1975, plastics have cannibalized market share from the paperboard industry for specialty packages like milk cartons and beverage carriers, as well as for high volume products like disposable diapers (for which a plastic bag is one-quarter the cost of a paperboard carton). Powdered laundry detergent moved from small, to big, to gigantic cartons, to the highest volume paperboard item in the converting industry, only to lose almost all of the market to liquid in plastic bottles.

Throughout the last three decades the paperboard industry has suffered from denial, failing to respond with creative alternatives, like combining plastic and paperboard, which could better meet the needs of the market. For example, Coors (a trademark of Adolph Coors) had to build its own converting plant to make a light-caliper paperboard/plastic film lamination for its multipacks, a package the paperboard industry long refused to admit existed. And the paperboard industry shunned Tetra Pak as not a real member of the industry. Once the leaders in consumer packaging, the paperboard industry's survival strategy based on mill tonnage has endangered its ability to grow and provide new packaging solutions (Brody 2004).

1.2.4 Corrugated and Solid Fiberboard/Fibreboard[3] Shipping Containers

The paperboard inventions of the 1800s gave birth to the corrugated fiberboard shipping container in the early 1900s. Corrugated boxes were to play an essential role in developing mass distribution throughout the 1900s.[4]

The first patent was issued for making corrugated paper in 1856 in England. It was formed on the same kind of fluted irons used to make ruffled collars in Elizabethan times. Like the earliest paperboard, it was used in men's hats. The fluted band of paper served as cushioning for the hat's sweat band.

The first patent issued for corrugated paper as a packaging material was granted in 1871 to American Albert L. Jones as shown in Figure 1.15. It was a textured cushioning material for wrapping glass bottles to protect against breakage. The flute profile was the original *A-flute*.[5] In 1874, Oliver Long's patents added the single and double facings to prevent stretching.

The British hatband patent was used as evidence of prior art in a patent in-

[3]The spelling of "fibre" board once reflected the British, French, and Latin spelling at least until the 1940s, but is no longer used (except inexplicably by the Fibre Box Association) because government standards and other documents use the American form.
[4]*Reference note:* The history of corrugated fiberboard boxes is well documented; the primary sources for this section are Bettendorf (1946) and Howell (1940), unless otherwise noted. They were eyewitnesses to some the industry's early development. Koning (1995) presents a more comprehensive bibliography.
[5]A-flute was the first flute size, and several other sizes are named for the order in which they were introduced. B-flute, the second, is smaller; C-flute is in between A and B. E and F flutes, introduced in the 1990s, are also known as *microflutes* because they are so small.

A. L. JONES.

Improvement in Paper for Packing.

No. 122,023.

Patented Dec. 19, 1871.

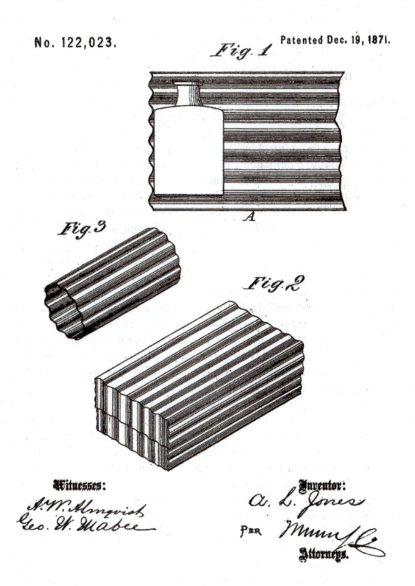

FIGURE 1.15. A. L. Jones' patent. Improvement in Paper for Packing. No. 122,023, patented December 19, 1871.

fringement settlement that led to an alliance of three entrepreneurs: Robert H. Thompson, Henry D. Norris and Robert Gair. They maintained a manufacturing and sales monopoly on corrugated bottle-wrapping paper for the life of the patents until the1890s. The Thompson Norris Company and Gair developed machinery, processes, and various forms of wrappers.

The early corrugated wrapping materials were made from thin straw sheets, like the straw wrapping paper of the period. A sheet was dipped in water and then passed through heated fluted rollers. A typical bottle wrapper was single-faced with the facing extending past the corrugated paper to fold in neatly around the neck and at the bottom. Some were gummed for convenience, and some were pre-made into cylinders. The material was also used for nests and pads as dunnage (void filler) inside wooden boxes.

When the bottle wrap patents ran out in the 1890s, Thompson and Norris developed the first double-faced board. Single-faced board would be unrolled, glue was applied to the second liner by a series of brushes, and the plies would be combined as they were pulled through the machine, which was stopped to cut off the sheet; in later machines, glue would be applied only to the tips of the flutes. They invented a rudimentary process for setting the glue under pressure: the machine operators piled up several sheets, covered them with boards and "would tramp around this board to give the necessary weight to make the liners adhere," (W. G. Chapin, quoted by Howell 1940).

In 1894, Thompson and Norris produced the first double-faced corrugated boxes for light express deliveries in New York City. The new box was tested by a Wells Fargo office that was "pleased to say has borne without damage, such handling as it would probably be called upon to stand in ordinary transportation," and their agent recommended them to other Wells Fargo shippers (W. B. Lindsay, quoted by Bettendorf 1946).

For the boxes, Thompson and Norris used a thicker and stiffer strawboard corrugated medium, close to the thickness (0.009 inches) of the standard medium material used ever since. The first liners were 0.016–0.018 inches thick and were made from "boiled wood pulp." Several box styles were designed, including the economical one-piece regular slotted container (RSC), which is still the most popular box style today.

Experimentation with different liner stocks continued until about 1906 when jute liners were first made. From then until 1936, jute linerboard was most common, made from wastepaper reinforced at first with jute and later with kraft fibers.

After 1895, corrugator machinery developments quickly multiplied. The first independent machinery manufacturer was S. F. Langston in Philadelphia; his first machine, in 1895 was a singlefacer. The first machine to corrugate the medium and affix both faces was invented in 1895 by Jefferson T. Ferres for the Sefton Manufacturing Company in Anderson, Indiana; he improved it in

1900 by adding steam-heated hot plates for drying the board and setting the glue. His machine, although it operated at only 10 feet/minute, is the basis for the corrugating machines used today.

RSC box-making equipment was also developed in the early 1900s. Rotary slitters and scorers were developed in 1905 by George Swift for making the blanks into tubes. Slotting was mechanized using saws in 1902, by the Sefton Manufacturing Company, a dusty practice that continued into the 1920s along with the up-and-down slotter developed in about 1905.

1.2.4.1 Rail Classification Cooperation

Throughout history, the makers of shipping containers have always had a strong relationship with the transport industry. But the marriage between the U.S. fiberboard shipping container industry and the railroads, which came to ensure the success of corrugated board, did not start out as a love affair.

The first serious effort to use corrugated fiberboard boxes for rail transport was about 1903, when a cereal manufacturer secured an exception to the wooden box requirement of the railroads in the Central Freight Association territory. This motivated nine manufacturers to unite in 1905 in a remarkable cooperative effort between competitors.[6] The first association was called The Progress Club and became the Corrugated Paper Patents Company. Their goal was to standardize the material and make it fully acceptable to all of the railroads. Through a series of trade associations (culminating in today's Fibre Box Association) they drafted specifications and in 1906 they applied to the Western Classification Committee of the railroads to permit the use of corrugated fiberboard boxes.

At the same time (1904), the solid fiber shipping container was being developed by the Illinois Fibre Box Company in Chicago. Solid fiber boxes were made from three or four sheets of flax or jute board pasted together with animal glue. The first ones had wooden frames because the material was hard to crease until George Swift developed a creasing machine in 1909. The W. K Kellogg Toasted Corn Flake Co. was the largest user of the boxes produced by the Illinois Fibre Box Co., which moved its manufacturing operations to Battle Creek, Michigan, in 1907 in order to better meet Kellogg's demand.

The Illinois Fibre Box Company also petitioned the Western Classification Committee in 1906, and joined the corrugated manufacturers in a July hearing at Frankfort, Michigan. As a result, the Western Committee accepted both kinds of boxes, as did the Eastern Committee in the same year, but they also authorized a 10% higher freight rate.

---

[6]The members were Thompson & Norris, Hinde & Dauch, J. W. Sefton, McPike Paper, Charles Boldt, Hunt & Crawford, J. N. Hahn, Modes-Turner Glass, and Lawrence.

More and more types of goods began to be shipped in fiberboard boxes. Mass production and distribution were multiplying markets. Besides cereal, RSCs were used in 1906 for glass-packed goods, starch, sugar, baking powder, candy, hardware, housewares, drugs, stationery, rubber goods, shoes, and soap. In 1910, the U.S. Bureau of Explosives approved fiber boxes for packaging Strike Anywhere matches. By 1916 they were also used for canned foods, cigarettes, and other tobacco products, blankets, clothing, chewing gum, chemicals, kitchen cabinets, and other furniture. In the 1920s their use would extend to products like radios, paint, and department store goods (Browder 1935).

But the railroads continued to charge a higher freight rate and argued that it was justified because fiberboard quality was considered to be mostly poor, and the railroad was liable for damage in transit. They also argued that they would lose freight revenue because of the corrugated box's lighter weight and smaller size.

In defense, quality standards were developed by the Corrugated Paper Patents Company and the Illinois Fibre Box Company, which joined other solid fiber box makers in 1909 to form the Fibre Shipping Container Association. Thirteen years later, the two industries joined to form one organization: The National Association of Corrugated & Fibre Box Manufacturers.

They standardized corrugated board properties and box weight limits, based on the thickness of facings and bursting strength. Bursting strength was based on the Mullen tester. It had been developed in 1887 by John W. Mullen for testing paper,[7] and by 1907 a jumbo sized version had been adopted by the Government Printing Office for testing book cover boards (Bettendorf 1946). It was also used by the textile industry for testing knit fabrics.

The Western railroads were especially resistant, charging on an "exception" basis as much as a 400% premium for shipments eastbound from California. They were concerned about more than damage. Due to federal land grants, the Western railroads had enormous timber holdings and investments in box-making sawmills, and therefore had a financial interest in favoring the use of wooden boxes in the West.

So an angry Pacific Coast boxmaker, R. W. Pridham, sued the eastbound railroads for discrimination. The Interstate Commerce Commission's landmark *Pridham Decision* in 1914 was a complete victory for fiberboard boxes: the ICC found that there were no transportation differences between the wood and fiber boxes, and it prohibited all tariff discrimination. The decision moved

---

[7]After serving as superintendent of several paper mills, Mullen operated the Mullen Brothers Paper Company board mill in St. Joseph, Michigan, until his death in 1915. The first tester was sold to the Parsons Paper Company in 1887 and is now in the Ford Museum. The test method still bears his name. B.F. Perkins and Sons (now owned by Standex International) acquired the trademark rights to manufacture the Mullen Burst Strength Tester in the early 1900s and has been the sole manufacturer since that time (Standex 2004).

fiber boxes out of the position of being a substitute container. Howells (1940) calls it "the Fourth-of-July of the industry.".

The ensuing cooperation gave the corrugated box the status it has today. The shipping container manufacturers and the rail, and later trucking, associations standardized corrugated fiberboard, and the standards appeared in all three railroad freight classifications in 1910, and when the three were merged in 1919, it was published as *Rule 41*, which exists to this day. In 1968, motor carriers adopted Item 222, similar to the railroads' Rule 41, for items carried by truck. The cooperative relationship between the corrugated board and transport industries continues today, and the implications are discussed in Chapters 14 and 15.

In the 1920s, corrugated boxes became the most prevalent form of shipping containers (FBA 1999). Except for a short increase in demand during WWII, solid fiber and wooden boxes gradually conceded markets to boxes made from the now ubiquitous corrugated board.

1.2.4.2 The Commodity RSC

From its outward appearance, the corrugated RSC that is still the most common box style used today has changed little from that used in 1914. But in the following decades there were many changes in materials and technology and, as a result, changes in the structure of the industry. There were new sources of pulp, changes in the process for making the containerboard, and new processes for combining the board and converting it into boxes.

The first board-combining adhesives were simple cooked starches and flours, which were capable of producing a good bond, but limited the corrugator speed. Between 1910 and 1920, the industry switched to a quicker-drying adhesive based on silicate of soda, which required less water and heat to cure than starch. Machine speeds increased to over 300 feet/minute. But silicate of soda is abrasive and over time ruined corrugating rolls, so in 1934, Jordan C. Bauer of the Stein-Hall Company developed a starch-based adhesive that would cure more quickly. The *Stein-Hall Formula* uses cooked starch as a carrier agent to keep the remaining raw starch suspended as heat at the glue line solidifies it and creates an instant bond; it is the basis for the adhesive used for corrugated board today.

In 1910, the invention of the printer-slotter by James Jones and Henry Gores, employees of the American Paper Products Company, simplified the box-making process. The concepts were based on rubber printing plates (developed by John Kerr in 1900), a "long way" slitter and scorer that also printed the blank (developed by George Swift in 1905), and Samuel Langston's 1908 printing press, which for the first time fed blanks the short way. Feeding blanks in the "short" direction would come to simplify folding and gluing, too.

Automatic tapers were invented in the 1920s for making the manufacturer's joint, but the blanks were still manually folded and fed into the machine. Once folding machines were developed and glues were improved, a gluing operation was added in the 1950s. Up until then, most manufacturers' joints were taped or stitched (stapled). One of the earliest folder-gluer machines was the Universal Comet (Shulman 1986).

The first boxes were closed with string, which was replaced by the 1920s with case sealing glue made from silicate of soda. But as it dried, it seeped out and hardened with sharp edges that cut the fingers of workers. Later, vegetable based adhesives were used that could be automatically applied, but the slow setting speeds required 50-foot long compression belts (Personal Touch 1977).

The first printing on corrugated board was with oil-based inks by letterpress. The development of flexography made possible the use of faster drying, low viscosity inks. Water-based inks have the advantage of being quickly absorbed by the porous board. Flexography also had the advantage of not crushing the flutes as much as letterpress.

In 1957–1960, a flexographic printing press was first added to a folder-gluer machine under the direction of Henry Kulwicki at Hooper-Swift (later Koppers). By 1970, there were approximately 2,000 flexo folder-gluers in use worldwide. Flexo folder-gluers greatly improved productivity, and were able to run at speeds up to 3-1/2 times faster than letterpress (Shulman 1986). Just as the development of lithography led to an extravaganza of color on labels and cartons, the development of flexography has led to an increasing emphasis on printing for corrugated fiberboard containers. Today, most corrugated boxes are formed and printed in flexo folder-gluer machines, as described in Chapter 14.

1.2.4.3 Corrugated Board Evolves

Containerboard pulp and papermaking processes have changed, too. In 1924, all-kraft cylinder machine liners first appeared in Rule 41. The jute linerboard and strawboard medium had been produced on multi-ply cylinder machines, and so was the first kraft linerboard. It was made in a Bogalusa, Louisiana, juteboard mill in 1915, and in 1923, the Hummel-Ross Fibre Corporation in Hopewell, Virginia, first produced kraft linerboard commercially.

In 1922, a variation on the kraft process was developed in the South that made the pulp stronger, less expensive, and easier to drain, making it possible to form board on a fourdrinier machine. In 1927, the Brown Paper Mill in Monroe, Louisiana, was the first mill to produce kraft board on a fourdrinier machine. The invention of the double flow headbox for the fourdrinier machine, in 1934 by John Sale of Hummel-Ross Fiber Corp., made it possible to put a smoother (or white) surface on the linerboard or a layer of virgin kraft atop a recycled layer (as had been done by cylinder machines for many years), and this is the process still used today. A smoother and/or whiter surface improves print quality.

The introduction of fourdrinier kraft board ignited a paper industry expansion in the South that fiercely competed with the cylinder-made jute linerboard producers in the North. From 1925 to1952, as linerboard production overall quadrupled, annual juteboard production remained relatively steady and kraft's share increased to 80% (Gates 1954).

Kraft linerboard, since it was stronger than juteboard, could be used at a lower basis weight. Whereas the original jute liners needed an 80-pound basis weight to pass the 200-pound burst test, "200 lb test" board made from today's kraft liners has a basis weight of half that.

The standard 0.009-inch corrugated strawboard medium began to be replaced in 1927 by one made from semichemical hardwood pulp. The first source was another waste product, spent chestnut chips that had been leached to extract tannin for the leather tanning industry. Chestnut pulp, like that made from straw, made a good stiff medium. This would prove to be true for other hardwoods as well, and neutral sulfite semichemical (NSSC) hardwood pulp is still used in some corrugated medium.

Once there were enough old corrugated containers (OCC) with high kraft content in service, it began to make economic sense to recycle them to reclaim their long fibers. In 1930, *bogus* medium made from OCC began to be used (the term *bogus*, though rarely used today, refers to paper made from recovered pulp). The recycled pulp was also used in juteboard, to replace imported virgin kraft and some of the wastepaper.

World War II was instrumental in the development of wet-strength board. There was a shortage of wooden boxes, and the regular fiberboard boxes that were sent as substitutes failed in the South Pacific where they were stored outdoors. In 1942, the paperboard mills came up first with "V" (for "Victory") water-resistant solid fiberboards: V1, V2, and V3. When demand quickly outstripped the capacity for solid fiberboard, the Southern kraft board mills and box makers developed *V3c* corrugated board, with heavy kraft medium and liners, glued with water-resistant urea formaldehyde resin. Its lighter weight cousin, *W5c*, was first used for the inner packing of wooden boxes, but later came to be used for lighter duty water resistant boxes (Lincoln 1945). After the war, the water-resistant boxes were found to be too costly for most domestic uses (Daly 1971).

The ballooning demand for corrugated board during and after the War stimulated the building of more virgin kraft mills. This new kraft supply ultimately crowded out the juteboard and solid fiber box suppliers. For the next 50 years, corrugated board in the United States had a high virgin kraft content.

The ascension of kraft linerboard dramatically changed the structure of the industry. The kraft manufacturers, corrugators, and box-making industries grew increasingly integrated. Further breakthroughs in the pulping of hardwoods and pine, and the ability to make containerboard on a fourdrinier machine, brought corrugated board into the heart of the forest industries.

1.2.4.4 Competition and Collusion in the Corrugated Industry

There has always been a powerful incentive to fix prices in the industry. Corrugated board is largely a commodity, and when the market is left to pure competition and inventory levels are high, prices can fall below variable costs. The integrated kraft/board/box producers have, since the 1930s, attempted to use their power and affiliation to maintain prices of linerboard, boxes, or both. They have been accused in anti-trust actions several times.

In 1939, the integrated producers were first accused of violating the Sherman Antitrust Act. They were sued for price fixing and allocating markets among themselves. In the 1930s, most of the producers did not know how to use their accounting systems to determine variable costs, and much of the early collusion was claimed to be based on attempts to standardize accounting principles. The suit was settled by a Consent Decree in which the independents agreed, without admitting guilt, to abstain from collusion (Daly 1971).

Mergers in the 1950s further increased the market power of the large producers.[8] The incentive to fix prices continued, and corrugated box salesmen developed the informally agreed upon practice of asking their customers for a competitor's price—and meeting it, because beating it caused the retaliatory threat of reciprocal price cutting. In 1967, the government again clamped down, suing the industry under the Clayton Antitrust Act. The result was an order to divest and the industry was ordered not to communicate between competitors about prices (Daly 1971).

In 1976, one of the largest anti-trust class action civil cases in history (brought by 58 companies on behalf of a class of over 200,000 purchasers of sheets and boxes) was brought against 37 corporations for exchanging price information, resulting in a record $550 million in settlements. The most certain violations were by the integrated producers who allegedly conspired to fix the price of linerboard, although they were also accused of fixing prices for their largest multi-plant customers (Goldberg *et al.* 1986). This coincided with an anti-trust indictment of 23 producers in the folding carton industry (Arzoumanian 1986).

In the 1990s, when the price of linerboard doubled in two years, it happened again. A class action suit was brought by a group of independent converters and end users. The suit alleged that the president of Stone Container, the largest

---

[8]This is the last we see of Robert Gair's company, as it merged with Continental Can. Other survivors were Owens Illinois (glass maker), Container Corp. of America, Inland Container, Weyerhaeuser, Crown Zellerbach, St. Regis, International Paper, Mead.

of the integrated producers, had orchestrated a plan in which his fellow liner-board manufacturers would close plants and idle mills in order to drive up the price of linerboard. Then they could charge the independents an inflated price for the linerboard as well as undercut the independents' box prices in the market. By 2003, the case's combined settlement was $210 million, not including several of the end users who opted out of the class action to pursue independent lawsuits (Duffy 2003). At this writing, another class action suit is pending.

Much of this litigation stemmed from the fact that the integrated producers control kraft linerboard production. During the 1950s, they perpetuated the legend of the virtues of virgin kraft linerboard, which has been emphasized by the industry ever since.

The increase in recycling in the 1990s has changed the materials again. Now most U.S. corrugated fiberboard is still kraft, but the OCC content varies depending on the state of recycling. Chapter 14 discusses the significance of corrugated board recycling today.

By the 1970s, almost every product in the United States was shipped in a corrugated fiberboard shipping container. Wax impregnation, introduced in the 1960s, even made them suitable for agricultural products with high moisture contents, like fruits, vegetables, fish, and meat.

1.2.4.5 Transportation Carriers and Test Grades of Corrugated Board

Some of the first tests for corrugated board and boxes were developed by the fiberboard associations from 1916 to 1927 through the Mellon Institute of Industrial Research in Pittsburgh. In 1913, the Forest Products Laboratory in Madison, Wisconsin, started research and developed the revolving drum test, which was a little like kicking a box down an endless set of stairs (Plaskett 1935). In 1950, Technical Association of the Pulp and Paper Industry (TAPPI) developed a Corrugated Containers Division that has been responsible for developing most of the material test methods used today, some of which are discussed in Chapter 8 (Coleman 1990).

In 1991, due to pressure from package buyers (notably members of the Institute of Packaging Professionals), the carriers and the Fibre Box Association (FBA) issued an alternative to the Mullen test specification for corrugated board based on edge crush test values (ECT). Package users had been dissatisfied with the burst test because, especially for palletized shipping, stacking strength had become a more useful criterion. Since ECT is related to compression strength, and compression strength is related to stacking strength, ECT was the property of corrugated board considered to be most related to performance. The ECT and Mullen tests are still the prevalent bases for corrugated board specification; their procedures are described in Chapter 8.

In the late 1990s there was yet another change to the classification, at the request of package users/shippers, and this one was more radical. Under the slogan "Do a 180," leaders in IoPP, ASTM, and International Safe Transit Association (ISTA) proposed and prevailed in adding the alternative *Item 180* to the *National Motor Freight Classification*.

Item 180 has no necessary relationship to corrugated fiberboard's material properties. In fact, a distribution package does not need to be made from corrugated board to pass the tests.

The low cost and light weight of corrugated boxes enabled more producers to economically employ wider distribution than ever before in history. Indeed, the exponential expansion of distribution throughout the 1900s would not have been possible without the help of the self-effacing brown corrugated RSC.

But, like other wood and paper based packaging, the market for corrugated boxes is also suffering from the competition of plastics. Deregulation of the transportation industry in 1980 and the phase-out of the Interstate Commerce Commission in the 1990s, coupled with new materials and more robust distribution testing, have reduced the barrier to new shipping container forms. Chapter 15 explores the details of this historic shift.

1.3 WOOD AND PAPER PACKAGING STATISTICS TODAY

Trees are still used to make the most commonly used packaging materials in the world: wood, paper, and paperboard. Table 1.1 shows world packaging use (by total value) for 2012. (Wood is not listed separately, rather, it is included in the "other" category.) Figure 1.16 shows the proportions for 2012.

In the United States, wood and paper-based packaging is a higher percentage of all packaging used than the world average. The reasons for this include our plentiful forest resources and mature, successful converting industries. The United States has a historical reliance on paperboard cartons, corrugated fiberboard boxes, and wooden pallets. As stated above, the proportional estimates vary depending on whether they are based on weight or value.

**TABLE 1.1. World Use of Packaging Materials
(by value in $ million) (Savinov 2013).**

	2008	2009	2010	2011	2012
Paper and paperboard	248,605	224,641	248,743	272,489	276,283
Plastic	249,418	226,513	248,400	276,888	280,007
Metal	118,006	106,153	112,224	122,027	121,938
Glass	49,297	47,735	51,985	56,802	56,211
Other	41,558	34,092	37,245	39,723	37,841
Total	**706,883**	**639,135**	**698,597**	**767,930**	**772,279**

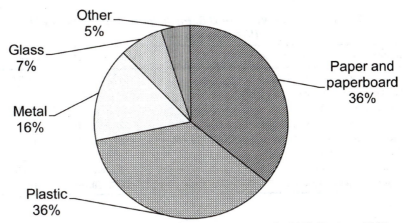

FIGURE 1.16. *Worldwide use of packaging materials, 2012 (Savinov, 2013).*

The U.S. Environmental Protection Agency measures *weight*, because it is concerned with waste disposal and recycling. As shown in Table 1.2, paper and paperboard are about one-half of the weight of all U.S. packaging waste; wood is 13%. Table 1.2 also shows that paper and paperboard have the highest percentage of materials recovered in recycling; over 75% of all paper-based packaging is recycled. Much of it is recycled right back into new packaging materials.

The U.S. Census of Manufacturing shows that paper-based packaging is about one-third of the value of all packaging, as shown in Table 1.3. The reason for the difference in weight verses value is that paper-based packaging materials are heavier and less expensive than other (especially plastic) materials. The difference is also reflected in the lower converting cost of wood and paper-based packaging, compared to plastic, glass, and metal. Also shown in Table 1.3, plastic packaging is nearly one-half of the value of all U.S. packaging.

TABLE 1.2. Weight (in Millions of Tons and Percent of Generation) of U.S. Packaging in 2012 (U.S. EPA 2014).

Container Type	Weight Generated	% of Weight	% Recovered
Paper and paperboard	38.01	51%	76.1%
Wood	9.61	13%	25.1%
Glass	9.38	12%	34.1%
Steel	2.23	3%	72.2%
Aluminum	1.87	2%	38%
Plastics	13.78	18%	13.8%
Total	**75.23**		

TABLE 1.3. Annual Value of U.S. Packaging Shipments in 2011 (U.S. Census 2013).

Container Type (NCAIS numbers)	Value ($1,000)	% Value
Paper and paperboard (see Table 1.4)	$43,906,825	32%
Wood containers and pallets (321,920)	$ 6,122,884	4.5%
Glass containers (327,213)	$ 4,870,286	3.5%
Metal containers (332,341, 332,115, 332,439, 32,225)	$20,622,916	15%
Plastic packaging (326,112, 326,111, 326,160, 326,140, 326,150, 3,261,994)	$61,378,823	44.7%
Textile bags (314,910)	$ 518,564	0.4%
Total U.S. Packaging Value	**$137,420,298**	**100%**

Table 1.4 shows the relative size of each of the U.S. wood and paper-based packaging industries that produce the forms featured in this book. Corrugated fiberboard boxes are over half of the value because they predominate the shipping container market. Chapters 14 and 15 focus on corrugated fiberboard packaging. Folding cartons are 19% (featured in Chapter 12), wood pallets and containers are 11% (discussed in Chapters 4 and 5), paper bags are 6%, and the other package forms (most of which are explored in Chapters 11 and 13) make up the remaining 7%.

Packaging is, of course, only one use for wood and paper. Worldwide, half of the wood harvested is used for fuel (IIED 1996). The highest volume (and value) use of non-fuel wood is for building and construction.

In the United States, 37% of the wood harvested is used to make pulp (U.S.

TABLE 1.4. Value of U.S. Wood- and Paper-Based Packaging Shipments, 2011 (U.S. Census, 2013), Comparing Package Forms in This Book (CC&CB).

Material (NAICS Codes included)	Value ($1,000)	% CC&CB
Corrugated fiberboard packaging (322,211)	$32,144,540	64%
Folding cartons (322,212)	$10,906,650	22%
Wood containers & pallets (321,920)	$ 6,122,884	12%
Uncoated paper and multiwall bags (322,224)	$ 2,157,240	4%
Plastics, foil, and coated paper bags (32,223)	$ 767,516	2%
Fiber cans, tubes, drums (322,214)	$ 611,787	1%
Liquid-tight paper cartons (3,222,151)	$ 821,889	2%
Other sanitary food cartons (3,222,155)	$ 1,545,726	3%
Molded pulp (3,222,991)	$ 505,966	1%
Total value, packages in CC&CB	**$50,029,709**	**100%**

National Agricultural Statistics Service 2012). Of that, about half is used in packaging (U.S. Census 2014).

This book focuses next on wood: the chemistry and properties of wood are examined in Chapters 2 and 3. Design principles for wood boxes, barrels, and crates are covered in Chapter 4, and wood pallets in Chapter 5.

The remaining chapters deal with paper and paperboard. Chapters 6 and 7 describe the wood pulping and paper-making process. Chapter 8 describes the properties of paper and paperboard, as well as the most common tests. Chapter 9 explores the types of paper and paperboard used for packaging. Chapter 10 is about printing. Chapter 11 discusses paper packages such as labels, bags, and wraps. Folding cartons, milk cartons, and set-up boxes are covered in Chapter 12 and 13. Chapters 14 and 15 cover corrugated board, the shipping containers made from it, and distribution package testing. Chapter 16 finishes the book by speculating on the future of wood and paper-based packaging.

1.4 REFERENCES

Abbott, D.L. 1989. *Packaging Perspectives,* Dubuque, IA: Kendall-Hunt.

Abramovitz, J.N. and A.T. Mattoon. 1999. *Paper Cuts: Recovering the Paper Landscape.* Washington, DC: Worldwatch Institute, Paper #39.

Ainsworth, J.H. 1959. *Paper, the Fifth Wonder.* Kaukauna, WI: Thomas Printing & Publishing.

American Forest and Paper Association. 2001. www.afandpa.org

Arkell and Smiths. *The Oldest Name in Paper Bags: Ninety Years of Know How.* Canajoharie, NY: Arkell & Smiths.

Arzoumanian, M.. 1986. Has Industry Learned how to Police Itself? *Paperboard Packaging,* v. 71, n. 8, August, 32–34, 37.

Bettendorf, H.J. 1946. *Paperboard and Paperboard Containers: A History.* Chicago: Board Products Publishing. Originally published in Fibre Containers, 1945–46.

Bobart, H.H. 1936. *Basketwork Through the Ages.* London: Humphrey Milford Oxford University Press.

Bower, P., Ed. 1996. *The Oxford Papers: Proceedings of the British Association of Paper Historians Fourth Annual Conference.* London: British Association of Paper Historians. Held at Oxford, Sept 17–19, 1993.

Brody, A. 2004. Personal correspondence, April 10.

Browder, G.R. 1935. A Story of the Fibre Box and History of its Developments. *Fibre Containers and Paperboard Mills,* vol. 20, no. 8, Reprinted in ed. G. Maltenfort, *Performance and Evaluation of Shipping Containers.* Plainview, NY: Jelmar, 1989, pp.1–7.

Bükle, J. and D. Leykamm. 2001. *The Manual of Labelling Technology.* Germany: Volker Kronseder.

Castle, J. 1962. Packaging of Pottery, in ed. Frank A. Paine, *Fundamentals of Packaging.* UK: Institute of Packaging.

Coleman, M.J., Ed. 1990. Forty Years of Corrugated, *Tappi Journal,* vol. 73, no. 11, 112A 1–32.

Conservatree, Making Paper. 2000. Available online, www.conservatree.com

Cornucopia of Processes. 1977. *Paper and Film Converter,* October, 64-71, 119–129.

Coyne, F.E. 1940. *The Development of the Cooperage Industry in the United States, 1620–1940.* Chicago: Lumber Buyers Publishing Company.

Daly, G.J. 1971. *The Corrugated Container Industry—A History and Analysis.* Oak Park, IL: Harry J. Bettendorf.

Davis, A. 1962. Print on Show in 1862," *British Printer,* August, 71–72.

Davis, A. 1967. *Package and Print: The Development of Container and Label Design.* London: Faber & Faber.

Doyle, J.J. 1947. The Coopers Union History as Recalled by Official, *Barrel and Box and Packages,* November, 14–16).

Drasner, R. 2004. *The History of Bags and Baggers.* Unpublished manuscript (Langston, Alabama).

Duffy, S.P. 2003. House of Cardboard, *The Legal Intelligencer,* November 212, vol. 229, no. 95, 1.

Elkington, G. 1933. *The Coopers: Company and Craft.* London: Sampson Low, Marston & Co.

FBA (Fibre Box Association). 1999. *The Fibre Box Handbook.* Rolling Meadows, IL: Fibre Box Association.

Forrest, D. 1973. *Tea for the British: The Social and Economic History of a Famous Trade.* London: Chatto & Windus.

Garoche, P. 1944. *Stowage, Handling and Transport of Ship Cargos.* New York: Cornell Maritime Press.

Gates, J.E. 1954. The Kraft-Jute Containerboard Controversy, *Fibre Containers and Paperboard Mills,* v. 39, no. 1, January. Reprinted in ed. G. Maltenfort, *Performance and Evaluation of Shipping Containers.* Plainview, NY: Jelmar, 1989, pp. 13–22.

Gilding, B. 1971. *The Journeymen Coopers of East London: Workers' Control in an Old London Trade with Historical Documents and Personal Reminiscences by One Who has Worked at the Block and an Account of Unofficial Practices down the Wine Vaults of the London Dock.* Raphael Samuel, ed., Ruskin College History Workshop Pamphlet. Oxford, UK: Ruskin College.

Goddard, R. 2000. One Thousand Years (and more) of Packaging. *Profit Through Innovation.* Leatherhead, UK: Pira International, pp. 53–54, 57.

Goldberg, R.C., C.S. Hakkio, and L.N. Moses. 1986. Competition and Collusion, Side by Side: The Corrugated Container Antitrust Litigation, in ed. Ronald E. Grieson, *Antitrust and Regulation.* Lexington, MA: Lexington Books.

Goodwin, J. 1991. *A Time for Tea: Travels Through China and India in Search of Tea.* NY: Alfred A. Knopf.

Hanlon, J.F., R.J. Kelsey, and H.E. Forcinio. 1998. *Handbook of Package Engineering,* 3rd ed. Lancaster, PA: Technomic.

Hankerson, F.P. 1947. *The Cooperage Handbook.* Brooklyn, NY: Chemical Publishing.

Hayes, E. P. 1929. History of the Dennison Manufacturing Company, *Journal of Economic and Business History,* v. 1, no. 4, 467–502.

Henderson, C.R. 2013. *Composite Cans, in Handbook of Paper and Paperboard Packaging Technology,* 2nd ed., edited by Mark J. Kirwan, pp 183–204. Chichester: John Wiley & Sons.

Hills, R.L. 1988. *Papermaking in Britain, 1488–1988.* London: Athlone Press.

Hine, T. 1996. *The Total Package.* New York: Little, Brown and Co.

Hollander, S.C. 1956. *History of labels: A Record of the Past Developed in the Search for the Origins of an Industry.* New York, N.Y.: Allen Hollander Co.

Howard, M. 1996. Coopers and Casks in the Whaling Trade, *Mariner's Mirror* (Great Britain) 82 (4), 436–450.

Howell, W.F. 1940. *A History of the Corrugated Shipping Container Industry in the United States.* Camden, New Jersey: Samuel M. Langston.

Hunter, D. 1947. *Papermaking: The History and Technique of an Ancient Craft.* 2nd ed. New York, Alfred A. Knopf.

Hunter, D. 1951. American Paper Labels, *Gutenberg Jahrbuch.* 30–33, tavel I-V.

IIED (International Institute for Environment and Development). 1996. *Towards a Sustainable Paper Cycle.* Geneva: World Business Council for Sustainable Development.

Impact Marketing Consultants. 2006. *The Marketing Guide to the U.S. Packaging Industry.* Manchester Center, VT: Impact Marketing Consultants.

Kelsey, R.J. 1989. *Packaging in Today's Society,* 3rd ed., Lancaster, PA: Technomic.

Kleiger, E.Fox. 2001. *The Better Bag, Invention and Technology,* v. 16, no. 3, 64.

Koning, J. 1995. *Corrugated Crossroads: A Reference Guide for the Corrugated Containers Industry.* Atlanta: TAPPI Press.

Koops, M. 1800. *Historical Account of the Substances Which have been Used to Describe Events and to Convey Ideas, from the Earliest Date to the Invention of Paper.* London: T. Burton.

Kouris, M., Ed. 1996. *Dictionary of Paper.* 5th ed. Atlanta: TAPPI.

Kovel, R. and T. Kovel. 1998. *The Label Made Me Buy It: From Aunt Jemima to Zonkers(The Best-Dressed Boxes, Bottles and Cans from the Past.* New York: Crown.

Kriebel, B. 2003. Box board plant has 101-year History, *Journal and Courier,* Lafayette, Indiana: March 2.

Labarre, E.J. 1952. *Dictionary and Encyclopedia of Paper and Paper-making.* 2nd ed, revised from a 1937 edition which included samples, reprinted 1969. Amsterdam: Swets & Zeitlinger.

LeBlanc, R., and S. Richardson. 2003. *Pallets: a North American Perspective.* Cobourg, Ont: PACTS Management.

Lewis, N. 1974. *Papyrus in Classical Antiquity.* Oxford: Clarendon Press.

Lincoln, W.B. 1945. Development of V' and W' Board Weatherproof Corrugated Shipping Containers, *Fibre Containers and Oaperboard Mills,* vol 39, no. 9, Reprinted in ed. G. Maltenfort, *Performance and Evaluation of Shipping Containers.* Plainview, NY: Jelmar, 1989, pp.8–12.

Lisiecki, R.E. 1997. Cartons, Gabletop. In *The Wiley Encyclopedia of Packaging Technology,* ed. Aaron L. Brody and Kenneth S. Marsh. New York: John Wiley & Sons, 187–89.

Locci, J.P. and M. Baussan-Wilczynski. 1994. *Memoires du Cartonnage de Valreas. Avignon,* France: Editions Equinoxe et Association pour la Sauvegarde et la Promotion du Patrimoine Industriel en Valcluse.

Long, R.P. 1964. *Package Printing.* Garden City, NY: Graphic Magazines Inc.

Lynn, C. 1981. *Bandboxes, Art & Antiques,* v. 4, no. 2, 50–55.

MacGregor, D.R. 1972. *The Tea Clippers.* New revised impression of 1952 impression; London: Conway Maritime Press.

Marquis, R.W. 1943. *The Wood Box Industry at War, Barrel and Box and Packages,* April, 8–10.

McCully, E.A. 2006. *Marvelous Mattie: How Margaret E. Knight Became an Inventor.* New York: Farrar, Straus and Giroux.

Meggs, P.B. 1998. *A History of Graphic Design,* 3rd ed. New York: John Wiley & Sons.

Milogrom, J. and A.L. Brody. 1974. *Packaging in Perspective.* Cambridge, MA: Arthur D. Little.

O'Brien, J.P. 1990. Straw Paper and Pulp—The World Market, *Straw - Opportunities and Innovations: 2nd Annual Conference,* May 15–17. Leatherhead, UK: Pira International.

Oka, H.. 1975. *How to Wrap Five More Eggs: Traditional Japanese Packaging.* Photographs by Michikazu Sakai. New York: Weatherhill.

Opie, R. 1987. *The Art of the Label.* Seacaucus, New Jersey: Chartwell Books.

Opie, R. 1989. *Packaging Source Book.* London: McDonald & Co.

Paine, F.A. 1991. *The Package User's Handbook.* Glasgow: Blackie (published in U.S. by AVI, New York).

Personal Touch, The. 1977. *Paper and Film Converter,* October, 57-63, 90–118.

Plaskett, C.A. 1930. *Principles of Box and Crate Construction,* Technical Bulletin No. 171, United States Department of Agriculture, Washington DC.

Plaskett, C.A. 1935. Forest Products Laboratory Launches Investigation of Fibre Boards, *Fibre Containers,* vol 20, no. 6, 6–8.

PSSMA: Paper Shipping Sack Manufacturers' Association. 1991. *Reference Guide for the Paper Shipping Sack Industry.* Tarrytown, NY: Paper Shipping Sack Manufacturers' Association, Inc.

The Pocket Pal. 2000. 18th ed. Michael H. Bruno. Memphis, TN: International Paper Co.

Raycraft, D. and C. Raycraft. 1980. *Shaker: A Collector's Source Book,* Des Moines, AI: Wallace-Homestead.

Rickards, M. 2000. *The Encyclopedia of Ephemera: A Guide to the Fragmentary Documents of Everyday Life for the Collector, Curator and Historian.* New York: Routledge.

Robbins, T. 1976. *Even Cowgirls Get the Blues.* New York: Houghton Mifflin, p. 103.

Robertson, G.L. 2002. The Paper Beverage Carton: Past and Future, *Food Technology,* vol. 56, no. 7, 46-52.

Roth, L. and G.L. Wybenga. 2000. *The Packaging Designer's Book of Patterns.* 2nd ed. New York: John Wiley and Sons.

Sacharow, S. N.d. *The Package—Its Art and History.* Unpublished manuscript received from the author in 2001.

Savinov, V. 2013. *The Future of Global Packaging to 2018.* Leatherhead, UK: Smithers Pira.

Shulman, J.J. 1986. *Introduction to Flexo Folder-Gluers.* Plainview, NY: Jelmar Publishing.

Sentance, B.. 2002. *Art of the Basket: Traditional Basketry from Around the World.* London: Thames & Hudson.

Shorter, A.H. 1993. *Studies of the History of Paper-making in Britain,* ed Richard L. Hills. Hampshire, UK: Variorum/Ashgate Publishing.

Soroka, W. 1995. *Fundamentals of Packaging Technology.* Herndon, VA: Institute of Packaging Professionals.

Smith, H.A. 1939. *Robert Gair: A Study,* with an Introduction by Lewis Mumford. New York: The Dial Press.

Steen, H. 1963. *Flour Milling in America.* Westport, CT: Greenwood Press.

Twede, D. 2012. The Birth of Modern Packaging: Cartons, Cans and Bottles. *Journal of Historical Research in Marketing,* vol 4, no 2, 245–272.

Twede, D. 2003. The Packaging Technology and Science of Ancient Transport Amphoras, *Packaging Technology and Science,* vol 15, no 4, 181–195.

Twede, D. 2005. The Cask Age: The Technology and History of Wooden Barrels, *Packaging Technology and Science,* vol 18, no 5, 253–264.

Twede, D. 1997. Uneeda Biscuit: The First Consumer Package? *Journal of Macromarketing,* 17 (2), 82–88.

Twede, D. and B. Drasner. 2013. Marketing Dust: The Effect of Packaging Technology on the Marketing of Cement and Carbon Black, *Proceedings of the 16th Biennial Conference on Historical Analysis and Research in Marketing,* ed. Leighanne Neilson, pp. 272–277. http://faculty.quinnipiac.edu/charm/cumulative_proceedings.htm

U.S. Census. 2013. *Annual Survey of Manufactures: Value of Products Shipments: Value of Shipments for Product Classes.* Washington, DC: U.S. Bureau of Census.

U.S. Census. 2014. *U.S. Statistical Abstract.* Proquest. http://cisupa.proquest.com/ws_display.asp?filter=Statistical%20Abstract

U.S. EPA. 2014. *Municipal Solid Waste Generation, Recycling and Disposal in the United States: Facts and Figures for 2012.* EPA530-F-14-001. Washington, D.C.: U.S. Environmental Protection Agency. http://www.epa.gov/epawaste/nonhaz/municipal/msw99.htm

U.S. National Agricultural Statistics Service. 2012. *Agricultural Statistics Annual.* http://www.nass.usda.gov/Publications/Ag_Statistics/

Van Der Molen, H. 2002. Short History of the Strawboard Industry in the Province of Groningen, The Netherlands. Accessed 12/20/03, page dated 3/6/02. http://home-1.worldonline.nl/~molen/scripophily/straw.html

Von Hagen, V.W. 1944. *The Aztec and Maya Papermakers.* New York: J.J. Augustin.

Weeks, L.H. 1916. *A History of Paper-Manufacturing in the United States, 1690–1916.* New York: Burt Franklin, reprinted in 1969.

Wells, D.A. 1899. *Recent Economic Changes and Their Effect on the Production and Distribution of Wealth and the Well-being of Society.* New York: D. Appleton.

Wever, R. and D. Twede. 2007. The History of Molded Fiber Packaging; A 20th Century Pulp Story, *22nd IAPRI Symposium Proceedings* (on disk). Leatherhead, UK: Pira (Windsor, UK September 3–5, 2007).

World Around Us. 1977. *Paper and Film Converter,* October, 72-75. 160–168.

1.5 REVIEW QUESTIONS

1. Explain why trees are a plentiful, natural, renewable, and biodegradable resource.

2. How do basket shapes differ for different uses?

3. For what kinds of products and distribution were wooden barrels used for over 2,000 years? What is meant by a "slack barrel?"

4. What are the advantages of barrels for material handling and transport?

5. What were the roles of the medieval coopers' guild? How did an apprentice's Trusso ceremony differ from the final exams and a graduation of today's packaging students?

6. Why did barrels lose market share with the coming of railroads and mechanized woodworking?

7. What wood material's earliest use was for tea chests?

8. Upon what factors were the size and weight of tea chests based?

9. What did tea chests symbolize in America? What is their greatest reputation in the United Kingdom?

10. How is the use of wooden boxes tied to the history of railroads?

11. Which war was responsible for the introduction of pallets?

12. Who invented paper? When and where?

13. How is paper different from papyrus?

14. For the 17 centuries (105–1850) after the invention of papermaking, what was the primary source of fibers?

15. What three key inventions in the early1800s enabled the mass production of paper based packaging?

16. What are the names of the two kinds of paper making machines and what is the difference? Which one has a history of being used to make thick paperboard?

17. Which was invented first: the paper-making machine or the method for pulping wood? Did this surprise you?

18. Why wasn't wood used in paper until after 1850? What advantages and disadvantages does it have compared to rags?

19. What was the source of jute fibers and why were they used to make paper?

20. Why was straw used to make packaging-grade paper and paperboard? What factors led to it being replaced by wood fiber?

21. What does the word "kraft" mean? In what language?

22. Which printing process brought color to packages in the early 1800s?

23. For what purpose was the paper bag first developed?

24. For what purpose was flexographic printing developed?

25. What package forms are the Dennison family known for?

26. How was Robert Gair inspired to use a single die ruled for cutting and scoring a paperboard folding carton blank in a single impression?

27. Why is it a great challenge to package a liquid in paper?

28. For what packaging purpose was single-faced corrugated fiberboard first used?

29. During the early 1900s, what kind of pulp was used to make corrugated linerboard and medium? What is used now?

30. What obstacles did corrugated fiberboard have to overcome in order to be accepted as a shipping container material?

31. What was the significance of the Pridham Decision in 1914?

32. Approximately what percentage of U.S. packaging usage is represented by paper and paperboard? Is this by value or by mass? What percentage is wood? Is this by value or by mass? Why do they differ?

CHAPTER 2

From Trees to Lumber

Wood has many advantages as a packaging material. It has high strength for its weight, high stiffness, good durability, and reasonable versatility for design. It maintains its strength well when it gets wet. Compared to other rigid materials like steel and glass, it is relatively light weight. Wood packaging offers excellent rigidity, stacking strength, and physical protection.

Wood is a low cost commodity, easily available everywhere in North America. Trees are a natural resource, making wood universal, abundant, and inexhaustible given proper forest management. Wood is an environmental packaging material because it requires little energy to process, stores carbon, does not pollute, and it is renewable and biodegradable.

But wood is not necessarily an inexpensive packaging material when the system-wide costs are considered. The cost to make wood into a container is relatively high. Sawmills are relatively slow. Wooden packages require time and skill to assemble. They are not very compatible with high speed automated packaging operations. Furthermore, more mass is required to make a package, compared to other materials like fiberboard or plastic. Wood is bulky, heavy, and may present storage problems.

Wood excels when it is made into packages that require great strength or are expected to convey a natural or furniture-like effect. Wooden packages are suited to small scale production and can be manufactured with simple equipment in a variety of forms, including boxes, crates, pallets, and barrels. For packages in which mass is an asset (like reusable pallets), wood can be less expensive than plastic or steel. Wood can be combined with metal to obtain benefits from both materials. Wood is used in pallets and crates for its strength and relatively low cost compared to other high-strength packaging materials. It is chosen for its appearance and tactile feel in presentation boxes for premium items ranging from fine liquors to silverware.

TABLE 2.1. Hardwoods and Softwoods.

Type of Wood	Softwoods (evergreen)	Hardwoods (deciduous)
Examples of Species	Pine, spruce, larch, fir, hemlock, cedar, redwood, juniper	Oak, maple, aspen, ash, beach, cottonwood, cherry, hickory, sycamore, walnut, willow, alder, birch
Non-packaging Uses	Building construction Printing paper (blended with hardwood)	Furniture Printing paper
Major Packaging Uses	Kraft paper Corrugated board liners Light-duty crates, boxes, pallets	Barrels, pallets Heavy-duty crates, boxes Corrugating medium
Characteristics	Long fibers High lignin content Easy to work with, but lower in strength	Short, stiff fibers Harder to work with, but stronger

Since it is a natural material from a living tree, wood continues to undergo a natural lifecycle after harvesting. It is especially affected by water and is not a good moisture barrier. It is subject to sorption of water vapor and liquid water, and the presence of water can change its dimensions. Because of the way trees grow, the properties of wood vary with direction (a condition called *anisotropy*). In the end, wood will ultimately biodegrade.

The properties and cost of wood vary considerably with the type of tree. The major two classifications of trees are softwoods (evergreen conifers) and hardwoods (deciduous with broad leaves) shown in Table 2.1. They both belong to the botanical division of *spermatophytes* which means produced from seeds.

2.1 HARDWOODS

Botanically, hardwoods are *angiosperms*, trees whose seeds are enclosed in the ovary of the flower. They are deciduous trees, with broad leaves, and, with few exceptions, in the temperate region lose their leaves in fall or winter. Some examples are oak, maple, aspen, ash, beech, cottonwood, cherry, hickory, sycamore, walnut, willow, alder, and birch.

Hardwood is commonly used for making furniture. In packaging, hardwood, most commonly oak, is used to make reusable pallets, barrels, and heavy-duty boxes and crates. It is, on average, denser and harder to work with than softwood, and it has higher strength.

Hardwoods and softwoods can be pulped to make paper. Hardwood fibers are shorter, stiffer, and less strong than softwood fibers. Hardwood fibers better fill in a sheet and make a smoother paper that is more opaque and better for printing. Most printing paper is made from a blend of both hardwood and softwood fibers.

Short, stiff hardwood fibers also make a paper that is easy to corrugate and yet provides a rigid structure. Hardwood fibers are often used to make the corrugated medium layer of corrugated fiberboard.

2.2 SOFTWOODS

Softwoods are *gymnosperms*, trees whose seeds lack a covering layer and are not enclosed in the ovary of the flower. They are evergreens, cone-bearing plants (conifers) with needles or scale-like evergreen leaves. Some examples are pine, spruce, larch, fir, hemlock, cedar, redwood, and juniper. It should be noted that some softwoods, such as longleaf pine and Douglas fir, are actually harder than some hardwoods, such as basswood and aspen.

Softwood lumber is commonly used in framed building and construction. In packaging, softwood, most commonly pine, is used to make light-duty crates, boxes, and pallets. Only the harder softwoods with a density higher than 30 lb/ft^3 (450 kg/m^3), like dried Southern Pine, are used to make pallets. In general, softwood is easier to work with than hardwood but has lower strength.

For papermaking, softwood is preferred because it has longer, stronger, and slender fibers. Softwood pulp is the source of the strongest kraft paper, used to make bags, high-strength paperboard, and the facings for corrugated fiberboard.

2.3 RESOURCE BASE AND WOOD PRODUCTION

The type of wood used for packages and paper pulp varies depending on the species of trees that are regionally available. Since wood is bulky and expensive to ship, most construction and pulp wood used is based on local natural resources.

The United States has abundant forest resources, covering one-third of the land area, about 1.2 million square miles (751 million acres) of the country's 3.5 million square miles (2.3 billion acres). In 1630 during the European settlement, U.S. forest area was estimated at one billion acres. Forest land was mainly in the North and Southeast when forests were cleared for agriculture and housing (USDA Forest Service 2011). A forested area is classified as *forestland* if it is at least one acre in size and if 10% of it is covered by trees. It is estimated that more than 30 million acres in urban settings also satisfy this definition but are not included in the U.S. forestland (Alvarez 2007).

Today, about two-thirds of the U.S. forestland, 514 million acres, is classified as *timberland*, defined as productive forests that are capable of producing at least 20 ft^3 per acre of industrial wood per year, and are not legally reserved from timber harvest. Seventy-five million acres are reserved for non-timber

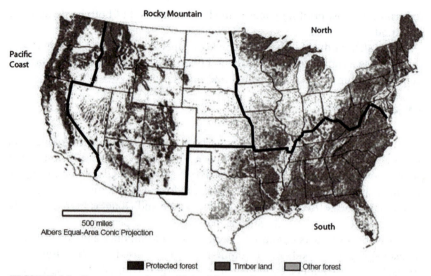

FIGURE 2.1. *Forest areas in the United States showing North, South, Rocky Mountain, and Pacific Coast regions. Alaska and Hawaii are also part of the Pacific Coast region but are not shown. Adapted from U.S. Forest Service, 2011.*

uses in the form of parks or wilderness areas. Primarily in the South and West, including Alaska, 162 million acres are classified as *other forest* because, while they produce some wood and tree products, they produce little industrial roundwood (US Forest Service 2011).

Industrial roundwood is defined as wood in the rough, intended for processing into lumber, panels, pulp, paper, or similar products, but not for energy production. Production of roundwood is primarily from growing stock, but some production also comes from sources that are not counted as growing stock such as dead trees, deformed cull trees, very small trees, trees from seldom used species, and trees from non-forest land.

The U.S. Department of Agriculture's (USDA) Forest Service divides the United States into four main regions: North, South, Rocky Mountain, and Pacific Coast. The map in Figure 2.1 shows the forested areas.

The most productive forest land in the United States is located in the South. Of the 179 million acres that are in the two highest productivity classes (capable of producing more than 85 ft^3 of industrial wood per year per acre), 98 million acres are located in the South, and the Pacific Coast region accounts for an additional 38 million acres. The South also has 106 million acres of lower productivity land (20–85 ft^3 per year per acre). The North has slightly more, 128 million acres of the total of 335 million acres in this category (Smith *et al.* 2009).

Most of the forest land in the "other forest" category, only capable of producing under 20 ft³ per year, is in the Pacific Coast region, 94 million acres of the 162 million acres in the category. The Rocky Mountain region has 59 million acres in this category. The Pacific Coast also has the largest fraction of the forest land in the reserved class, 46 million of the total 77 million acres. The Rocky Mountain region is next, with 20 million acres (Smith *et al.* 2009).

The types of timberland in the United States vary widely. In the eastern half of the United States, the majority of the forested land is classified as eastern hardwood forest, containing a number of species. Oak-hickory forests are most widespread. Oak-pine and oak-gum-cypress forests are found in the South, and maple-beech-birch and aspen-birch forests in the North.

Softwood forests are common in the South. The most important group are pines in most of the East, although they cover a smaller surface area than hardwood forests. Longleaf-slash pine and loblolly-short leaf pine forests are the major important softwood species in the South.

Western forests are primarily softwood, with Douglas fir the most common and useful. Other forest types include ponderosa pine in the Rocky Mountains, east of the Cascade Range in the Northwest, and in the Southwest. Fir-spruce forests are very common at higher elevations but most are not productive timberland. Hemlock-sitka spruce forests are found in Oregon, Washington, and Alaska. Lodgepole pine forests are found in the Rocky Mountain area.

Hardwoods in the West include oak in California, aspen in the Rocky Mountains, and red alder in the Northwest. Figures 2.2 and 2.3 show the predominant types of timberland in the Eastern and Western United States (Smith *et al.* 2009).

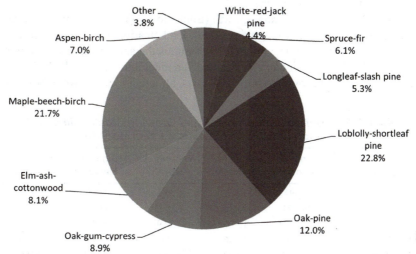

FIGURE 2.2. *Types of forests in the Eastern United States (Smith et al. 2009).*

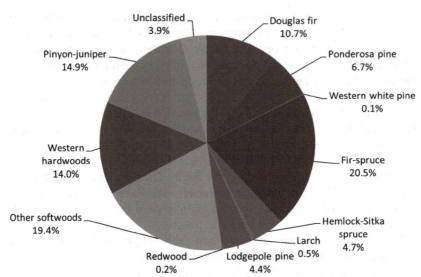

FIGURE 2.3. *Types of forests in the Western United States (Smith et al. 2009).*

The United States is rich in forest resources, and the resource base has slightly increased by 1% (8 million acres) in the last decade. Growth of timber in the United States as a whole has exceeded losses due to harvesting, mortality, and changing land uses for the last several decades, though concern has been expressed that harvesting has been increasing at a faster rate than growth in some regions. While overall growth exceeds production, in some areas and for some species this is not the case.

Table 2.2 shows net annual growth (minus loss to mortality, fire, etc.), annual harvest (harvesting, logging residues, land clearing, etc.), and the net annual gain, which is the net annual growth minus the annual harvest by region. About 62% of all timber harvested came from the South region, which also has the lowest ratio of growth to harvest (1.5:1), although it is still well above the

TABLE 2.2. United States Timberland Growing Stock (million ft.³) by Region (U.S. Forest Products Laboratory 2006).

Regions	Annual Growth	Annual Mortality	Net Annual Growth	Annual Harvest	Net Annual Gain
North	7,694	1,789	5,905	2,711	3,194
South	14,171	2,138	12,033	9,696	2,337
Rocky Mountain	2,997	1,127	1,869	541	1,328
Pacific	6,786	1,586	5,200	2,474	2,726
U.S. Total	**31,647**	**6,640**	**25,007**	**15,423**	**9,585**

replacement level ratio of 1:1. The replacement level ratios of all other regions are higher than 2:1, confirming that growth considerably exceeds harvesting and mortality (U.S. Forest Products Laboratory 2006).

Of the kinds of wood *removed*, industrial roundwood is the highest percentage use. Overall, industrial roundwood production in the United States accounted for 11.3 billion ft³ (321 million m³) of timber removal in 2012. Logging residues totaled 0.45 billion ft³ (14.5 million m³) and wood fuel totaled 1.4 billion ft³ (40.4 million m³) (FAO 2014).

Of all the roundwood harvested in 2012, *sawlogs* accounted for 52% and *pulpwood* totaled 43%. Sawlogs accounted for 60% of all softwood harvested, but only 34% for hardwood.

Nearly 61% of the pulpwood harvested was softwoods.

In 2006, about 76% of the total sawlogs came from the South Central, Southeast, and Pacific Coast regions, and 98% of the pulpwood came from the eastern half of the United States, with 75% of the total production from the South. The pulp industry in the Pacific Coast region is dependent primarily on chips produced as byproducts from lumber manufacture (Smith *et al.* 2009).

While the United States is a significant net exporter of wood, it is also an importer, primarily importing from Canada. Overall in 2012, the United States exported 442 million ft³ (12,507 thousand m³) of industrial roundwood, and imported 5.2 million ft³ (148,000 m³) of the same. Production of industrial roundwood is dominated by North America (United States and Canada) and Europe with about 58% of the 1.7 billion m³ total as shown in Figure 2.4 (FAO 2014).

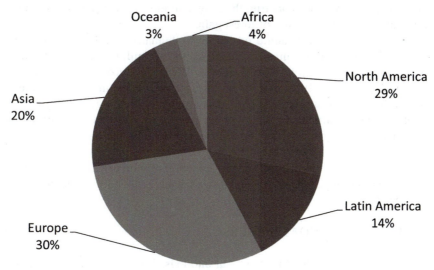

FIGURE 2.4. *World production of industrial roundwood in 2012 (FAO 2014).*

Fuelwood and charcoal are the largest use for wood worldwide, accounting for 1,870 million m^3 of consumption in 2012 compared to 1,657 million m^3 of industrial roundwood. About 90% of the wood used for fuel and charcoal is consumed in developing countries, mostly for household energy, while 70% of the industrial roundwood is used in developed countries (FAO 2012).

Deforestation around the world averaged 13 million hectares per year between 2000 and 2010, but the rate is slowing. The five countries with the largest annual net loss of forest area in the period of 2000–2010 were Brazil, Australia, Indonesia, Nigeria, and Tanzania. South America and Africa lost 4 million hectares and 3.4 million hectares of forest cover annually from 2000 to 2010, respectively. Brazil lost about 2.6 million hectares of forest annually in the same period but this was down from 2.9 million per year in the 1990s. Indonesia's forest loss went from 1.9 million hectares/yr. in the 1990s to 0.6 million/yr. between 2000 and 2010 (FAO 2010).

However, while there was a net loss of forests in some parts of the world, there were gains in other areas. The countries with the largest annual net gain in forest area in the past decade were China, the United States, India, and Vietnam, and Turkey and Spain tied for fifth. Costa Rica, Japan, and Thailand are among the countries that have recently recorded a change from having a net loss of forests to having a net gain in forest area (FAO 2010).

In percentage, the net world forest area decreased by 0.2% per year in the period from 1990 to 2000, by 0.12% per year between 2000 and 2005, and was up again to 0.14% per year between 2005 and 2010 (FAO 2012).

The major cause of loss of forests is the conversion of forests to agricultural land, while in developed countries the growth of forests on abandoned agricultural land is the largest cause of forest gain.

Forest fires, whether natural or caused by humans, also have an impact. In 1997 and 1998, forest fires consumed about four million hectares in Indonesia. Fires burned two million hectares of forest in Brazil and two million hectares in Russia in 1998. In 2003 in Portugal, nearly 380 thousand hectares, 6% of the total forest area, was burned in a series of fires. Fire activity in the United States is particularly high with an average of 3,730 thousand hectares consumed annually since 2000. Of course, regrowth on the burned lands can potentially replace the lost forest (FAO 2010).

Forest cover and production and consumption of lumber vary considerably in different regions of the world, and also vary markedly between countries in a given region.

2.4 PRODUCTION OF LUMBER

Trees are felled in the forest and cut into logs. The logs are transported by truck, rail, and/or water lines (such as rivers or purpose-built *flumes*) to the sawmill, where they are stored, sometimes in ponds, for a short period before processing. The logs are washed, and then enter the sawmill.

First, the bark is removed with a debarker. This is done to produce bark-free wood that will be used to produce strands for oriented strand boards, veneers for plywood and laminated veneer lumbers (LVL), or lumber for pallets, boxes, and crates. Debarking is essential to satisfy the international requirement of bark-free wooden packaging for export.

The most used technologies to remove bark include the ring debarker, the rosserhead debarker, and the drum debarker. In the ring debarker, a circular ring equipped with knives holds logs under pressure and then removes the bark by tearing it off. This debarker is suited for straight logs. The rosserhead debarker is a rotating cutterhead that travels along rotating logs to cut the bark on straight and crooked logs. The drum debarker is a large diameter drum where small diameter logs tumble on each other to abrade the bark away.

Once the bark is off, the primary breakdown of the log consists of using a *headsaw* to reduce the bark-free log to pieces called *cants* (the thickest pieces, sometimes with only the rounded bark edges removed) and *timbers* (more than five inches wide and thick). As each cut is made, the sawyer examines the log to determine the best way to cut it up to get the maximum grade and yield. In general, the grade of lumber is related to the strength for softwood and to the defects visible on the surface for hardwoods. The lumber yield is related to the amount of useable lumber that can be cut from a log. The primary breakdown is mostly performed with two types of saws: a circular saw and a band saw.

The secondary breakdown, also known as resaw, consists of cutting cants and timbers into lumber of specific thickness, width (edging), and length (trimming). *Lumber* is classified into three main categories: boards, dimensional lumber, and timbers, depending on the thickness and width. *Boards* measure less than two inches thick, with a width of one inch or more. *Dimensional lumber* is thicker, between two and five inches thick, with a width of two inches or wider; this category is easy to remember because it is used for thick three-dimensional objects. The thickness and width of *timber* is over five inches.

After sawing, the wood is sorted visually by grade and then dried. In order to help users select lumber meeting their needs, use categories, or grades, have been established based on the characteristics of the lumber. Features that are included in determining the grade are the size (width) of the piece and the number, character, and location of features that can lower the strength, durability, or utility of the lumber, such as knots, checks, and stains.

Wood is strongest and most stable in the *grain* direction, which follows the arrangement of cells in the tree, from bottom to top. This gives long boards and timbers their strength. But when a log is cut into lumber, pieces of varying quality result. Since the tree and its rings are cylindrical and the lumber is sliced flat, there is always a certain amount of curved cross-grain. There is also a certain amount of non-uniform grain from the tree's branches, knots, and other discontinuities. Such characteristics affect the properties and aesthetics of a piece of lumber.

The grading criteria have been developed by manufacturers' associations, and there is a different grading system for softwoods than for hardwoods. There are also differences in how hardwood and softwood lumber is dimensioned and how the dimensions are specified.

Lumber is also categorized by the degree of surface finishing. *Rough* lumber is simply sawn to a dimension. It shows saw marks on all four surfaces. Most of the hardwood lumber used for packaging is rough. Most softwood is surfaced on both sides because surfacing is a standard part of the softwood sawing process.

Dressed (or *surfaced*) lumber has been smoothed by a planning machine on one side (*S1S*), two sides (*S2S*), one edge (*S1E*), two edges (*S2E*) or any combination of these (*S2S1E, S1S2E, S2S1E,* or *S4S*). *Worked* lumber has been dressed and shaped to some pattern such as by notching, beading, or beveling.

2.4.1 Hardwood Lumber Grades and Dimensions

The grade of a lumber refers to the number, character, and location of features that may lower the strength, the volume, the durability, the aesthetics, and the use value of the lumber. It includes knots, checks, splits, shakes, pitch, and gums pockets, stains, wanes, and barky edges. The grade is determined by the proportion of the piece that can be cut into a certain number of pieces called *cutting units* measuring one inch by one foot by one foot, clear on one side and with the reverse face sound.

The criteria for hardwood grades are established by the *National Hardwood Lumber Association* (NHLA) (NHLA 2011). Hardwood lumber is mostly used for furniture manufacture, cabinetwork, flooring, paneling, molding, millwork, and pallets. Hardwood lumber is graded and marketed as factory lumber, dimension parts, and finished products.

The best grades with at least 83% of the piece clear of defects on the poorest face are termed *Firsts* and *Seconds* (*FAS*). *F1F* is FAS on one face. The third grade, Selects, have one face equal to a Second. *FAS* have a minimum width of 6 inches. Selects have a minimum width of 4 inches. The remainder of the grades are called *Common: No. 1 Common, No. 2 Common, Sound Wormy, No. 3A Common,* and *No. 3B Common.*

For pallets, *No. 3B Common and below* is used. A low grade of *factory lumber* is used as opposed to the hardwood *dimension* or *finished* products that are used for furniture and construction. A *3 common* hardwood lumber is at least four feet long, three inches wide, and is at least 25% defect-free cutting units (NHLA 2011).

TABLE 2.3. Selected Standard Thicknesses for Hardwood Lumber.

Rough (in inches)	Surfaced Two Sides (S2S) (in inches)
1	1-3/16
2	1-3/4
3	2-3/4
4	3-3/4

Hardwood dimensions are straightforward. The dimensions specified are the actual dimensions of the seasoned wood. There are some standard lengths: 4, 5, 6, 7, 8, 9, 10, 11, 12, 13, 14, 15, and 16 feet. There are no standard widths, but there are minimum widths for each grade. Thickness should be expressed in increments of one-quarter-inch. Surfacing one edge (S1E) or two edges (S2E) reduces the thickness. Some standard hardwood thicknesses are shown in Table 2.3.

2.4.2 Softwood Grades and Dimensions

Softwood can be graded by visual inspection or by machine. The size and location of knots, slope of the grain, decay pockets, and other defects are evaluated on the surface of the wood. A machine is used to stress-rate softwood lumber in terms of stiffness which is related to bending strength.

The two most important softwood grades are *select* and *common*. *Select softwood* is high quality, suitable for finishing, and used for construction and carpentry. It comes in Grades A (free from all defects) and B that are suitable for natural finishes, and C and D that have more blemishes but are suitable for painting.

For most packaging applications, common grade, intended for general construction, is suitable. *Common softwood* is not high enough quality for finishing, and is categorized by number. *Number 1 common* is sound, tight-knotted, and watertight, and is used as a low priced finish lumber. *Number 2 common* is grain tight, with coarser knots, used for subflooring and unfinished cover lumber. *Number 3 common* is the grade used for boxes and crates; it has knotholes and greater amounts of shake and decay. In some species, there are also *Grades 4 and 5 common*, which are very low quality. The American Softwood Lumber Standard sets the grades for softwood (U.S. National Institute of Standards and Technology 2010).

Boards under two inches thick are generally not stress-graded. The grade is determined by the shape and size of the pieces, along with the visual requirements of the grades. Depending on the wood species and the manufacturing association setting the standards, three to five different grades are generally established, using either numbers or descriptive terms.

In the United States, the width and thickness of softwood boards are expressed in *nominal* dimensions, which means that a standard "two by four" is neither two nor four inches, but is actually 1-1/2 by 3-1/2 inches. The actual widths and thicknesses are specified by the American Softwood Lumber Standard (U.S. National Institute of Standards and Technology 2010), and common dimensions are shown in Table 2.4.

One explanation for the difference between nominal and actual dimensions is an assumption that the rough green wood was really two by four inches when it was first cut, but that shrinking and surfacing made it thinner and narrower. Although this is not strictly true (because improvements in processing have made it possible to reach the targets shown in Table 2.4 by beginning with thinner stock), it is a significant factor. The width and thickness of softwood decrease, because the wood shrinks radially and tangentially as it dries.

A second explanation is that softwood is always sold surfaced on four sides, and such planning removes some of the surface, reducing the dimensions. A third explanation is that since wood is sold by the *board foot*, keeping the nominal dimensions in whole numbers simplifies transactions.

Standard softwood board nominal widths are 2, 3, 4, 6, 8, 10, and 12 inches. Nominal thicknesses are in half-inch increments. Softwood lumber is manufactured in lengths of multiples of one foot, as specified in the grading rules, but in practice two-foot (and even number) multiples are the rule.

The thinnest wood is called *veneer*. It is less than three-eighths inch thick. It can be sliced or sawn, but is usually rotary cut in a lathe that rotates the log against a knife. Veneer is usually glued to an engineered wood product like fiberboard and is used for inexpensive furniture and shelves. It is glued in plies to make plywood. It can be used alone for thin box walls, as in wire-bound boxes.

TABLE 2.4. American Softwood Lumber Standard Widths and Thicknesses for Construction Lumber (Forest Products Laboratory 2010).

Thickness (dressed, in inches)			Width (dressed, in inches)		
Nominal	Dry	Green	Nominal	Dry	Green
1	3/4	25/32	2	1-1/2	1-9/16
1-1/4	1	1-1/32	3	2-1/2	2-9/16
1-1/2	1-1/4	1-9/32	4	3-1/2	3-9/16
2	1-1/2	1-9/16	5	4-1/2	4-5/8
2-1/2	2	2-1/16	6	5-1/2	5-5/8
3	2-1/2	2-9/16	8	7-1/4	7-1/2
4	3-1/2	3-9/16	12	11-1/4	11-1/2
5 and ≥ 5	1/2 < nominal	1/2 off	≥ 12	1/2 off	1/2 off

2.4.3 Seasoning and Moisture Content

Seasoned lumber refers to wood that has been dried to a moisture content lower than 19%. It is dimensionally stable and strong. *Green*, or unseasoned, wood still has too high a moisture content to have reliable properties for all uses. If wood is not properly dried before use, it will shrink, warp, bend, and release its grip on nails and other fastening devices. It can also cause rust, biodegradation, and other humidity-induced damages to products packed in it.

Wood, as it exists in a living tree, contains a considerable amount of moisture, from 30–200% *moisture content*. This moisture is necessary to the life and growth of the tree. It carries nutrients from the roots to the leaves. It is normal for wood in a living tree to be very wet, especially the *sapwood*, the outer rings of the log. The difference in moisture content between the heartwood and sapwood is more pronounced in softwoods.

When the tree is cut down and converted into lumber, the wood immediately begins to loose moisture through evapotranspiration, but not necessarily in a predictable manner. It needs to be dried slowly enough to control the formation of drying defects so it won't shrink too quickly and crack, called *checking*, and yet quickly enough to minimize the in-process time and inventory. One of the most important processes of lumber manufacture is to control and accelerate the process of drying green wood.

The moisture content of wood in the forest products industry is typically specified on a dry weight basis, as a percentage of the weight of moisture over the weight of moisture free or oven-dry wood:

$$MC\% = 100 \times \frac{W_g - W_{od}}{W_{od}}$$

where $MC\%$ is the moisture content in percentage, W_g is the weight of green wood, and W_{od} the weight of oven dried wood.

The equation above can be manipulated to calculate the oven-dried (OD) weight of wood if the moisture and the weight of green wood are known:

$$W_{od} = 100 \times \frac{W_g}{100 + MC\%}$$

if the oven-dried weight and the moisture content are known:

$$W_g = W_{od} \frac{100 + MC\%}{100}$$

A piece of wood with 40% moisture content is holding the equivalent of an additional 40% of its dry weight in water. If the weight of a board is 1.4 pounds and its moisture content is 40% MC, the equation above can be used to calculate the oven dried weight and the amount of water as follows:

$$W_{od} = 100 \times \frac{1.40 \text{ lb}}{100 + 40} = 1 \text{ lb}$$

the oven dried weight is 1 lb and the weight of water is 0.4 lb

Aside from the structural issues, the use of green wood in packaging adds to shipping cost.

Consider 50 wooden boxes made with 10 m³ of green wood with a moisture content of 40%. If it is assumed that an oven dried cubic meter weighs 500 kg and the shipping cost is 10¢/kg, calculate the cost of shipping boxes made of wood as-is with 40% moisture content and after drying down to 19% moisture content.

Answer:

Total oven dried weight of wood needed to build 50 boxes:

500 kg per cubic meter multiplied by 10 cubic meters = 5,000 kg

Weight at any specific *MC%* is calculated using the equations for the green weight when oven dried weight and moisture content are known:

At 40% MC: 5,000 kg × (100+40)/100 = 7,000 kg
Cost of shipping: 7,000 kg × 10¢/kg = $700
At 19% MC: 5,000 kg × (100+19)/100 = 5,950 kg
Cost of shipping: 5,950 kg × 10¢/kg = $595

Cost saving for going from 40% to 19% MC for this shipment of 50 boxes = $105.

Lumber with a moisture content of 20% and higher is susceptible to deterioration from mold, mildew, sap-stain, and decay fungi. The effect and measurement of moisture in the wood is discussed in more detail in Chapter 3.

The rate of drying depends on a number of factors including wood anatomy, density, microstructure, size, and initial moisture content. Drying is accelerated by high temperature, low relative humidity, and good air circulation.

The three major methods of seasoning wood are air drying, accelerated air drying, and kiln (oven) drying. Other techniques are available for special applications. One such special application is the heat treatment of pallets for export in order to eliminate infestation.

Air drying exposes the wood to air at ambient conditions. It is, therefore, an economical technique, but is inherently variable, with the speed of drying dependent on weather conditions. It can be used alone, or as a preparatory step before kiln drying. The wood is usually air dried in piles, and proper stacking practices with adequate ventilation are necessary for adequate results. The average moisture content of air-dried softwood and hardwood lumber ranges from 15–25%, depending mainly on wood density. Extended air drying can bring down the moisture content slightly more, depending on how dry the climate is; the lowest that air-drying can accomplish is a minimum moisture content of 12–15% in much of the United States.

Accelerated air drying involves using forced air circulation and protecting the lumber from the weather. Fans are used to blow air, which may be heated somewhat, through the lumber piles which are housed in a shed or provided with some other form of weather protection. This can reduce the time required for drying by 50–75%, compared to conventional air drying.

Kiln drying uses more rapid air circulation and higher temperatures to speed up the drying process. Of course, the temperatures used must not be so high that the wood burns or is degraded. Also, drying too fast can cause the wood to split. In some cases, equalizing and conditioning treatments are used to relieve stresses. Ordinary kiln drying uses temperatures between 110–180°F (43–82°C). Elevated temperature kiln drying uses temperatures of 180–212°F (82–100°C), high temperature kiln drying uses temperatures above 212°F, and thermal modification uses temperature between 300°F and 470°F. Lower moisture contents between 8% and 14% can be achieved with kiln drying than with air drying (18–25%). The average moisture content of kiln-dry lumber is generally 12% or less, and 19% for air-dry lumber.

For most packaging applications, air drying is sufficient to reduce the lumber moisture content to close to 19%. Sometimes relatively unseasoned green wood with moisture above 25% is used for boxes, crates, and pallets when the application does not demand the added strength, dimensional stability, dryness, and cost of fully seasoned wood. There is a greater need to dry softwood for pallets in order to increase its strength and to control wood pests such as nematodes and beetles. ASTM D6199 (Standard Practice for Quality of Wood Members of Containers and Pallets) specifies that container and pallet members should have a moisture content at the time of fabrication of no greater than 19% and no less than 9% of their dry weight basis.

Unseasoned wood is rarely used for building construction or manufacturing since it has a tendency to shrink excessively and it may warp, curl, or twist as it dries. For such applications, it is best to use wood that has been seasoned to approximate equilibrium with the conditions under which it will be used. For indoor applications in most of the United States, this is approximately an 8% moisture content, on a dry weight basis. However, there are significant differ-

ences associated with climate. In desert areas of the west and southwest, seasoning to 6% moisture is recommended, while in the southeastern coastal regions of
the United States with high relative humidity, an equilibrium moisture content of
11% is recommended by the U.S. Forest Products Laboratory (2010).

2.4.4 Heat Treatment for Pest Control

When wood is used for export packages, there are special requirements for
bark-free wood and heat-treatment to destroy insects and other pests. Export
shipments can expose a recipient country to pests that are not indigenous and
pose special risk of spreading in the ecosystem.

An example is the *pinewood nematode*, a tiny worm that eats and causes
disease in wood, which has been found in softwood pallets, boxes, and crates
exported from the United States, Canada, Japan, and China. There are fears
that it, and other wood-borne pests, including beetles and emerald ash borer,
could spread around the world by means of international shipments and could
devastate forest ecosystems.

Heat-treatment regulations from the United Nations have been developed
with the cooperation of European and U.S. organizations (including the U.S.
Department of Agriculture). They require wood to be heat-treated beyond the
level ordinarily reached by some low temperature kilns. The International
Standards for Phytosanitary Measures, *Guidelines for Regulating Wood Packaging Material in International Trade*, require treating wood used for export
shipments. The requirement is to heat the wood to a minimum core temperature of 133°F (56°C) for a minimum of 30 minutes (IPPC 2002).

Records are required to be kept, and the boards or pallet must be marked
by an agency accredited by the American Lumber Standards committee. The
mark has an HT symbol showing compliance, as shown in Figure 2.5 (ASTM).

Although fumigation with methyl bromide has been accepted in the past, it
is expected to be abandoned in favor of heat treatment in all cases in the future.
Other treatments under consideration are microwave, radiofrequency, vacuum
treatment, infrared, chemical preservative impregnation, and irradiation.

The regulation applies to all solid wood packaging used for pallets, crates,

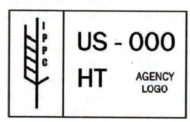

FIGURE 2.5. Heat-treatment mark for export wood.

boxes, and dunnage used for separating or balancing loads. It does not apply to engineered wood products like plywood, laminated lumber veneers, particleboards, fiberboards, and oriented strand board, which are sufficiently processed by heat and pressure to kill any pests.

The regulations began with similar requirements for shipments to China, the EU nations, Australia, and some South American countries because of wood beetles. Compliance was required beginning March 2002 for all shipments to the European Union. A number of other countries have also adopted these rules.

The packager must understand the country of destination's import packaging requirements. Up-to-date information for American shippers can be found at http://www.aphis.usda.gov/wps/portal/aphis/ourfocus/importexport. Customs inspectors look for the accredited mark, and shipments found not to comply can be delayed, refused entry, destroyed, or treated there. If the shipment is marked and live pests are found anyway, the same penalties apply.

2.5 ENGINEERED WOOD PRODUCTS

Engineered wood products like plywood, LVL, fiberboard, particleboard, wood plastic composites, and oriented strand board (OSB) are primarily used for building construction, but can also be useful packaging materials. Packaging has not been a major concern for the engineered wood product industry over the years, but it is becoming more important now as a second significant market.

Plywood, LVL, and OSB make strong and easy-to-assemble box walls at a cost that is much lower than equivalent structures made from lumber. They have revolutionized the field of crate-making by making it easier and less expensive to build a cleated panel box than a traditional framed wooden crate. They can provide a flat surface for pallet and skid decks. Sold in sheets, they offer more dimensional stability than a series of boards and keep the crate or pallet framework from racking out of square. These uses are described in Chapters 4 and 5.

Because of the way they are made, engineered wood panels are never "green". They are formed using dried wood under high temperature and pressure. The processing temperature corresponds to the curing temperature of the thermoset or thermoplastic adhesive used for bonding, which is higher than the 133°F temperature recommended for export wood packaging. This temperature improves the dimensional stability and it also gives favored status for international shipments because it renders engineered wood products exempt from the export requirements for heat treatment. As a result, the use of engineered wood products in packaging continues to grow. There is particular interest in using engineered wood to replace boards in export packages.

2.5.1 Plywood

Plywood was the first type of engineered wood panel. Tea chests (described in Chapter 1) were the first large scale use of plywood.

Plywood is formed from several thin sheets of wood veneer called *plies* that have been glued together under heat and pressure to form a rectangular panel. Each ply consists of a single thin sheet, or of two or more sheets laminated together with the grain direction parallel. Each ply is generally wood veneer, although the core layers may also be lumber or particleboard. The wood grain in adjacent layers is arranged perpendicular to each other with the outside plies having the grain running in the long direction of a typical eight-by-four-foot sheet. To balance the plywood against warping, it is nearly always assembled with an odd number of layers so the structure is symmetrical around the interior layers. One surface, called the *face*, is usually better quality than the *back*.

Because of this layered construction, strength and other properties of plywood are more uniform than in solid wood. The properties are, in general, between those of wood in the grain and the cross-grain directions, weaker than lumber's grain direction, but stronger than lumber's cross-grain direction, and sheets are balanced in the eight-by-four-foot dimensions. Engineered wood leaves the manufacturing process at moisture content levels well below 10% and has more dimensional stability when exposed to moisture variations than solid wood.

In North America, most plywood is made from softwood lumber, although it can be made from over 70 species of wood. The plies are generally glued together with thermoset phenol formaldehyde adhesives, and then the structure is sent into a hot press, where the adhesive is activated and cured.

The weakest part of the plywood generally is the wood, not the adhesive material. The glue acts to limit water absorption, so plywood tends to be less susceptible to gain and loss of moisture than solid wood. Plywood panels with the grade of either *Exposure 1* or *Exterior* on their trademarks have been manufactured with water resistant resins like phenol formaldehyde. Plywood panels for interior use are made with urea formaldehyde and soybean based resins.

Plywood is characterized by excellent split and puncture resistance, stiffness, and fastener holding power. It resists racking, shape distortion, and impacts. It is used extensively in building construction. In packaging, it is used for crate sheathing and light duty bases. U.S. Department of Defense (1999) tests show that it is stronger than lumber for these applications and less expensive than lumber.

A wide variety of grades are available. The Engineered Wood Association (APA), (formerly the American Plywood Association) is a non-profit trade association that serves the industry as a source for research and development as well as education. All APA trademarked products are subject to verification through APA audits that are based on standards, including the PS1-09 Voluntary Product Standard, (APA 2010). The grade is marked on each panel. Typical sheets are four by eight feet with thickness of one-quarter, three-eighths, five-eights, and three-quarters inch.

The span rating applies more to decks than walls of packages, since it is the deck that supports the weight. Plywood is used when a flat smooth surface is needed, such as for solid deck pallets, large bins, and cleated panel boxes. A major use is as a stiff decking platform to aid stacking and spreading loads during shipment, for example used on top of a pallet load to make a flat protective surface for stacking a second pallet load.

2.5.2 Oriented Strand Board

OSB is made of compressed strands or flakes of wood that are arranged according to their grain direction, then in layers of alternating grain direction similar to layers of veneer in plywood. OSB is stronger than particle boards because its strands are purposefully aligned to make the panel stronger and stiffer. It is considered structural and subject to standards published by the APA.

This cross-lamination yields the same characteristics obtained in plywood with better dimensional stability, stiffness, and cross-panel strength. As with plywood, the resins used to manufacture OSB are urea formaldehyde (for interior furniture) and water-resistant phenol formaldehyde cured with heat and pressure. In the construction industry, plywood and OSB are almost always considered to be interchangeable depending on equivalent thicknesses and required structural properties.

But for packaging applications, the difference in the performance of plywood versus OSB is more pronounced. OSB has lower impact resistance than plywood. Because the wood fibers in plywood are in a longer continuous ply, impacts are distributed across a larger area, eventually resulting in the uneven splitting of the wood. Impacts to OSB result in a quicker fracture of the panel, leaving a removed section close in size to the impacted area. For example, the impact of a forklift tine will leave an uneven, elongated gash in plywood, whereas the same impact to an OSB panel will leave a hole completely through the panel in the size and shape of the fork tine. Similarly, strips of OSB are commonly not as strong as strips of plywood.

The amount of OSB used for packaging is growing, primarily for cleated panel boxes.

2.5.3 Particle Boards

Particle board is made from residues of wood milling. Wood shavings, chips, wafers, flakes, or strands are glued together with a thermoset resin in a hot pressing operation.

Particle board is less expensive than plywood, but is also less strong. It has more glue and less wood than plywood, generally 6% or more glue by weight, resulting in a resin content about four times as high as plywood. It is also heavier than plywood, so boxes made from particleboard are heavy and more expensive to ship.

The mechanical properties of particle board are influenced by the geometry of the particles and the type and amount of adhesives. It has no grain direction, and consequently is weak in bending. It does not hold nails or screws well, but is reasonably dimensionally stable, although it does tend to absorb water readily. Because particle board can be made from wood that would otherwise likely be waste, it is a relatively inexpensive material.

The strength of the board is determined primarily by the strength of the glue bonds, rather than that of the wood components. Urea formaldehyde adhesives are most common, blended with melamine formaldehyde resins when greater water resistance is required. Urea formaldehyde-based adhesives can be a source of formaldehyde emissions, so alternatives like phenol formaldehyde and isocyanate adhesives are increasing in popularity (Walker 1993).

Particle boards may be of single-layer or multilayer construction. In multilayer boards, coarser particles are used in the core, and thinner, smaller particles in the upper and lower surfaces. Single-layer graded density boards are also available in which air classification is used to control the distribution of wood particle sizes in the mat.

Particle board is often used as flooring; floor, wall, and roof sheathing; and covering. Packaging applications include the manufacturing of inexpensive, relatively low quality wood boxes and crate sheathing, and as a pallet deck covering to provide a level surface.

2.5.4 Fiberboard

Fiberboard is a manufactured wood product that is in essence very thick paper. It is not often used in packaging but in furniture manufacturing, underlayment for flooring and carpet, and insulation. It is made in a way similar to paper, by a thermomechanical pulping process that is less complete than that used for paper-making, as some fiber bundles are acceptable.

Hardboard is a high density fiberboard with density values higher than 800 kg/m^3 (50 pcf), with a specific gravity near 1, which is sometimes called *Masonite*, after the originator of the material. It is made in a wet forming process, bonded by the wood's own lignin. *Medium density fiberboard* (mdf) is formed in a dry process that adds synthetic resin between 6–7% on a dry weight basis, rather than using lignin bonding. The density of mdf ranges from

500–800 kg/m^3 (30–50 pcf). Low density fiberboard or insulation board has a density less than 500 kg/m^3 (30 pcf). The term presswood is sometimes used for these materials, which can be molded into pallets.

2.6 WOOD COST ESTIMATION

Lumber is measured, priced, bought, and sold by the *board foot*. A board foot (BF) is defined as the volume of wood in a board measuring nominally one inch thickness by one foot wide and one foot long.

$$BF = \text{Thickness in inches} \times \text{Width in feet} \times \text{Length in feet}$$

For comparison, a cubic foot (CF, or ft^3) is the volume of a piece of wood measuring one foot in thickness by one foot in width and one foot in length, so there are 12 BF in a CF.

$$CF = \text{Thickness in feet} \times \text{Width in feet} \times \text{length in feet}$$

$$CF = BF/12$$

One of the best ways to calculate BF is to express the thickness in inches, and the width and length in feet. Unfortunately, lumber width is always expressed in inches, and therefore the volume in board feet will be calculated using the following equations, using nominal dimensions:

$$\text{Volume (BF)} = \text{Thickness in inches} \times \frac{\text{Width (inches)}}{12}) \times \text{length in feet}$$

Example 1:

How many board feet are in a piece of pine lumber measuring 3/4″ × 5-1/2″ × 10′?

Answer:

Since pine is a softwood, the nominal dimensions of this piece of lumber are 1″ × 6″ × 10′. Therefore the number of board feet is:

$$\text{Volume in BF} = \frac{1'' \times 6'' \times 10'}{12} = 5 \text{ BF}$$

Example 2:

If pine is selling for $200/1,000 BF, how much will it cost for the wood to make a crate that consists of four pieces with nominal dimensions of 4″ × 3″ × 2′ and 12 pieces with dimensions 1″ × 6″ × 4′?

Answer:

$$\text{Volume in BF} = 4\frac{4'' \times 3'' \times 2'}{12} + 12\frac{1'' \times 6'' \times 4'}{12} = 32 \text{ BF}$$

$$\text{Cost} = 32 \text{ BF} \times \frac{\$200}{1000 \text{ BF}} = \$6.40$$

The price per board foot depends on the type and grade of wood, the volume ordered, the season, and other factors. The price per unit is lower when larger quantities are purchased. Since lumber cannot usually be purchased in the exact sizes required, scrap is generated, which needs to be figured into the cost. On the other hand, if smaller pieces are required, they can sometimes be made from someone else's scrap, called *random lengths and widths* (thickness is specified in one-quarter inch increments) and purchased at a bargain. It is important to remember that allowance must be made for saw cuts. An eight-foot long two-by-four cannot be cut into four pieces exactly two feet long.

Small quantities of pre-cut boards are frequently priced by the linear foot (LF) or by the square foot (SF). Long thin pieces like moldings are usually priced by the linear foot. A linear foot of wood measures one inch thick by one inch wide and one foot long. One board foot contains 12 linear feet of wood.

$$1 \text{ LF} = \text{one inch thick by one inch wide and one foot long} = \frac{\text{BF}}{12} = \frac{\text{CF}}{144}$$

At a given thickness, flat pieces of wood such as paneling materials, flooring planks, plywood, particleboard, and fiberboard are often priced by the square foot, which is the width times the length, both measured in feet. Pieces may also be priced by the unit. The price for lumber also varies with width.

While in the United States quantities of lumber are specified by the number of board feet based on nominal dimensions, in Europe and most other countries lumber is specified by its actual size, using the volume in cubic meters of the wood when dry. If the lumber is green, a correction factor is applied to account for shrinkage. One cubic meter of dry lumber equals 424 BF, determined from the actual, rather than nominal, dimensions. For international trade in green lumber, a conversion factor of 450 BF/m^3 is often used (Bowyer *et al.* 2002).

2.7 TREES, WOOD AND SUSTAINABILITY

Wood has distinct sustainability advantages compared to other packaging materials. Trees filter the water we drink and produce the oxygen we breathe. They sequester carbon, are renewable, recyclable, and ultimately biodegrade.

Trees help to slow global warming because wood is an effective *carbon sink.* As trees grow, they absorb some of the CO_2 that has been added to the atmosphere by the burning of fossil fuels. CO_2 is the most important greenhouse gas that contributes to global warming. The growing tree converts natural energy from sunlight during photosynthesis to biomass and thus stores the carbon.

Therefore, the conversion of trees into useful forest products such as lumber, panels, and paper is highly energy-efficient, and gives wood a very low-to-neutral carbon footprint in comparison to materials made from finite resources such as fossil fuels, metal, or glass that require a large amount of energy to extract or convert.

Trees are a renewable resource. In the last century, great progress has been made in the field of forest management to grow short-rotation plantation species with high yields such as Southern Pine. Sustainable and responsible forest management is one of the most viable strategies to combat climate change through the absorption of CO_2 and other greenhouse gases.

To fully implement and take advantage of the environmental benefits, it is strongly recommended to use wood from forests that have been certified to follow the guidelines and principles of sustainable forest management.

Non-government organizations have developed sustainability certification standards that may be labeled on forest products. *The Forest Stewardship Council (FSC)* is an international organization for forest management that tracks and certifies that wood and paper products come from responsible sources that are environmentally appropriate, socially beneficial, and economically viable (https://us.fsc.org/). The Sustainable Forestry Initiative (SFI) targets large industrial forestry operations in Canada and the United States (http://www.sfiprogram.org/). The *Canadian Standards Association (CSA)* tracks and labels certified products in Canada. The *Programme for the Endorsement of Forest Certification (PFEC)* is a recognition framework for national forest certification standards (http://www.pefc.org/).

On the other hand, the conversion of wood to paper requires a high use of energy and water, and the environmental implications are discussed in Chapter 9.

At the end of their service life, wood and paper are recyclable, biodegradable, or the energy can be recovered by incineration. The effects of recycling are also discussed later, in terms of recyclability and the effect of recycled content related to pallets, pulp, paperboard, and corrugated boxes. Paper and wood biodegrade over time, but it takes a long time and in landfills produces methane gas, another greenhouse gas, which can be captured to produce energy. Paper and wood can be also be incinerated to produce energy, subject to strict emission standards.

2.8 REFERENCES

Alvarez, M. 2007. *The State of America's Forests.* Bethesda, MD: Society of American Foresters.

APA (Engineered Wood Association). 2010. PS 1-09 Structural Plywood Voluntary Product Standard. Tacoma, WA: APA (Engineered Wood Association). http://www.apawood.org/

ASTM (formerly American Society for Testing and Materials, now ASTM International). No date. *Annual Book of ASTM Standards.* West Conshohocken, PA: ASTM International.

Bowyer, J.L., R. Shmulsky, and J.G. Haygreen. 2002. *Forest Products and Wood Science.* 4th ed. Ames, Iowa: Iowa State University Press.

Brown, N.C. and J.S. Bethel. 1958. *Lumber.* 2nd ed. New York: John Wiley & Sons.

Bush, R., J.J. Bejune, B.G. Hansen, and P.A. Arman. 2002. Trends in the Use of Materials for Pallets and Other Factors Affecting the Demand for Hardwood Pallets, *Proceedings of the 30th Annual Hardwood Symposium,* May 30-June 1, 2002. Memphis: National Hardwood Lumber Association.

Dombeck, M. 2000. Intelligent Consumption: The Forest Service Role. Presented at *The Intelligent Consumption Forum,* Madison, WI, July 19, 2000; available online www.fs.fed.us/dombeck_intelligent_consumption.htm.

FAO. 2010. Global Forest Resources Assessment 2010, Main Report, FAO Forestry Paper #163. Rome: Food and Agriculture Organization of the United Nations. http://www.fao.org/docrep/013/i1757e/i1757e.pdf

FAO. 2012. State of the World's Forests 2012. Rome: Food and Agriculture Organization of the United Nations. http://www.fao.org/docrep/016/i3010e/i3010e.pdf

FAO. 2014. Forest Products Statistics: Facts and Figures 2012. Rome: Food and Agriculture Organization of the United Nations. http://www.fao.org/forestry/statistics

Forest Products Laboratory (US). 2010. *Wood Handbook: Wood as an Engineering Material.* Gen. Tech.Rep. FPL-GTR-190. Madison, WI: U.S. Department of Agriculture, Forest Service, Forest Products Laboratory. http://purl.fdlp.gov/GPO/gpo11821

Haack, R.A., T.R. Petrice, P. Nzokou, and D.P. Kamdem. 2007. Do Insects Infest Wood Packaging Material with Bark Following Heat-treatment? In *Alien Invasive Species and International Trade,* edited by H.F. Evans and T. Oszako, pp. 145–149. Warsaw, Poland: Forest Research Institute.

IPPC (International Plant Protection Convention). 2013. *Guidelines for Regulating Wood Packaging Material in International Trade,* IPSM #15. Rome Italy, www.ippc.int

NHLA. 2011. *Rules for the Measurement and Inspection of Hardwood & Cypress.* Memphis: National Hardwood Lumber Association. http://www.nhla.com/assets/1603/2011_rules_book.pdf

Smith, W., B. Visage, P.D. Miles, C.H. Perry, and S.A. Pugh. 2009. *Forest Resources of the United States, 2007,* General Technical Report WO-78, USDA Forest Service.

U.S. Forest Service. 2011. *National Report on Sustainable Forests-2010.* Report FS 979. Washington, DC: U.S. Government Printing Office.

U.S. Department of Defense. 1999. Wooden Containers and Pallets and Crates, *Packaging of Materiel,* Chapters 3 and 6. FM 38-701/ MCO 4030.21D/ NAVSUP PUB 503/ AFPAM(I) 24-209/ DLAI 4145.2, United States Departments of the Army, Navy, Air Force and the Defense Logistics Agency. Washington, DC.

U.S. National Institute of Standards and Technology. 2010. *American Softwood Lumber Standard,* Voluntary Product Standard PS 20-10. Washington: U.S. Government Printing Office.

Walker, J.C.F. 1993. *Primary Wood Processing: Principles and Practice.* London: Chapman & Hall.

2.9 REVIEW QUESTIONS

1. What is the difference between hardwoods and softwoods? Give examples of trees in each category. How do their wood and uses differ?

2. Name two common species of woods used in packaging.

3. Which is most often used for reusable pallets: hardwood or softwood?

4. What proportion of the U.S. land area is covered with forests? Of that, how much is harvestable timberland?

5. Why is the amount of U.S. timberland increasing, despite harvesting?

6. Where is the most productive forest land in the United States (north, south, east, or west)?

7. What is the difference between dimensional lumber and boards?

8. Would a four inch by four inch by eight foot piece of lumber be classified as timber, board, or dimensional lumber?

9. What grade of hardwood is used for pallets?

10. What is meant by "nominal dimensions" and what kind of wood does it refer to?

11. What are the actual dimensions of a standard two by four?

12. What is the maximum thickness of veneer?

13. How and why is wood seasoned?

14. What is the name of the tiny worm that has been found in softwood pallets, boxes, and crates?

15. What special treatment is required for wood used in export packaging? How hot and for how long?

16. Which engineered wood products are used in packaging, and what are they used for?

17. Does plywood usually contain an odd or even number of layers? Why? Is the strength of a sheet of a four foot by eight foot sheet of plywood greater along the four foot or the eight foot direction?

18. Why is oriented strand board superior to types of board made from wood particles?

19. How does the performance of OSB differ from plywood for packaging applications?

20. How many board feet are in a piece of wood with nominal dimensions 1-1/2″ × 4″ × 20′?

21. If pine sells for $200/1,000 BF, how much will the wood cost to make a pine crate consisting of four pieces with nominal dimensions measuring 4″ × 4″ × 5′ and 12 pieces measuring 2″ × 6″ × 5′?

22. If oak sells for $1/BF, what is the cost for the wood to make an oak pallet consisting of the following materials:
    a. Twelve deckboards that measure 1″ × 4″ × 40″
    b. Three stringers measuring 12″ × 4″ × 48″

23. If one inch by one inch strips of wood are priced at $0.10/LF, what is the equivalent price per BF?

24. List five reasons or advantages of the use of wood or paper packaging as one of the most sustainable packaging materials.

CHAPTER 3

Structure and Properties of Wood

The chemical and physical structure of wood determines its properties as well as the properties of the paper-based materials made from it. Wood has a fibrous nature, with long, hollow, sap-filled cells and *grain* arranged parallel to the trunk of the tree. The characteristics and arrangement of the fibrous cells affect its strength and reaction to changes in humidity.

The walls of the fiber-shaped cells are made from cellulose and hemicellulose. The fibers are surrounded by lignin, an intercellular material that "glues" the cells together. This structure makes trees tall and strong. It also gives strength to wood.

Wood fibers are good for papermaking because the process of wood pulping frees the fibers from the inter-cellular lignin so that they can be rearranged into another useful shape, like a sheet. Although pulping causes considerable changes in the physical macrostructure of the wood, there is much less change in the structure of the fibers themselves.

3.1 CHEMICAL STRUCTURE OF WOOD

The wood cell walls contain *lignin, hemicellulose*, and *cellulose*. Hemicellulose and cellulose are carbohydrates and are the major chemical components. Cellulose predominates, with about 50% by weight. Lignin is not a carbohydrate. As shown in Figure 3.1, there is over twice as much cellulose and hemicellulose as lignin.

The amount of cellulose, hemicellulose, and lignin varies with species, location within the tree, climate, age, and exposure of the wood. The relative proportions of these three constituents also differ with the position within the cell wall. In general, softwoods have more lignin than hardwoods, as shown in Table 3.1.

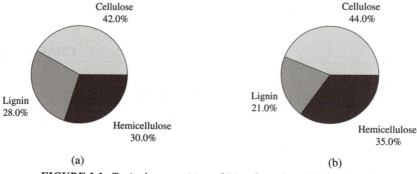

FIGURE 3.1. *Typical composition of (a) softwood and (b) hardwood.*

Wood contains 49% elemental carbon, 44% oxygen, 6% hydrogen, less than 0.5% nitrogen, and traces of other elements (Fengel and Wegener 1983). In addition to cellulose, hemicellulose, and lignin, wood also contains a variety of other chemical components known as extractives and inorganic non-combustibles known as ash, most in very small amounts (Table 3.2).

3.1.1 Cellulose

Cellulose gives wood its strength. It is a linear homopolymer of glucose residues joined by β-1-4 linkages (β-D-glucopyranose), as shown in Figure 3.2. It has an extremely high molecular weight. Cellulose can be represented by the chemical formula $(C_6H_{10}O_5)_n$, where n is thought to be about 10,000 (Bowyer *et al.* 2002).

Cellulose has very strong intramolecular (within the molecule) and intermolecular (between adjacent molecules) hydrogen bonding, and exhibits a high degree of crystallinity. Because of the extent of hydrogen bonding, cellulose also tends to interact strongly with water, and hence is primarily responsible for the tendency of both wood and paper to sorb water and change dimensions on exposure to changes in humidity.

As a tree grows, the cellulose molecules are arranged into long oriented,

TABLE 3.1. Chemical Composition of Softwoods and Hardwoods.

Component	Softwoods (mass percent)	Hardwoods (mass percent)
Cellulose	40–44	40–44
Hemicellulose	20–32	15–35
Lignin	25–35	18–25
Extractives	0–5	0–10

**TABLE 3.2. Elemental Composition of Wood
(Fengel and Wegener 1984).**

Element	% Dry Weight
Carbon	49
Oxygen	44
Hydrogen	6
Nitrogen	> 0.1
Ash content*	0.2–0.5

*Ash content in some tropical species is close to 5% (Nzokou and Kamdem 2006).

thin crystalline strands called *fibrils*, which in turn make up the cell wall. Cellulose's crystalline nature (estimated to be up to 70%) makes it resistant to chemical attack during the pulping process as it is liberated from the lignin "glue" leaving fibers that can be reformed into a sheet of paper. Cellulose gives paper its strength and flexibility, as well as its sensitivity to water.

3.1.2 Hemicellulose

Hemicellulose plays a more important role in papermaking than it does in wood properties. Like cellulose, it is a condensation polymer formed from sugar residues. However, hemicelluloses are actually copolymers, rather than a pure homopolymer like cellulose. Rather than a linear crystalline structure, it is irregular, commonly containing short side-chains, and is sometimes heavily branched. It also has a much lower molecular weight than cellulose.

The monomeric units commonly found in wood hemicellulose are glucose, mannose, galactose, xylose, arabinose, 4-O-methylglucuronic acid, and galacturonic acid residues. Hemicellulose is predominantly amorphous, with a degree of polymerization ranging from 100 to 400 (Walker 1993).

Because of its lower molecular weight and amorphous nature, hemicellulose is much more readily hydrolyzed and soluble than cellulose. It plays a less significant role in wood properties than either cellulose or lignin.

But in paper, hemicellulose forms some of the "flesh" that fills out the paper between the longer cellulose fibers. It makes a smoother, more printable

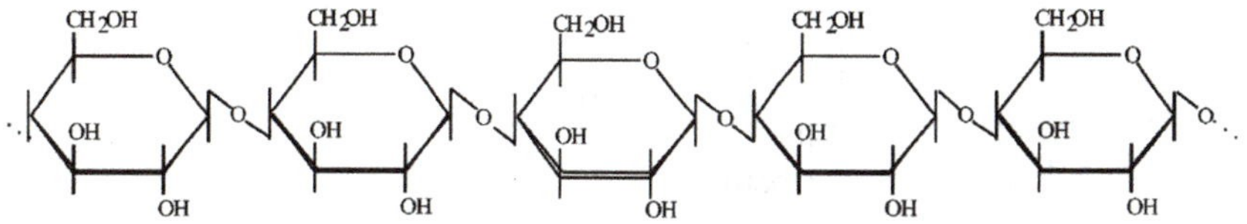

FIGURE 3.2. *Cellulose.*

surface. It increases the tensile, burst, and fold strength of paper, and increases pulp yield. The starch that is often added to pulp as a binder has a very similar effect to hemicellulose (Bierman 1996).

3.1.3 Lignin

Lignin acts to protect the cellulose and imparts compressive strength, rigidity, brittleness, and some degree of resistance to various types of physical, chemical, and biological degradation. Half of the lignin is in the cell walls and is linked to hemicellulose and cellulose; the other half can be regarded as the glue that holds the fibers together. Lignin makes the cell walls hard, renders wood indigestible, and makes trees resistant to cold.

Lignin is a highly complex, three-dimensional aromatic condensation polymer of very high molecular weight. It is a totally amorphous polymer, formed by non-selective free radical addition and condensation processes that yield a random structure.

Lignin is formed from hydroxy- and methoxy-substituted phenylpropane units (Figure 3.3). Softwood lignin has a molecular weight estimated at 90,000 or more. Hardwood lignin has somewhat lower molecular weights (Walker 1993).

The main goal of chemical pulping is to remove lignin, freeing and purifying the fibers for the papermaking process. Since lignin is susceptible to UV degradation, mechanical pulping methods that do not remove it result in paper that yellows upon exposure to light due to the oxidation of the residual lignin.

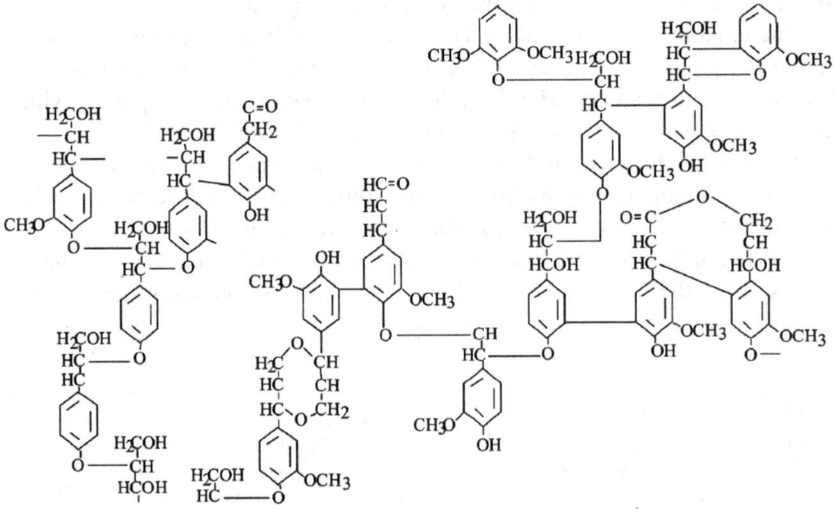

FIGURE 3.3. *Portion of a lignin molecule.*

3.1.4 Other Components

A variety of other components are found in wood in small amounts. These are not structural materials.

Organic components such as tannins, phenolic compounds, terpenes, fatty acids, resin acids, waxes, gums, and starches contribute to color, odor, decay resistance, density, hygroscopicity, and flammability. These are classified as extractives because they can be removed from wood by extraction with solvents. There are also trace amounts of non-combustible compounds containing inorganic calcium, sodium, potassium, silicon, and magnesium, known as ash components.

The total amount of extractives typically ranges from about 1% to 18% dry weight, and is dependent on species, growth conditions, time of year, and several other factors. In some cases like tropical species, extractives content can be as high as 30%. Many extractives serve to protect wood against biological degradation, by their toxicity to certain biological organisms. The higher concentrations of extractives in heartwood compared to sapwood are largely responsible for the greater decay-resistance of heartwood. The extractive content within a tree generally decreases with height. The chemical composition and concentration of extractives varies markedly with species, and can serve as an aid in species identification.

3.2 TREE GROWTH AND THE PHYSICAL STRUCTURE OF WOOD

The elongated fibrous cells in wood are hollow. Most are arranged parallel to each other along the trunk and branches of a tree. But all cells in the tree are not the same because of the way a tree grows.

The outside of the tree is covered with *bark*. The wood of the tree trunk is divided into two zones—*sapwood* and *heartwood*—with different and distinct functions. The sapwood is closest to the bark, and contains active conducting and storage cells that are light in color. The heartwood is older, darker in color, and farther from the bark. The very center of the tree, and of each twig and branch, is the *pith*, the soft tissue about which the first wood was formed in the newly growing twig.

The sapwood has high moisture content, because it transports the water or sap. It has an important storage and conduction function for starch, sugar, and other biochemical essentials for tree growth. The cell walls of sapwood are thinner to facilitate the storage of tree nutrients. Its high starch and sugar content make it more susceptible to microbial attack, chemical staining, and biodegradation than heartwood, but it is easier to make into pulp. Sapwood typically ranges from 1-1/2 to 2 inches in thickness, though it may be one-half inch or less thick in species such as catalpa and black locust. It can be 6 inches or more in thickness in fast-growing species like maple and Southern pine, which is why these are so useful for papermaking.

As cells age and die, sapwood is gradually converted into heartwood. Heartwood consists of inactive cells that store chemicals known as extractives. It does not function for either conducting sap or nutrient storage. The extractives are formed in the area between the sapwood and heartwood through a complex biochemical transformation of starch and sugar that is deposited in heartwood dead cells under conditions of little to no oxygen. These extractives are responsible for heartwood's high density, dark color, distinct odor, lower hygroscopicity, and fiber saturation point. It is more difficult to penetrate with liquid, more difficult to pulp, and is more resistant to decay and termites than sapwood.

Growth rings are produced because, in temperate climates for most species of trees, there is enough difference between the wood produced early and late in the growing season to produce distinct rings. It is possible to determine the age of a tree by counting the number of rings in its trunk. The wood produced in the spring is called *earlywood*. Produced at a time of rapid growth, earlywood is characterized by cells with relatively large cavities and thin walls. It is typically lighter in weight, softer, and weaker than *latewood*, which is produced during the summer. Latewood cells, produced at a time of slower growth with decreased photoperiod and lower temperature, have smaller cavities, thicker walls, and a higher density than earlywood.

Bark typically constitutes between 5% and 30% of the volume of a log. As the tree grows, the existing bark is pushed outward, becoming stretched, cracked, and ridged, and eventually sloughs off. In contrast, the old wood cells are covered over by the new wood cells, as the diameter of the tree increases. As branches die and fall off, the portion of the branch that was projecting from the tree is encapsulated, and becomes a knot in the wood of the tree.

It is useful, in discussing the structure of wood and tree, to talk about three mutually perpendicular directions. The *longitudinal* direction, from the ground up, is the *grain* direction in the tree, the direction in which most of the cells are oriented. The *radial* direction is the direction from the center of the tree horizontally outward. The third direction, *tangential*, is perpendicular to the radial direction and approximately tangent to the growth rings in the tree (Figure 3.4).

3.2.1 Cellular Structure

Most wood cells are long and pointed at the ends. There are also cells of different sizes and shapes. The cell walls surround a hollow central area called the *lumen*, which contains sap and extractives. Small pits in the cell walls allow sap to flow laterally from one cell to the other. The cells are joined to-

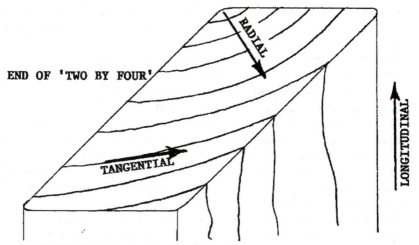

FIGURE 3.3. *Longitudinal, radial, and tangential directions in wood.*

gether through the *middle lamella* which is rich in lignin. The main objective of chemical pulping is the removal of lignin to obtain discrete fibers.

Within the tree, the cells are arranged and cemented together predominantly in straight radial parallel rows and straight spoke-like rays. Softwood and hardwood contain several different types of cells. The cellular structure of softwoods is much more uniform than that of hardwoods. Hardwood cells are more specialized.

The most common type of cells in softwoods are longitudinal fiber cells, called *tracheids*, which make up 90–95% of the volume, and are responsible for the strength and rigidity of the tree, as well as for longitudinal conduction of sap. A typical softwood fiber cell is 3–4 mm long and 25–45 µm in diameter, for an aspect ratio (ratio of length to diameter) of about 100:1. Tracheids have a hollow center (lumen) and are often rectangular in cross section. The general shape is described as blunt or rounded radially and pointed tangentially, much like a soda straw pinched shut at both ends (Bowyer *et al.* 2002).

The other major type of cell found in softwoods is *ray cells*. Rays serve to transport materials in the horizontal direction within the tree. Ray cells can be either tracheids that are positioned horizontally rather than vertically, or *parenchyma cells*. The main function of parenchyma cells is food storage and transport. Parenchyma cells are the longest living wood cells and remain functional throughout the sapwood; their death marks the transition of sapwood to heartwood. Softwood rays are narrow and typically comprise about 5–7% of softwood volume. While softwood can contain other types of cells, these are present in only minor amounts. Other softwood features include resin canals to distribute resin and epithelial cells where resins are stored.

Hardwoods have four major types of cells, each of which may comprise 15% or more of the wood volume. These are vessel elements, longitudinal tracheids, parenchyma cells, and ray cells.

Longitudinal sap conduction in hardwoods occurs in specialized cells called *vessel elements*. These cells are found in virtually all hardwoods, but only rarely in softwoods. They are short cells with very large diameters that are lined up and linked end-to-end along the grain to form long tube-like structures known as vessels. Viewed in cross-section, the vessels look like holes, and are often referred to as pores. In some hardwood species, the hollow lumen of the vessels, and thus the access for liquid transport, can be blocked by outgrowths from other cells (such as parenchyma cells) known as tyloses. The presence of tyloses in the vessels of white oak is the reason that oak is used to make barrels that are impervious to liquid.

The presence of these large vessels is a primary distinguishing characteristic between hardwoods and softwoods. In some species, the pores in the earlywood are much larger than those in the latewood; these species are classified as ring-porous, since the large early-wood vessels form a visible ring in a cross-section of the tree. In the majority of hardwood species, classified as diffuse-porous, the pores are more uniform in size and are distributed fairly evenly across the growth ring. In addition to adding a different type of cell, the large size of the vessels disrupts the orderly arrangement of the remaining cells, causing the pattern of cell arrangement in hardwoods to be much less uniform than in softwoods.

Hardwoods also have longitudinal fiber tracheids, but these cells are considerably shorter than softwood tracheids, and have thicker walls. Hardwood fibers average less than 1 mm in length, and tend to be rounded rather than rectangular in cross-section. Since hardwood fibers do not need to provide for sap flow, they are specialized for strength.

In many hardwood species, longitudinal parenchyma cells make up a significant portion of the wood volume, up to 24% in some domestic hardwoods and greater than 50% in a few tropical hardwoods, while some hardwood species do not contain any at all. They function primarily for food storage and transport.

As is the case for softwoods, hardwoods also contain ray cells. However, hardwood rays are often much larger than those in softwoods. Also, they are seldom arranged in the straight spoke-like patterns characteristic of softwoods.

3.2.2 Paper Fibers

Once wood has been subjected to pulping, all the recovered material is classified as *fiber*, regardless of the type of cell. The long fibrous wood cells are good for making paper because of their ability to interlock. The original di-

TABLE 3.3. Typical Fiber Dimensions (Peel 1999).

Wood Type	Length (mm)	Width (μm)	Thickness (μm)
Hardwoods: Birch, Beech, Eucalyptus, Poplar, Aspen	0.9–1.3	10–35	2–5
Softwoods: Balsam fir, Scots pine, Lodgepole pine, Norway spruce, Black spruce, White spruce	3–3.5	28–35	2.1–3
Softwoods: Douglas fir; Loblolly pine, Slash pine, Longleaf pine, Shortleaf pine, Radiata pine, Western hemlock	3.5–4.2	35–40	2.6–4.2

mensions of the wood cells are important for determining the dimensions of the fibers recovered from the tree by pulping. The pulping process itself has an influence on fiber dimensions, but it can only act to shorten, not to lengthen, the fibers.

Since softwood fibers are, on average, longer than hardwood fibers, and since the strength properties of paper are enhanced by longer fibers, softwood is more desirable for making many kinds of packaging paper, especially strong kraft paper. The shorter and thicker fibers in hardwood, however, make a smoother paper that has better printability and stiffness. Most printing papers have a combination of hard and soft wood fibers. The corrugated layer in corrugated board also benefits from short hardwood fibers, which makes it stiffer and easier to form. Typical fiber dimensions are presented in Table 3.3.

3.3 INTERACTION WITH WATER

A substantial portion of the weight of a living tree consists of water, from 30% to 200% of the dry weight. Moisture content varies by species, climate, and even within a single tree, depending on whether the wood is from the heartwood or sapwood layers.

Once the tree is harvested, the wood begins to lose moisture and weight. As it dries, it gains strength, shrinks, and can warp or crack. These changes can affect wooden pallet and container performance.

Throughout this book (especially when the subject turns to pulping and papermaking in Chapters 6 and 7), the subject of water will return. Moisture content also affects the lifecycle of the fibers in paper and paperboard as they undergo significant, repeated wetting and drying episodes. Water is a significant ingredient in wood, paper, and paperboard packaging and helps to determine its properties.

3.3.1 Moisture Content

The forest products industry usually specifies the moisture content of wood on a dry weight basis. The moisture content (MC_d) of wood on a *dry weight basis* is defined as the percentage of water in the moist wood, based on the amount of dry wood:

$$MC_d = 100 \times \frac{W_m - W_d}{W_d}$$

where W_m is the initial moist weight and W_d is the dry weight of the wood.

Example:

What is the moisture content of a sample of wood, on a dry weight basis, if its initial weight is 50 g, and after oven-drying its weight is 37 g?

$$MC, \% = 100 \times \frac{50g - 37g}{37g} = 35\%$$

Freshly cut wood can have moisture content on a dry weight basis as high as 200%, which means that it initially held twice as much water as it did dry wood. A moisture content of 100% indicates that the wood has equal amounts of water and wood, by weight.

Green wood is defined as wood with a moisture content greater than 19%, while seasoned or dry wood has a moisture content under 19%. Most wood used for packaging has a moisture content between 9% and 19% (ASTM D 6199).

The usual method for measuring the moisture content of a wood sample is to weigh the (wet) wood, and then dry it in an oven at a temperature of $103 \pm 2°C$ (214–221°F, ASTM D2016 standard conditions) until a constant weight is reached. This above-boiling temperature evaporates the water. A two-inch piece of lumber will reach a constant weight in 12–24 hours. This constant weight is considered to be 0% moisture content. The moisture content is calculated from the initial and final weights, using the formula above.

Several types of hand-held moisture meters are available to determine moisture content by measuring the electrical resistance of the wood. The advantage of the electrical method is speed and convenience, although it can be less accurate than the weighing method. A moisture meter is used only if the moisture content is within the range of 2–30%.

If the moisture content and the wet weight (W_m) or the moisture content and the oven dry weight are known, the moisture content can be used to calculate the unknown parameters following the equations below:

$$W_m = W_d \times \frac{(100 + MC, \%)}{100}$$

$$W_d = 100 \times \frac{W_m}{100 + MC}, \%$$

Moisture content can also be calculated on a wet weight basis (MC_m), as it is in the paper industry (further discussed in Chapter 8). This is the percentage of the total weight of the wet wood which is moisture:

$$MC_m = 100 \times \frac{W_m - W_d}{W_m}$$

Moisture content on a wet weight basis is always lower than the moisture content on a dry weight basis, because wet weight in the denominator is always higher than the oven dry weight. This explains why the moisture content of solid wood can be higher than 200%. For the example above, with 35% moisture content based on the dry weight, the wet basis moisture content is only 26%. The difference between the two values increases with increasing moisture content, and it is very important to distinguish between the two methods when reporting moisture content.

The dry basis is commonly used for wood and the wet basis is more common for paper. This book, in accordance with industry practice, bases the moisture content of wood on the dry weight, and of paper on the wet weight, unless otherwise indicated. It is important to remember the difference and to use the correct calculation for the right industry, and also to specify which basis is used for a given measurement.

The moisture content of freshly harvested wood is highly variable. It depends on species, season of the year, geographic location, and position within the tree. The difference between the heartwood and sapwood is more pronounced in softwoods than in hardwoods, as shown in Table 3.4. It should be recognized that variations of 15–25% from the mean are common.

TABLE 3.4. Average Moisture Content of Freshly Harvested Green Wood (Forest Products Laboratory 2010).

Tree	Heartwood (MC,%)	Sapwood (MC,%)
Birch, yellow	74	72
Elm, American	95	92
Oak, White	64	78
Fir, Douglas	37	115
Redwood (old growth)	32–41	106–140
Pine	86	210
Spruce, Sitka	41	142

3.3.2 Equilibrium Moisture Content

Once the tree is harvested, it begins to lose moisture. The wood dries as water moves to the surface of the wood and evaporates. Eventually if the wood is exposed to constant relative humidity and temperature, it will reach equilibrium with its surroundings. Below about 30% moisture content (the fiber saturation point, discussed in the next section), evaporation is limited by the amount of moisture in the atmosphere.

The *equilibrium moisture content* is that which a piece of wood will have when it remains in an atmosphere with a constant relative humidity and temperature for some time. Additional moisture cannot evaporate unless the atmospheric temperature rises and/or the relative humidity falls. The moisture content of wood at equilibrium is a function of the relative humidity and temperature of the surrounding air, and is relatively independent of wood species.

Relative humidity (RH) means the actual humidity *relative* to the maximum moisture that the air can hold at a given temperature. It is expressed as the percentage of actual moisture in the air compared to the amount of moisture that saturates air at that temperature.

For example, as temperature goes up, so does the amount of moisture that the air can hold. Therefore, if the temperature goes up and the amount of moisture in the air remains constant, the relative humidity decreases. Similarly, when air cools, the relative humidity goes up. When it gets cold enough to reach saturation, 100% RH, that temperature is the *dew point*, below which further cooling will lead to condensation (dew).

When wood or paper is exposed to conditions that vary in both temperature and RH, moisture gain and loss is ongoing. The moisture content of a board that is not at equilibrium with its surroundings is not uniform throughout. If the board is losing moisture, the ends and surfaces will tend to be dryer than the more central areas. The converse is true if the board is gaining moisture.

The *sorption and desorption* behaviors of wood are not identical. Wood, as is the case for many other products, exhibits *hysteresis*, an effect of previous conditions on equilibrium behavior. So when wood is drying from a moist condition, the equilibrium moisture content is generally higher than when it is gaining moisture from a dry condition.

The effect of hysteresis on paper properties, discussed in Chapter 8, is so significant that test methods under prescribed environmental conditions require for the paper or paperboard to be preconditioned to a drier state before it is conditioned at a higher RH.

3.3.3 Fiber Saturation Point and Dimensional Stability

There are two kinds of water in wood: free and bound. Free water is contained in cell cavities, in much the same way that a cup holds water. This water, located in the lumen and intercellular spaces, is also referred to as *capillary held water*. Free water moves relatively easily through a living tree in the form of sap flow, and is also relatively easily removed by drying when the wood is harvested. The absence or presence of free water does not have much effect on wood properties.

On the other hand, *bound water* is in the cell walls. It is held much more tightly than is free water. It is chemically attracted (adsorbed) to the cell walls, held by secondary forces, primarily hydrogen bonding, with a binding energy of about 20 kJ/mol (Walker 1993).

Bound water is much more difficult to remove because of these physiochemical forces. During drying, the free water always leaves the wood before the bound water can be removed. The *fiber saturation point (FSP)* is the moisture content at which the cell walls are completely saturated with bound water (all the binding sites are occupied) but there is no free water left in the cell cavities.

The FSP is about 30% moisture content. It varies between and within species, ranging from 25% to 35%. Above the FSP, changes in moisture content do not affect the properties of wood (other than weight), since the cell walls do not change.

But below the FSP, both the mechanical properties and the dimensions of wood change with changing moisture content of the bound water in the cell walls. As discussed in Chapter 2, the wood gets stronger as it dries.

As the wood dries below the FSP, it also shrinks. As it gains moisture, it expands. (This phenomenon is familiarly experienced with the changing fit of wooden doors on a humid or dry day.) Nearly all dimensional change occurs between the fiber saturation point and zero percent moisture. As moisture content increases above the fiber saturation point (30%), there is no further dimensional change.

Swelling and shrinkage are expressed in terms of percentage of increase or decrease of dimension or volume before the change (Bowyer *et al.* 2002):

$$\text{Swelling in percentage} = 100 \times \frac{\text{increase in dimension or volume}}{\text{original dimension or volume}}$$

$$\text{Shrinkage in percentage} = 100 \times \frac{\text{decrease in dimension or volume}}{\text{original dimension or volume}}$$

The change in dimensions is not uniform. Wood is an anisotropic material. It swells or shrinks differently in the longitudinal, radial, and tangential directions. When the cell walls lose moisture, the cells get thinner but do not change much in length. So the change in the longitudinal dimension is very slight.

TABLE 3.5. Typical Shrinkage Values for Common Woods, From Green to Oven-Dry Moisture Content (As Percentage of Green Dimension; Ranges Represent Different Species of The Same Group).

Species Group	Radial	Tangential	Volumetric
Aspen	3.3–3.5	6.7–7.9	11.5–11.8
Birch	4.7–7.3	8.6–9.9	14.7–16.8
Cottonwood	3–3.9	7.1–9.2	10.5–13.9
Maple	3–4.8	7.1–9.9	11.6–14.7
Oak	4–6.6	8.8–12.7	12.7–19
Fir	2.6–4.8	6.9–9.2	9.4–13
Pine	2.1–5.4	6.1–7.7	7.9–12.3

(Forest Products Laboratory 2010)

Shrinkage in the tangential direction is greatest. The change in the radial direction is lower than that in the tangential but higher than in the longitudinal. Thus, a piece of wood does not change much in its dimension along the grain (longitudinally) as it gains and loses moisture. It shrinks most in the tangential direction, around the log. The effect can also be seen in paper, where wet paper fibers swell markedly more in width than in length.

The amount of shrinkage during and after wood drying is directly related to the density of the wood. Woods with higher densities generally have proportionally more cell wall and less lumen, so they tend to shrink and swell more with changes in moisture content. In general, wood from hardwood species shrinks more than softwoods. Typical shrinkage values for selected woods are shown in Table 3.5.

A good rule of thumb for estimating the amount of shrinkage between 0% moisture and the approximately 30% FSP is to assume 10% shrinkage tangentially; half as much, 5%, radially; and negligible shrinkage longitudinally. Shrinkage in volume is about 15%. The amount of shrinkage/swelling is species dependent.

The amount of shrinkage from drying is approximately the reverse of the swelling, subject to hysteresis between absorption and desorption (Suchsland 2004). The amount of shrinkage is obtained by dividing the total change in dimension (green versus oven dry dimension) by the original green dimension when the wood moisture content is close to the FSP (30%), and then expressed in percentage as in the equation below. The total shrinkage value (S) in percent is calculated as follows:

$$S_{R,T,L}, \% = 100 \times \frac{\Delta D_{R,T,L}}{D_{1_{R,T,L}}}$$

where R, L, and T subscripts represent the dimension in the radial (R), transverse (T), and longitudinal (L) directions, D_1 the original dimension, D_2 the oven dry dimension, and ΔD the absolute difference between the green (D_1) and oven dry (D_2) dimension in inches.

This approximation generalizes three assumptions that are not absolutely accurate. A 30% FSP is not accurate for many species. Sawn board does not have its dimensions oriented squarely in the radial and tangential direction because it is cut from a cylindrical log. And to simplify the estimate, shrinkage is assumed to be linear with decreasing moisture content, which it is not. Nevertheless, this calculation can roughly approximate the shrinkage effect of drying. The dimensional change is computed by the following equation:

$$\Delta D_{R,T,L} = \frac{S_{R,T,L},\% \times D_{1R,T,L}}{100}$$

The equation above can be used to estimate dimensional changes from green to oven dry and vice versa. In practice, wood will seldom be dried from green to oven dry, but more likely to an equilibrium moisture content of 8–25% depending on the drying method and the storage conditions. Once the equilibrium moisture content and the fiber saturation point are known, the dimensional changes will be calculated following the equation below:

$$\Delta D_{R,T,L} = \frac{S_{R,T,L},\% \times D_{1R,T,L} \times \Delta MC}{100 \times FSP}$$

where FSP is the fiber saturation point average value (typically 30%) and ΔMC is the decrease in moisture content of wood below the FSP value, expressed in percent.

Example:

A saw mill cuts 8 foot-long two-by-fours from green wood (100% MC) to near exact dimensions of 2″ × 4″ × 8′; the four inch dimension is in the radial direction (D_{1R}), the 2 inch dimension is in the tangential direction (D_{1T}). The wood is seasoned (dried) to a moisture content of 12%. Estimate the final dimensions assuming that the swelling-shrinkage coefficients in percentage are 5% in the radial (S_{1R}), 10% in tangential (S_{1T}), and almost zero in the longitudinal (S_{1L}), and that swelling-shrinkage is negligible above the fiber saturation point (FSP) which is 30% in this case.

Answer:

The swelling–shrinkage coefficient given in the example in percentage is the amount of swelling-shrinkage of a dimension when the moisture goes from the FSP to 0% moisture content.

However, we want to know the lumber dimension at 12%, not at 0%. The other concept to consider is the fiber saturation point above which the lumber dimension does not change; it can be assumed that the lumber dimension at 200% or 100% or FSP is the same and does not change. Therefore, it is safe to assume that the lumber dimension in this case at 100% will remain the same if lumber moisture content is reduced from 100 % to FSP (30%).

The question is now to calculate the lumber dimension if the lumber is dried from 30% to 12% moisture content.

The difference in moisture content (ΔMC) from 30% to EMC of 12% is equal to 18%.

The following equation is used to compute the dimensional changes in the radial (R), Tangential (T), and longitudinal (L) direction:

$$\Delta D_{R,T,L} = \frac{S_{R,T,L},\%}{100}) \times D_{1R,T,L} \times \frac{\Delta MC}{FSP}$$

where,

$\Delta MC = 30\text{--}12\% = 18\%$
$FSP = 30\%$
$D_{1R} = 4 \text{ in}$
$D_{1T} = 2 \text{ in}$
$D_{1L} = 8 \text{ in}$
$S_R = 5\%$
$S_T = 10\%$
$S_L = 0$

$$\Delta D_R = \frac{5}{100} \times 4 \text{ in} \times \frac{18}{30} = 0.12 \text{ in}$$

$$D_{2R} = 4 \text{ in} - 0.12 \text{ in} = 3.88 \text{ in}$$

$$\Delta D_T = \frac{10}{100} \times 2 \text{ in} \times \frac{18}{30} = 0.12 \text{ in}$$

$$D_{2T} = 2 \text{ in} - 0.12 \text{ in} = 1.88 \text{ in}$$

$$\Delta D_L = 0.0; \ D_{2L} = 8 \text{ in} - 0 \text{ in} = 8 \text{ in}$$

The final dimensions are approximately $1.88'' \times 3.88'' \times 8'$

Shrinkage results in warping and loose fasteners. Wood that is restrained during shrinkage can crack and split under tension, in the effort to free itself

TABLE 3.6. Problems Associated With Changes in Moisture Content of Wood.

Problem	Common Cause
Warping of pallet deckboards	More shrinkage on one side than the other
Loosening of banding in palletized load	Shrinkage of green wood in pallet
Loosening of nails in boards	Shrinkage of green wood
Checks and splitting in deckboards	Restrained shrinkage
Warped paperboard jamming cartoner	Differential swelling of paperboard plies
Biological degradation	Moisture higher than 20%

from the restraint. An example is splitting of deck boards in a pallet during drying. Common problems associated with wood shrinkage are presented in Table 3.6.

Shrinking contributes to warping because the difference in shrinkage in radial and transverse directions is combined with the curvature of the annual rings. Warping can also be caused by different parts of the wood being exposed to different conditions. For example, a house door in winter may deform because the depletion of moisture from the warm indoor side causes the inside surface to shrink and the door to warp inward.

Use of improperly seasoned wood in construction of pallets can cause loosening of the banding in palletized loads as the green wood shrinks, corrosion of metal fasteners, and growth of bacteria, mold, and mildew on the surface. Shrinkage can also loosen nails, causing them to push out of the wood.

Too rapid drying of lumber will cause surface and end *checking* or splits in the lumber. *Checks* are long, v-shaped splits directed toward the log's center. Checks occur because the wood shrinks more tangentially than radially under ordinary circumstances, and accelerated drying makes the outer part of the log shrink more quickly than the interior.

Cracking, splitting, and buckling problems, as shown in Figure 3.5, arise when wood is restrained from expansion or contraction. If moisture content increases and the wood is not allowed to expand, it is placed in compression and may buckle if the forces are large enough. Examples are the buckling of wood paneling in a humid basement, buckling of a seasoned board that is attached to a green one as it dries, or buckling of one board that is nailed to a second board with grain perpendicular to the first.

3.3.5 Other Problems Associated with Moisture Gain and Loss

If pallets containing large amounts of moisture are used for shipping products, they may act as a source of moisture to be absorbed by the corrugated

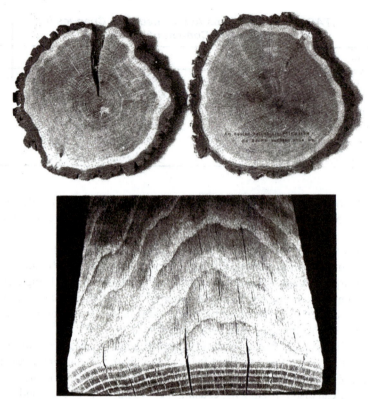

FIGURE 3.5. *Splitting and cracking in wood. Reprinted with permission of John Wiley & Sons, from Kubler,* Wood as a Building and Hobby Material *(1980) 51, 56.*

boxes that rest upon them. Moisture that evaporates from pallets can condense on products if they are contained in an enclosed space such as a trailer or box-car that travels to a cooler climate

Moisture weakens corrugated fiberboard and is a source of a great deal of damage. Moisture evaporating from pallets may contribute to the rust and corrosion of metal products, like fasteners, coils of steel, or appliances. Moisture can contribute to rot and other biological problems.

Strength is also affected by moisture content. Dry wood is stronger but less flexible than green wood, as discussed in Section 3.5.1.

3.4 DENSITY OF WOOD

Different types of wood differ substantially in density. In general, the density of temperate and tropical wood ranges from 15.6–62.4 lb/ft^3 (250–1,000 kg/m^3). The weight of dry lumber per thousand board feet varies from about 1,800 pounds for very light species to over 4,000 pounds for very heavy ones.

Density affects strength and the cost of transport. Dense woods provide higher nail withdrawal resistance but they split more easily due to nailing, and shrink more than lighter woods. As a rule, it is easier to work with lighter-weight woods.

There are a number of variables that can affect the density of wood. In general, hardwoods are denser than softwoods. But this varies from tree to tree, and even for the wood within a tree.

Green wood is denser than seasoned wood. Since both the weight of the wood and its dimensions change as wood gains and loses moisture, density is specified in terms of nominal density at a given moisture content, defined as:

$$\text{Nominal density at } x\% \text{ moisture content} = \frac{\text{Weight at } x\% \text{ moisture content}}{\text{Volume at } x\% \text{ moisture content}}$$

Wood density is always reported on the basis of moisture content due to the hygroscopic nature of wood.

Aside from moisture, the primary influences on wood density are the thickness of the wood cell walls compared to the size of the lumen, and to a considerably lesser extent, the amount and type of extractives that are present. Wood cell walls are made of a combination of cellulose, hemicellulose, and lignin. The density of a wood cell wall without water and not including any void space (lumen) is estimated to be 1.5 g/cm^3.

3.4.1 Specific Gravity

Specific gravity is defined as the ratio of density of a substance such as wood to the density of water at a specific temperature and pressure where the density of the water is 1 g/cm^3. Tabulations of the specific gravity of wood are often based on the oven-dry weight of wood and its green volume, referred to as "basic specific gravity." This calculation provides the lowest possible specific gravity, as it is based on the minimum weight and maximum volume. Other common bases for what volume to use in this calculation are the oven dry volume, and the volume at 12% moisture content. Of course, the resulting specific gravity depends on this choice.

The basic specific gravity in combination with the volume and moisture content can be used to estimate the weight of green wood products as follows (Bowyer *et al.* 2002):

$$\text{Weight} = \text{specific gravity} \times \text{volume} \times \text{density of water} \times \frac{100 + MC\%}{100}$$

Example 1:

Estimate the weight of 20 m³ of red pine to be shipped green at a moisture content of 50%. The average basic specific gravity of red pine is 0.39.

Answer:

$$\text{Weight} = 0.39 \times 20 \text{ m}^3 \times 1000 \text{ kg.m}^{-3} \times \frac{100 + MC\%}{100}$$

$$\text{Weight} = 0.39 \times 20 \text{ m}^3 \times 1000 \text{ kg.m}^{-3} \times \frac{100 + 50}{100} = 11{,}700 \text{ Kg}$$

The porosity of wood products is defined as the proportion of voids. The porosity can be calculated from the specific gravity of the wood and the specific gravity of the oven-dry cell walls, as expressed in the equation below where 1.5 is the specific gravity of oven-dry cell wall and *SG* is the specific gravity of the wood.

$$\text{Percent of void} = 100 \times \frac{1.5 - SG}{1.5}$$

Example 2:

If the oven dry specific gravity of a piece of oak is 0.75, what is the percentage of voids when the wood is oven dry? (*Note:* for this calculation to be meaningful, we must use the specific gravity calculated from the volume when the wood is oven dry.)

Answer:

$$\text{The percentage of voids} = \frac{1.5 - 0.75}{1.5} \times 100 = 50\%$$

3.5 STRENGTH

Wood is strongest in the longitudinal (grain) direction, and weakest in the radial direction. Strength is determined by the wood species and density, but is also affected by moisture content, grain slope and defects like knots, bark, checks, and splits.

In packaging, the most important strength property is *bending strength* because the primary use of wood is for bearing loads across a span of wood, as in pallet decks. Bending strength and *stiffness* are primarily related to *thickness*, although the characteristics of the wood species is also a contributing factor.

Bending strength and stiffness will also be relevant in later chapters dealing with paperboard and corrugated board. All of the materials in this book have this in common: for a given material, the thicker it is, the stiffer it is.

In most packaging applications, wood is subjected to bending stress, rather than to tensile or compressive stress. In bending stress, portions of the wood are in tension and other portions are in compression. A *modulus of elasticity (MOE)* can be calculated from the results of a bending test of a beam loaded at the center of its span and supported at its ends, in the elastic region where deformation is totally recoverable, as shown in Figure 3.6. The MOE is calculated from the slope of the linear relation between the stress below the proportional limit (σ) in pounds per square inch or in SI units and the strain (ε) using Hooke's law:

$$MOE = \frac{\sigma}{\varepsilon}$$

For a rectangular beam with a uniform cross section, *MOE* can be calculated using the span, the cross sectional area, the load and the deformation within the elastic zone and below the proportional limit as:

$$MOE = \frac{PL^3}{4Dwh^3}$$

where:

MOE is in psi or N/m^2
P = load in pounds or N
D = deflection at midspan in in. or m
L = span, the distance between supports in in. or m
w = width in in. or m
h = thickness in in. or m

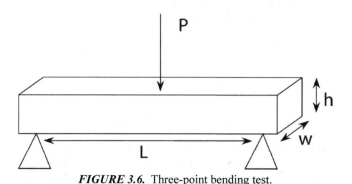

FIGURE 3.6. Three-point bending test.

TABLE 3.7. Mechanical Property Values for Selected Woods (Forest Products Laboratory, 2010).

Species	Moisture Condition	Specific Gravity*	Bending MOE (million psi)	Bending MOR (psi)
Aspen, bigtooth	Green	0.36	1.12	5,4,00
	Dry	0.39	1.43	9,100
Douglas fir, coast	Green	0.45	1.56	7,700
	Dry	0.48	1.95	12,400
Larch, western	Green	0.48	1.46	7,700
	Dry	0.52	1.87	13,000
Maple, red	Green	0.49	1.39	7,700
	Dry	0.54	1.64	13,400
Oak, northern red	Green	0.56	1.35	8,300
	Dry	0.63	1.82	14,300
Pine, eastern white	Green	0.34	0.99	4,900
	Dry	0.35	1.24	8,600
Spruce, white	Green	0.33	1.14	5,000
	Dry	0.36	1.43	9,400

*Specific gravity green is based on green volume; specific gravity dry is based on volume at 12% moisture.

The amount of deflection can be predicted from the MOE of a given type of wood. Some average MOE values are given in Table 3.7. This calculation is the basis for theoretical models that predict the deflection of the deckboards that span a pallet or help it to span supports in a rack.

Example:

Calculate the deflection in a 20-inch span of four-inch wide by 1-1/2-inch thick board of eastern white pine subjected to a load of 200 lb in a three-point bending test, if the MOE is 1×10^6 psi?

The following equation should be used to calculate D:

$$D = \frac{P \times L^3}{4 \times MOE \times w \times h^3}$$

Answer:

Deflection is 0.030 inch

The *bending strength* of solid wood and wood-based products is usually expressed in terms of the *modulus of rupture (MOR)*. The MOR is used to design wooden pallets and crates so that the lumber will not break during handling.

The MOR is calculated from the maximum load (load at failure) in a bending test:

$$MOR = \frac{1.5P_{max}L}{wh^2}$$

where the terms are as defined above.

Example:

The 20-inch span of four-inch wide by 1-1/2 inch thick board of eastern white pine described above is subjected to a bending test to determine the failure point. If the *MOR* is 4,900 psi, what is the breaking load in pounds?

$$P_{max} = \frac{MOR \times w \times h^2}{1.5 \times L}$$

Answer:

1,470 pounds

Example:

A 40-inch span of board needs to support a 400 lb load. The board is eight inches wide. Calculate the minimum thickness of the board, if the MOR of the wood is 5,000 psi?

$$h = \sqrt{\frac{1.5P_{max}L}{MOR \times w}}$$

Answer:

0.77 inch

Given a species with the MOE and MOR listed, design properties such as width, thickness, and span can be estimated when the maximum load and the acceptable deflection are known, using the equations above.

The MOE and MOR are greater for dry wood than for green wood. But hardwoods do not necessarily have greater bending strength than softwoods. The values in Table 3.7 are based on the average of extensive samples, and the Forest Products Laboratory (2010) has developed them for all wood species. They note, however, that there is wide variation in MOE, MOR, and density for a given species.

The flexural modulus of elasticity described above is similar in value to the modulus of elasticity that would be obtained in a tensile test for paper or plastic. The modulus of rupture is similar to the tensile strength.

3.5.1 Effect on Strength of Moisture Content and Temperature

Dry wood is stronger than green wood, but has less flexibility. The strength of wood generally increases as wood dries, provided that drying is not done too quickly at too high a temperature, which can cause the wood to crack or check.

Once the FSP is reached, the fibers stiffen and strengthen. Drying from the FSP to 12% moisture results in a 30–100% increase in strength. But since dry wood will not bend as easily as green wood, shock resistance and toughness may decrease.

In general, the strength of wood decreases in warmer temperatures, and increases when it is cold. For temperatures below about 150°C, the change in mechanical properties is approximately linear with temperature, if moisture content is constant. Below about 100°C, the changes are reversible. Above this temperature, the wood undergoes degradation, which results in some permanent loss of weight and strength. The degree of change depends on a variety of factors, including moisture content, temperature, and exposure period and conditions. The relative change in bending strength of wood with temperature is shown in Table 3.8.

3.5.2 Effect on Strength of Duration of Load

Time under load also affects the strength of wood, since wood is subject to creep and relaxation. Wood is able to support more weight over a short period than it can for a longer period of time. A wood member can support continuously for 1 year only two-thirds of the load that will cause failure in a standard strength test that only lasts for a few minutes.

Under continuous load, as might be imposed on a loaded pallet deck or a stack of crates, wood tends to deform. In typical use environments, the additional deformation resulting from creep after several years is approximately equal to the amount of initial instantaneous elastic deformation, doubling the total deformation.

Creep and relaxation are affected greatly by changes in temperature and

TABLE 3.8. Percent Change in Bending Strength, Compared to Strength at 20°C (Forest Products Laboratory 2010).

Moisture Content	−50°C	+50°C
≤ 4%	+18	−10
11–15%	+35	−20
18–20%	+60	−25
Above FSP	+110	−25

relative humidity. An increase in temperature of about 50°F can increase creep by two to three times. Green wood drying under load may creep by four to six times the initial deformation. After unloading, typically about half of the creep deformation is recovered. Relaxation results in a decrease in stress at constant deformation to 60–70% of the initial value within a few months.

When wood carries a load for a long period of time, it is also subject to fatigue, which reduces the load required to produce failure, often to 60% or less of the initial value after 10 years loading. When the loading is intermittent, the effect is cumulative.

Age appears to have little effect on the strength of wood, provided it is protected from decay and similar influences. Normal aging appears to result in significant loss in strength only after several centuries, so is certainly not an issue in packaging applications.

3.5.3 Effect on Strength of Physical Characteristics

Clear, straight-grained wood is used for determining the properties shown in Table 3.7. But every tree is different, and there are natural growth characteristics of a tree that reduce the strength of its wood.

Bending strength is dramatically reduced by characteristics like knots, grain slope, annual ring orientation, pitch pockets, and stresses to which the tree was exposed (like leaning, bending, and bird pecks) during its life.

A *knot* affects bending strength. A knot interrupts the direction of the wood grain and causes localized cross grain. The effect of a knot on strength depends on how much of the board the knot occupies, and how far the cross grain extends into width of the board. A knot's location affects bending strength the most when located near the middle of a span or at the top or bottom edges.

The *slope of grain* refers to the direction of the wood fibers in relationship to the longitudinal axis of the board. A board is stronger when the grain is parallel to its length; the steeper the cross grain, the weaker the board. A slope of one inch in an eight inch member will result in a board with only half of the bending strength of a piece without the grain slope. Boards with cross grain also tend to twist with changes in moisture content.

Decay resulting from fungi reduces strength as well as nail withdrawal resistance. *Wane* is bark or a lack of wood on the edge of a piece of lumber, and reduces strength because the cross sectional area is reduced. *Shakes*, separations along the grain between the growth rings, and *checks*, lengthwise v-shaped splits, reduce bending shear strength. *Warping* does not affect strength, but makes fabrication more difficult.

Specifications for load-bearing members in crates, pallets and boxes generally limit the extent and position of knots, grain slope, decay, and defects like checks, splits, and warping. Sheathing and other non-structural parts can usually bear such defects without resulting in problems due to decreased strength.

The working stress for a crate or pallet can be estimated by reducing the MOR of the wood by factors related to variability, impact loading, and strength-reducing characteristics like knots and cross grain. For example, the following factors are recommended for crate wood: 75% for variability, 33% for impact loading, and 25–75% for knots or cross grain, based on the green MOR under the assumption that it will be wet at some time. Cumulatively, these factors reduce the strength to 5–25% of the original dry MOR (Anderson and Heebink 1964).

3.6 COEFFICIENT OF FRICTION

The coefficient of friction (COF) between two surfaces is the ratio of the force required to move one over the other to the total force pressing the two surfaces together. Packages are often pushed or pulled into position by sliding, which benefits from a low COF. On the other hand, when packages are stacked in storage, handling, or transit, a high COF improves safety and stability.

As is the case for most materials, the coefficient of static friction is generally larger than the coefficient of sliding friction, and the coefficient is dependent on the surface roughness.

In the case of wood, the coefficient of friction is also dependent on the moisture content. In general, the coefficient of friction increases as the moisture content increases from oven-dry to fiber saturation, and then stays relatively constant as the moisture content increases further, until a considerable amount of free water is present. When the surface is flooded with water, the coefficient of friction decreases. Common values for the coefficient of sliding friction for smooth, dry wood against hard, smooth surfaces are 0.3–0.5. Near fiber saturation, the values are 0.7–0.9 (Forest Products Laboratory 2010).

The coefficient of friction for wood varies little with species, with the exception of species that contain large amounts of oily or waxy extractives. Some typical values of static friction coefficients for wood are presented in Table 3.9.

TABLE 3.9. Coefficient of Static Friction.

	Dry	Wet
Wood on Wood	0.25–0.5	0.2
Wood on Metal	0.2–0.6	0.2
Wood on Brick	0.3–0.4	

3.7 RELEVANCE OF WOOD PROPERTIES TO PACKAGING

Wood is not a homogenous material. It does not have precisely predictable engineering properties. It varies by species, weather, moisture content, and even by the position within the tree. There is an art and craft to working with wood, more so than for other packaging materials.

Nevertheless, there are predictable relationships, and the science of wood technology is well developed. The chemistry of wood and its relationship to structural properties of strength, density, and interaction with moisture is more complex than discussed in this chapter, and the reader who wants to know more is directed to the literature of the field.

Packaging uses for wood are limited to relatively crude forms, compared to fine woodwork and structural building. But the basic concepts of strength and shrinkage, and their relationship to moisture content, do apply to boxes, crates, and pallets. And the properties of wood's fibrous cells will prove to be a useful consideration throughout this book, especially in the chapters dealing with paper.

3.8 REFERENCES

ASTM (formerly American Society for Testing and Materials, now ASTM International). No date, annually updated. *Annual Book of ASTM Standards.* West Conshohocken, PA: ASTM International.

Anderson, L.O. and T.B. Heebink. 1964. *Wood Crate Design Manual.* Agriculture Handbook No. 252 (U.S. Department of Agriculture, U.S. Forest Products Laboratory).

Bierman, C.J. 1996. *Handbook of Pulping and Papermaking,* 2nd ed. San Diego: Academic Press.

Bowyer, J.L., R. Shmulsky, and J.G. Haygreen. 2002. *Forest Products and Wood Science.* 4th ed. Ames, Iowa: Iowa State University Press.

Brown, N.C. and J.S. Bethel. 1958. *Lumber.* 2nd ed, New York: John Wiley & Sons.

Fengel D. and G. Wegener. 1984. *Wood: Chemistry, Ultrastructure, Reactions.* Berlin: deGruyter.

Forest Products Laboratory (U.S.). 2010. *Wood Handbook: Wood as an Engineering Material.* Gen.Tech.Rep. FPL-GTR-190. Madison, WI: U.S. Department of Agriculture, Forest Service, Forest Products Laboratory. http://purl.fdlp.gov/GPO/gpo11821

Hoadley R.B.1992. *Understanding Wood. A Craftsman's Guide to Wood Technology.* Newtown, Connecticut: The Taunton Press.

Kubler, H. 1980. *Wood as a Building and Hobby Material.* New York: Wiley.

1. Panshin, A.J. and C. deZeeuw. 1980. *Textbook of Wood Technology.* 4th ed. New York: Mc-Graw Hill.

Peel, J.D. 1999. *Paper Science and Paper Manufacture,* Vancouver, Canada: Angus Wilde.

Nzokou P. and D.P. Kamdem. 2006. The Influence of Extractives on the Photodiscoloration of Wood Surfaces Exposed to Artificial Weathering. *Color Research & Application* 31(4):425–434.

Rowell R.M. 2005. *Handbook of Wood Chemistry and Wood Composites.* New York: Taylor & Francis.

Suchsland, O. 2004. *The Swelling and Shrinkage of Wood.* A Practical Technology Primer. Madison, WI: Forest Products Society.

Walker, J.C.F. 1993. *Primary Wood Processing: Principles and Practice.* London: Chapman & Hall.

3.9 REVIEW QUESTIONS

1. What are the three major chemical constituents of wood cell walls? Which one predominates? Which one "glues" the other two together?

2. What is it about wood cell structures that make them good for papermaking?

3. Which chemical constituent is removed in the pulping process?

4. What part of the tree has the highest moisture content and is easier to pulp? Why?

5. How is moisture content determined and calculated for wood? Is this the same for paper?

6. If a piece of wood weighing 35 pounds dries to 12 pounds in an oven, what is its dry basis moisture content?

7. What is meant by the "equilibrium moisture content" of wood?

8. What is "relative humidity" relative to? What is meant by the "dew point"?

9. What is meant by "hysteresis?"

10. What is the difference between free and bound water (in terms of wood cells) and which affects the dimensional stability of wood?

11. What is the fiber saturation point, and why is it significant? What percent moisture content is the fiber saturation point (in general)?

12. If a board at the fiber saturation point weighs 30 pounds, how much will it weigh when thoroughly dry?

13. If a piece of wood cut in the tangential direction is 10 inches wide when its moisture content is 100%, how wide will it be when the moisture content is 50%? How wide will it be when it is oven dry?

14. A piece of green wood (100% MC) measures $1'' \times 4'' \times 6'$. The one-inch dimension is in the radial direction, the four-inch dimension is in the tangential direction, and it is six feet long. If the wood is seasoned (dried) to a moisture content of 10%, estimate the final dimensions.

15. What two assumptions are used in this chapter's shrinkage estimates that are not absolutely accurate?

16. What is meant by "checks" in wood and why do they occur?

17. What causes buckling and splitting of wood?

18. What factor most affects a board's stiffness and bending strength?

19. What strength property is measured by a wood species' modulus of rupture?

20. How do knots and cross grain affect wood's bending strength?

21. Calculate the amount of carbon locked in 1,000 pounds of wood used to build packaging.

22. Calculate the weight of carbon dioxide stored in 1,000 pounds of wood.

CHAPTER 4

Wooden Containers

Wood has long been used to make shipping containers. Barrels, boxes, crates, and pallets all rely on wood's strength, availability, and simple construction methods. Strong wooden containers are especially useful for military and other extreme operations. The custom-built dimensional versatility of wooden crates is an advantage for shipping machinery and other large, low volume, awkward-sized items. Wood is also used on a limited basis to make shipping containers for bottles of high-value wine and for gift premium presentation packages for products like bottles of liquor and cigars.

Wooden containers are a relatively small part—about 1%—of the packaging market today. The industry is the most fragmented of all packaging industries. There are four types of producers: large integrated timber companies, captive suppliers, independent producers, and co-ops. Most are small, with sales between $15 and $50 million annually.

Wooden boxes, crates, and pallets are designed to minimize the use of material while providing adequate protection from internal and external mechanical forces. This chapter describes the types of woods and fasteners that are used, standard box styles, and the construction of crates, barrels, and baskets.

4.1 COMMERCIAL BOX WOODS

The strength of a box, crate, or pallet depends on the type of wood, the design characteristics, and the strength of the fasteners and joints. Boxes can be made from softwoods or hardwoods. Hardwood is preferred for high stress applications, although thicker softwood can often give equivalent bending strength.

The type of wood also affects the strength of the joints: the harder the wood, the greater the nail-holding power. But harder wood has a greater tendency to split when nailed, which in turn, reduces joint strength. For this reason, hardwoods and softwoods have been further categorized by the U.S. Forest Products Laboratory and ASTM into four groups, based on their nail-holding power and tendency to split, as shown in Table 4.1.

Group I is mostly softwoods that have little tendency to split as a result of nailing; they have moderate holding power, moderate beam strength, and moderate shock resistance. They are soft, lightweight, low density, easy to work, hold their shape well, and are easy to dry.

Group II woods are denser softwoods with softer springwood rings between harder summer rings. They have greater nail-holding power and more tendency to split than Group I.

Group III woods are medium density hardwoods. They have about the same strength and nail-holding characteristics as Group II but are less inclined to split or shatter on impact. These are the most useful woods for box ends and cleats and furnish most of the veneer for wirebound boxes.

Group IV comprises the dense hardwoods. They have great shock resistance and nail-holding power, but are difficult to nail because of their hardness and have a tendency to split at the nails. This group is generally used in manufactured wooden containers made on high-speed equipment, since mechanization can somewhat overcome the nail driving problems. Group IV is often fastened with bolts or pre-drilled screws.

The grain of the wood usually runs horizontally in a box, lengthwise on the face. This makes for a stronger package than if the grain runs in the short vertical direction of the panels, because the walls when stressed are better able to flex instead of pulling out nails.

TABLE 4.1. Commercial Box Woods.

Group 1	Group 2	Group 3	Group 4
Aspen	Douglas fir	Ash (except white)	Beech
Basswood	Hemlock	California black oak	Birch
Buckeye	Western larch	California maple	Hackberry
Cedar	Southern yellow pine	Soft elm	Hickory
Chestnut	Tamarack	Soft maple	Hard maple
Cottonwood		Sweetgum	Oak
Cypress		Sycamore	Pecan
Fir (true firs)		Tupelo	Rock elm
Magnolia			White ash
Redwood			
Spruce			
Yellow poplar			
Willow			
Red alder			
Pine (except Southern yellow)			

ASTM D 6199-97

The wood must be of sufficient grade to do the job. Often the outer wood surface of a box has a smoother finish while the inside is rough. This provides a smooth surface for communication, marking, and handling. The width and thickness of softwood lumber are specified and sold in nominal dimensions (actual versus nominal dimensions were shown in Table 2.4). Most specifications limit the number and size of defects present on lumber. For presentation and gift boxes, all surfaces may be polished, stained, varnished, or finished to highlight the wood grain.

For container applications, it is important to use seasoned dry wood, in order to prevent shrinkage, warping, nail loosening and rust, and biodegradation. The moisture content may be specified. Per *ASTM D6199: Standard Practice for Quality of Wood Members of Containers and Pallets*, the moisture content at the time of fabrication should be no greater than 19% and no less than 9%. The U.S. Department of Defense requires that wood for crates should have moisture content between 12% and 19% of its dry basis weight (U.S. DOD 1999).

ASTM D6199 also gives a description of allowable defects in wood members, with the caveat that they must occur in positions where they do not interfere with fabricating the package. These include checks, warp, knots, and non-straight grain.

4.2 FASTENERS AND FASTENING

A wooden package will be stronger if it is designed so that the wooden members, rather than the nails, bear the load. The fastening devices and methods used to join lumber into shipping containers are an important contributor to package strength. The primary fasteners used are nails, staples, corrugated fasteners, screws, bolts, dowels, and strapping.

4.2.1 Nails and Nailing

Nails are classified by size and the shape of the head and body. The characteristics that affect joint quality are nail length, diameter, bending resistance, and withdrawal resistance. Most nails used in packaging are made from steel.

Nail size may be simply referred to by the type of nail and the shank length. But nail size is often still specified in the traditional *penny* system (designated by d).[9]

[9]The origin of this designation is obscure; there are at least two theories. The most cited is that in England, nails were made a certain number of pounds to the thousand for many years, and so 1,000 "ten-pound nails" weighed 10 pounds. By abbreviation and slang, the word "pound" came to be pronounced "pun," which came to stand for "penny," and the letter d came to stand for both "penny" and "pound" (Park 1912). On the other hand, the Oxford English Dictionary says that "penny" referred to a price per hundred at some point in time, and that 100 five-penny nails cost five pennies.

There are many shapes of nails; *ASTM F1667: Standard Specification for Driven Fasteners: Nails, Spikes, and Staples* specifies the materials, dimensions, and profiles of 62 styles. It also provides an ASTM identification code as an alternative to the penny designation. The shape of the nail body can be smooth, like *common nails*, or it can be coated or textured for added holding power.

The types of nails used most often for boxes, crates, and pallets are *common* and *box nails* with a plain (also called bright) or coated shank, or textured like *annularly threaded* (also called *ring shank*) or *helically threaded* nails.[10] Various types of nails are shown in Figure 4.1. Threaded nails can better resist dynamic forces in distribution and they do not loosen easily if the wood shrinks.

The fact that two types of nails have the same penny designation does not mean that they are the same dimensions. For the same penny size, most nails have the same length but diameter varies by type. Standard dimensions for common, box, and threaded nails are given in Table 4.2.

All styles used for packaging have a secure flat head as opposed to *finish* nails that have small heads that can easily pull through the wood. Nails that are applied by pneumatic tools may have heads that are shaped to better feed through the nailing gun, for example, flat on one side.

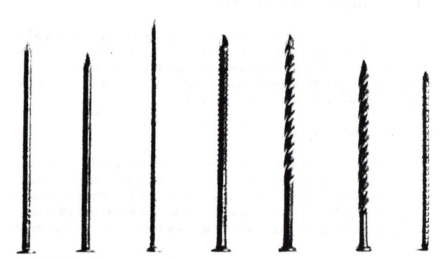

FIGURE 4.1. *Various types of nails (left to right); bright smooth wire nail, cement coated, zinc coated, annularly threaded, helically threaded, helically threaded, and barbed (Forest Products Laboratory 2010).*

[10]The names and styles of nails have changed over time. The nails that were used in the 1900s for boxes had more colorful names: *cooler, sinker,* and *corker* nails.

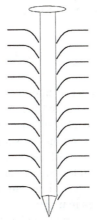

FIGURE 4.2. *Wood fibers (side grain) resisting the withdrawal of a nail.*

The most important measure of fastener effectiveness is *nail withdrawal resistance*. When a nail is withdrawn from wood, the wood fibers act like tiny wedges that resist the withdrawal of the nail (see Figure 4.2). As anyone who has pulled a nail will understand, the hardest part is the initial pulling. The maximum load occurs at relatively small values of initial displacement.

Withdrawal resistance depends in part on the density and dryness of the wood. Dense, dry wood is best for holding nails.

Withdrawal resistance also depends on the grain direction of the two wood pieces to be joined. Whenever possible, wooden pieces should be nailed together into their *side grain*, because nails into the *end grain* have poor withdrawal resistance and are more likely to work loose. The withdrawal resistance of nails clinched into the end grain (F_{eg}), direction perpendicular to the wood

TABLE 4.2. *Length (L) and Diameter (D) in inches of Different Styles of Nails.*

	Common Nails		Box Nails		Threaded Nails	
Penny (d)	L	D	L	D	L	D
6d	2	0.113	2	0.098	2	0.120
8d	2.5	0.131	2.5	0.113	2.5	0.120
10d	3	0.148	3	0.128	3	0.135
16d	3.5	0.162	3.5	0.135	3.5	0.148
20d	4	0.192	4	0.148	4	0.177
40	5	0.226	5	0.162	5	0.177

Forest Products Laboratory 2010 and ASTM F1667.

fibers is estimated to be about 75% of that of nails driven into the side grain (F_{sg}):

$$F_{eg} = 0.75 \times F_{sg}$$

Time loosens most nails. Nails driven into green wood and pulled out before it is dry, and nails driven into seasoned wood and pulled out soon thereafter, exhibit about the same withdrawal resistance. But if the moisture content of the wood changes between the time the nails are driven and the time they are removed, either through seasoning of the wood or due to cycles of wetting and drying, the nails may lose up to 75% of their initial withdrawal resistance. Relaxation of the wood fibers over time can also decrease nail withdrawal resistance, even without changes in moisture content. On the other hand, if the wood fibers deteriorate or the nail rusts under these conditions, the resistance may be regained or even increased.

It is 15–30% easier to withdraw nails from plywood than from solid wood at the same thickness, because fiber distortion is less uniform in plywood. However, plywood also has less tendency to split than solid wood, especially when nails are driven near an edge of the wood, and for thicknesses less than one-half inch, this tends to offset the lower withdrawal resistance. The direction of the plywood face grain has little effect on nail withdrawal resistance.

The most significant determinant of withdrawal resistance is the length and diameter of the nail. A rough estimate of the force required to remove a common nail from side grain is 300 pounds per inch of nail diameter per inch of penetration in the solid wood member:

$$F = 300 \text{ lb/in}^2 \times L \times d$$

Example 1:

One of the most common nails used for construction is a 16d spike (3-1/2 inches long, 0.162 inches diameter). Estimate the force required to pull out a fully-seated nail in the side grain.

$$F = 300 \text{ lb/in}^2 \times 3.5 \times 0.162 = 170 \text{ lb}$$

Example 2:

What weight can the box shown below (upside down) carry? Each nail is 2-1/2 inches long and 0.131 inches diameter. The bottom of the box is three-quarters inch-thick and the wood on the box sides is oriented so that the bottom is nailed into the side grain.

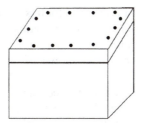

$$F = 14 \times 300 \text{ lb/in}^2 \times (2.5 - 0.75) \times 0.131 = 963 \text{ lb}$$

Of course, this is only a rough estimate. The actual holding power is a function of the wood's density, moisture content, grain direction, and age. The effect on withdrawal resistance of the wood density, the length, and diameter of the nail as expressed in the equation below:

$$F = k \times G^a \times L \times d$$

where,

F = the withdrawal resistance or the maximum load in lb or N (newtons)
G = the specific gravity of wood based on oven-dried weight and volume at 12% EMC
L = the penetration of nail in the member holding the nail point) in in. or m
D = the nail diameter in in. or m (Forest Products Laboratory 2010)

The constants k and a vary with the form of fasteners (spikes, screws, bolts, and staples) and wood members (plywood, particle board, oriented strand board, and solid wood). For more precise estimation, the reader is directed to the *Wood Handbook* (Forest Products Laboratory 2010) and the *Wood Crate Design Manual* (Anderson and Heebink 1964).

Withdrawal resistance is also affected by the shape of the nail. Common nails have a diamond-shaped point. A nail with a long, smooth round, sharp point would have greater withdrawal resistance, but is more likely to split the wood. A blunt point would reduce splitting, but has less withdrawal resistance since driving it breaks the wood fibers.

Withdrawal resistance can be increased by using nails with surface treatments for softer and greener woods. Surface treatments include cement or resin coatings and threading.

Cement coatings, despite their name, contain no cement. They contain resin that is applied to the nail to increase its withdrawal resistance by increasing the friction between the nail and the wood. In hardwoods, most of the coating is removed in the process of driving in the nail, so they provide no significant advantage. Cement coated nails are used for boxes, crates, and other containers made out of softer woods and intended for relatively short service, since it can add about 40% greater initial withdrawal resistance. The short service life is significant because the increase in withdrawal resistance drops off as the wood continues to dry.

If still greater holding power is required, nails that are *threaded* or *barbed* can be used. ASTM F1667 specifies several styles, most notably *pallet nails* that have a *helically threaded* shank. Threaded nails have the greatest resistance, especially over time and as wood dries, because they have a larger surface area. Furthermore the fibers stiffen between the threads, and so the nail is held by a greater force than simply friction. Compared to the same size bright nails, a threaded nail has 40% greater initial withdrawal resistance, and the difference can increase to about four times greater resistance over time. *Ring shank* (*annularly threaded*) nails also have good withdrawal resistance and are widely used for cleated panel crates.

The withdrawal resistance of nails and staples can also be increased by *clinching*, bending over the protruding points of the nails. Clinched nails have 45–170% more withdrawal resistance than unclinched nails. The relative increase in withdrawal resistance varies with species, moisture content of the wood and its variation, the size of the nail, and the direction of clinch with respect to the grain of the wood.

Bending resistance is another useful measure of a nail's effectiveness. A stiffer nail makes a more durable joint. Nail-bending resistance is measured by the angle in degrees that a nail bends when hit from the side with a test weight. The measure is called the *MIBANT angle*. For example, nails with MIBANT angles of 8°–24° are considered to be suitable for pallet manufacturing.

The number of nails and their spacing affects the strength of a joint. Nails should be no closer than one inch to each other or to the edge of the board, whenever possible, to avoid splitting the wood. Every board should have at least two nails at each end, in order to give sufficient rigidity to the structure.

Nail heads should be driven flush. Over-driving weakens the wood grain, and under-driving reduces the holding power. Furthermore, protruding nail heads are a potential injury and damage hazard.

4.2.2 Other Fasteners

Screws and bolts are used especially when the package is meant to be easy to open or re-closable, and for boxes made from dense Group 4 woods. Bolts are also used for mounting objects to bases in crates.

Common wood screws are available in many different diameters, lengths and shapes. Most wood screws are available with either slotted or *Phillips heads* (with a recessed cross). The Phillips heads offer a production advantage over slotted heads, because the screw driver is less likely to slip and more likely to stay centered.

Holes for bolts and screws are pre-drilled, especially when using dry dense hardwoods, to make the screw easier to drive and to prevent the wood from splitting. The hole should be shorter and narrower than the screw. The holes for softwood should be narrower than those for hard woods.

Staples, like some nails, can be driven mechanically. They should be driven with their crowns across the grain of the lumber or face-grain of the plywood. When two pieces are joined with their grain forming a right angle, the staples should be driven so that the crowns bridge the two at a 45° angle.

Corrugated fasteners and glue are used to connect *built-up faceboards* (a panel made from joined pieces). Corrugated fasteners also provide a security advantage at box closure points. Since a corrugated fastener has no head, it cannot be removed without great effort and leaves signs of prying.

4.2.3 Strapping

Metal strapping is often used to reinforce boxes and crates. The use of strapping on crates and boxes relieves stress on the nails, and also acts to minimize bulging. Strapping placed around the box, at some distance from the ends, can dissipate shocks. So can strapping that runs lengthwise on the box, perpendicular to the grain of the ends.

For very heavy crates, shorter strapping that extends around a corner is used for reinforcement. The legs of a *corner strap* are generally eight inches long, and the ends are nailed into the wood. Longer *tension strapping* is used to secure the top to the sides of a crate, with anchor plates on each side that permit tightening the strap as it is applied.

Strapping can be made from steel or plastic, but metal is most often used for crates. It can be wire-like or flat. *ASTM D4675, Standard Guide for Selection and Use of Flat Strapping Materials*, compares strap types based on their properties, ability to withstand outdoors, tension, and unit load type. It also provides advice on the number and placement of straps for different wood and corrugated box configurations.

Strapping should always be perpendicular to the grain of the wood, and should be applied tightly, sinking into the wood at the edges. For best performance, strapping should be applied just prior to shipment, since any wood shrinkage will result in loose strapping.

4.3 WOODEN BOXES

Three kinds of wooden boxes are described in the next sections: standard nailed box styles, wirebound boxes and pallet bins. These differ from the crates discussed later (in Section 4.4), which depend on their frame for strength, and may not even have walls. It should be noted, however, that in common jargon, the word "crate" is often used to refer to any wooden package.

Strictly speaking, wooden boxes rely on the walls as their primary source of strength, although cleats and diagonal members can be added to prevent distortion.

When the contents of boxes are not mounted to the base, dunnage material like plastic foam, corrugated blocks, wood, or other material is used to block and brace the product(s) inside. If a water vapor barrier is required, the box can be lined, or the product encapsulated in polyethylene film that may be impregnated with a vapor corrosion inhibitor to further prevent moisture damage.

Paper or foam impregnated with a vapor corrosion inhibitor (VCI) is also often used to protect metal products from corrosion. VCI materials do not provide a moisture barrier, but the chemicals they contain act to prevent or reduce corrosion.

4.3.1 Standard Nailed Wooden Box Styles

Wooden boxes can be built in a sawmill or carpentry shop. Nailed wood boxes and cleated panel crates account for 39% of all wooden containers, and about half of that is shipped as shook, thin boards or engineered wood products that are sawn to size at mills and later assembled in the customer's *box room* (Impact Marketing Consultants 2006).

There are nine standard styles of nailed wooden boxes, the first six developed in the early 1900s by the National Association of Wooden Box Manufacturers and the U.S. Department of Agriculture's Forest Products Laboratory. The principal difference in their construction is the design of the ends, as shown in Figure 4.3. (The following discussion is adapted from Plaskett 1930.)

Styles 1 (uncleated end) and *6 (locked-corner and dovetail)* are simple constructions that are less strong than the styles with reinforced ends. They have a tendency to split along the grain when dropped, and are stronger when the ends are made from a single piece of wood. The nails in style 1 have low holding power because they are driven into the end grain of the box.

Weight limits have been developed for each style, but strapping or other reinforcement can be added to increase the weight bearing ability. Style 1 is used to ship bottles of premium French Bordeaux wines, and depending on the materials can hold loads weighing up to 60 pounds. The locked corner and dovetail in style 6 are secured by glue between the series of tenons, which is more rigid than nailing, but can fail by breaking or pulling apart. Style 6 can carry loads up to 100 pounds. Variations of style 6 are favored for presentation boxes used to encase an expensive bottle of liquor, often with a coffin-type lid.

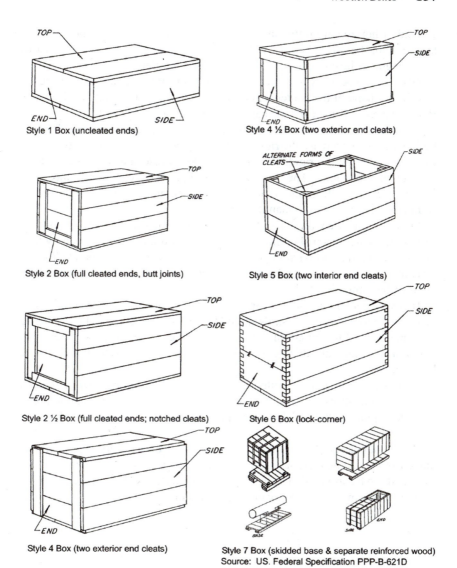

FIGURE 4.3. *Standard nailed wooden box styles.*

The other styles have *cleats*, reinforcing members that strengthen the joints and secure the ends. Some of the nails are driven through the sides into the cleats and some are driven into them from the ends. The use of cleats avoids end-grain nailing. This increases the nail-holding power and adds rigidity to the box. It also reduces the possibility of the nails splitting the wood.

Styles 4, 4-1/2, and 5 have ends that are reinforced by two cleats, either vertical or horizontal on the outside or vertical (rectangular or triangular) on the inside of the box. The use of cleats across the grain of each end increases the strength of the container and reduces the need for end-grain nailing. Styles 4 and 4-1/2 can be made to carry up to 600 pounds. Putting the cleats inside the box, if the shape of the contents allows space, reduces the length of the sides. But it also reduces the carrying capacity to less than 200 pounds.

Styles 2 and 2-1/2 are further strengthened with four cleats per end. The two horizontal cleats increase the rigidity of the top and bottom, and the two vertical cleats increase the rigidity of the box sides. These styles can be made to carry up to 600 pounds. The two styles differ in how the cleats are joined. The cleated end with butt joints, Style 2, is most commonly used since less labor and lumber are required to manufacture it.

Style 7 has a base with *skids*, used to facilitate material handling by forklift. The contents are attached to the base that has, at minimum, three inch by four inch boards (called skids) along each of two bottom edges. A separate hood is assembled from top, side, and end panels that have been reinforced with framing members and diagonals.

Box styles other than Style 7 can also have skids attached to the bottom. The general rule for skids is that skids should be no less than 2-1/2 inches nor more than one-sixth of the length of the box from the end. When four-way entry is required, the skids are thicker and notched to permit entry from the side or lower rubbing strips are added.

Reinforcements include battens, diagonals, and additional cleats. Battens are reinforcing strips mid-wall that are either inside the box or encircle it and are held in place with the help of strapping. Outer battens can also function as skids.

Additional cleats can be added to the ends. Diagonal reinforcing members, made from the same material as the cleats, are sometimes used on the ends or sides and can be arranged in a zigzag pattern, with a peak at the center of a long panel where it meets a center batten, as shown in Figure 4.4.

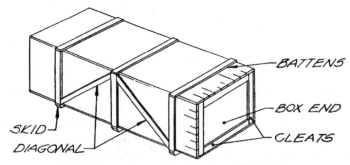

FIGURE 4.4. *Example of reinforced Style 2 box.*

The box standards specify the thickness of the wood in each of the four wood groups to be used for various weight products shipped either domestically or overseas. They specify how to set up boxes of each style and how to choose wood thickness, nail and screw sizes, and fastener spacing patterns.

In general the standards reveal:

- The thicker or softer the wood, the bigger the nails need to be.
- The smaller the nail, the closer should be the spacing of the nails.
- Nails driven into end grain should be more closely spaced than those in side grain.
- It recommends using 25% fewer coated nails, compared with bright ones.

The standards can be found in *ASTM D6880/D6880M Standard Specification for Wood Boxes*.

All U.S. government wooden container standards like this have in the past been managed by the U.S. Government's General Services Administration (GSA). Some have been prepared by the Department of Defense (DOD) through a mutual agreement. As a result of Circular Number A-119, titled "Federal Participation in the Development and Use of Voluntary Consensus Standards and in Conformity Assessment Activities," GSA and DOD adopted ASTM standards in place of the government's wooden container standards.

In the new ASTM standards, many parts were left out or rewritten to allow more liberal fabrication procedures and reflect commercial practices. Mandatory standard statements were changed to allow more liberal fabrication procedures. The requirements that are unique to the military are added in a supplementary section and do not form a part of the voluntary standard unless cited in government contracts.

But military specifications are unique to the military's logistics operation and site requirements such as the degree of protection, military marking, and shelf life. Military standards are based on extreme logistics situations. Most are designed for a 10-year life expectancy for purposes of military preparedness, in circumstances ranging from desert to swamps, from blazing hot to freezing cold. Testing, described in Mil-Std-2073, is much more stringent than for commercial packages (U.S. DOD 2011).

For products shipped in modern civilian logistics systems, the military standards may be excessive, but the standard styles represent a set of successful principles to guide commercial wooden package design. Most commercial wooden boxes are still based on the standard styles.

4.3.2 Wirebound Boxes

Wirebound boxes are manufactured from thin veneer reinforced by steel

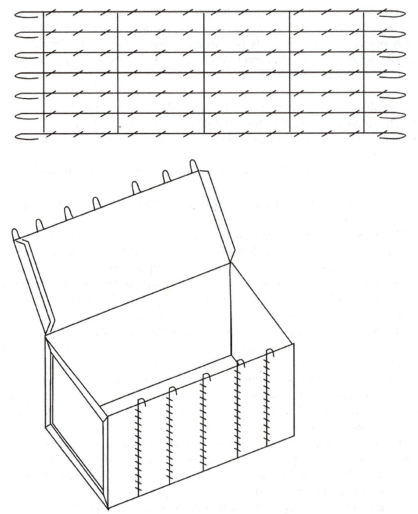

FIGURE 4.5. *Wirebound box, wire configuration on knocked-down form and assembled box.*

binding wires that are fastened to the wood by staples, and by wooden cleats on the ends of the box, as shown in Figure 4.5. There are a number of different styles. Most are closed by putting wire loops on the cover through mating loops on the body and bending them over.

Wirebound boxes became popular in the early 1900s as a substitute for nailed wooden boxes. They offered many advantages. They are inexpensive and can be produced at high speeds. They have very high strength-to-weight ratios, can carry up to 500 pounds, and can support heavy stacking loads. In contrast to wooden boxes and crates, they are easily shipped flat, in bulk knocked-down form, which significantly reduces transportation costs for empty containers.

Wirebound boxes are primarily used now only for shipping heavy, bulky products like machinery and automotive parts. They are also used for fresh produce like broccoli that is shipped wet or on ice, since they have better stacking strength when wet than do corrugated fiberboard boxes, allow the melted ice to run off, and provide good ventilation and absorbency. However, they have largely been replaced by waxed corrugated boxes for such wet uses, including shipping of meat and poultry.

Large wirebound boxes are used for overseas shipments and military applications. The U.S. military has developed standards for wirebound boxes, which have been adopted as ASTM standards: *ASTM D6573: Specification for General Purpose Wirebound Shipping Boxes* and *ASTM D6254/D6254M Specification for Wirebound Pallet-Type Wood Boxes.*

4.3.3 Pallet Bins, Collars, and Build-Ons

There are many cases in which walls are built onto a pallet, or assembled from a folded and hinged collar. The walls give puncture protection and provide some stacking strength. They can be built from OSB, LVL, plywood, wirebound veneer, or corrugated board. The Engineered Wood Products Association (APA) provides some guidance in choosing panel products for these uses (APA 1995).

Demountable bins (where the walls can be taken off again) are reusable, and the walls can be folded down for shipment when empty. There are a number of patented systems in use, such as those offered by Foldy-pak and Nefab. Nailed pallet bins usually employ cleats inside or outside to aid nailing.

One of the major uses for pallet bins is for hardware and parts shipped to assembly factories. They are also used in agribusiness for products ranging from granular materials like fertilizer to fresh produce shipped on ice.

4.4 CRATES

Traditionally, a wooden *crate* differed from a box because a crate's cleat-like *frame members* (also called *struts*) provide the strength and define the shape. Usually the item inside a crate is mounted to the base, and the walls (if there are any) are mere sheathing.

On the contrary, in the package called a box, the walls provide the main structural strength, and the cleats are mere reinforcement. A *cleated panel box*, also called a *cleated panel crate*, is a hybrid between the two.

The word *crate* is also used in the fruit industry for wooden boxes (as in *field crates*), and some people use the word to refer to all wooden boxes. It comes from the Latin word *cratis*, which referred to wicker baskets.

Crates and cleated panel boxes are used for large items like equipment, airplane parts, engines, vehicles, and sculptures. They are manually custom-built, although some sawing and nailing can be automated for high volume uses. The following sections show the difference between traditional framed crates and cleated panel boxes.

4.4.1 Traditional Framed Crates

A traditional framed crate may or may not have walls. Open crates are used when the contents can be safely exposed to the atmosphere and other distribution hazards. Sheathed crates are used when complete puncture resistance is required. Crates can be nailed with the intention of a single use or bolted to be demountable and reusable.

Products shipped in crates are usually large and awkward to move without the crate. Equipment, vehicles, sheet metal, and glass are often shipped in framed crates. Glass sheets, in particular, benefit from being tightly held in an open crate, because the lift truck driver can see the glass's vulnerability and take adequate care to avoid damage. Glass sheets are also easier to pack and unpack in an open crate.

The type of wood chosen depends on the weight of the product, the needs of the distribution system, and the bending strength of the wood (discussed in Chapter 3). Knots and cross-grain are avoided for weight-bearing frame members.

Most of the lumber used for domestic crate frames is softwood (usually pine), nominally two inches-thick, although load-bearing struts and skids are usually thicker. Hardwood is used when more strength is required. Heat treated wood is used for export crates because of regulations regarding infestation, described in Chapter 2. Rough lumber can be used for the base and framing, but some surfaces must be dressed to provide a smooth surface for marking.

The strongest crates employ *three-way corners* (shown in Figure 4.6) wherever possible, so that every nail is driven perpendicular to the grain. This creates a very strong structure. Since each of the three pieces is locked in by the other two, the nails are not likely to work loose. For very heavy crates, the corners may be reinforced by corner straps that are nailed to the wood. Three-way corners in sheathed crates especially improve the security of the top, since it is interlocked between two sides, nailed on from two directions.

Diagonal members are added to prevent distortion and strengthen crate frames. Nail clusters in joints alone will not prevent the *racking* that can occur in frames with only end-to-end joined members. Figure 4.7 shows how diagonal reinforcement prevents racking.

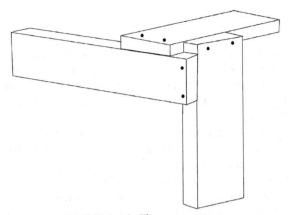

FIGURE 4.6. *Three-way corner.*

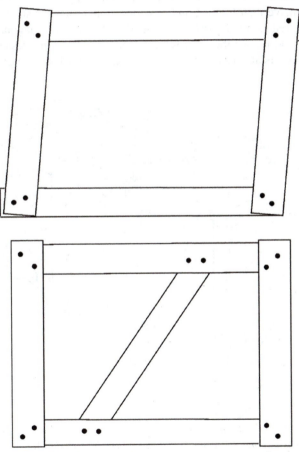

FIGURE 4.7. *Diagonal reinforcement prevents racking.*

Most crates incorporate triangles as basic building blocks. Diagonals placed at or near 45° significantly improve the rigidity of a crate, especially when it is impacted on a corner. Two diagonals crossing each other add even more strength. Figure 4.8 shows a typical crate with three-way corners and diagonal members for bracing.

Traditional framed crates are used extensively by the military due to the large size of military equipment, the small customized orders, and the extreme nature of their logistical systems. The U.S. military has set standard specifications for wooden crates, found in *ASTM D6039: Specification for Crates, Wood, Open and Covered*, for five categories of content weights from 200 to 2,500 pounds. Higher weight contents require thicker bases, wider frame members, and more reinforcement from diagonal members and cleats.

When sheathing is required, plywood or OSB are generally used. Plywood is less expensive than lumber, and provides more structural strength and stability. It can reduce total costs by eliminating the need for diagonal members and a liner, and reduce the crate weight. OSB is heavier and less strong than plywood, but it is also less expensive. The least expensive and lightest weight

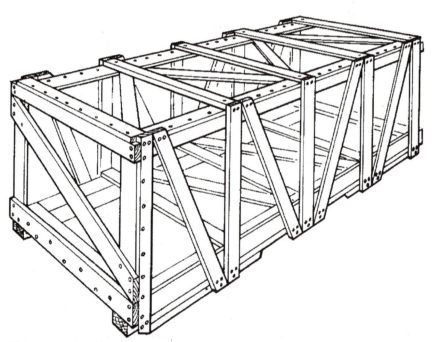

FIGURE 4.8. *Crate with three-way corners and diagonal members for bracing. U.S. Forest Products Laboratory, from Anderson and Heebink (1964) Wood Crate Design Manual.*

sheathing material is corrugated fiberboard, but this provides no strength beyond minimal puncture resistance and dust cover.

As the strength of the sheathing increases, the distinction between crates and boxes decreases. Developed by the U.S. military, ASTM has published standards for some of these hybrid types of containers. Three examples are:

- *ASTM D6251/D6251M: Specification for Wood-Cleated Panelboard Shipping Boxes*
- *ASTM D6254/D6254M: Specification for Wirebound Pallet-Type Wood Boxes*
- *ASTM D6256/D6256M: Specification for Wood-Cleated Shipping Boxes with Skidded, Load-bearing Bases*

Steel crates can also be sheathed with plywood or other engineered wood products, as illustrated in *ASTM D6255/D6255M: Standard Specification for Steel or Aluminum Slotted Angle Crates.*

When the item is moisture sensitive or vulnerable to rust, the item is first thoroughly dried and waterproof liners may be used, ranging from asphalt-covered paper to polyethylene, often in combination with desiccant packets to pull the moisture from the air inside the package. Such liners should hang free from the item to permit some air circulation. If water trapped inside the package tends to be a problem, ventilation and drainage holes are used to allow moisture from condensation to escape; they can be covered with screens to prevent the entry of animals or insects. VCI paper is a common way to provide protection from rust and other corrosion for metal products. VCI foam can also be used.

To the extent possible, crates use the item itself to provide strength and structure. A sturdy steel item with a flat top may be packed with no head clearance, in order to take advantage of the strength of the product. But when surfaces and elements could be damaged by contacting the wood during rough handling and transit, there should be a minimum one-inch clearance to allow for distortion and surface protection. Additional clearance is required for fragile or shock mounted items.

In many cases the item is mounted to the base. It is important to identify the best points for mounting and contact. Mounting methods include steel strapping, carriage bolts, brackets, and anchor blocks and plates. Shock mounts or cushions may be used, but they should be tested to ensure that they do not cause vibration problems since they lower the natural frequency of the crated product.

A skid base is most common, with pallet-like stringers or blocks, capable of being handled from any of the four sides. Whenever possible, the center of gravity should be low so that it does not tip easily, and centered so that the crate is well-balanced.

Long crates must often be handled from their short ends in order to load and unload trailers, and the base should provide a means for doing so even if the lift truck's forks do not extend long enough to fully pick it up from the end. Stringers that run parallel to the length, with chamfered ends like sleigh runners, will facilitate sliding. Long, thin crates also need a base wide enough to stabilize it and keep it upright during transit and handling. A 22° tip test can be used to judge the stability of such packages if they have a high center of gravity.

Crates can be fastened with nails, screws, bolts, patented metal connectors, and some kinds of staples. Two inch to three inch ring shank and cement-coated common nails are used. Staples are used for connecting sheathing to frames, but not for connecting the frames together. Screws and bolts are used when the crate is designed to be reusable. Sometimes just one panel is connected by screws in order to facilitate easy opening.

The panels of the frame are assembled first, including connecting sheathing. Then the frame is nailed together. It is common to use *jigs* and *templates* during crate construction in order to ensure consistent quality, because the dimensions and assembly require precision.

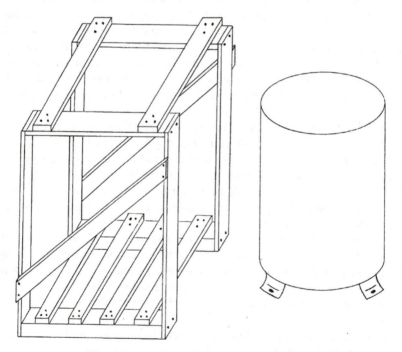

EXISTING DESIGN: 400 LB WATER HEATER BOLTED TO DECK OF PALLET BUILD-ON
STACKED TWO HIGH, NON RETURNABLE

FIGURE 4.9. Crate design change example.

Crate Design Change Example:

A 400 pound water heater, packed in the crate shown in Figure 4.9, was experiencing a great deal of damage: the product was shifting and the crate was distorting in transit and handling. The water heater was bolted to the deck of the pallet, and the crates are stacked two-high. How could you redesign this package to be sturdier?

Improvements:

1. Add more boards to the corners and use three-way corners.
2. Move top boards to the sides and upright boards to the adjacent side to make three-way corners, which will stiffen them and eliminate the need for end-grain nailing. This also gives the present diagonals a larger nailing area.
3. Add diagonals on two more sides creating corner triangles.
4. Change nailing pattern to move nails farther from the edge.
5. Add a center stringer and four-way entry to the skid (making it more like a pallet) to ease handling and stiffen the deck.

4.4.2 Cleated Panel Boxes/Crates

Cleated panel crates are actually more like boxes than they are like traditional framed and sheathed crates, even though they look similar. The government calls them boxes (*cleated panel boxes*) because they rely on the walls, reinforced by cleats, for strength, rather than a frame. They have gained use due to the increasing use of engineered wood products for packaging.

Cleated panel boxes are made by stapling wood cleat reinforcements to the perimeter of sheets of plywood or oriented strand board (or even corrugated board), like a picture frame, to form panels. The panels are then nailed together at the cleats to form a container mounted on a skid. The cleats reinforce and stiffen the edges of the composite sheet material to provide support and aid fastening. They are often used to give a side grain lumber surface for nailing (although in practice a great deal of cleated panel end-grain is nailed). Cleated panel crates may be reinforced with straps. An example is shown in Figure 4.10.

Cleated panel boxes have become popular alternatives to solid wooden boxes and sheathed crates for commercial uses. They differ from traditional crates because most do not have three-way corners, and some do involve end-grain nailing. The tops are simply nailed straight down to the load-bearing cleated walls, without the benefit of interlocking joints.

Most commercial "crating" companies are in the business of producing

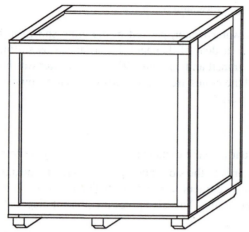

FIGURE 4.10. *Cleated panel box without interlocking corners. Used by permission of Deploy Services LLC (2004).*

such boxes, and thus the name "crate" has come to include cleated panel boxes. Most cleated panel boxes are produced for commercial use, rather than military.

Cleated panel boxes are widely used for equipment, engines, metal stampings, large machinery parts, computer hardware, artwork, and personal belongings. They are used when a factory or production line must be moved. They are easily handled by an ordinary forklift, and can be custom built to the right size for the product.

Cleated panel boxes are lightweight and inexpensive, compared to solid wood. They are easier to make than traditional sheathed crates because the panels can be prefabricated. Their dimensions and construction methods are easy for workers to understand. They have good resistance to diagonal distortion and corner damage.

They can withstand the forces of normal distribution, especially since packages in commercial transport are not stacked very high or handled as roughly as they are in extreme military situations. Even commercial ocean transport is now primarily containerized, which greatly reduces the stresses compared to break-bulk shipping.

There are also a number of reusable panel boxes that are joined at the edges by patented locking devices. Some examples are systems made by Klimp, Hardy-Graham, Shure-Lok (trade name of Pak-Rite), and Nefab. Such panel boxes clip together easily, and are knocked down once they are emptied to be shipped to another location to be re-erected and re-filled. Some of these do not even have cleats.

The U.S. government has developed two specifications that apply to cleated panel boxes. They are now published as *ASTM D6251/D6251M: Specification for Wood-Cleated Panelboard Shipping Boxes* and *ASTM D6256/D6256/M: Specification for Wood-Cleated Shipping Boxes with Skidded, Load-bearing Bases*. D6251 provides specifications for 11 standardized styles of cleated panels, four of which are all wood and the rest use corrugated board as sheathing. The variations include cleat number and placement, and whether the box has two-way or three-way corners. D6256 deals with the choice of base depending on the weight of the load.

But the vast majority of cleated panel boxes/crates are not built strictly according to any published standards. Wood crates are best suited for custom packaging, and one of the great benefits of wood construction is its adaptability.

4.5 BASKETS

Baskets, through the ages, have proven to be an exceedingly versatile package form. Most are formed by weaving natural materials. They can be woven tightly enough to hold liquid or coarsely to be more lightweight. There is a wide range of sizes and shapes, each suited to a specific purpose.

Most baskets are made by hand, although mold-like forms are used to guide shapes. Baskets are not used much commercially because of the high labor input. A notable exception is the bushel basket.

The traditional "bushel" basket (Figure 4.11) is formed from many (25–100) thin strips of wood called *splints* that radiate from the basket bottom cen-

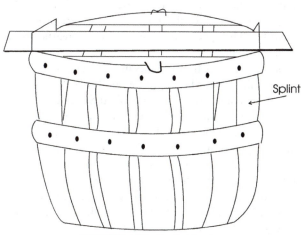

FIGURE 4.11. *Bushel basket.*

ter, overlap, and are bent into shape. They are held together by hoops made from wood strips that are stapled into place. The term "bushel" has been used for various units of capacity over time. In the United States it is a dry capacity measure equal to 2,150.42 in^3 and it is a British unit of dry or liquid capacity equal to 8 imperial gallons or 2,219.36 in^3.

Baskets are still used widely for gathering fresh fruits and vegetables from the field and distributing them to local markets. In less developed countries, baskets are still used extensively for transporting and handling many kinds of products ranging from live chickens, to grain, to gravel.

4.6 BARRELS

Barrels, also known as casks, employ one of the strongest building principles in engineering: the principle of the double arch. When a barrel is viewed from the head, the width of each stave acts as a keystone in the arch construction, supported by the other staves as a base. When a barrel is viewed from the side, the length of the stave is the keystone, supported by the two heads as a base. Because of its shape, the barrel is "one of the strongest containers known to man" (Hankerson 1947).

Its shape also endows it with unique handling characteristics: the ability to easily roll, making it like a container with a built-in wheel. Although a single person can typically carry less than 100 pounds, he or she can roll a large weight, up to several hundred pounds. Because of the bilge construction, only a small surface comes in contact with the floor, reducing friction to a minimum so that it will roll and pivot easily, especially down an incline. Figure 4.12 shows an example of barrel handling and transport.

There are two basic kinds of barrels. *Tight barrels* are designed to hold liquids and *slack barrels* are for dry products. They are produced in different production facilities. The species of wood and the construction are related to their expected performance.

Tight barrels are used mostly for wine and whiskey aging, valued for their traditional role and their contribution to the mellow flavor of aged beverages. Eighty percent of the barrels made in the United States today are tight barrels made from hardwoods, produced near Tennessee and Kentucky, and used for bourbon whiskey. Some of the larger distillers, like Brown-Forman, make their own casks (Impact Marketing Consultants 2006). The Nevers and Limousin regions of France have the greatest reputation for making wine barrels.

Making a barrel "tight" is a skill requiring great precision. Hardwood species like white oaks (*quercus rubra, quercus petraea, quercus alba*) are most often used to make barrels because they resist soakage and are hard enough to resist abuse and yet soft enough to form into the barrel shape. The parts of a tight barrel are shown in Figure 4.13.

A tight barrel is made from *staves* that are radially or *quarter sawn*, from the wood between the heart and sapwood. They are cut parallel to the log's medullary ray, a hard fibrous membrane radiating out from the center of the tree that makes the wood more dimensionally stable and impervious to seepage due to the presences of tyloses in the pores. Beer barrels, designed to be reused many times, have especially heavy staves.

The edges are cut so that the staves are slightly wider in the center and so that one face is slightly narrower than the other. They are partially dried, leaving just enough moisture to keep them flexible. They are then set in a form that is tight on one end and are steamed to make them pliable. The staves are then drawn together into the conventional barrel shape (with a slightly bulging *bilge*) and held in place with heavy iron *hoops*.

The formed barrel is then fired to dry and shrink the wood and set the staves in shape. Some barrels for whiskey and wine are also slightly charred to impart a special taste. After cooling, the hoops are hammered towards the bilge in order to press the staves more tightly together. Next are cut the *chime*, a bevel on the end of the staves, and the croze, a groove into which the *heads* will fit. The circular heads are cut with a beveled edge and inserted into the crozes. The last step is driving on the *head hoops* that complete the *tight head* (Hankerson 1947).

The process of shaping was mechanized during the Civil War, although it produced barrels of inferior quality. Leakproofness was improved with glue

FIGURE 4.12. *Barrel handling and transport in Portugal.*

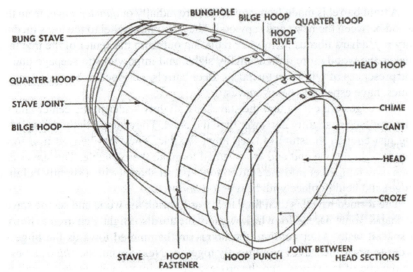

FIGURE 4.13. *Tight barrel with parts labeled. Reprinted with permission of Chemical Publishing, from Hankerson,* The Cooperage Handbook *(1947).*

and linings. As late as the 1940s, articles on the topic of linings and how to join the staves to better prevent leakage were still common in the trade press (e.g., in *Barrel and Box and Packages* 1943–1946).

One of the staves has the *bunghole* through which the barrel will be filled. The *bung* closure is a cork-shaped piece of wood or rubber that is wedged into the hole. It is applied and removed by using a mallet. Tight barrels when full are always stored on their sides with their *bungstave*, the stave containing the bunghole, up. In some cases the bunghole is in the head.

Slack barrels are more diverse. They can be made from a number of different woods such as gum, elm, cottonwood, basswood, and pine. Slack barrels are formed in the same way as tight barrels except that the staves are thinner and do not need to be cross sawn. Some have tongue-and-groove joints, designed to be siftproof for holding powdered products. The heads may be made from several pieces. They can be bound with hoops made from steel, wire, or wood. The choice among these variables depends on the contents' weight and consistency, as well as the rigors of distribution. Slack barrels are opened by removing one head (carefully driving the top hoop halfway off and tapping the head into the barrel) that is carefully replaced after filling.

Slack barrels are rarely used today, but in its prime the cooperage industry had a number of standard styles for specific commodities like dry powders and flour, meat and poultry, glass and pottery, potatoes, apples, fish, vegetables, nails and spikes, heavy weights, and explosives (Hankerson 1947).

4.7 CORKS

Cork closures were first used by the Romans to close amphoras, barrels, and glass bottles. The tight friction-fitting cork kept the "stars" in Dom Perignon's champagne beginning in the 1600s. After the Industrial Revolution, their use as bottle closures grew with the growing use of glass bottles for foods, beverages, and patent medicines; in the mid-1800s cork was the predominant closure material for bottles. With the introduction of standardized continuous thread caps and plastic closures in the 1920s, cork use declined.

Corks are made from the bark of cork oak trees (*quercus suber*) in Portugal, Spain, Italy, France, Tunisia, Morocco, and Algeria. The trees are unusual in that their bark is so thick and resistant that it can be stripped from the trunk without hurting the tree. The structure is unique in that it has very small, closely packed 14-sided cells that make it light, elastic, inert, and impermeable. It also has low thermal conductivity, which is why it can be used as an insulating material.

Cork farming is a long-term business. Cork suitable for wine stoppers cannot be obtained until the tree is 25 years old, at which time the cork bark is stripped from the trees and allowed to re-grow. Thereafter, the trees are re-stripped at intervals of no less than nine years (by Portugese law). The older the tree, the larger the harvest; the average tree life is about 170 years. About 150,000 tons of cork per year, which represents 50% of world production, is harvested in Portugal.

After air drying in stacks for six months, like other wood is seasoned, the cork is boiled for about 90 minutes to make it more flexible and to destroy insects and mold. The planks rest for another three weeks and then they are cut into strips (horizontal around the tree) as wide as the length of the final corks. The strips are turned on their sides and the corks are punched out so that any knots or fissures in the wood are perpendicular to the length of the cork, to minimize leakage due to discontinuities. The ends are then polished to present a smooth surface. They are bleached to achieve a more uniform color, branded with the name of the winery, and sometimes coated with wax or silicone to make them easier to insert and to remove using high speed equipment (Robinson 1999).

Today, cork closures are used almost exclusively for closing wine bottles. Natural cork wine stoppers are the most profitable form of wine closure, accounting for approximately 15% of the total production by weight but two-thirds of the revenue (Natural Cork Quality Council 2002).

Cork closures are valued because they give a good positive seal. Cork length ranges from one inch to 2.3 inches (25–60 mm), with longer corks being used for more prestigious wines that are meant to be cellared for longer periods. But contrary to popular myth, the length of even a short cork does not allow for any exchange of air, and the cork does not need to be kept moist in order to provide a tight seal. The traditional closure is part of an image that is cherished for wine.

However in recent years, the wine industry has increasingly switched to artificial corks made from plastic and composite materials and to threaded closures in response to two cork-related problems. One is the insufficient supply of cork to serve the growing wine market.

But the bigger problem is that some corks have been found to be infected with a particular species of bacterium that changes the taste of the wine (which is said to taste "corked"). It has proven difficult to detect which corks have the cork taint, since it occurs randomly, but it has been estimated that 2–5% of wine has this problem. It has proven difficult to prevent since its source is not clear, but likely contributors include the bleaching method and storage conditions (Margalit 1997). Cork taint has been attributed to the presence of 2,4,6-trichloroanisole (TCA), and/or 2,4,6-tribromoanisole (TBA) (Chatonnet *et al.* 2004).

Corks are a matter of great controversy in the wine industry. While they lend a traditional cachet to the wine package, the cork taint drawback has begun to outweigh the aesthetic advantages. The fine wine industry in New Zealand was the first to switch to screw caps, once only thought suitable for low priced wine. Consumer acceptance is growing for substitutes to traditional wine bottles closed with corks.

4.8 REFERENCES

Anderson, L.O. and T.B. Heebink. 1964. *Wood Crate Design Manual.* Agriculture Handbook No. 252 (U.S. Department of Agriculture, U.S. Forest Products Laboratory).

APA (The Engineered Wood Products Association). 1995. Materials Handling Industrial Use Guide. Washington, DC: APA.

ASTM (formerly American Society for Testing and Materials, now ASTM International). No date, annually updated. *Annual Book of ASTM Standards.* West Conshohocken, PA: ASTM International.

Barrel and Box and Packages. Trade publication from the early 1900s. MSU has the War years, 1943–1946, vols 48–51.

Chatonnet P., S. Bonnet, S. Boutou, and M.D. Labadie. 2004. Identification and Responsibility of 2,4,6-tribromoanisole in Musty, Corked Odors in Wine, *Journal of Agricultural and Food Chemistry 52* (5): 1255–1262.

DeployTech. 2004. *Crate Pro.* CDROM. Available from www.cratepro.com.

Forest Products Laboratory (U.S.). 2010. *Wood Handbook: Wood as an Engineering Material.* Gen.Tech.Rep. FPL-GTR-190. Madison, WI: U.S. Department of Agriculture, Forest Service, Forest Products Laboratory. http://purl.fdlp.gov/GPO/gpo11821

Friedman, W.F. and J.J. Kipnees. 1977. *Distribution Packaging.* Malabar, FL: Robert E. Krieger Publishing Company.

Harvey, J. 1986. Manual on wooden packaging. International Trade Center UNCTAD/GATT, Geneva, 114p.

Hankerson, F.P. 1947. *The Cooperage Handbook.* Brooklyn, NY: Chemical Publishing Inc.

Impact Marketing Consultants. 2006. *The Marketing Guide to the U.S. Packaging Industry.* Manchester Center, VT: Impact Marketing Consultants.

Margalit, Y.. 1997. *Concepts in Wine Chemistry.* San Francisco, CA: The Wine Appreciation Guild.

McKinlay, A.H. 1998. *Transport Packaging.* Herndon, VA: Institute of Packaging Professionals.

NWPCA. 2012. *Uniform Standard for Wood Containers.* Alexandria, VA: National Wooden Pallet and Container Association. http://www.palletcentral.com/images/files/uniform%20standard%20for%20wood%20containers%202012.pdf

Natural Cork Quality Council. 2002. The Cork Industry Worldwide and U.S. Production Estimates. www.corkqc.com/cqfstat.htm, accessed March 10, 2002.

Park, J.C. 1912. *Educational Woodworking for Home and School.* New York: The Macmillan Company.

Plaskett, C.A. 1930. *Principles of Box and Crate Construction,* Technical Bulletin No. 171, United States Department of Agriculture, Washington DC.

Robinson, J., Ed. 1999. Corks, *The Oxford Companion to Wine.* 2nd ed. Oxford: Oxford University Press.

The Wooden Crates Organization. 2009. www.woodencrates.org

U.S. Department of Defense. 1999. Wooden Containers and Pallets and Cra*tes, Packaging of Materiel,* Chapters 3 and 6. FM 38-701/ MCO 4030.21D/ NAVSUP PUB 503/ AFPAM(I) 24-209/ DLAI 4145.2. Washington, DC: United States Departments of the Army, Navy, Air Force and the Defense Logistics Agency.

U.S. Department of Defense. 2011. Mil-Std-2073: Standard Practice for Military Packaging. Washington, DC: U.S. Government Printing Office.

4.9 REVIEW QUESTIONS

1. On what basis do ASTM and the U.S. Forest Products Laboratory further categorize hard and soft woods?

2. Where can military specifications for wooden containers be found?

3. Which way should the wood grain run on a box: horizontal or vertical? Why?

4. How is the size of nails traditionally designated?

5. What factors *related to the wood* affect nail withdrawal resistance?

6. Why is it not desirable to connect wood boards by nailing into the end grain?

7. What are the two most significant factors *related to the nail* that affect withdrawal resistance?

8. What is the estimated force required to pull out an 8d common nail?

9. What estimated weight can be carried by a panel that is three-quarters-inch thick, nailed in place with 12 16d common nails?

10. How many 10d common nails are required to hold a solid two inch bottom on a box that holds 600 pounds (assume that the nails go into the walls' side grain)?

11. The simplified withdrawal resistance estimation ignores the effect of other factors related to the nail and wood. What are five of these?

12. How can the surface of a nail's shank be modified to make it harder to remove?

13. What is vapor corrosion inhibitor paper and when is it used?

14. What is the strongest way to make a wooden box or crate corner? Why?

15. What are cleats in a wooden box and why are they used? What are battens?

16. What is the function of diagonal members in a wooden crate?

17. What are "staves" and "splints" and what kinds of packages are made from them?

18. What is a bung, and how is it used?

19. What is the difference between a tight and slack barrel?

20. What kind of wood is used to make tight barrels? Why?

21. In what ways are staves cut differently from ordinary lumber?

22. What are the advantages and disadvantages of using cork as a wine bottle closure?

CHAPTER 5

Wooden Pallets

For most of human existence, material handling was a manual activity. Humans lifted, carried, and stacked packages, and loaded and unloaded vehicles by hand by rolling barrels and wheeled hand trucks. The invention in the 1930s of a system of easily-maneuvered mechanized forklifts and rugged pallets greatly reduced the expense, time, and back-breaking work of material handling.

A pallet is a portable, horizontal, rigid platform used as a base for assembling, storing, stacking, handling, and transporting shipping containers as a unit load. Pallets are also used as platforms for intermediate bulk containers.

Most of the material handling in the world today takes place with the use of wooden pallets and forklifts. Wooden pallets are the highest volume category of wood packages in use today.

It is estimated that more than two billion pallets are used every day in the United States with more than 500 million introduced yearly newly made or re-manufactured/repaired (Bush and Araman 2008; Clark 2004). About 90% are made of wood, 2% from plastics, 2–4% from wood panels, 1% from metal, and 1% are paper-based (Baumeister and Beaulieu 2009; NWCPA 2014).

Two-thirds by volume of wood pallets are from hardwood species and one-third are from softwood material. Oak is the most commonly utilized species with almost 26.9%, followed by maple at 7.2%, and other hardwood species at 4.7%. Within the softwood species, in 2006 the southern pine group amounted to 53.5%, SPF (spruce-pine fir) for 35.5%, and Douglas fir for 3.8% (Bush and Araman 2009).

The use of pallets and other unit load systems dramatically improves the productivity of material handling. Whereas it may take one worker all day to manually unload a *break-bulk* trailer package-by-package, he/she can do it in a matter of minutes with a forklift when the shipping containers are palletized.

Pallets are handled with various kinds of equipment, but the most predominant are forklifts and pallet jacks. *Forklifts* are motor powered, maneuverable vehicles (also called *fork trucks* and *hi-los*) with forks that penetrate the pallet's opening to lift and move the load. Forklifts can also raise the load from the floor to a high rack or stack. There are many different weight capacities and lifting heights, based on the capability and counterweight of the forklift.

A *pallet jack* does not raise the load to any significant height, but simply puts it on wheels. The pallet jack's wheels are driven into the pallet's entry opening and a jack slightly lifts the pallet from the floor, enough to permit it to be pulled or pushed into position, usually with a motorized assist.

Pallets were first used commercially after they proved their value in military applications during World War II. Wood is a good material for pallets because of its relatively low cost and high strength. Most wooden pallets are strong enough to be reused. And when they do get damaged, they are easy and inexpensive to repair.

Over the years, there have been variations in pallet structure, strength, and quality, prompting the need for standards. Damaged or poor quality pallets can cause problems throughout a supply chain, damaging products, tipping stacks, and causing injuries. Consistency and quality are especially necessary for *automated storage and retrieval* or *automated guided vehicle systems*. Sloppy tolerances, missing boards, and protruding nails can cause a system to jam or to not recognize a palletload.

Furthermore, most pallets are intended to be reused and repaired within a specific supply chain, like those supplying groceries or assembly parts. But since the company that buys a new pallet is not necessarily the one who will reuse it, there is little incentive to purchase a strong one, although that would be the wisest choice from the supply chain's point of view. As a result, it has been in the best interest of specific supply chains to set standards, for example the specifications set by the Grocery Manufacturers Association (GMA).

Comprehensive standards that apply across user industries were first developed in the United States by the National Wooden Pallet and Container Association (NWPCA) in 1994. The members of NWPCA are manufacturers, recyclers, and distributors of wood packaging materials or supply products and services to the industry (www.palletcentral.com). The NWPCA has established a Uniform Standard for Wood Pallets that is used by most pallet manufacturers in the United States.

The uniform standard for wood pallets contains pertinent information on the prescriptive, performance, and phytosanitary standards. The prescriptive standard includes terminology and definitions, classifications, materials to be used, and information on manufacturing, remanufacturing and repairing pallets. The performance standard defines the testing procedures and minimum requirements for specific pallets. The phytosanitary standard specifies heat-treatment requirements (NWPCA 2012).

5.1 PALLET TERMINOLOGY AND TYPES

The parts of a wooden pallet are shown and named in Figure 5.1. Two types of pallets are pictured: a *stringer* design and a *block* design. The stringer type is most common; in 2006, approximately 80% of those produced were stringer type (Bush and Araman 2009).

The *deck* is the top or bottom surface of a pallet. It is comprised of a number of *deckboards*. The number of deckboards is a trade-off between cost and a level surface: the fewer deckboards, the lower the cost, but wide spaces between deckboards make a poorer platform and a less-level stack.

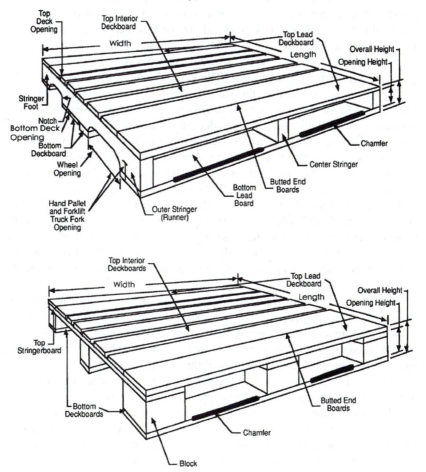

FIGURE 5.1. *Typical stringer and block pallets with parts labeled. Reprinted with permission of the National Wooden Pallet and Container Association, copyright 2012.*

The *top deck* is the upper surface of the pallet, upon which the shipping containers rest. A pallet often has more boards on the top deck than on the bottom deck in order to provide a more level surface for the packages stacked upon it. Discontinuities in the surface can cause damage to containers that do not have sufficient stiffness to bridge the gap. A solid deck pallet is constructed with no spaces between deckboards.

A *double-deck (double-face)* pallet has both top and bottom decks. The *bottom deck* is the load bearing surface of the pallet. If the bottom deck has too few boards, when stacked atop another load, it concentrates pressure on a smaller area of the top of the load below it, which can crush packages at the pressure points. This is the reason that *single-deck* pallets, also called *single-face* pallets or *skids*, are not used in stacks without the aid of a sheet of plywood or other sufficiently stiff material to spread out the load beneath the stringers. In a *reversible* pallet, the top and bottom decks are identical.

The *deckboard span* is the distance between the supports, which may be stringers, stringerboards, or blocks. The strength and stiffness needed in the deckboards is determined by the weight of the load supported by this span when the pallet rests on the floor or in a stack. However, if palletloads will be stored on racks that only support two edges, the full span of the pallet must support the entire weight of the load, reducing the amount that a given design can safely support. The NWPCA's Pallet Design System (PDS) provides structural analyses that give the safe load for each design under specified conditions of racking and stacking, based on the bending strength described in Section 3.5.

The *lead deckboards* are those at the front and back. They are the first deckboards that the forklift encounters. These are sometimes made wider than the interior deckboards, because they take more abuse. *Chamfered deckboards* are lead deckboards with the leading edge of the bottom face beveled to ease the entry of pallet jack wheels. Protecting the lead board will increase the durability of a wood pallet.

Wings are deckboards that extend beyond the stringers (not shown on the examples in Figure 5.1). On a top deck, wings are useful for interlocking stretch wrap under the load and tying it to the pallet. *Double wing* pallets are the type used in seaports, where pallets are hoisted from above by a crane using spreader bars and slings that are hooked under the wings.

Stringers are the continuous, longitudinal, solid or notched beam components of the pallet that are used to support the deck. A stringer is a piece of wood with a rectangular cross section, with good bending strength and loaded on the narrow face, and the word "stringer" is also used to indicate a longitudinal member in other kinds of construction. Pallets usually have three stringers, identified by their location as the outer or center stringers.

Block pallets have *stringerboards* that extend the length of the pallet perpendicular to the deckboards, securing the deckboards to the blocks.

A *two-way entry* pallet has un-notched solid stringers that allow entry only from the ends. A *partial four-way entry* stringer pallet has notched *fork* openings in the stringers for pallet jacks or forks, although some pallet jacks will not fit (hence the *partial* designation). Although they make a pallet easier to pick up since the forks can approach from any side, the notches reduce a pallet's strength and durability.

A *block* pallet has blocks, instead of stringers, separating the pallet decks. Blocks are rectangular, square, or cylindrical deck spacers. A block pallet is always *four-way entry*. Block pallets are stronger than stringer pallets, and are preferred by pallet rental companies because of their durability. Stringer pallets are generally less expensive to make because the process can be more easily automated.

Pallet size varies, but there are standards within industries and countries. Standard dimensions facilitate pallet exchange programs and provide a common footprint for other dimensional aspects of a supply chain such as rack sizes, automated handling systems, conveyors, vehicles, and assembly line layout.

The *length* of a pallet is defined as the length of the stringers. It is the first dimension given in describing a pallet. Thus a 48" × 40" pallet has 48-inch stringers and 40-inch deckboards. The height of the opening needs to be sufficient to accommodate forks or pallet jack wheels, usually 3.75 inches.

The most standard size in the United States is the GMA 48" × 40" pallet. This is the standard for dry groceries. The automobile industry has a 48" × 45" standard.

The most common international standard dimension is 1,200 mm × 1,000 mm (47.2" × 39.5", which is close to the U.S. standard). The U.S. electronics industry, for example, has adopted the international standard dimensions (IoPP 2000).

Throughout history, standard dimensions of shipping containers have been chosen to match the dimensions of vehicles, in order to maximize cube utilization. The standardization in the United States of the 48-inch, 45-inch, and 40-inch dimensions of pallets reflects this fact, since they have been chosen to fit two abreast in a trailer and three abreast in a railcar. The older Europallet's 800 mm × 1,200 mm (47.2" × 31.5") standard was chosen because it best fits into European sea containers. There are also international air freight pallet dimensions designed to fit into air freight containers (most common is 8' × 10').

But the dimensions of some products do not fit into a standard footprint, and wooden pallets offer the additional advantage of being easy to dimensionally customize. A specific size can easily be built for products with a smaller footprint (such as appliances) or a larger footprint (such as furniture and industrial machinery). Unlike the manufacturing of plastic pallets which requires a different mold for different sizes, it is relatively inexpensive for a wood pallet manufacturer to customize size.

5.2 PALLET CONSTRUCTION

The strongest wood pallets in the United States are made of hardwoods, often oak. Hardwood deckboards provide more strength and stiffness to resist bending and breaking under the weight of the load. Some pallets have oak deckboards but softwood (e.g., pine) stringers.

While for most carpentry applications it is undesirable to use wood that is still green, it is not uncommon for hardwood pallets. Drying increases the cost of wood, and pallets are generally made from the least expensive wood practical.

But a dry pallet is better. Below the FSP the wood's strength and stiffness increase, and the pallets' fastener withdrawal resistance and shear resistance increase. Green pallets have a tendency to shrink as they dry, resulting in instability and loose nails and bands. High moisture can also cause rust, rot, and contamination, and can weaken corrugated boxes stacked on a pallet. *ASTM D6199, Standard Practice for Quality of Wood Members of Containers and Pallets* specifies that pallet members should have a moisture content at the time of fabrication no greater than 19% and no less than 9% of their dry weight basis. However, the NWPCA Standard does not limit the moisture content of pallet components.

Softwoods like southern pine may be used for pallets, but it is more important to dry softwood to avoid shrinkage problems like nail loosening and dimensional changes. The NWPCA (2003) ranks southern pine high in strength when it is dry (near the strength of the eastern hardwoods) but very low when it is green.

The choice of wood type and thickness depends on the strength required and cost (which depends on availability). Table 5.1 shows the NWPCA classification used for pallet woods and ranks them by their relative strength and stiffness. It is interesting to note that NWPCA ranks woods a little differently than the box wood classification in Chapter 4.

Both classifications agree that eastern hardwoods like red and white oak, birch, maple, ash, elm, hickory and beech are the strongest. These woods are preferred in high stress applications, such as where wood is subjected to bending forces (as in pallet deckboards), because of its higher modulus of rupture. And they both agree that some hardwoods like basswood and aspen are less strong than some softwoods.

But the pallet wood classification gives a higher strength ranking to "dry southern pine" and douglas fir than to some of the medium density hardwoods, and it rates poplar and red alder higher than they are rated for box making. Both classifications agree that some of the softer hardwoods like aspen and cottonwood have low strength comparable to pine, fir, and cedar, although these are all used for lighter duty pallets.

TABLE 5.1. Wood Species Classes (NWCPA 2012).

Class Number	Class Name	Species
North American Hardwoods		
1	High Density Eastern Hardwoods	Hickory, Birch (Yellow, Sweet), Maple (Sugar, Black, Red), Ash (Green, White), Elm (Rock, Slippery), American Beech, Black Locust, Black Cherry, Tan Oak, Dogwood, Persimmon
2		Bigleaf Maple, Oregon Ash
3	Medium Density Eastern Hardwoods	Sweetgum, Tupelo, Paper Birch, Ash (Black, Pumpkin), Hackberry, Sycamore, Maple (Silver, Striped), Magnolia
4	Western Hardwoods	Oregon White Oak, California Black Oak, Cascara, Chinquapin, Myrtle, Madrone
6		Red Alder
7	Low Density Eastern Hardwoods	Aspen (Bigtooth, Quaking), Catalpa, Buckeye, Butternut, American Basswood, Cottonwood (Black, Balsam), Eastern Poplar
21	Eastern Oaks	Red Oak, White Oak
29		Yellow Poplar
North American Softwoods		
11	Douglas Fir	
12	Hem-fir	Hemlock (Western, Mountain), Fir (California Red, Grand, Noble, Pacific Silver, White)
13	SPF	Spruce (White, Black, Red, Engelmann, Sitka), Pine (Sugar, Western White, Lodgepole, Ponderosa, Monterey, Jack, Norway, Eastern White), Southern Pine (Pitch, Pond, Spruce, Virginia), Fir (Subalpine, Balsam), Bald Cypress, Eastern Hemlock, Western Red Cedar, Redwood
14	Low Density Softwoods	Cedar (Alaska, Incense, Port Orford, Atlantic White, Northern White, Eastern Red)
22	SYP	Southern Yellow Pine (Loblolly, Longleaf, Shortleaf, Slash)
European Species		
31	Imported Hardwoods	Kapur, Keruing, Mengkulang
32	Dense European hardwoods	Ash, Beech, Oak, Plane
33	Dense European Softwoods	Douglas Fir, Larch (European, Japanese Pine (Jack, Maritime, Scots)
34	Medium Dense Woods	Dutch Elm , Hybrid Larch, Pine (Corsican, Lodgepole), Poplar (Black Italian, Grey) Redwood, Silver Fir
35	Whitewood	English Elm, Sitka Spruce (Canada), Whitewood
36	Common European Softwoods	Radiata Pine, Spruce (Black, Norway, White, Sitka), White Willow
37		Hybrid Poplar

(continued)

163

TABLE 5.1 (continued). Wood Species Classes (NWCPA 2012).

Class Number	Class Name	Species
	South America and Other Species	
41		Radiate Pine (Chile)
42		*Gmelina arborea* (Costa Rica)
43		*Pinus caribaea* (Venezuela)
44		*Pinus elliottii*
45		*Pinus taeda*
46		*Eucalyptus grandis* (Uruguay)

Wood pallets used for export must comply with the international regulations for heat treatment discussed in Chapter 2. Some manufacturers prefer to heat-treat hardwood pallets after manufacturing, because the heat treatment makes the wood harder to nail.

The wood in pallets is generally fastened with nails. The nails are usually helically or annularly threaded for extra holding power. The most common pallet nail is at least 2 inches long, which embeds the nail 1-1/4 inch (32 mm) in the stringer after passing through and attaching to a typical three-quarters-inch deckboard. Its bottom 1-1/4 inch is threaded so that it can hold tight to the stringer.

Pallets intended for light loads and short-time uses may be fastened with long-pronged wire staples. Two to six fasteners should be used per joint to retain holding power after the wood dries. Wider deckboards require more nails. They are usually staggered to reduce the tendency to split the stringers. Screws or bolts can be used as fasteners, but they increase labor requirements and cost.

Pallets can also be made from engineered wood products. Plywood and OSB decks are used when a flat continuous deck is required. These are more expensive than ordinary deckboards, but provide an added measure of protection from stacking damage because the platform is smooth, warp-resistant, and provides uniform support, although it is less stiff when spanning the width of a rack. A solid deck can carry a heavier load, and it offers superior impact resistance to forklift tines. Furthermore, there is no need for additional heat treatment for export.

The stringers or laminated nailed blocks can also be made from plywood or OSB. Since the materials were formed using heat, there is less nail protrusion due to shrinkage. Pallets made from plywood or OSB can be repaired, but repairs can be more costly than with ordinary wooden pallets because the full deck must be replaced. A common form of damage is block failure when the nails are not long enough to firmly secure the blocks from both sides.

As shown in Chapter 4, large wooden crates and boxes are commonly attached to pallets or skids. Pallets can be also attached to large intermediate bulk containers or integrated into other material handling systems. For example, the assembly lines for many appliances such as refrigerators begin with a pallet upon which the refrigerator is built.

5.2.1 Pallet Specification

In general, the selection of a specific pallet depends on the load-carrying capacity, resistance to deformation under load, resistance to damage during storage and handling in various environments, compatibility within the supply chain, and cost.

The most common pallet specification system used in the United States is the NWPCA's *PDS*. The PDS provides information on the specification, structural analysis, durability and physical properties of pallets. It can compare wood species, wood dimensions, grades, physical properties, and moisture content of lumbers, as well as types of fasteners and connections. The program provides structural analyses for each specification. It estimates a safe load capacity based on deflection at maximum load in the service environment (wet or dry) required for the handling, stacking, and racking system (NWPCA 2013).

Pallet specification of materials and dimensions include the following:

- Pallet dimensions (stringer $L \times$ deck W)
- Lumber dimensions, wood species class, grade, and moisture content
- Type of fasteners, their lengths, and the number per joint
- Drawings showing each deck and stringer configurations
- A bill of materials

Lumber used to manufacture wood pallets is graded in a different system than that used for lumber appearance grades. The grading of lumber used to manufacture wood pallets is based on the size and location of tight or loose knots and holes, slope of the grain, and the presence of bark, splits, shakes, and checks.

Accordingly, PDS combines lumber standards into five grades:

- Select
- Premium
- Standard
- Utility
- Economy

Select is the grade with the fewest defects while *economy* is the grade with the most defects that reduce the strength of wood. A wood pallet made of select grade lumber will be stronger than one build with economy grade lumber.

The most critical wood pallet strength property is bending strength which is related to the maximum carrying load without permanent deformation. For a rectangular beam, such as lumber, the slope of the relation between the stress (σ) and strain (ε) applied uniformly at mid-span and below the wood's proportional limit is known as the *modulus of elasticity* (*MOE*). According to Hooke's law:

$$MOE = \frac{\sigma}{\varepsilon}$$

where σ is the applied stress and, in the case of a loaded pallet, is equal to the load (P) divided by the area (S) of the top of the pallet, since this is the area that resists the load.

$$\sigma = \frac{P}{S}$$

$$MOE = \frac{P}{S \times \varepsilon}$$

$$P = MOE \times S \times \varepsilon$$

The load can be increased by increasing the *MOE* (which depends on the wood species) and/or dried wood, high surface values, and high strain ε. This may partially explain why the use of plywood or particleboard as the deck platform supports a higher load in comparison to the traditional spacing of deck boards which results in a reduced surface area.

The strain ε is defined as:

$$\varepsilon = \frac{\Delta L}{L} = \frac{L_f - L_i}{L_i}$$

where L_i and L_f are the sample length before and after the load is applied during a static bending test.

For a rectangular beam with known span, width and thickness, load, and deflection, the static bending modulus of elasticity of the beam is obtained using the equation below:

$$MOE = \frac{PL^3}{4Dbd^3}$$

where P is the load, L is the span length, d is the thickness, b is the width, and D is the deflection.

This equation is valid only in the region up to the linear proportional limit as described above. The proportional limit is also known as the allowable stress without permanent deformation. Above the proportional limit, the additional deformation cannot predicted from this equation, and may even be permanent. The load that results in failure can be used to calculate the *modulus of rupture* (*MOR*).

To avoid rupture and damage, it is strongly advised to use the proportional limit to select the maximum load in the design of pallets by using the published values of *MOE*. One of the most important aspects of the design of a wood pallet is the maximum load allowable to stay below the proportional limit to avoid any permanent deformation:

$$P_{\text{max}} = \frac{4 \times MOE \times D \times b \times d^3}{L^3}$$

where P_{max} is the maximum load capacity

MOE = modulus of elasticity
  D = deflection limit
  b = width
  d = thickness
  L = span, distance between the supports

From the equation above, it is evident that P_{max} will vary with wood species, the moisture content of wood, the thickness and the width of wood, and the length of the span. Typical racks in the grocery industry support only two edges of a pallet. Clearly, racking across a width of 40 inches versus racking across length of 48 inches will result in different maximum loads according to the equation above. A short span will support a higher load than a longer span.

Pallets may be marked according to *ASTM D6253, Standard Practice for Marking of Pallets*. This standard recommends identifying the manufacturer, the specification with which the pallet conforms, the year of manufacture or repair, and the service designation to indicate whether they are limited use (marked with an *L*), repaired (*R*), or multiple use (*M*).

Tolerances of more or ± 1/8″ are common for the wood dimensions that do not affect the height of the pallet. The thickness of boards (± 1/32″), and stringer or block heights (± 1/16″) have less tolerance since they affect the levelness of the platform.

5.2.2 Pallet Reuse, Repair, and Recycling

Most pallets are reused many times. A survey by *Material Handling* magazine found that 15% of pallets were used more than 20 times, 5% between 11 and 20 times, 10% between 7 and 10 times, 41% between 2 and 6 times, and 29% were used only once (Trebilock 2010).

One of the biggest advantages of reusing wooden pallets is how easy they are to repair. In the 1990s, there was a large increase in recovery, repair, and recycling of pallets. In 1999, 69% of the recovered pallets were repaired and reused, and this number is growing as pallet recycling increases (Bejune 2001; Bush *et al.* 2002). By 2013, it was reported that 75% of wood pallets in the United States were reused. The increase from 69% to 75% was attributed to the creative approach of some recyclers, the ban on wood pallets in some municipal landfills, the higher cost of virgin pallets, the high cost of disposal, and the acceptance of the idea of using recycled pallets (Howe *et al.* 2013).

It is inevitable that sooner or later a pallet will become damaged due to routine handling; rough handling makes it happen sooner. Pallets should always be inspected before reuse to sort out those needing repair.

Broken deckboards, stringerboards, and blocks can easily be replaced with new or used components of similar material. An occasional wider or narrower deckboard is acceptable as long as the gaps are not excessive. Loose components are removed or refastened. Steel connector plates are used to repair horizontal breaks in stringers, and wooden companion stringers are nailed onto stringers with broken notches. The NWPCA Standard (2012) specifies repair methods and workmanship; for example deckboard thickness should not vary by more than three-sixteenths inch (5 mm).

Another way to recycle damaged pallets is remanufacturing. A *remanufactured pallet* is newly assembled from recycled components. The damaged pallets are mechanically *de-nailed*, and the parts are sorted and re-assembled. When pallets have a mixture of recycled and new parts, they may be called *combo pallets*.

The service life, or number of uses in a pallet's life, is dramatically increased by the ability to be repaired. PDS estimates a pallet's service life, employing a number of assumptions that illustrate the advantage of being able to repair wooden pallets. It assumes the following:[11]

* Connections can be repaired once without replacing the board.
* A replaced board is restored to 100% of its original damage resistance but its connections lose 10% with each replacement.
* The number of times a board can be replaced depends on the stringer width:
  — Twice for stringer widths of 1-1/2 to 2 inches
  — Once if stringer width is less than 1-1/2 inches
  — Three times if stringer is greater than two inches.

[11]The PDS simulates the 15-handling cycles with equally distributed forces and impacts. When one member is repaired or replaced, all members of that type (e.g., stringers or lead deck boards) are repaired or replaced at once. The simulation proceeds until a specific component fails that has already been replaced the allowed number of times.

- Stringers can be repaired twice without having to be replaced, and they can be replaced once. When repaired, they are restored to 65%, and when replaced, restored to 100%, of their original damage resistance.

This degree of repair is considered normal for wooden pallets, until the cost of repair exceeds the value.[12]

Besides repair and remanufacture, pallets can be recycled for many purposes. The production of landscape mulch is the leading use for ground or chipped pallets. They can be used for animal bedding, fuel, pulped for paper or chipped to be added to composite materials. The nails are problematic for some recycling uses.

In many places, however, there has been a lack of infrastructure to support pallet recycling. The pallet industry is more fragmented than other packaging industries, made up of many small to medium sized businesses, competing fiercely to sell within their region. Pallet and skid producers are found in every state, with the heaviest concentration in the industrial regions of Ohio, California, Pennsylvania, Wisconsin, Texas, and Michigan. Most have fewer than 20 employees and serve customers within a 100-mile radius. They are mostly small wood-products companies with annual sales under $10 million (Impact Marketing Consultants 2006).

But since pallets are used to ship outside of a region and are usually discarded somewhere far away from where they are produced, there has traditionally been little cooperation between suppliers in different regions. Although it lacks the coordination of the corrugated recycling industry, there is a successful business model for collecting and repairing pallets to be resold.

Pallet recycling is increasing due to a greater understanding of the benefits. Pallet producers benefit from repairing and salvaging material from used pallets because they are a low cost source of wood. Retailers and other customers at the end of a supply chain benefit by avoiding the cost of pallet disposal and landfill, which is expensive because of their mass.

5.3 ALTERNATIVE PALLET MATERIALS

An average 48″ × 40″ wooden pallet weighs 40–50 pounds and costs $10–$20, depending on its strength and quality. But there are a number of competing pallet materials, as well as alternative material handling methods that do not require a pallet platform.

The most common alternative pallet material is plastic, generally high density polyethylene (HDPE). Plastic pallets can be thermoformed or injection molded, or molded from structural foam. They are typically lighter weight and

---

[12]PDS also analyzes repair costs to find the number of cycles that results in the lowest estimated cost per cycle.

stronger than wood, but are more expensive and less environmentally positive.

Thermoformed plastic pallets are made by thermoforming one or more heavy HDPE sheets together. Thermoformed pallets are light weight, about 20 pounds, nest well together and cost about $15. *Structural foam pallets* are heavier, about 25 pounds, and are more expensive ($40–$100), but they are also stronger than thermoformed pallets. *High pressure injection molded* HDPE or PP pallets are also used.

Plastic pallets, especially the expensive structural foam ones, are generally used many times in closed-loop distribution systems. For example, they are widely used in the automobile assembly industry and by the U.S. Postal System. Plastic pallets are very durable and can last longer than wooden pallets, although some styles are stronger than others. They eliminate the need for costly repairs and inspections, but when plastic pallets break, they cannot be repaired. They can be recycled, however, and the recovered material used in manufacture of new pallets.

Steel and *aluminum* reusable pallets are used mainly by the military for supplies like explosives and equipment parts. Metal platforms and racks are also used for some assembly parts, especially large sheet metal parts. They are very durable, but expensive, about $75–$200. Aluminum is the most expensive material used in pallets, but it also has a high recycling value.

Presswood pallets are made of a molded material similar to particle board. Presswood pallets are heavy (about 42 pounds) and inexpensive, about $10–$25 per pallet because they are made from waste materials, but they are relatively weak. There are also some pallets made from molded pulp. Engineered wood products offer the advantage of not being vulnerable to infestation, overcoming the need for heat treatment described in Chapter 2.

Corrugated fiberboard can also be used to make pallets. Several layers of corrugated board are glued together, with built-up blocks or other structures such as tubes separating the decks. They are used for light-weight loads and are, themselves, lightweight (about 10 pounds) and inexpensive, $5–$25 per pallet. They are considered to be disposable, used only once and then recycled along with other corrugated board.

Slipsheets and clamp handling are alternative unit-load handling methods that do not require a rigid pallet at all. They are lifted not with forks, but with more specialized material handling equipment. *Slipsheets* made from corrugated or solid fiberboard are handled by a lift truck that pulls the slipsheet and the packages stacked upon it onto a solid platform that carries it until it is pushed off. Fiberboard slipsheets are described in more detail in Chapter 15. *Clamp handling* uses a lift truck that squeezes the unit load between two platens. This is best for lightweight products, such as breakfast cereal cartons, and products that do not permanently deform from the pressure, such as rolls of toilet paper or insulation. Appliances are handled by still another method that uses the box's top cap as a lifting device for a *baseloid* lift truck.

5.4 PALLETIZATION AND UNITIZATION

A primary function of pallets is to assemble goods in shipping containers to be handled, transported and stored as a single unit. The integrity and stability of a palletload, as well as its cubic efficiency, depends on the pattern in which items are stacked on the pallet.

The *pallet pattern* footprint is mostly responsible for cubic efficiency. Cubic efficiency is valuable because the more products that fit on a pallet, the lower the transport and handling cost will be. Thus, shipping container dimensions are often designed to fit together onto a 48″ × 40″ pallet with no wasted space, and also with no overhang that can be a source of damage.

Specialized computer programs can provide the best pallet pattern from given container dimensions, or can generate best fit dimensions for a shipping container and its contents. The two most popular pallet pattern programs are named CAPE and TOPS. One of their measures is *pallet utilization*, which is the percentage of the available volume used. For example, loads with 80% utilization waste 20% of the space, resulting in higher transportation cost. It also requires more pallets, labor, and handling.

The *stacking pattern* and unit load restraint method contribute to the load's strength and stability. Although *column stacking* gives the best compression strength (as discussed later in Chapter 14), an *interlocking pattern* is more stable. Stability prevents damage and facilitates handling productivity as lift truck drivers and transport vehicles whip around corners and bounce over bumps. Stability can also be added by stretch-wrapping or strapping loads, or by applying temporary adhesives in between the containers in a stack.

Palletization and stretch-wrapping can be an automated or manual operation. Palletizing equipment has the ability to vary pallet patterns according to case dimensions, stack in an interlocking manner, and convey a load directly into a stretch-wrapper.

5.5 PALLET EXCHANGE AND RENTAL PROGRAMS

There are a number of options for pallet ownership. The products loaded onto a pallet will change ownership as a result of business transactions, and the pallet will ultimately no longer be needed for those products. But since most pallets are intended for multiple uses (and from a total cost point of view, sturdier pallets are more economical, but only if they are reused), there must be a shared responsibility in the supply chain for returning the pallet to useful service. The question of who buys, owns, and repairs pallets is negotiated within each supply chain.

A *captive pallet* is a pallet that is intended for use within a single facility or system. Some examples are slave pallets and stevedore pallets. A *slave pallet* is a pallet, platform, or a single thick panel that is used in rack-storage facilities or production systems. A *stevedore pallet* is designed for use on seaport shipping docks, and is usually of heavy-duty double-wing construction; pallet size varies from port to port.

An *exchange* pallet is a pallet intended for use among a designated group of shippers and receivers. Ownership of the pallets is transferred with the truckload in exchange for empty pallets, which are, in turn, exchanged for a full load. Pallets used for dry groceries are sometimes exchanged. However, this practice has become less popular because it generally leads to a deterioration of the common pool when no one party has an incentive to invest in new pallets.

Pallets can be part of the *transaction*. In this case, empty pallets are sold by the receiver to a pallet recycling company that inspects and repairs it, to sell it to another nearby shipper.

A *rental pallet* is a pallet that is owned by a third party, not one of the actual pallet users. CHEP and IFCO are two companies with a long history in the pallet rental business beginning in Europe. They provide pallets to customers for a fee, and manage the return, repair, redistribution, and reuse of the pallets. Rental pallets are the very strongest, most well-constructed pallets, designed for many uses. Some providers, like iGPS, offer plastic pallets with built-in radio frequency identification (RFID) tags. Pallet rental is increasing because the higher quality pallets prevent damage and avoid quality problems associated with pallet exchanges.

5.6 PALLET PERFORMANCE PREDICTION AND TESTING

There are two kinds of damage associated with palletloads. The most common damage occurs to products (and shipping containers) in the load due to poor stacking strength, leaning, and crushing. The design and testing of corrugated shipping containers to provide stacking strength are further discussed in Chapter 14.

Damage to the pallet itself is usually caused by forklifts in the form of punctures, collapsed stringers, and broken deck boards. The load itself can damage the pallet if it is too heavy or unstable. Damaged pallets, in turn, can cause damage to products stacked upon them, and can injure the people working nearby, when they tip, stab, or simply fail to support loads.

The strength of a pallet made according to a given material specification can be predicted by the PDS program described in Section 5.1. A structural analysis is generated, based on research conducted at Virginia Tech. The safe load capacity, deflection load limit, and critical member, stacking height, durability, and dimensional change based on drying can be analyzed. PDS also estimates the pallet service life, given a number of repairs.

There are standard tests to evaluate pallet strength. The test methods have been standardized by *ASTM D1185, Standard Test Methods for Pallets and Related Structures Employed in Materials Handling and Shipping*, which describes static and dynamic tests.

The static test methods determine a pallet's strength and stiffness. Compression testing determines the safe working load. Bending tests determine the structural stiffness of the pallet when supported by fork tines, rack, or sling.

The dynamic tests determine the stability of the pallet when exposed to a series, or cycle, of hazards. A free-fall drop test from 40 inches (1 m) on corners and edges evaluates strength and diagonal rigidity. Vibration and impact tests of loaded pallets, although they are used in this standard to evaluate the pallet alone, are similar to standard tests for unit load integrity. Incline impact tests with fork tines mounted to the impact surface simulate fork impacts and misaligned forks and test the leading deckboards' separation resistance.

Single-use pallets need to survive one *cycle* of tests. Multiple use pallets are expected to survive at least 10 cycles. The test standard's disclaimer that the tests are not intended to exactly simulate pallet performance in distribution is followed by the recommendation to couple laboratory testing with field tests.

Field tests to determine a pallet's actual life span can be useful, but require a system for tracking the number of trips to failure. Actual use situations are more realistic than tests, but they do not always provide the data necessary for strict comparisons. Laboratory tests are used in specifications and for comparing pallets. The laboratory tests are necessarily more severe than reality to provide a margin of safety.

In addition to pallet performance testing, there are two tests that are used to provide relative measures of pallet joint strength. The *fastener withdrawal test* and *fastener shear test* are used, as well as the *MIBANT angle* test for impact resistance. These are found in *ASTM F-680, Standard Test Methods for Nails*.

5.7 REFERENCES

APA (Engineered Wood Association). 2001. Engineered Wood Pallets: A Solid Value. Washington, DC: APA.

ASTM (formerly American Society for Testing and Materials, now ASTM International). No date, annually updated. *Annual Book of ASTM Standards.* West Conshohocken, PA: ASTM International.

Baumesiter, R. and G. Beaulieu. 2009. Markets and attributes trends. Pallets trends Canada. http://www.solutionsforwood.com/_docs/reports/PalletTrends09.pdf

Bejune, J.J. 2001. *Wood Use Trends in the Pallet and Container Industry: 1992–1999.* MS Thesis, Blacksburg, VA: Virginia Tech.

Bush, R.J. and P.A. Araman. 2008. Final Report—Updated Pallet and Container Industry Production and Recycling Research. Blacksburg: Virginia Polytechnic Institute.

Bush, R.J. and P.A. Araman. 2009. Pallet Recovery, Repair and Remanufacturing in a Changing Industry: 1992 to 2006. *Pallet Enterprise,* August 2009:22–27.

Bush, R., J.J. Bejune, B.G. Hansen, and P.A. Arman. 2002. Trends in the Use of Materials for Pallets and Other Factors Affecting the Demand for Hardwood Pallets, *Proceedings of the 30th Annual Hardwood Symposium,* May 30–June 1. Memphis: National Hardwood Lumber Association.

Clark, J.W. 2004. Pallets 101: Industry Overview and Wood, Plastic, Paper and Metal Options. Pallet Enterprise 2004. http://www.palletenterprise.com, http://www.ista.org/forms/Pallets_101-Clarke_2004.pdf

Conway, J.G. and M.S. White. 1999. Comparative Performance of Plywood and Timber Pallet Designs for Pallet Pooling. http://www.apawood.org/pdfs/unmanaged/Pallet%20comparative.pdf

Howe, J., S. Bratkovitch, J. Boyer, M. Frank, and K. Fernholz. 2013. The Current State of Wood Reuse and Recycling in North America and Recommendations for Improvements. Minneapolis: Dovetail Partners. http://www.dovetailinc.org/files/ReportWoodRecyclingUSCanadaReport0513.pdf

Impact Marketing Consultants. 2006. *The Marketing Guide to the U.S. Packaging Industry.* Manchester Center, VT: Impact Marketing Consultants.

IoPP (Institute of Packaging Professionals). 2000. Electronics Industry Pallet Specification, EIPS 2000. Ed. Bob Sanders. Naperville, IL: IoPP.

NWPCA. 2013. *Pallet Design System,* version 5.1 Alexandria, VA: National Wooden Pallet and Container Association. On disc.

NWPCA. 2012. *Uniform Standard for Wood Pallets.* Alexandria, VA: National Wooden Pallet and Container Association. http://www.palletcentral.com/images/files/nwpca_uniform_standard_for_wood_pallets_2012.pdf

Trebilcok, B. 2010. Modern Materials Handling's pallet webcast Results of the 2010 Pallet Usage and trending Study. September 2010:p. 24–29. http://www.mmh.com/issue_archive/2010/mmh_10_09.pdf

5.8 REVIEW QUESTIONS

1. What is the highest volume category of wood packaging used today?

2. What is the difference between a forklift and a pallet jack?

3. When did pallets begin to be commercially used?

4. What is the difference between a single-use and reusable pallet?

5. What is the name for the boards on the top surface of a pallet?

6. Why are the lead deckboards sometimes wider than the interior ones?

7. What is the advantage of using only a few deckboards with wide spaces between them? What kind of problems can this cause?

8. What are stringers? How many do most stringer pallets have?

9. Which pallet design, in general, is stronger: stringer or block?

10. From how many sides can you pick up a pallet with a forklift?

11. Upon what factors are the dimensions of pallets based?

12. What is the size of a standard GMA pallet?

13. Are softwoods or hardwoods used most commonly for pallets?

14. What special treatment is required for export pallets?

15. What factors determine how many times a wooden pallet can be used?

16. What is done with a wooden pallet that is damaged?

17. What is the difference between pallet remanufacturing and repair?

18. Can pallets be recycled? For what uses?

19. In the pallet industry, to what do the following acronyms refer: NWPCA, PDS, GMA?

20. From what materials other than wood can pallets be made?

21. What are the advantages of using wood for pallets?

22. What are some methods for handling unit loads using no pallets at all?

23. What is meant by a "pallet utilization" of 80%? What extra costs are incurred for poor pallet utilization?

24. What are the methods for improving palletload stability?

25. How are pallets statically and dynamically tested?

CHAPTER 6

Making Pulp

It was not until a process was invented for pulping wood in the 1850s that paper became the inexpensive commodity that it is today. Although paper can be made from many kinds of fibers, wood pulp is now the most prevalent, economical, and abundant source of supply in most parts of the world.

The primary disadvantage of using wood, as compared to other fiber sources, is that pulping wood is energy-intensive and uses a lot of trees. But trees are a renewable resource and have many advantages as a fiber source in places like the United States where forests and water are abundant, reforestation is practiced, and energy is inexpensive (for example, located near a hydroelectric generator).

The use of paper and paperboard in packaging not only helps promote a high standard of living in rural communities that are close to forest resources through generation of sustainable jobs that help to keep communities vibrant, but it also provides environmental, social, and economic benefits through the ability of sustainable forest management to capture and store carbon to combat climate change. The use of wood from forests certified by SFC, SFI, the Rain Forest Alliance, and other certification bodies is paramount for the future of paper and paperboard packaging.

Wood can be made into a wide variety of paper types, from toilet tissue to writing paper to fiberboard. The properties of paper depend largely on the pulping process used. Today's papermaking industry benefits from the past 170 years of developments in pulping since the invention of that first mechanical pulping machine.

In order to make logs into paper, the cellulosic wood fibers first need to be set free in a series of operations that break down the wood into pulp. The logs are first debarked and chipped mechanically. The chips are made into a wet fibrous pulp in either a chemical, mechanical, or hybrid process. The pulp is further processed by beating or refining to modify the fiber surfaces and en-hance bonding, and by bleaching for whiteness or by incorporating additives to modify the paper properties.

The pulp-making process has an ultimate goal of harvesting long refined fibers with strong intermolecular forces. The fibers must have an affinity for bonding together chemically and mechanically into a strong, smooth, flexible sheet of paper.

The pulp-making processes are described in this chapter: debarking, chip-ping, pulping, and refining. Chapter 7 describes the processes for making pulp into paper, and Chapter 8 discusses the properties and tests for paper and pa-perboard. The three chapters are inter-related because the properties of paper depend on the raw materials, the process used for pulping, and the papermak-ing process.

6.1 DEBARKING AND CHIPPING

Before wood can be pulped, the bark must be removed. Bark does not con-tain the types of fibers desirable for paper-making, and does contain a number of substances such as dark pigments that are undesirable, and it is hard to bleach. But the bark serves as an important fuel source in many lumber and pulp facilities, reducing the need to purchase energy.

The most common method for debarking wood in pulp mills is using a *drum debarker*, which consists of a rotating drum into which whole logs are placed. Drum dimensions vary, but a typical one is 50 feet long and 12 feet in diameter (15.2 m × 3.66 m). The logs go into the center of the drum, which has a slight incline and rotates slowly. The bark is sheared off by friction during the rub-bing and tumbling as the log travels down the drum. In some drum debarkers, water sprays are used to help loosen the bark. In others, the wood is debarked dry, reducing the cost to process the wastewater. The bark falls through slots in the drum wall, and is collected for use as fuel.

Ring debarkers pass the log through a rotating ring with scraping tips that shear off the bark. *Rosserhead debarkers* rotate the log under a rotating cutter head that removes the bark in a helical pattern. *Hydraulic debarkers* use jets of high-pressure water to blast the bark loose.

The wood is next *chipped* to small pieces of a uniform size in preparation for pulping, about seven-eighths inch long, seven-eighths inch wide and one-quarter inch thick (22 mm × 22 mm × 6 mm). Most wood is chipped using a *disk chipper*, which consists of a rotating disk with blades. It works like the grater on a food processor. Wood pieces or whole logs are fed lengthwise into the chipper. They are cut at an angle across the grain by the knives and split along the grain due to shearing action. Wood can also be chipped using *drum chippers*, or *disc, drum,* or *knife-ring flakers*.

Next, chips of an acceptable size are sorted out, using screens, from those that are too large or too small. The first screen retains the oversized material for further processing, and successive screens remove undersized chips and sawdust destined to be used as fuel. Non-wood contaminants that are not removed by screening are removed by magnets, chip washers, and pneumatic cleaners.

Rather than chipping after harvesting and debarking, it is also possible to use whole-tree chipping, usually done in the forest as the wood is harvested. In this process, the whole tree, including bark, branches, twigs, and leaves, is converted into chips. The primary advantage is increased yield of fiber. The disadvantage is increased contamination, which requires a greater degree of attention to cleaning and screening. Usually, it is impossible to remove all the bark, so pulping processes and end uses that can tolerate the presence of some bark must be selected.

The chips are conveyed into a storage area where they await pulping.

6.2 PULPING

The pulping process frees the cellulosic wood fibers from the chips by breaking down the wood's inter-cellular lignin glue, using a mechanical, chemical, or hybrid process. Once the fibers are separated from each other, they are suspended in water, bleached (if desired) and refined. This slurry may be used directly in the paper manufacturing process, or the pulp may be dried and then resuspended in water for paper-making at a later time or at a different location. An overview of the pulping process is shown in Figure 6.1.

The process has to be well controlled to ensure that the fibers are adequately refined but not simply pulverized into dust. The yield and properties of the pulp depend on how much lignin is removed. Cellulose (which is about 45% of the wood) is the most desirable for paper; it gives fibers their strength and is light in color.

Any kind of wood and many kinds of vegetation can be pulped. The properties of the fibers vary by species. Prior to the invention of the mechanical method for pulping wood in the 1850s, almost all paper was made from linen rags or straw and was pulped in a large mortar and pestle. Some alternative sources of fibers are discussed on page 197.

Most packaging paper today is made from softwoods because they have long, strong fibers. This is the opposite of the relative strengths of the wood itself. Hardwood has short fibers, but the lignin and structure of the tree give it more strength than softwood (on average).

On the other hand, the strength of paper relies on the network of interlocking fibers. Long, flexible, and slender softwood fibers interlock better than do short and stiff hardwood ones. Softwood has fibers that, on average, are more than three-fold as long as those of hardwoods (1 cm versus 3 cm on average).

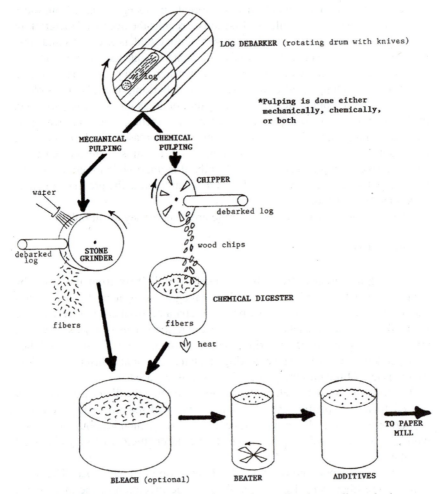

FIGURE 6.1. *Pulping is done either mechanically, chemically, or both.*

The longer the fibers, the stronger the paper. Long fibers contribute to tensile and puncture strength, tearing resistance, folding endurance, and stiffness.

But long softwood fibers have a drawback: they can give the paper a rough or coarse surface with the density varying across the sheet, which is unsuitable for high quality printing. Therefore, softwood pulp is only used alone when strength is needed and graphic demands are low, for example in kraft paper bags and liners for corrugated fiberboard.

Hardwood has shorter fibers and produces weaker paper. But because the shorter hardwood fibers better fill in the sheet, they make a smoother paper

FIGURE 6.2. *Long silky softwood pulp fibers. Reprinted with permission of TAPPI, from Parham and Gray,* Practical Identification of Wood Pulp Fibers *(1982), iii.*

FIGURE 6.3. *Shorter stiffer hardwood pulp fibers. Reprinted with permission of TAP-PI, from Parham and Gray,* Practical Identification of Wood Pulp Fibers *(1982), iii.*

that is more opaque, and is therefore better for printing. Most printing paper is made from a blend of both hardwood and softwood fibers.

The fluted middle layer of corrugated board is generally made from hardwood because the short fibers help improve the thickness and the stiffness properties. Corrugated medium is used for its structure, not its tensile strength.

Fiber made from logs and wood chips is called *virgin* fiber. Virgin fibers are longer, stronger, and more expensive than recycled fiber.

An increasing amount of recycled fiber is being used in packaging. This *secondary fiber* is recycled from wastepaper. It is less expensive than virgin fiber and it is believed to be an environmentally favorable alternative. Secondary fiber is distinguished from *broke*, which is trim or off specification paper that is reused within the same mill. But repulping produces a weaker paper than paper made with virgin fibers (from the same species of tree). For this reason, many papers are made from a mixture of virgin and secondary fibers.

There are three categories of pulping processes. Mechanical processes simply tear apart the fibers from logs and wood chips. Chemical processes rely on chemical treatment to dissolve the lignin glue and separate the cellulose fibers, resulting in pulp with longer fibers and therefore stronger paper. Hybrid processes are a combination of mechanical, physical, and chemical means to remove the lignin and separate the fibers.

Chemical pulping, because it can produce relatively undamaged fibers, makes stronger paper, but the yield is lower than 75% because of the removal of the lignin (20–25%) and part of the hemicellulose and extractives. Mechanical pulping produces higher yields because it retains most of the lignin and hemicellulose; the yield can be as high as 95%. Table 6.1 compares these three processes, which are discussed later in this chapter.

6.2.1 Mechanical Pulping

Mechanical pulping literally grinds and shreds the wood into pulp. It was the first method invented for breaking down wood into fibers for papermaking. Mechanical pulp was first produced commercially in Europe in 1852 and in the United States in 1867.

Because the process is mechanical, the lignin is not removed, and so the fibers are shorter than those produced in chemical processes. The paper is less strong and more prone to yellowing than that produced from chemical pulp. Newsprint (and the recycled paperboard made from it) is made from mechanical pulp.

Yields are high, typically 90–95%, which is why mechanical pulping produces the lowest cost virgin fiber. But mechanical pulping is energy-intensive, so mechanical pulp mills are often found near relatively inexpensive sources of electrical power, such as hydroelectric power facilities.

TABLE 6.1. Comparisons of Mechanical, Semi-chemical, and Chemical Pulping.

	Mechanical	Hybrid	Chemical
Methods	Stone groundwood Refiner mechanical Thermomechanical	Neutral sulfite semi-chemical (NSSC) TMP, CTMP, High yield kraft	Kraft Sulfite Soda
Pulping	Mechanical energy, heat, steam (little or no chemicals)	Chemical, biological pretreatment, then mechanical	Chemicals and heat (little or no mechanical)
Yield	High (85–95%)	Intermediate (55–90%)	Low (40–75%)
Type of Fiber	Short • Weak • Unstable Stiff, thick	Intermediate (some unique properties)	Longer • Strong • Stable Flexible, slender
Printing	Good print quality Opaque and smooth	Intermediate	Poorer print quality Rough
Bleaching	Difficult to bleach	Intermediate	Easier to bleach
Examples of Packaging Uses	Newspapers (ONP) recycled into paperboard	Corrugating medium (NSSC) Linerboard (high yield kraft)	Bag paper, SBS and SUS board (kraft) Greaseproof, glassine (sulfite)

6.2.1.1 Stone Groundwood Pulp

The oldest mechanical pulping method is the *stone groundwood* process. The abrasive surface of a revolving grindstone generates heat and shear in the presence of water. It also cuts and shortens fibers. The frictional heat generated (120–130°C) raises the temperature of the wood surface, softening the lignin. The fibers are washed away from the grinding stone's surface with water, the slurry is screened to remove slivers and large particles, and the water is removed leaving the *pulp stock*.

Pulp from the stone groundwood process is commonly designated *SGW*. The SGW process can use either hardwood or softwood as the fiber source, but softwood is used most often.

The efficiency of the simple groundwood process depends on careful control of the stone surface roughness, pressure against the stone, water temperature, and flow rate. For example, when the stone has been freshly roughened, production rates are high and energy consumption low, but because the sharp grits cut the fibers, the pulp quality is poor. As wear accumulates, cutting of the fibers decreases causing an improvement in the quality of the pulp, but also decreasing production and increasing the energy needed. A pulp mill commonly adjusts sharpening cycles for individual grindstones so that the overall pulp quality is consistent.

A variation on this process is *pressure groundwood* (*PGW*), which is carried out at a higher temperature and higher than ambient pressure. Since the temperature at the grinding zone is higher, less power is required. It results in paper with improved tear strength compared to SGW, though not as good as refiner mechanical pulp.

6.2.1.2 Refiner Mechanical Pulp

Refiner mechanical pulp (*RMP*) was developed in the 1970s to use wood residues from sawmills and sources that convert wood into chips. Most mechanical pulping mills have switched to this process because it reduces the cost of materials and labor and retains longer fibers. This produces paper that is somewhat stronger than that from SWP, but lower in opacity and brightness.

In the RMP process, wood chips, in a slurry with about 70% water, are fed into a refiner where they are reduced to fiber fragments in a series of attrition mills, between rotating disks. The internal friction in the refiner heats the wood, softening the lignin and permitting efficient reduction of the chips to individual fibers. Since the mechanical treatment results in stiffening of the lignin, refiner pulps are sent to latency tanks where they are held at about 90°C and are slightly agitated to relieve internal stresses and allow the fibers to uncurl. The RMP process has undergone extensive development, and most new factories use thermal or chemical pre-softening of the chips.

6.2.1.3 Thermomechanical and Chemi-Thermomechanical Pulp

Thermomechanical pulp (*TMP*) is significantly stronger and contains less screen reject material than simple mechanical processes. TMP preheats the chips with steam followed by refining at higher temperatures and pressures than are used for standard mechanical pulping. Typically, the chips are presteamed at 120–140°C for about four minutes, and then ground through a pressurized disc refiner in a slurry of water. In some cases, the high pressure steam generated during refining is used for drying in the paper machine, reducing the overall energy use.

The higher temperatures and pressures cause more lignin softening, and therefore fibers are less damaged and can make paper with higher strength. Figure 6.4 compares strength properties of pulp made by the SGW, refiner groundwood, and thermomechanical processes, and shows how strength has increased with each innovation, making it less necessary to mix in some long-fibered chemical pulp. For example, SGW was traditionally mixed with 15–

25% chemical pulp to make newsprint, its largest market. Now some newsprint is made from 100% TMP.

Chemi-thermomechanical pulping (CTMP) uses pretreatment with sodium sulfite at 130–150°C for up to 1 hour, followed by pressurized refining. Yields are high, 85–90%, but lower than pure mechanical processes since some lignin is removed. The pulps have good strength and can be bleached, but produce paper with less opacity than SGW. They are also absorbent, and so CTMP is used to produce pulp for tissue.

These mechanical processes produce an inexpensive high-yield printing paper. The pulp is not as bright as bleached chemical pulp, but is more opaque (due to the contribution from lignin, fines, and broken fibers). It also is more affected by light and aging, becoming yellow and brittle on exposure to UV light that chemically degrades the lignin. It is not possible to permanently bleach the pulp.

Mechanical pulp is weaker than chemical pulp. The fibers are shorter and, because they are not chemically altered, are stiff and resist consolidation in the paper sheet. Therefore, the bonding between the fibers is relatively poor, and the sheet has relatively low density, tending to be bulky and porous. Mechanical pulping produces a large quantity of *fines* (non-fibrous particles) some of which are lost along with water-soluble components of the wood. The presence of fines in the pulp aids bonding between fibers in the papermaking process and adds opacity.

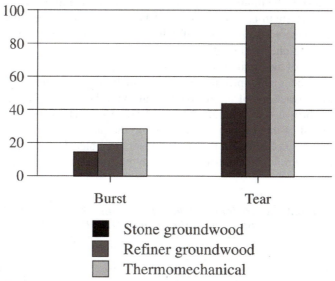

FIGURE 6.4. *Relative strength of spruce mechanical pulps (MeadPaper 1983).*

Mechanical pulp forms a highly opaque paper with good printability. It is used to make newspapers, catalogs, and magazines. Its greatest use in packaging is secondary, in grades of paperboard such as chipboard, which contain recycled newsprint. It is also added to chemically pulped fibers to reduce the cost of some packaging papers.

The characteristics of newsprint and other papers made from mechanical pulp make them ideal for use as recycled stock for paperboard. It is inexpensive and high-yield. It has weak fiber-to-fiber bonds that are easy to disrupt during recycling. And its low density makes a high quality, light weight, thick (and therefore stiff) board.

6.2.2 Chemical Pulping

With the primary exception of paperboard made from recycled mechanical newsprint, most paper for packaging is made from chemically processed pulp. Seventy percent of all pulping is chemical, and most of that is the kraft process. Chemical pulp produces higher quality, stronger paper that if bleached, is also whiter.

Chemical pulping cooks the wood chips in a chemical solution that is either acid or alkali. The chemicals, high temperature (140–190°C) and high pressure (0.6–1 MPa) dissolve much of the lignin, thus freeing the fibers rather than mechanically tearing the wood apart.

Yields are much lower than for mechanical pulping, typically 40–55% lower, since most of the lignin and a great deal of the hemicellulose are lost in the process. But strength is higher due to the largely intact long cellulosic fibers. Table 6.2 compares the retention of cellulose and other wood components in mechanical and chemical pulps.

Since most of the lignin is removed, chemical pulp is easy to bleach in order to make a higher quality printing substrate. The lignin removal is not total, and the chemical reactions of the residual lignin with the pulping chemicals turn it the dark brown color characteristic of kraft paper. The bleaching process removes most of this residual lignin, leaving the pure white color of the cellulose. The absence of lignin is also the reason why bleached chemical pulp is less prone to ultraviolet light-induced discoloration, compared to mechanical pulp.

TABLE 6.2. Approximate Yields and Percent Retention of Wood Components During Pulping (Peel 1991).

	Cellulose	Hemicellulose	Lignin	Solubles	Overall Yield
Mechanical	100	100	100	25	90–95
Chemical, unbleached	90–95	32	9–25	5	40–55

Chemical processing, in the past, has been responsible for significant pollution problems. A great deal of water is used in the process, and the wastewater contains pulping chemicals, sometimes bleach, and also organic materials that are high in biological oxygen demand (BOD). Since sulfur-based chemicals are used, resultant emissions of organic sulfide gases tend to have strong noxious odors which can be offensive even at concentrations as low as one part per billion. Due to environmental legislation, the paper industry now recycles and cleans the water and has emission controls for the air. The pulping chemicals are generally recovered by burning the residual cooking liquor to recover metal salts, and then chemical treatment is used to regenerate the pulping chemicals. Many mills are switching from chlorine to alternative bleaching systems, reducing the toxicity of the effluent. While these changes have not eliminated all environmental problems, emissions have been greatly reduced in both amount and environmental impact.

The soda process was the first chemical pulping method, invented in 1851. It used a strongly alkaline solution of sodium hydroxide to dissolve the lignin in straw. There are still a few *soda mills* operating, using this process to produce pulp from hardwoods and non-wood plant materials. But most were converted to kraft mills once the kraft process was developed.

The following three sections describe the three primary kinds of chemical pulping used today, in the order that they were commercialized: sulfite, kraft, and semi-chemical processes. The most-used chemical pulping method for packaging paper is the kraft process because of its better chemical recovery and higher strength paper.

6.2.2.1 Sulfite Process

The sulfite process, first commercialized in Sweden in 1875, was the most common chemical pulping method around the world for many years. In the 1930s the kraft process became more dominant because it can utilize woods that are not suitable for sulfite processing. Since then, the number of sulfite mills has decreased and no new ones have been built in North America since the 1960s. Less than 3% of pulping now uses this process.

The sulfite pulping process is based on the use of sulfurous acid (H_2SO_3) and bisulfite (HSO_3-ion) to attack and solubilize the lignin. Hydrolysis is followed by sulfonation of the lignin, fragmenting and eventually dissolving it, freeing the cellulosic fibers. Some species of trees are not suitable for sulfite pulping.

It is an acidic process, so it operates at a low pH compared to the kraft and soda processes that are alkaline. A wide range of pH can be used depending on the relative proportions of sulfur dioxide, bisulfite ions, and sulfite ions.

The *acid sulfite* process is carried out at a pH of 1.5–2. Typical conditions are temperature of about 110°C for up to 3 hours, followed by a final 2-hour treatment at 140°C. Originally, calcium sulfite was used along with sulfur dioxide, but calcium cannot be recovered economically. Therefore, more modern processes typically use ammonium hydroxide, magnesium hydroxide, or sodium carbonate to absorb the sulfur dioxide, rather than using limestone. This results in ammonium, magnesium, or sodium sulfite in the pulping liquor, rather than calcium sulfite. A typical concentration is 5–6% sulfur dioxide and 1–1.5% Na_2SO_3. The acid sulfite process is suitable for non-resinous softwoods and some hardwoods.

Acid bisulfite pulping is carried out at a pH of 1.5–4, depending on the ratio of salt to free SO_2. True *bisulfite pulping* is carried out at a pH of 4–5, and a temperature of 170–175°C for 2-1/2–3-1/2 hours. It is suitable for hardwoods and softwoods with low extractive content.

Recent research in sulfite pulping has concentrated on alkaline processes. One promising new development is the *alkaline sulfite-anthraquinone* pulping (*ASAM*) process that can also be used on more kinds of woods. The yield is about 2–3% higher than for kraft pulp, and tear strength of the resultant paper is about 20% less. It results in improved lignin removal and easier bleaching.

Sulfite pulps tend to have lower strength than kraft pulps, but are easier to refine. They are light in color, easier to bleach, and produce a higher yield of bleached pulp. A few mills still exist because of the advantage offered by the sulfite process of easier delignification, which produces pulp with more hemicellulose. On the other hand, long cook times result in nearly complete removal of both hemicellulose and lignin. Some cellulose is also lost at long cook times.

Sulfite pulp makes a less porous sheet than kraft pulp. It is still used somewhat for making fine writing paper and specialty packaging papers like greaseproof, glassine, and waxed paper. However, the proportion of wood pulp made using the sulfite process has declined substantially over the last few decades.

6.2.2.2 Kraft Process

Kraft pulping is also known as the *sulfate process*. The word *kraft* comes from the German word for "strong." It produces the highest strength pulp of any process, and can process a wide variety of species, hardwoods and softwoods, although in the United States young southern pine predominates.

Kraft pulp has the longest fibers, which form a strong network. The fiber surfaces have higher free energy and more carbohydrates, compared to other types of pulp. Kraft pulping increases the size and volume of submicroscopic pores in the cell walls where the lignin has been removed. It increases the flexibility and conformability of fibers so that they more readily flatten into a ribbon-like form, which form stronger inter-fiber bonds. Refining further "activates" the long fibers by unraveling fibrils and delaminating the cell walls.

Kraft is by far the most common pulping process for packaging papers and paperboard. *Kraft paper* is the plain brown paper used to make shopping bags, shipping sacks, and the liners for corrugated fiberboard, applications that take advantage of its strength. Bleached kraft pulp is used to make the *solid bleached sulfate* (*SBS*) board used in high-quality white folding cartons, especially for food applications where the board is exposed to wet conditions.

The kraft process was invented by C.F. Dahl in Germany in 1884 by adding sodium sulfide to the cooking liquor in the soda pulping process. This dramatically accelerated delignification and produced a pulp with stronger fibers. Since it was easy to convert existing soda mills to kraft mills, they were competitive with the sulfite mills, although sulfite pulp predominated for many years longer.

Today, 95% of all chemical pulping uses the kraft process. It surpassed the sulfite process in the 1950s once a better system was developed for recovering the chemicals and energy, and chlorine dioxide bleaching was developed.

The kraft process uses an alkaline solution, so it operates at a high pH, in contrast to the sulfite process, which is acid. The *cooking liquor* changes chemical composition and changes colors from white to black as it moves through the processing and recovery process. Although most waste products are recovered and reused, the kraft process is also responsible for bad smells, primarily organic sulfide gases, which cause environmental concern.

Kraft pulping uses a solution of sodium hydroxide (NaOH) and sodium sulfide (Na_2S), known as *white liquor*. The alkali attacks and fragments the lignin molecules into segments with salts that are soluble in the cooking liquor. Pulping is carried out at a maximum temperature of 170–175°C, with a total pulping time of 2-1/4 to 3 hours. A typical digester is 45–55 feet (14–17 m) high and 11–12 feet (3.4–3.7 m) in diameter. In the initial part of the cook, while the temperature is still increasing, most of the alkali is consumed by the carbohydrates and little lignin is removed. Lignin removal occurs during the 1–2 hours the pulp is held at its maximum temperature.

After the cooking is completed, the pressure is released and the softened chips are discharged into a large *blow tank*. The sudden release in pressure reduces the chips to fibers. Continuous digesters are also available, in which there is no sudden pressure release in the tank. Instead, chips are fed continually into one end of the large 200 foot (60 m) tall digester, and cool cooked chips, at temperatures below 100°C, or washed pulp comes out the other.

Then the pulp is screened to remove fragments and washed to remove the residual pulping chemicals, dissolved lignin, and other materials. Kraft pulps tend to be dark brown in color, and the spent cooking liquor is known as *black liquor*.

The recovery of the pulping chemicals from the black liquor involves evaporation to a highly viscous solution, followed by burning in the recovery furnace, using lignin as the fuel. The inorganic smelt of sodium carbonate (Na_2CO_3) and sodium sulfide, recovered at the base of the furnace, is dissolved to form *green liquor*, which is transformed back into *white liquor* by being causticized with reburned lime (CaO). Sodium sulfate (Na_2SO_4) is added to compensate for the sulfur lost in the process, and gives the kraft process its other name, the *sulfate* process.

Along with 80% of the lignin, about half the hemicellulose in the wood and a small amount of the cellulose are also dissolved and lost in a typical kraft pulping process.

A number of different kraft pulp grades are produced. The unbleached grades used for paper bags and linerboard are cooked less for a higher yield and have more lignin than grades that are bleached to make white paper and paperboard.

6.2.3 Semi-chemical Pulping

Semi-chemical processes combine chemical and mechanical methods. Wood chips are partially delignified (to a lesser extent than in chemical pulps) by being first soaked in the cooking chemical and then fed into one end of the digester vessel. There, chemical treatment, pressure, and high temperatures soften the chips. Very little lignin is dissolved. Next, the mechanical action in a disk refiner breaks down the chips into fibers. The pulp is then washed and refined.

The distinction between semi-chemical and chemi-thermomechanical pulping depends on the extent of the chemical treatment. A variety of process variations form essentially a continuum between purely mechanical and purely chemical pulping processes. In general, processes with more chemical treatment will have lower yields, as they involve greater removal of lignin and other wood components.

Semi-chemical paper is generally stronger than that from mechanical pulp, but not as strong as paper from chemical pulp, and has an intermediate yield. The pulps have a number of end uses and some unique properties. For example, the higher yield pulps produce stiff paper, ideally suited for the fluted medium in corrugated board.

Three common variations are neutral sulfite semi-chemical, high-yield sulfite, and high-yield kraft. The *high-yield sulfite* and *high-yield kraft* processes, as the names imply, involve less cooking and more mechanical processing than the ordinary sulfite and kraft processes. High-yield kraft is most commonly used to make linerboard for corrugated board.

The *neutral sulfite semi-chemical process* (*NSSC*) is used on hardwoods and is the most widely used semi-chemical process. It produces a high yield stiff paper. The main application for the NSSC process is the manufacture of corrugating medium, which benefits from the paper's stiffness, and the short hardwood fibers make it easy to shape into the corrugated form.

The NSSC process involves digestion for about 1 hour with 10–15% sodium sulfite and 4% sodium carbonate at temperatures of 170–180°C. The sodium carbonate neutralizes wood acids produced during hydrolysis, maintaining the pH at the desired level of 8–10. Next, the partially delignified wood chips are mechanically refined. Yields are high, typically 75–85%.

To avoid sulfur, there is also a semichemical process that uses a mixture of sodium carbonate and sodium hydroxide. Pulp for corrugating medium can also be produced in this process. Another semichemical process uses green liquor, a mixture of sodium carbonate and sodium sulfide.

6.3 DEFIBERING, WASHING, AND SCREENING

Before the pulp is made into paper or bleached, it must be defibered, washed, and screened. *Defibering* and *deknotting* are mechanical processes to separate fibers and chunks. Disc refiners are the most common, rubbing the pulp between rotating discs. Large chunks, known as *knots*, are screened out and reprocessed or discarded.

Washing is particularly important for chemical pulps to remove the pulping chemicals. It refines the pulp and recovers the chemicals in the black liquor for reuse. A series of washers are arranged so that soiled water from a cleaner pulp passes to the showers for the next dirtier pulp in a counterflow arrangement that minimizes water requirements and dilution of the pulping chemicals.

There are a number of different processes used for washing, but the most typical is the *rotary drum washer*, in which the stock is introduced into a tank under the washing drum, which is covered with a screen. The water is sucked through the screen into the drum by use of vacuum pressure from inside the drum causing the fibers to accumulate on the surface. *Rotary pressure washers* are similar, but employ air pressure on the outside of the drum to force the liquor into the drum. As the pad of fibers rotates out of the tank, it is further washed by showers located above the drum. The liquid inside the drum is removed, and is replaced with fresh water from the showers.

Displacement washers form a dense pulp sheet on a screen and shower it with clean water, displacing dirty water from the sheet by means of a vacuum. *Diffusion washers* submerge the pulp for a long period to diffuse the liquor from the fibers.

The washed stock is then mixed with water and sent to another screening operation. Screening and cleaning remove oversize particles such as slivers (called *shives*), fiber bundles, and other debris. Typically, large pieces are returned to the pulper for additional processing. Centrifugal cleaners and screens that employ pressure are typical. The pulp is thickened to prepare it for further processing.

6.4 BLEACHING

The color of natural pulps ranges from creamy to dark brown. Kraft paper is darkest, NSSC is slightly brighter, and the brightest are sulfite pulps and groundwood, depending on the species.

But for most printing applications, it is desirable for paper to be white in color and to consist predominantly of cellulose, with little lignin (or hemicellulose) remaining. Bleaching helps to accomplish both of these objectives. It can be regarded as a continuation of the pulping process, with the aim of removing the residual lignin that imparts a brownish color to the pulp. Sulfite and hardwood kraft pulps are the easiest to bleach because they have less lignin. Pure cellulose is white.

Bleaching is a sequence, generally with five stages using different chemicals during each stage, usually with washing between stages. The chemicals used for bleaching are chosen because of their ability to react with the residual lignin and remove it from the pulp.

Chlorine and hypochlorite were commonly used until the 1990s, because they are low in cost and effective. However, it was discovered that chlorine bleaching results in the production of small amounts of highly toxic chlorinated dioxins that pollute the wastewater and can migrate into food products (discussed further in Chapter 9). Regulation of emissions of these chemicals has caused this method to go out of favor.

If chlorine is used for bleaching at all, it is likely in the form of chlorine dioxide, which produces a much smaller amount of dioxin. Increasingly, nonchlorine bleaching is being used, especially for milk cartons. Pulps bleached without chlorine are referred to as *elemental chlorine free (ECF) pulps*. Today, the most common bleaching sequence uses chlorine dioxide and alkaline extraction with oxygen, peroxide, or ozone.

To meet increasingly stringent wastewater standards, some mills in Scandinavia and North America have switched from chlorine dioxide as well, using primarily hydrogen peroxide, in processes referred to as *totally chlorine free (TCF)*. *Biobleaching enzymes* are another alternative, using enzymes that dissolve lignin and hemicellulose but not cellulose.

The paper industry in the United States and Europe has converted to bleaching methods that are more expensive but are more environmentally favorable. Most bleaching processes now recycle the waste water from a later stage to an earlier one, reducing the use of fresh water and the production of effluents. In the same way that the industry has worked to clean up the sulfur smell from the air, it has worked to clean up the water.

It should be noted that not all paper that looks white is bleached. As will be discussed in Chapter 7, paper is sometimes coated with white clay to produce a white background for printing.

6.5 RECYCLING AND REPULPING

Most paper mills use some dried and/or secondary (recycled) pulp. Since the fibers are already free from the wood and lignin, repulping is a simple mechanical process. Pulp that is dried and sold to another mill is called *market pulp*. It is formed into a thick mat, dried in an *air float dryer*, and cut into sheets and baled.

Dried pulp is rehydrated into a slurry in a *furnish pulper*. This works like a kitchen blender, with a turning rotor at the bottom of a large tank that breaks up the material and separates the fibers. There are a number of variations on the design, such as the *Betonniere* cement-mixer type with two rotors turning in opposite directions, causing thorough mixing in the tank.

6.5.1 Recycling Process

The use of secondary pulp is increasing. But all waste paper cannot just be mixed together, because the mixture would be highly variable. It is important to have pure sources of homogenous paper to feed into the papermaking process. These are usually sorted at the source.

There are five basic grades of wastepaper in the United States (although these can be further subdivided, and in other countries the grades may be defined differently). *Pulp substitutes* are the highest value grades because they are most pure: unprinted uncoated white sheets, cuttings, and trim. They can be used directly in the papermaking process, and have always been recycled.

Other wastepapers must be more thoroughly cleaned to remove contaminants. *Mixed Paper* is of varied quality such as office waste and boxboard cuttings. *Old Newspapers* (*ONP*) and *Old Corrugated Containers* (*OCC*) are recycled in the highest volume because of efficient collection systems. *High grade Deinked* is recycled in smaller quantities. These recycled stocks have slightly different classifications in other countries (McKinney 1995).

Sorting packaging waste increases its value. Sorting can take place at the origin or at a material recovery facility (MRF). Sorting at origin requires separate bins for accumulation, collection, and transport, but it saves by putting responsibility on those who generate the waste. Comingled (mixed) waste is more efficient to collect, but requires a sorting operation in the MRF. Most systems (like on-campus recycling) are a hybrid, with separation at origin of specific streams such as corrugated board, paper, glass, plastic, and metal, with finer sorting taking place in the MRF.

In many places, paperboard cartons are not recycled because of the low quality contaminated fiber that results, especially when the heavily coated and printed board is made from ONP. Other drawbacks are that mixing various types of paperboard results in a variable mixture, and low volume makes collection costly. However, recycling of cartons is increasing.

The process for repulping waste paper is similar to that used to rehydrate dry pulp, but requires more extensive processing. Adhesives, coatings, inks, and other contaminants need to be removed. The recycling pulper processes at a low consistency (more water than for simple repulping). Gross contamination is removed with a *junker*, which separates heavy objects by centrifugal force, and a *ragger*, which forms a thick rope from the wires, rags, plastic and other stringy debris, that can be pulled from the slurry. *Drum pulpers*, introduced in the late 1980s, are a gentler alternative, consisting of a horizontal rotating drum that tumbles the paper into fibers in a higher consistency slurry, then washes and filters the fibers.

The recycled pulp is further screened and cleaned to remove smaller bits of contamination. The most problematic are packaging glues, coatings, and hotmelt adhesives, called *stickies*. Centrifugal, vortex, and flow-through cleaners are used. Then the pulp is deinked with the help of a soapy surfactant and bleached if intended for use in newsprint, tissue, and other bright grades.

Last, the fibers are re-refined and combined with fibers from other sources in the slurry that takes its next ride on the papermaking machine. A multi-layer paperboard may have more recycled fiber in the middle layers, combined with virgin or other higher- quality fiber on the surface.

6.5.2 Significance of Paper-Based Package Recycling

Recycled pulp is a significant input for packaging, especially for corrugated board and paperboard. These industrial uses can accept more variation, contamination, and a generally lower grade of fiber than can writing papers.

Approximately 75% of all secondary fiber in North America is used to make paperboard and corrugated board, more for the corrugated medium than in the liners (Smook 2002). It is also used in paper bags, labels, and wraps. The opportunity to use recycled content in paper adds to its benefits as a package material.

Used corrugated fiberboard, with its strong kraft fibers, is the source of a great deal of recycled fiber that is reused to make new corrugated fiberboard. Corrugated boxes are ubiquitous and easy to collect, sorted and baled, from retailers. Other packaging papers and board have much lower recycling rates, since they are more expensive to collect and sort, often contain fibers of lower quality, and tend to be contaminated.

Paper recycling has increased markedly worldwide since the late 1980s. There is an increased environmental consciousness of the fact that recycling conserves trees and provides a lower cost input than wood. Recycling also conserves land by reducing the amount of municipal waste going to landfills. (Although the biodegradability of paper is another environmental benefit, paper takes a very long time to degrade in sanitary landfill conditions.)

In countries with fewer forest resources, paper is more highly recycled. Heavily forested countries like the United States, Canada, and Sweden have less incentive to use recycled fiber.

Corporate responsibility has led product manufacturers to increasingly specify recycled content in their packages. It has led package manufacturers (especially those making corrugated board) to set up a collection infrastructure and processing operations. Paper-based packaging has been a big winner in many environmental debates because of its recyclability.

Just as they can be designed for strength or good graphics, packages can be designed for reincarnation. Recyclable packages should use a single material that has some recycled content. Coatings, inks, and adhesives that would contaminate the recycled fiber are avoided. The material should have a high recycling rate so as to increase the chance that it will be recycled into a new single material package that has some recycled content. The Sustainable Packaging Coalition has published guidelines and opportunities for using more recycled fiber (GreenBlue 2011).

Environmental legislation is growing. In Europe and Japan, where there is less space for landfills, the political response has been most active. Legislation in the United States has been limited to increasing municipal collection of recyclables and mandating recycled content standards for government purchases.

In 1994, the European Union issued a waste directive mandating that each country ensure that packaging materials are collected and recycled at established minimum rates. The guiding philosophy is producer responsibility, meaning that the producers of the packaged product are financially responsible for its collection, recycling, or disposal (and specific recycling targets must be met). Generally, these systems are run by industry organizations that tax participants on the basis of amount and type of packaging generated, in order to fund the collection and reprocessing systems. For example, the Green Dot in Germany (*grün punkt*) designates that the package manufacturer participates in the industry collection system, and has paid the fee. Every EU country has slightly different requirements (fee amounts and whether they are paid by volume or weight), which increases the complexity of export, but increases the probability of recycling. The system offers an economic incentive for reducing the amount of material in packages, as well as for choosing materials that are more easily recycled.

6.5.3 Properties Affected by Recycling

For all its advantages, there is a disadvantage to the use of recycled pulp.

Recycling, especially in the case of chemically pulped fibers like kraft, is usually associated with a loss of strength.

Recycling damages fibers and weakens the between-fiber bonds. Hot drying and calendaring, during the initial papermaking process, close the small pores in the fiber walls in a phenomenon called *irreversible hornification*. This makes the fibers resist swelling during rewetting, reduces their external surface area, and makes them stiffer and less conformable. Factors that improve bonding of the fibers during the virgin phase, like refining, also tend to reduce the capability of the fibers to form strong inter-fiber bonds when recycled. Furthermore, their brittle nature (due to hornification) renders them more susceptible to breakage (Hubbe *et al.* 2007).

The extent of the effect of recycling on pulp quality depends on the original pulping method and the amount of refining, the original papermaking process and printing, as well as the collection history and recycling process (Howard 1995).

As discussed earlier, the initial *virgin* kraft process "activates" long softwood fibers with strong bonds. The long fibers become flexible and ribbon-like; the fiber surfaces have higher free energy in comparison to mechanical pulp; the removal of lignin leaves surfaces that are rich in carbohydrates; and the size and volume of submicroscopic pore spaces are increased during refining. But once the virgin kraft fibers are dried in the papermaking process, they lose swelling ability and wet-flexibility (Hubbe *et al.* 2007).

Each time the fibers are recycled, the tensile, elongation, burst, folding strength, and density are reduced (although the tear strength, stiffness, and opacity slightly increase). The greatest change occurs the first time that fiber is recycled. After five times, most physical properties have stabilized (Nazhad 2005).

On the other hand, mechanical pulp does not necessarily become weaker with recycling, and can even show small gains in tensile strength, stiffness, and density when recycled, but a reduction in tear strength. This is because the wiry fibers are flattened and are made more flexible during repulping, giving a thinner denser sheet. The bonds are not weakened because the fiber walls do not swell and delaminate as much as with chemical pulp (Hubbe *et al.* 2007). This adds to the stiffness of corrugated medium and paperboard made from ONP, which contain semi-chemical and mechanical pulp.

Optical properties are also affected by recycling. For white paper, brightness is lower and there may be specks of ink or dirt. ONP gives cereal carton board its characteristic gray color. Recycled paper is more porous, so more ink is required for printing, which can result in poor resolution or smearing. For these reasons, paperboard made from wastepaper is usually coated with clay or a bleached virgin fiber layer to give a smooth white surface for printing.

Some virgin fiber is usually mixed with recycled paper fiber to make up for the reduction in strength, and so a piece of recycled paper may have fibers having a wide range of age and previous processing. The reduction in the compression strength of corrugated board after repeated recycling is primarily due to the dilution of the fiber stock with the shorter fibers from the corrugated medium. The effect of recycling on corrugated board is further explored in Chapter 14.

Recycled content in paper-based packaging is increasing. As more paper of all kinds is collected, packaging is a good market for the recycled pulp.

But there are some uses that are not appropriate for recycled pulp. For example, the Tetrapak juice box depends on long virgin Swedish softwood fibers to give it strength and stiffness. Heavy duty multiwall bags used for food aid and hazardous materials shipments, and corrugated fiberboard boxes for the military, have extreme requirements. For these, the strength of the strongest virgin kraft pulp is essential.

Post-consumer recycled paper is rarely used for direct food contact because there are various levels of contamination in post-consumer waste. The following types of papers are considered unsuitable for inclusion in food contact paper (Robertson 2013):

• Contaminated paper and board waste from hospitals
• Paper and board that is sorted from comingled garbage
• Used sacks that have contained chemicals

A primary source of contamination is from mineral oil in the ink of the recycled paper which can migrate from paperboard into food (Biedermann *et al.* 2011). These factors are discussed further in Chapter 9.

The subject of paper recycling is a very active research subject as scientists work to learn how fibers change in use and recycling and experiment with ways to overcome the technical obstacles to paper being a true cradle-to-cradle multiple-use renewable resource (Hubbe *et al.* 2007).

6.6 WOOD VERSUS NON-WOOD PULPS

The United States and Canada are the largest producers of wood pulp because of abundant resources of timber, energy, water, and technology, as well as its markets. But there are plenty of other bio-based resources that can be used to make paper and board, and the use of non-wood pulp is increasing. China is the largest producer (FAO 2013).

A variety of non-wood plants can yield significant quantities of usable fiber. Bagasse from sugar cane production is among the leading materials. Various types of straw, including wheat, rice, oats, and barley, are also used. Other fiber sources include kenaf, jute, hemp, sisal, reeds, bamboo, corn stalks, cotton lint-

TABLE 6.3. Average Yields of Selected Non-wood Fibers (Pande 1998).

Raw Material	Fiber Yield (ton/hectare/yr.)	Pulp Yield (ton/hectare/yr.)
Sugar cane bagasse	9	4.2
Kenaf	15	6.5
Wheat straw	4	1.9
Bamboo	4	1.6
Hemp	15	6.7
Rice straw	3	1.2

ers, and even elephant dung! Table 6.3 shows estimated annual fiber and pulp yields for selected non-wood materials.

Pulping of non-wood fibers is similar to wood pulping, but alkaline processes (especially soda) are most common. Table 6.4 shows appropriate pulping processes and pulping yields.

Until relatively recently, foreign producers were not a significant factor in the U.S. paper market, a position enjoyed by few other industries. However, this situation is changing. An increasing amount of paper is imported. Short-rotation pine and hardwood plantations are being developed in sub-temperate regions around the world; countries like Brazil and Chile are already important producers. Russia, with over half of the world's softwood timber resources, has the potential to produce twice the amount of pulp produced in the United States Furthermore, the United States is losing its papermaking technology lead to Europe and Japan. An increasing number of domestic producers have merged with foreign ones, and there is no U.S. company building papermaking machinery (Smook 2002).

TABLE 6.4. Common Pulping Processes and Pulp Yields for Selected Non-wood Materials (Kocurek and Stevens 1983).

Raw Material	Pulping Process	Pulp Yield in % Unbleached	Bleached
Bagasse	Soda, kraft	50–70	45–48
Kenaf	Soda, kraft	45–51	40–46
Mixed cereal straw	Lime, soda, kraft	55–67	50
Bamboo	Soda, sulfite	44–47	40–43
Jute	Soda	62	58
Hemp	Soda, kraft	45–54	43–52

6.7 STOCK PREPARATION AT THE PAPER MILL

Paper mills are often integrated with pulp mills, but they may be separate facilities. When they are separate, the pulp mill dries, compresses, and bales the pulp sheets (after bleaching, for white paper) and ships it to the paper mill.

The first part of the paper mill is the *stock preparation* area. Here the bales are repulped or, in integrated mills, the high density (15%) stock coming from the *storage chest* is diluted. The pulp is beaten and refined. It is blended with wet end additives in the *blend chest* to form a uniform consistency. Accurate proportioning and the sequence of additives depends on the properties of the paper and the mill.

6.7.1 Beating/Refining and Cleaning

Beating or *refining* are used to modify the surface of the fibers so that they cling together more tenaciously. Refining makes the fibers soft and flexible and increases the surface area of the fibers for better bonding. They are literally "beat into a pulp."

The pulp is agitated in water to swell and soften the fibers. The agitation fibrillates and ruptures their thin outer walls, partially collapsing the wood fibers, allowing water to penetrate the cells and causing them to become more flexible. *Fibrils* unfurl along their surface, as illustrated in Figure 6.5, increasing the fiber-to-fiber contact surface area. The hairy fibrils and the fiber debris also increase the hydrogen bonding that is required to make strong paper.

The degree of refining has a large effect on the properties of the paper that is produced. Unrefined fibers tend to be stiff, springy, and resistant to collapse, and paper made from them tends to be bulky and porous. It also tends to have low tensile strength, since there is little inter-fiber bonding. For example, absorbent tissue paper products are made from pulp that has been lightly refined.

Tensile, burst, and tear strength all improve with some refining and decline with too much. This is because at low levels of refining, the improvement in hydrogen bonding increases all three kinds of strength.

FIGURE 6.5. *Refining causes fibrils to unfurl on the surface of the fiber.*

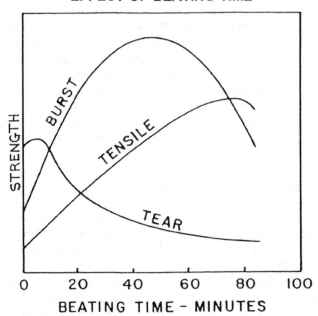

FIGURE 6.6. *The effect of refining on paper properties.*

Longer refining time is better for tensile and burst properties than it is for tear strength. It damages and shortens the fibers, and so tear strength, which is influenced markedly by the length and the strength of individual fibers, begins to decline. As refining time continues, bursting strength likewise improves, plateaus, and then drops off while tensile strength continues to grow. Tensile strength is the last to decline. The effect of refining time on paper properties is summarized in Figure 6.6.

Refined fibers make a sheet denser, less porous, and less opaque. The improved inter-fiber bonding makes the material more uniform, and thus light penetration is easier since differences in refractive index are reduced.

If refining is carried out to a very high degree, paper properties change markedly. For example, glassine and waxed paper are made from fibers that have been very heavily beaten to a jelly-like pulp. Most grease-resistant papers are highly refined, which limits the absorption of oil.

A number of types of beaters and refiners are available. The traditional *hollander beater* (a Dutch invention dating from the mid-1600s and still sometimes used, especially in laboratories) rolls a bar-covered cylinder through a trough fitted with a textured bedplate and filled with the pulpy suspension. In this batch process, the fibers move perpendicular to the *flybars*.

Modern refining equipment employs a continuous process, moving the pulp through disks or cones where the fibers move parallel to the flybars. The *Jordan and Claflin refiners* employ a conically shaped plug with longitudinal flutes, which is rotated inside a ribbed conical housing. Moving the plug farther in or out of the housing controls the spacing between the flutes and ribs, and thus affects the degree of refining. The paper stock enters the refiner at the small end of the cone and is discharged at the large end.

Disk refiners are a more recent development. A disk refiner is similar to the refiner used for mechanical pulping. It consists of two fluted metal disks, closely spaced and rotated in opposite directions, or with one or two fixed and one that rotates. The pulp is subjected to a grinding action as it passes between the disks. Disk refiners offer the advantages of lower cost, lower energy usage, better efficiency, and more versatility.

Papermakers have found that refining decreases the *freeness* (drainability) of the fiber because the water is more tightly bound to the fibers. The *Canadian Standard Freeness* (*CSF*) test is used for process control to determine when a pulp meets the required qualities and to ensure consistency. It measures the rate at which water drains from a given amount of fiber suspended in water through a wire onto which the fiber is allowed to settle.

After the refiner, the pulp is cleaned. There are various cleaning systems, but they all employ the principle of swirling the slurry and drawing out the contaminants by force that is *centrifugal*, *vortex*, or *pressure* (*flow through* and *reverse cleaners*).

At this point the pulp stock is ready to blend with other pulps and additives to become the *furnish*. The blend of hardwood, softwood, market pulp, recycled pulp, and/or broke is metered and mixed in the desired proportions in *blending chests*, and fillers, binders, sizing, and coloring are added.

6.7.2 Binders, Fillers, Sizing, and Coloring

To improve properties, non-fibrous minerals and chemicals are added to the fibers before the paper is made. Additives can make paper brighter, smoother, stiffer, and/or water resistant. About 10% of the cost of making paper is the cost of the additives.

These are called *wet end additives*, and their use depends on the desired properties of the finished sheet. Not all additives are added to the wet stock; some are coated onto the paper sheet at the dry end of the papermaking machine.

Fillers are used, as their name implies, to fill in the gaps between fibers. This makes the paper smoother and improves its printability, opacity, and brightness. Most fillers are white. White paper for printing usually contains from 10% to 30% filler.

White clay is the most commonly used filler. Its low cost and good printing quality are prime assets. Clay is also a common white coating for papers and paperboard.

Other fillers are titanium dioxide, calcium carbonate, and talc. Titanium dioxide gives excellent brightness and opacity to paper, but is relatively high in cost. Calcium carbonate is brighter than clay, and its excellent ink absorption promotes fast drying of printing inks. It is primarily used in coated paper for magazines, books, and other printed materials. It is a much less expensive filler than titanium dioxide, but cannot be used with acidic papers. Talc is used to give a silky feel to the paper.

The purpose of *binders* is to enhance inter-fiber bonding, and thus increase the strength of the paper. Unlike refining, which also increases inter-fiber adhesion, binders also make the paper stiffer. Binders improve burst, tensile, and folding strength. They make paper crisp and give it "rattle." They decrease surface fuzz and increase hardness, stiffness, and durability. They also reduce the rate of water penetration. Binders are natural or synthetic polymers.

The most commonly used binder is starch. Like cellulose, starch has many OH groups that can participate in hydrogen bonding. The starch can get between adjacent fibers, bind strongly to both, and thus bind the fibers to each other. Starch is a natural polymer of glucose derived from plants like corn, tapioca, potatoes, and wheat. Other binders are gums (derived from bean and guar seeds) and synthetic polymers.

Starch is the third most important ingredient in the paper furnish, based on weight, after pulp and clay. It plays many roles in papermaking, from binder to adhesive to external sizing. In paperboard, it is commonly used in amounts of 2–4%.

Sizing agents increase the liquid resistance of paper. *Internal sizing* agents are those used as additives in the furnish. Liquid penetration is retarded by the non-polar regions of the size molecules that are anchored to the fibers. Papers with a high level of sizing are called *hard sized* and are very water and ink resistant. *Slack sized* paper has little or no ability to resist liquid. *External sizing* agents are applied to the surface of the formed sheet and are discussed in Chapter 7.

The most commonly used internal sizing is rosin, produced from pine. The rosin is dissolved in an alkaline solution and added to the pulp. Alum (aluminum sulfate) is added to precipitate the rosin onto the fibers in a finely divided state. Once the formed sheet passes through the dryers, the fine rosin particles melt, coating the fibers with a water resistant coat.

Because of the presence of alum, rosin-sized paper is acidic, and is therefore more subject to deterioration than non-acidic papers. Alkaline sizing agents have also been developed. These are more expensive, but produce paper with better permanence (retention of properties such as color over time), and also permit the use of calcium carbonate filler rather than titanium dioxide.

For many packaging uses in wet environments, sizing alone does not give adequate water resistance. Some examples include paper bags exposed to wet conditions and corrugated fiberboard boxes for fruits and vegetables. Liquid attacks the fiber to fiber bonds and makes the paper lose its strength. Wet paper typically retains only about 5% of its tensile strength.

Wet-strength resins are thermosetting resin additives that are fixed onto the fibers. Curing in the drying process leads to forming covalent fiber-to-fiber bonds. Wet-strength additives can make wet paper that retains from 15% to 50% of its dry strength, depending on the grade. *Wet-strength kraft* paper is often marked with a faint yellow or green stripe to identify it.

Coloring agents, such as pigments or dyes, can be added to the paper furnish. Pigments are insoluble coloring materials, while dyes are soluble. Pigments function by becoming mechanically entrapped in the paper. Dyes become chemically attached to the fiber. The three most common fillers, clay, calcium carbonate, and titanium dioxide, also act as white pigments.

A variety of other chemicals may be added to the paper furnish. Specialty additives include *corrosion inhibitors*, *flame proofing*, and *anti-tarnish chemicals*. Many additives are used not for their effect on the finished paper, but rather to facilitate the paper manufacturing process. *Defoamers* and *fiber flocculants* are added to eliminate foam and encourage deposition of fibers in the paper. Chemicals are added to fix pitch (tacky resins from the wood) onto the fibers so it does not build up on the papermaking machinery. *Slimicides* are commonly added to prevent the growth of bacteria and fungi in the paper machine.

The furnish is fed into the *machine chest* at a consistency of 3–5% solids. Agitation is maintained to keep the stock uniform and prevent settling. Before being fed into the papermaking machine, the stock is diluted to 1% or less, depending on the grade and basis weight of paper that is being produced. Generally, the lower the basis weight of paper being produced, the lower is the consistency of the stock.

The pulp is now ready to take its ride on the papermaking machine, described next, in Chapter 7.

6.8 REFERENCES

Ainsworth, J.H. 1959. *Paper, the Fifth Wonder*. Kaukauna, WI: Thomas Printing & Publishing.

Biedermann, M., Y. Uematsu, and K. Grob. 2011. Mineral Oil Contents in Paper and Board Recycled to Paperboard for Food Packaging, *Packaging Technology and Science*, 24 (2): 61–73.

Bierman, C.J. 1996. *Handbook of Pulping and Papermaking*, 2nd ed. San Diego: Academic Press.

FAO. 2013. FAO Yearbook of Forest Products 2011. Food and Agriculture Organization of the United Nations. Rome. http://www.fao.org/forestry/46203/en/

GreenBlue. 2011. *Guidelines for Recycled Content in Paper and Paperboard Packaging*. Charlottesville, VA: GreenBlue.

Howard, R.C. 1995 The Effects of Recycling on Pulp Quality. In *Technology of Paper Recycling,* ed. R.W.J. McKinney, 180-203. London: Blackie Academic and Professional Press.

Hubbe, M.A., R.A. Venditti, and O.J. Rojas. 2007. How Fibers Change in Use, Recycling, *BioResources,* 2 (4): 739–788.

Kline, J.E. 1991. *Paper and Paperboard; Manufacturing and Converting Fundamentals,* 2nd ed. San Francisco: Miller Freeman.

Kocurek, M.J. and C.F.B. Stevens. 1983. *Pulp and Paper Manufacture, Volume 1, Properties of Fibrous Raw Materials and their Preparation for Pulping.* Atlanta: TAPPI.

McKinney, R.W.J. 1995. Manufacture of Packaging Grades from Wastepaper, in *Technology of Paper Recycling,* ed. R.W.J. McKinney, 244–295. London: Blackie Academic and Professional Press.

MeadPaper. 1983. *The Manufacture of Printing Paper,* The Mead Corp.

Nazhad, M.M. 2005. Recycled Fiber Quality—A Review, *Journal of Industrial and Engineering Chemistry,* 11(3): 314-329.

Pande, H. 1998. Non-wood fibre and global fibre supply. Unasylva, No. 193, Vol. 49. http://www.fao.org/docrep/w7990e/w7990e08.htm

Parham, R.A. and R.L. Gray. 1982. *The Practical Identification of Wood Pulp Fibers.* Atlanta: TAPPI.

Peel, J.D. 1999. *Paper Science and Paper Manufacture,* Vancouver, Canada: Angus Wilde.

Robertson, G.L. 2013. *Food Packaging: Principles and Practice,* 3rd ed. Boca Raton, FL: CRC.

Smook, G. 2002. *Handbook for Pulp and Paper Technologists,* 3rd ed. Vancouver: Angus Wilde.

TAPPI. 1997. *How Paper is Made.* Atlanta: TAPPI, CD Rom.

Walker, J.C.F. 1993. *Primary Wood Processing: Principles and Practice,* London: Chapman & Hall.

6.9 REVIEW QUESTIONS

1. What advantages and disadvantages does wood have compared to other paper-making raw materials?

2. Why must the bark be removed before making pulp?

3. How does the pulping process free the cellulose fibers from wood chips?

4. What are the advantages and disadvantages of softwood versus hardwood pulp? Which produces stronger paper? Why? For what uses is one better than the other?

5. How does fiber length affect paper properties?

6. Describe the difference between the mechanical and kraft processes for making pulp and give examples of how and why the resulting paper and end uses differ.

7. Which pulping process has the highest yield? Which pulping process results in the strongest paper?

8. What combination of wood type, pulping method, and refining time produces the strongest pulp?

9. In the kraft process, what is the difference between white, black, and green "liquor?"

10. What is the sequence of operations involved in making pulp?

11. What is the problem with chlorine-based bleach?

12. What are the benefits of recycling paper-based packaging? How does recycled fiber affect the properties of paper? How can packages be designed for recycling?

13. Why is non-wood based pulp more commonly used in Asia than the United States?

14. What are some of the non-wood materials used to make pulp?

15. How does pulp beating/refining time affect the tensile, tearing, and burst properties of paper? What kinds of paper are most highly refined?

16. What is the purpose of adding the following wet-end additives: filler, binder, sizing, and wet-strength?

17. What is the purpose of sizing added to the pulp? From what is it made? What is the difference between hard sized and slack sized paper?

18. What is the most common filler? The most common binder?

Fabrication of Paper and Paperboard

Handmade paper is made just like Ts'ai Lun's ancient Chinese secret (Chapter 1). The pulp is suspended in water, and a sheet mold with a woven screen-like floor called the *wire* and a wooden frame called the *deckle* is dipped into the mixture. A mass of furnish is scooped up onto the mold, which is shaken to distribute the fibers. The water is then allowed to drain through the porous mold, leaving a fibrous mat. The molded paper is carefully removed from the wire in a process called *couching* (which rhymes with smooching, and means to *lay*) onto a felt blotter sheet. A stack of paper and felt sheets is pressed to remove the free water and create a physical bond between the fibers. The sheet is then removed from the felt and allowed to dry.

All paper was made in sheets, by hand, until 1804 when the Fourdrinier brothers financed, built, and sold the first papermaking machines that still bear their name. Mechanization is a continuous process that results in rolls, rather than sheets, and this made possible today's web-fed paper-based packaging conversion processes like printing, bag-making, and corrugating. Web-fed processing is faster and cheaper than sheet feeding.

Chapter 7 describes the two types of papermaking machines: the fourdrinier and the cylinder machines. The fourdrinier is used to make all kinds of paper and some kinds of paperboard. The cylinder machine is now used exclusively for multi-ply paperboard. These machines form the pulp slurry onto a continuous web, drain, couch, smooth, and dry it.

7.1 FOURDRINIER PROCESS

A fourdrinier machine is huge, housed in a factory the size of an entire city block three stories tall (Figure 7.1). It is considered in two parts: the *wet end* and the *dry end*. Fourdrinier machines are used to make most kinds of paper,

207

especially printing paper for books, magazines, and newspapers. It is also used to make kraft paper and some kinds of paperboard.

7.1.1 The Wet End

First, the furnish from the machine chest (see Section 6.7.2) is moved into the *stuff box*. It is diluted in proportions of about 99% water and 1% pulp by the *fan pump*, which uniformly distributes fibers in the stock slurry, with no lumps.

The slurry is pumped into the *headbox*. The headbox, pressurized to maintain adequate flow, and *flowspreader* distribute the slurry evenly across the width of the machine. This is a critical step that must be carefully controlled in order to avoid variation in the sheet properties such as basis weight, grammage, and thickness. The pulp is deposited through a long thin rectangular nozzle, called the *slice*, onto the moving screen, called the *wire* or *forming fabric*. The pressure in the headbox determines the volume of stock flowing through the slice, and the volume of stock determines the paper thickness.

The velocity of the stock flowing through the slice and the forward-moving wire orient a higher proportion of the fibers in the *machine direction* (*MD*), the direction in which the continuous web of paper runs through the machine. This is also called the grain. This fiber orientation is the reason why paper is stiffer and stronger in the machine direction than it is in the *cross machine direction* (*CD*).

The endless moving forming *wire* screen (also called the *table*), with its supports and transport system, extends from the breast roll to the *couch roll* and then loops back underneath to the breast roll again, as shown in Figure 7.2. While on the wire, the water is drained by gravity and vacuum suction.

The wire screen is a woven fabric mesh, usually made from polyester, formed into a continuous loop belt that is typically 7 m wide and 70 m long (24' × 230'). Polyester or nylon plastic mesh has replaced stainless steel and bronze mesh in most machines (although it is still called the "wire"). The wire is characterized by its mesh size (number of filaments per inch), which often differs in the machine and cross directions, and by the type of weave (the pattern in which filaments go under and over each other). The characteristics of the wire affect the smoothness of the paper.

The mesh belt covered with the pulp slurry passes first over the *breast roll*. *Deckle boards* prevent the stock from slopping over the sides. The pulp settles onto the wire, begins to drain, and the fibers begin to interlock. Turbulence is induced on the wire as the fiber slurry mat passes over it, which improves fiber distribution and causes more fibers to lay in the cross-machine direction. For example, the *Sheraton Roll* with a sawtooth surface vibrates the wire from side to side.

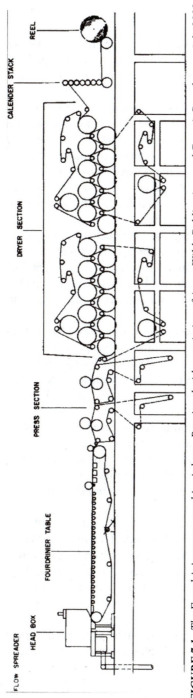

FIGURE 7.1. The Fourdrinier paper machine is huge. Reprinted with permission of Angus Wilde Publications and Gary A. Smook, copyright 2002.

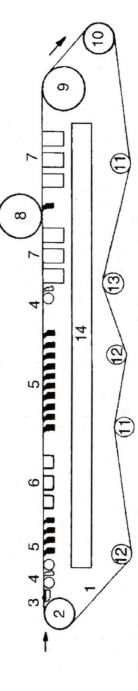

FIGURE 7.2. Fourdrinier Table: (1) forming screen/wire, (2) breast roll, (3) forming board, (4) table roll, (5) foil, (6) wet suction box, (7) dry suction box, (8) dandy roll, (9) suction couch roll, (10) wire turning roll, (11) guide roll, (12) stretch roll, (13) wire return roll, and (14) tray. Note that this is the left half of Figure 7.1, showing the direction of the moving screen; the paper sheet continues to move on to the dry end. Reprinted with permission of Angus Wilde Publications and Gary A. Smook, copyright 2002.

209

The balance of the wet end is devoted to draining water from the pulp mat to create physical and chemical bonds between the fibers. The wire is supported by a series of *table rolls, suction boxes, hydrofoil blades*, and *vacuum assisted foils and boxes*.

A *dandy roll*, wrapped with a wire cloth with about half the mesh count of the fourdrinier wire, sucks and pushes water out of the sheet, smoothing out the sheet. The dandy roll can also be used to make a *watermark* or a laid surface, used for fine writing papers. For example, a *bond paper* watermark is applied by a bronze letterpress-type dandy roll. (Watermarks can also be applied using a rubber stamp roll, called a Molette, in the wet press section of the paper machine.) *Laid paper* has a ribbed appearance, due to a dandy roll covered with parallel wires running in the cross-machine direction.

Much of the water is now drained, leaving the fibers behind in a smooth sheet as they leave the wire. The water removed through this point is called *white water*, and contains a significant amount of fiber (about 20% of that initially present), fines, and other furnish ingredients. It is continuously recycled in the paper-making process, so that the fibers in it will get another chance to become part of the paper mat.

At the end of the wire, the wide web of wet paper is transferred by the suction *couch roll* to the *wet press* section of the machine. At this point, the water content has been reduced to about 75–80%. Off-spec material, called *broke*, is removed into the couch pit where it is showered with water and agitated so that the fiber can be directly reused. (Dry broke is likewise reprocessed from the end of the machine.)

Next, in the *wet press* section, additional water is squeezed out and the fibers are pressed together. It irons out the lumps, further shaping the paper and interlocking the fibers. The web is transferred to a sheet of porous *carrier felt*, made of wool or synthetic. Like the wire, the felt is a continuous belt. The felt and paper are squeezed between a pair of *nip rolls*, which wrings water from the web and absorbs it in the pores of the carrier felt.

The top rollers are very hard and smooth; traditionally granite, but now usually steel covered with ceramic. The rollers that contact the felt are covered with an elastomer to control hardness, and are perforated with holes through which suction is applied to remove some of the water from the felt and help prevent it from being reabsorbed by the paper when the pressure is released. The paper leaves the felt immediately as it leaves the wet press, to prevent reabsorption of water.

One to three wet presses may be used, depending on the paper grade. There are alternative press designs in which grooved rolls, a shrunk fabric sleeve replacing a felt, fabric replacing a felt, or other designs are used. The thickness of the pulp sheet is partly controlled by the pressure applied in the wet press. Before the belt of carrier felt returns to collect new fiber, the water is removed from it to restore its ability to absorb moisture. When the paper web leaves the wet press section, it is down to 50–60% water, and still has little strength, being held together primarily by surface tension and frictional forces between the entangled fibers.

The top side of the paper is referred to as the *felt side*, and the bottom side as the *wire side*. Table 7.1 shows how the properties differ. The wire side is more rough because the coarse fibers settle fastest and provide a base on which fines can be more readily retained. It also takes on some of the texture of the mesh. The felt side is smoother since it has more small fibers and filler particles to fill in the gaps. If average filler content is 9%, for example, the filler content on the wire side may be as low as 4% while on the felt side it is as high as 14% (MeadPaper 1983). The smoother felt side is better for printing.

The wire side can be identified by using a low power microscope to look for the *wire marks*, a rectangular or diamond-shaped regular pattern. The felt side has a more irregular pattern. A carbon smudge (made by rubbing the surface with carbon paper) can intensify the pattern. Another way is to place a sample on a flat surface and tear it, starting along the machine direction and then curving to end up tearing across the grain; turn the sheet over and tear in the same manner. The tear that shows the most feathering, especially in the curved portion, is the one that was torn with the wire side up. This and other standard tests for distinguishing the wire side can be found in *TAPPI T455: Identification of wire side of paper.*

7.1.2 Twin-Wire and Hybrid Formers

A popular alternative to the fourdrinier former is *twin-wire forming* (also known as *gap forming*). Instead of using the long fourdrinier table, the jet of stock is fed into a converging gap between two continuous forming fabrics that squeeze and dewater it from both sides simultaneously.

TABLE 7.1. Comparison of the Felt (Top) and Wire (Bottom) Sides of Fourdrinier Paper.

Wire Side	Felt Side
More fiber interlocking	Less fiber interlocking
Wire mark	Felt or dandy mark
Large fibers	Small fibers and fines
Lower in filler	Higher in filler
High MD fiber orientation	More random fiber orientation
Rougher surface	Smoother surface
Worse for printing	Better for printing

Twin-wire forming has two advantages. The sheet sets faster, almost immediately, and the top and bottom sides of the paper are more uniform than in the traditional fourdrinier process. The first commercial twin-wire machine in the United States was installed in 1965. A number of different designs now exist, and most new machines being installed are of this type.

The on *top former* (also called the *preformer*) is a hybrid twin wire former that introduces the second wire after the slice has deposited the fiber onto the first wire and some drainage has occurred.

The web formed by twin wire or hybrid formers proceeds through wet and dry presses like those used in the fourdrinier process.

7.1.3 Multi-ply Fourdrinier Formers

Although most thick multi-ply paperboard is made from recycled pulp on a cylinder machine as described in Section 7.2, the fourdrinier machine is also used to make multi-ply board. This practice began in the 1970s, mostly in Europe.

The fourdrinier machine is limited in how quickly the water can drain through the successively applied layers. It is not possible to make thick board because the water will not drain quickly enough. Most fourdrinier multi-ply board is made from kraft pulp because it drains more freely than recycled fibers.

To make more than one ply, the fourdrinier machine is fitted with *multiple formers*. A series of headboxes and forming wires apply a sequence of plies on top of the still-wet first ply. Since the formers are in series, a higher grade fiber can be used in the ply that will be the printed surface. The multi-ply fourdrinier process is faster than cylinder forming.

A newer method is the *multi-layered headbox*, which discharges more than one layer in a single jet. Since its plies are more mixed, they are better bonded. But the surface coverage of the top and bottom plies is relatively poor, and sometimes a separate former is used for the outer plies.

Multi-ply fourdriniers are used for thinner grades of paperboard. The number of plies is limited to four. The fourdrinier is also capable of making a thinner, single-ply paperboard. The resulting board is higher quality than cylinder board because it is stronger in the cross-machine direction and contains kraft pulp.

This process is used to make the *duplex linerboard* commonly used for corrugated board. Multiple formers are also used to make the solid kraft boards called *solid bleached sulfate* (*SBS*) and *solid unbleached sulfate* (*SUS*).

7.1.4 The Dry End

Leaving the wet press, the paper enters the *dry end*, where hot drum dryers and hot air further evaporate the water, decreasing the water content in the paper to 5%. The drying operation accounts for a significant portion of the cost and energy usage of the paper-making process.

The web threads through a series of large steam-heated cylinder dryers with smaller diameter rollers in between them, as shown in Figure 7.3. The paper heats up as it passes around each dryer, and the moisture is flashed off in the pocket between the top and bottom cylinders before the paper goes around the next dryer. The side of the paper in contact with the hot dryer rolls alternates. The rolls are usually 4–6 ft (1.5–1.8 m) in diameter, and the dryer section is typically about 230 ft (70 m) in length.

The web is supported and held in contact with the dryers by a permeable fabric called the *dryer felt*. Most paper machines have three to five independently felted sections. Lightweight paper needs to be carried by felt on both sides for stability. On heavy weight paper and paperboard, the web is in direct contact with the dryer rolls.

The temperature of the dryers increases gradually through the section, reaching a maximum of 170–200°C (338–392°F) before being reduced again in the final section. Hot air is often also employed. Because the web is cooled by evaporation, its temperature stays at approximately 100°C (212°F) regardless of the roll temperature, as long as free water is present. In the final dryer section, the paper is dry enough that there is danger of overheating the web, so temperatures must be reduced.

In the manufacture of machine glazed (*MG*) paper, the still-damp web is pressed tightly to the surface of a *Yankee dryer*, which uses one large highly-polished drum roller to complete the drying. This results in a sheet with an extremely smooth glossy surface on the side that was in contact with the dryer. It

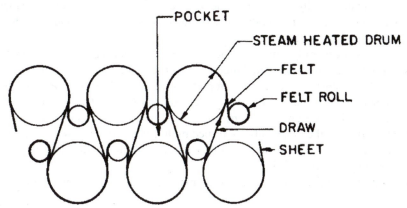

FIGURE 7.3. *Two-tier dryer configuration. Reprinted with permission of Angus Wilde Publications and Gary A. Smook, copyright 2002.*

is also the kind of dryer used to make crepe paper; the sheet is creped by scraping it from the dryer before it is completely dry. MG paper is used for making decorative tissue paper, crepe paper, and some kinds of food wrapping paper.

Similarly to crepe paper, *extensible paper* is produced in a micro-creping process that increases its strength and elasticity. Extensible paper is used in multiwall bags to increase their impact strength. The process involves pressing the moist web against a stretched rubber surface that is then allowed to shrink back, compressing the web on its surface.

Towards the end of the drying process, coating materials such as *starch sizing* may be added, if desired, to fill surface voids and add resistance to penetration by water. This external sizing differs from the internal rosin sizing that is added to the pulp. Sizing also increases resistance to handling and abrasion, aids in ink hold-out, and produces a firmer, stiffer sheet.

The size press applies a continuous film of starch on both sides of the paper, filling the capillaries so that they will resist liquid. The sheet, at 5–10% moisture, runs through a puddle of sizing agent and/or has sizing agent sprayed on, and then enters the nip of two rollers, which drives the size into the sheet and squeezes off the excess. The addition of sizing increases the moisture content of the sheet to about 20%, so another section of dryers is required to again reduce the moisture.

Next, most paper is sent through a *calender*[13] stack to smooth and compress the paper to a uniform thickness (*caliper*). In the papermaking machine, calendering is second only to the amount of fiber deposited by the headbox in determining thickness.

Calendering smooths the surface for printing. It usually slightly increases tensile strength and increases oil resistance; however, excessive calendering reduces tensile strength. *Machine-finishing* (*MF*) uses highly-polished rolls that are hydraulically loaded to squeeze and polish both sides of the paper. Since calendering is an operation performed on dry paper, it can also be done off-line, at a later time.

Calenders generally consist of one to three vertical stacks of rolls, each with two to nine very smooth heated metal rolls, ranging in diameter from 16 in to 36 in (40–91 cm). The paper is compressed between the rolls, which makes it thinner and smashes small fiber concentrations (called *flocs*) to the same thickness as lightweight spots. The bottom roll is larger than the others and is machine driven; the other rolls are turned by friction. Steam, water, and starch or other chemicals may be applied to the surface to achieve the desired *finish*, for printing or later adhesion.

Coatings or other surface treatments may be applied at the end of the ma-

---

[13]Note that this word is spelled differently from "calendar."

chine, or they can be applied off-line. The most common is clay coating. Section 7.4 discusses clay and other coatings.

Finally, the paper is trimmed and then rolled onto a reel for shipment. A roll of paper is known as a *log*. It is typically 25 ft (7.6 m) wide, weighs 28 tons, and contains 45 miles (72 km) of paper. A typical fourdrinier machine produces one log of paper every 1-1/2 hours—enough paper every month to circle the equator.

Since it is not desirable to stop the machine once a reel is full, the usual procedure is to make a *flying splice*, connecting the stream of paper moving at about 30 miles per hour (50 km/hr) to a new core, while removing the full roll at the same time without stopping the machine. The trimmed-off broke is repulped and returned to the papermaking process.

Winders and rewinders are used off-line to cut the large rolls of paper produced on the fourdrinier into smaller rolls more suitable for end users.

7.2 CYLINDER MULTI-PLY PAPERBOARD PROCESS

The traditional way to make multi-ply paperboard is on a *cylinder machine*. These make most of the recycled paperboard used today.

The cylinder former, patented in 1807, has been used for over 100 years to make paper, competing with the fourdrinier process. The essential difference is that instead of the headbox used to form the web in the fourdrinier process, the cylinder process uses a cylindrical former.

But the great advantage of the cylinder former was not realized until 1870 when a number of cylinder vats were combined in series to produce a multi-ply board. Previous to this, when a stiff board was required, layers of paper were pasted together to make *pasteboard.*

Today the cylinder process is used only to produce multi-ply board. It produces the board used to make most of the folding cartons for cereal, cookies, crackers, cake mixes, and many other dry processed foods. Cylinder board is also used as the blister backing material for cosmetics and hardware.

The raw material for a cylinder mill is usually recycled paper fiber, especially ONP made from mechanical pulp. Its short fibers make it weaker and coarser than the multi-ply board made in a fourdrinier mill. Most cylinder mills are located in metropolitan areas to be near the source of recycled material.

The cylinder process makes multi-ply board by laying down separate layers of pulp slurry on top of one another. This makes it possible to use different kinds of fibers in each layer. It is common to use higher quality fiber on the surface layers and lower quality fiber in the interior of the sheet. Each layer of material is typically about 0.003 in up to 0.01 in (3–10 points or 76–250 μm) thick.

In the traditional *vat cylinder* process, the furnish at about a 3% consistency is pumped into a series of vats. Very large screen-covered cylinders lie partially submerged in the vats and pick up the furnish as they rotate. A common counterflow vat is shown in Figure 7.4. The water is sucked into the cylinders, drained through the screens, and the pulp that remains on the rotating screen is transferred to the underside of a moving felt mat, pressed against it by a soft rubber couch roll, as the cylinder rotates.

The strong directional movement through the vat and around the cylinder orients the fibers markedly in the machine direction, more than in the fourdrinier process. Newer vat designs have tried to improve the cross-machine direction properties and make a more uniform sheet.

Each cylinder, in turn, contributes its layer to the mat as shown in Figure 7.5. The number of vats usually does not exceed eight.

Water removal from multi-ply board needs to be more gradual and gentle than from single-ply paper to bond the plies without crushing the board. Suction is used to assist in dewatering and to help orient more fibers in the cross machine direction. A large quantity of water is removed by suction and vacuum while the board is on the underside of the felt. Next, the web is joined by a second felt that helps carry it through the primary wet press section.

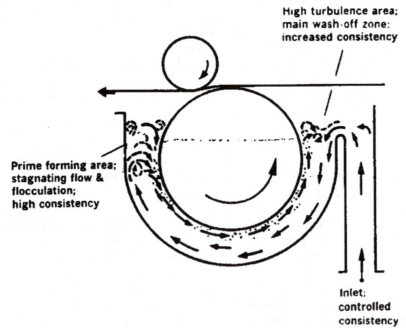

FIGURE 7.4. *Cylinder machine counterflow vat and couch roll. Reprinted with permission of Angus Wilde Publications and Gary A. Smook, copyright 2002.*

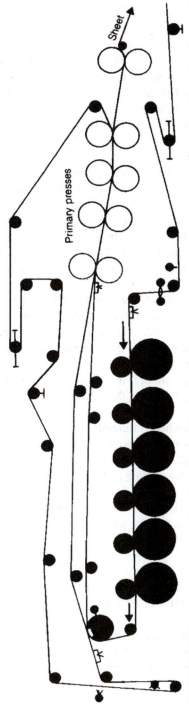

FIGURE 7.5. *Typical arrangement of multiple cylinder formers and felt. Compare this to the fourdrinier arrangement in Figure 7.2. Reprinted with permission of Angus Wilde Publications and Gary A. Smook, copyright 2002.*

217

The press section is designed to dry the board evenly throughout in order to prevent warping. The thickness and composite furnish affect the heat transfer rate from the surface to the center of the sheet and the steam diffusion rate from the center to the surface. The drying process is adjusted to best process each layer of the composite web. When one side of the board has different characteristics than the other, the dryer section needs to be able to control drying of each side independently.

A good ply bond (adhesion between layers) is essential for folding carton board. The board needs to be capable of being scored and formed well, and to hold its form throughout its use. The ply bond, stiffness, and inter-laminar shear strength depend on the mechanical interlocking of fines and fibrils in adjacent layers and the development of hydrogen bonds between fibers in the layers. The ply bond can be improved by increasing the fines and water content at the interfaces, further refining the pulp, giving better control of nip pressure throughout the process, and by decreasing the thickness of the plies (and increasing the number).

Some of the differences between drying in the fourdrinier and cylinder papermaking processes are that in the cylinder process the second and third presses are independently felted, evaporation rates are improved by more aggressive condensate removal, and larger steam cylinders are used. The temperature is progressively increased to avoid sheet blistering and the steam pockets that cause ply separation. If the surface evaporation is more rapid than evaporation from the board's center, the surface hardens, warping the sheet and reducing the drying rate.

Paperboard is also finished differently from the paper in the fourdrinier process to improve the surface smoothness. The two most common finishing treatments are the *calender water doctor* that applies starch to the surface, and a *yankee dryer machine glazing cylinder* that provides a better surface for subsequent coating.

Some paperboard is shipped to the convertor in sheet form rather than in rolls. This depends on whether the carton printing and conversion process is web fed or sheet fed, as discussed in Chapters 12 and 13.

7.3 MACHINE DIRECTION COMPARISON OF CYLINDER AND FOURDRINIER MULTI-PLY BOARDS

Both the fourdrinier and the cylinder processes produce paperboard with a *grain*, paper with properties that differ in the *MD* and the *CD*. This is because the papermaking machine tends to orient the fibers in the direction of web travel.

All paper is stiffer and has higher tensile strength in the MD. It is easier to tear, and naturally tears in a more nearly straight line, along the MD, following along the fibers.

When paper gets wet, the fibers swell more in width than length, causing the paper to swell more in the CD; therefore, it tends initially to curl into a cylinder with length parallel to the MD. These properties are the basis for tests to identify the machine direction, described in Section 8.4.1.

For most uses, a more balanced material is better. Many papermaking machine design improvements have focused on decreasing the directional difference by orienting a higher percentage of fibers in the CD. However, the directional properties can add value; for example, the MD stiffness of paperboard for folding cartons is oriented horizontally to prevent bulging caused by flowable contents.

There is a marked difference in the amount of fiber orientation in the fourdrinier and cylinder processes. The directionality is more pronounced in cylinder board because of the way the cylinders drag into the pulp slurry to collect fibers. New foudrinier processes are more successful at orienting fibers in the CD by creating turbulence across the table and modifications to the headbox and forming bed.

Directionality is usually measured by the ratio of tensile strength values, MD:CD. A ratio of 4:1 to 5:1 is typical for cylinder board, compared to less than 2:1 for fourdrinier paper. The cylinder process results in more fibers oriented parallel to the machine direction and to the plane direction of the sheet, which adds thickness and therefore stiffens the board.

7.4 PAPER COATING AND SURFACE TREATMENT

There are two categories of surface coatings: pigmented and functional. The pigmented coatings are used to provide a glossy, white, smooth surface for printing, and are usually applied in line with the papermaking machine.

The functional coatings provide a barrier of lacquer, varnish, or plastic, and are usually applied in a separate process from papermaking. Lacquer and varnish coatings are applied after printing to improve ink gloss, contrast, and image detail. Plastic extrusion coating gives gloss, and also can be used to add a water barrier and/or heat sealability, as it does for milk cartons and drink boxes. A plastic extrusion coating can be used as an adhesive for laminating layers of foil or other materials to the paper.

It should be noted that sizing could be considered a functional coating, but in papermaking it is classified separately, referred to simply as sizing rather than as a coating process.

Consistent coating coverage has always been a technical challenge. The first continuous coatings were applied with a brush. Now coatings are applied by means of rolls, fountains, and flooded nips. Counter-rotating rolls meter the thickness of the coating. It is then smoothed out with a blade or air knife, although there have been new developments in spray coating that may someday become widely used.

Paper and board can either be coated on one side or both. There may be two separate coating stations with drying in between. Clay-coating of paper is generally done on the papermaking machine.

7.4.1 Clay and Other Pigmented Coatings

Pigmented coatings are similar to fillers, and clay is used for both. A white clay coating fills in the void areas on the surface of the sheet. After it is dried and calendered, the coating provides a smooth, bright white printing surface.

The smoothness gives good printing resolution and reduces ink consumption. The brightness improves graphics by providing good color contrast. Coatings can also improve glueability by providing a smooth well-bonded surface

Clay-coating on one side is specified as C1S, and on both sides is designated as C2S.

It is not unusual for a coating to contain more than 10 ingredients. The major portion (70–85%) is the white mineral pigment. It is mixed with binders to hold it onto the paper. Additives (< 1%) are used to give specific properties to the coating or sheet. A typical coating contains mostly clay, some calcium carbonate, and a starch binder.

The most common white pigmented coating is *kaolin* or *China clay*. It is a natural, earthy, fine-grained material, composed primarily of aluminum silicate. It is plastic when wet, and hard when fired. It is classified by particle size, brightness, and particle shape.

Two other common pigments are calcium carbonate and titanium dioxide. *Calcium carbonate* ($CaCO_3$) is made from ground or precipitated chalk, limestone, or marble. Its use as a pigmented coating has been growing because it is compatible with alkaline-pulped papers, which are also growing in use, and it is low cost. *Titanium dioxide* (TiO_2) is high cost, and only used for specialty papers. *Plastic pigments* are used in combination with other pigments to provide high gloss.

Binders glue the particles to the paper and to each other. The dried coating is not a continuous film, but is rather a porous structure of pigment particles cemented together with binders. Binders improve the mechanical properties of coatings. They increase pick strength (the ability of the paper to resist lifting of the coating from its surface), flexibility, and the coating's resistance to dusting, fold cracking, abrasion, and water.

In the past, the most common binders have been natural water-soluble adhesives like starches and proteins. These natural binders, although inexpensive, are subject to spoilage, and are being replaced by synthetics like latex, acrylic, and PVA.

The coating can form a significant proportion of the thickness of a sheet. It is specified in terms of the number of pounds it contributes to the basis weight.

7.4.2 Plastic Coating and Laminating

Some of the most important paper and paperboard developments today are in the area of plastic extrusion coating and laminating. Paper and plastic both benefit from the combination. The plastic adds desirable properties, particularly for food applications, like moisture barrier, heat sealability, and a clean food-contact surface. The paper or paperboard adds low-cost stiffness, strength, and printability.

Most kinds of plastic can be *extrusion coated* onto paper. Extrusion coating is similar to all plastic film extrusion: thermoplastic pellets are melted in an extruder and a molten film is extruded through a pressurized thin slot die. The hot thin film is extruded onto the paper as it unwinds from a roll. The web then passes over chill rolls to firmly secure the paper to the plastic.

Plastic can also be used as an adhesive for combining another material with paper, such as foil, in a process called *extrusion lamination*. One substrate is extrusion-coated, and married to the second substrate while the plastic is still hot. The most common material laminated to paper is aluminum foil, which adds barrier properties to pouches for dry food products (like drink mixes) and candy. Lamination can also be used for decorative appearance; for example reverse-printed plastic film has been laminated to beverage carriers to give a glossy, water-resistant surface.

Low density polyethylene (*LDPE*) is the most common plastic extrusion coating. Its benefits include low cost, moisture barrier, and heat sealability. It is approved for food contact. LDPE is the coating that gives drink boxes and milk cartons their ability to hold liquid. It is used to make paper bags and pouches heat-sealable. LDPE and other plastic coatings for paperboard are discussed in more detail in Chapter 9.

There is no doubt that the use of plastic coatings will continue to increase. Plastic and paper have great complementary properties. Improvements in plastics' properties and in extrusion coating and laminating technology are increasing the number of applications for which paper-based packaging is appropriate.

7.5 REFERENCES

Ainsworth, J.H. 1959. *Paper, the Fifth Wonder.* Kaukauna, WI: Thomas Printing & Publishing.

Bierman, C.J. 1996. *Handbook of Pulping and Papermaking,* 2nd ed. San Diego: Academic Press.

Kline, J.E. 1991. *Paper and Paperboard; Manufacturing and Converting Fundamentals,* 2nd ed. San Francisco: Miller Freeman.

Kocurek, M.J. and C.B.F. Stevens. 1983. *Pulp and Paper Manufacture, Volume 1, Properties of Fibrous Raw Materials and their Preparation for Pulping,* Atlanta:TAPPI.

MeadPaper. 1983. *The Manufacture of Printing Paper.* The Mead Corp.

Peel, J.D. 1999. *Paper Science and Paper Manufacture.* Vancouver, Canada: Angus Wilde Publications Inc.

Savolainen, A. 1998. *Paper and Paperboard Converting.* Helsinki: Fapet Oy. Published in cooperation with TAPPI.

Smook, G.. 2002. *Handbook for Pulp and Paper Technologists.* Vancouver: Angus Wilde.

TAPPI. 1997. *How Paper is Made.* Atlanta: TAPPI, CD Rom.

Walker, J.C.F. 1993. *Primary Wood Processing: Principles and Practice,* London:Chapman & Hall.

7.6 REVIEW QUESTIONS

1. Define how the following terms apply to handmade paper-making: wire, deckle, and couch.

2. Define the following terms and explain their significance on a fourdrinier machine: headbox, wire, couch roll, dandy roll, wet end presses and carrier felt, dry end presses, dryer section, and calender stack.

3. What is meant by the "machine direction?" Which direction is stiffer? Which has higher tensile strength? Which has higher tear resistance? Why?

4. What is the difference between the felt side and the wire side of paper? How does it affect the properties of each side? Which side is better for printing?

5. What is a twin-wire (gap) former, and what are its advantages over the fourdrinier headbox and wire?

6. How has the fourdrinier machine been adapted to make paperboard?

7. Which two operations on a fourdrinier machine most affect the paper thickness?

8. How does calendering affect paper properties?

9. What is the purpose of external sizing? How does it differ from internal sizing discussed in Chapter 6?

10. Which papermaking process is used most often for the multi-ply recycled paperboard used in cereal cartons?

11. Which papermaking process is used most for newspapers? For kraft paper?

12. Which papermaking process orients the fibers more in the machine direction: fourdrinier or cylinder?

13. How do the outer and inner layers of multiply paperboard contribute to its stiffness?

14. What are the purposes of using clay and starch in the pulping and paper-making processes?

15. What is the purpose of clay coating on paper and paperboard?

16. What is meant by C1S and C2S, and to which material does it generally refer?

17. What is the difference between extrusion coating and lamination?

18. What material is most commonly laminated to paper and paperboard? What benefits does it add?

19. What plastic is most commonly used for coating paper and paperboard? What benefits does it add?

CHAPTER 8

Paper, Paperboard, and Corrugated Board Properties and Tests

Paper and paperboard are valued because of their printability, stiffness and strength. But such properties vary widely depending on the source of fiber and the process used for pulping and papermaking. The buyers and suppliers of paper-based materials use the results of standard tests as a common language for specifying properties and ensuring consistency.

Some of the properties of paper and paperboard that can be measured are weight, thickness, brightness, whiteness, gloss, porosity, tensile strength, tear resistance, coefficient of friction, and stiffness. Corrugated board tests are used to evaluate burst and compression strength. These properties vary depending on the moisture content of the paper. This chapter describes standard tests for measuring these properties.

The use of *standard test methods* ensures that everyone is testing in the same way. They aim for *precision*, with the hope that everyone testing the same thing will get the same results. They aim for *accuracy*, the measurement of true values, within the limits of the measurement methods that are clearly specified.

The use of standard tests ensures that the results will be interpreted in the same way by suppliers and buyers in their *packaging material specifications*. Specifications are an agreement between the buyer and seller. Material test requirements are used in buyers' specifications for specific critical properties and for consistency.

For example, when breakfast cereal manufacturers want to specify paperboard cartons that are stiff, with easy-to-bend scores to run on a cartoner, stiff enough to resist bulging, and white enough for good graphics, the properties of the paperboard are specified according to standard measurements of thickness, stiffness, score bend, and brightness.

Test standards are developed and updated periodically by committees of technical experts who routinely use the test. They work together to improve

225

the standard, including making it very clear how to perform the test in order to get reproducible results.

In the United States, the most important source for paper-based material test methods is *TAPPI, the Technical Association for the Pulp and Paper Industries* (*www.tappi.org*). TAPPI is an international membership organization for the paper industry that promotes research, education, and standardization of test methods. Its members make pulp, papers of all kinds, paperboard, corrugated board, and paper-based packages; suppliers to the paper industry and large buyers of paper-based packaging materials are also represented.

TAPPI committees update and revise the standards periodically. The primary TAPPI tests for paper and board are discussed in this chapter, but it should be noted that there are hundreds of TAPPI tests that are not discussed in this textbook, and for nearly every property of pulp, paper, and paper products that might affect their performance, a relevant test method can be found in the *TAPPI Test Methods* publication. These test standards are available through most university libraries.

There are also some packaging test standards that are maintained by *ASTM International* (formerly the American Society for Testing and Materials, referred to simply as ASTM throughout this book). ASTM is one of the largest standards organizations in the world and its technical committees develop standards for most sectors of industry, not just packaging. In packaging, ASTM focuses more on plastic material tests and package performance tests. ASTM has withdrawn its former standards that were specific to paper, in deference to TAPPI.

Other countries have their own standards organizations, like the British Standards Institution (BSI), the Technical Association of the Australian and New Zealand Pulp and Paper Industry (APPITA), and the Scandinavian Pulp, Paper, and Board Testing Committee (SCAN).

The national standards are harmonized by the *International Standards Organization (ISO)*. Most of ISO's paper standards derive from those developed in paper-producing countries like North America and Scandinavia, and so they strongly reflect standards maintained by organizations like TAPPI and SCAN. Many of the standards refer to equivalent standards in other countries, and all TAPPI standards refer to the applicable ISO standard.

There is a difference between material tests and package performance tests. As the name implies, *material tests* address the material in isolation, evaluating properties like thickness or tensile strength. They are used by suppliers for controlling their papermaking processes. Material tests are the primary focus of this chapter.

On the other hand, *package performance tests* are for the converted and filled package. These are tests like compression strength or drop height survival. When, in Chapter 15, performance tests for shipping containers are described, two other standards organizations will be introduced: ISTA (the International Safe Transit Association) and DOT (the U.S. Department of Transportation). Their tests judge the performance of a filled package in distribution.

The published standard documents are extremely detailed about testing equipment, fixtures, and procedures. Whenever there is a need to perform such tests, especially if there are profits or loss at stake, the standard, as published, should be followed exactly. There are five parts that are common to most test standards:

1. A legal disclaimer gives the validity and the scope of the standard.
2. Specific machinery to perform the test and calibration procedures are described.
3. The selection, number, and conditioning of test samples is specified.
4. The units of measurement are given.
5. There is an assessment of the accuracy, precision, and errors involved.

Some tests are inherently more precise—meaning that their results are more *repeatable* and *reproducible*—than others. The standards give values for each test method's repeatability, defined as an estimate of the maximum difference in results that can be expected to be obtained 95% of the time when measurements are done within a single laboratory on the same material, using the same equipment, by the same person, under the same environmental conditions, and within a short time frame.[14]

The standards cite the method's *reproducibility* as an estimate of the difference in results that could be expected 95% of the time when comparing tests performed by two laboratories.[15] And they show the number of test reports that were compared to determine it. These values assume that the apparatus used is properly calibrated, the environmental conditions are appropriately controlled, the personnel are correctly trained, etc. The test descriptions in this book reflect the TAPPI estimate of precision.

Some tests are more accurate—meaning that they more closely measure the actual value—than others. Test instruments and sample preparation impose their own errors. The standards therefore list and discuss the sources of bias and errors.

---

[14]For example, when TAPPI T411 notes the thickness measurement test as having a *repeatability* of 1.25%, this means that when one person measures the average thickness of a lot of paper at two different times, we can be 95% confident that the second result will not differ by more than 1.25% from the first. It also shows that the data source was 24 reports.

[15]Returning to the example, TAPPI's *reproducibility* value of 5.5% for paper thickness testing indicates that if another person in a second laboratory tests the same paper, using different (but appropriately calibrated) equipment, we can be 95% confident that the results will not be more than 5.50% different from the value in the first laboratory.

They also specify how the results should be reported in terms of the number of *significant figures*, giving a basis for rounding numbers. "Significant figures" are the number of digits that are known with some degree of certainty, that carry meaning. Using the correct number of significant figures avoids the appearance of greater precision than the test device can measure.[16]

This chapter summarizes the most common paper and paperboard tests. It describes why each test is used and how it is performed. Although the equivalent ISO number is given, the description is based on the TAPPI version of the standard. Metric measures, which are favored in the TAPPI standards, are given as well as the equivalent U.S. customary units that still predominate in the U.S. paper industry, and conversion factors are provided.

8.1 BASIS WEIGHT (GRAMMAGE)

TAPPI T 410: Grammage of paper and paperboard (weight per unit area)

ISO 536: Paper and board—determination of grammage

The most common way to specify paper (and some paperboards) is by *basis weight* in the United States, or by *grammage* elsewhere. It is the basis for pricing paper and paperboard, since prices are expressed in dollars per ton, and the area (and number of packages) per ton depends on the basis weight.

The strength and stiffness of paper-based materials, to a certain extent, are related to basis weight. However, a material with a higher basis weight is not necessarily stronger or stiffer, since these properties are also related to thickness, composition, and the properties of the fibers.

The U.S. basis weight designation is tricky, and a matter of historical tradition. It is defined differently for paper than for paperboard. For paper, like that used for bags, basis weight refers to the weight in pounds of a standard ream of 500 sheets of paper, measuring $24'' \times 36''$, for a total of 3,000 ft^2. A typical basis weight for paper is 16–90 lb. BW.

On the other hand, for paperboard and the containerboard used in corrugated board, the basis weight is the weight in pounds per 1,000 ft^2 (also known as *MSF*) of paperboard. This is based on a standard ream of 500 sheets measuring $12'' \times 24''$.[17] A typical basis weight for paperboard is 26–90 lb BW.

---

[16]For example, if a test device that measures in increments of 0.001 yields an average of several readings of 0.1234567, the standard would specify that the results be reported to three significant figures, and so the value reported would be rounded to 0.123. Likewise, readings that average 123.45678 would be rounded to 123.457. Significant figures do not include leading zeros, or trailing zeros that are used only to set the location of the decimal point. Zeros that follow a decimal point are always significant.

[17]For printing and writing papers, the basis weight still refers to the weight in pounds of a standard ream of 500 sheets, but the sheet size, and hence the total area, differs.

Paperboard is generally defined as thick paper, and thicknesses above 10 points are generally considered to be paperboard for the purpose of basis weight calculation. In practice, however, the range between 9 and 12 points is defined by convention and based on the use.

For example, all containerboard used to make corrugated board is specified by a basis weight based on 1,000 ft^2. This includes the corrugated medium, which is generally only 9 points, as well as the thinnest microflute components that are even thinner. On the other hand, paper used in bags is specified by a basis weight based on 3,000 ft^2, even up to 12 points.

Therefore, in the United States, when specifying basis weight in this 9–12 point range, it is necessary to know whether the material is considered to be paper or board. It is easy to see that these designations can be confusing. A piece of containerboard with a given basis weight is three times as heavy as a piece of bag paper with the same specified BW. If the basis weight is to be used in a price estimate, it would be a costly mistake to assume that the basis is 3,000 ft^2 if it is actually only 1,000 ft^2.

To add to the confusion, the paperboard used for folding cartons is sometimes not referred to by basis weight at all, since the industry more typically refers to paperboard by thickness (points). And for non-bending chipboard (used to make set-up boxes, discussed in Chapter 13), "basis weight" does not refer to weight at all, but to the number of sheets in a 50-pound bundle.

In other parts of the world, paper and paperboard are classified by *grammage*, which is much easier to calculate. It refers to the weight of paper or paperboard in grams per square meter, expressed as *gsm*. The thinnest tissue paper has a grammage of 12–30 g/m^2. Paper grammage ranges from 30–100 g/m^2, paperboard is over 224 g/m^2 (up to 800 g/m^2 or more), and in the intermediate range of 100–225 g/m^2 the products may be referred to as either paper or paperboard, depending on their use. But the grammage measure eliminates the possibility for misinterpretation in specifications or cost estimates.

For purposes of conversion between metric and English units, it can be remembered that 1 pound equals 454 grams, and 1 square foot equals 0.0929 square meters (derived from the fact that 1 $ft^2 = 12^2$ $in^2 = 144$ in^2; 1 $in^2 = 2.54^2$ $cm^2 = 6.45$ cm^2; and 1 $m^2 = 100^2$ cm^2). The weight basis for paperboard (1,000 ft^2) is expressed as *msf* (thousand square feet). Table 8.1 gives conversion factors from TAPPI T 410. Note that these calculations are based on the more accurate conversion factor of 1 lb = 453.5924 g.

Strictly speaking, pounds is a measure of force, not of mass. However, in the United States we routinely use pounds as a mass measurement. Consequently, we convert back and forth between pounds and grams or kilograms, without worrying about specifying, as an engineer might, that we mean lb_m (pounds mass) rather than lb_f (pounds force).

Basis weight is measured by cutting samples of given dimensions, averag-

TABLE 8.1. Basis Weight/Grammage Conversion Factors.

Kind of Paper	BW ft^2 area	BW to g/m^2	g/m^2 to BW
Paper (bag paper, labels, wrapping paper, tissue)	3,000 ft^2	1.627	0.614
Paperboard (folding cartons)	1,000 ft^2	4.882	0.205
Corrugated containerboard (liners and medium separately)	1,000 ft^2	4.882	0.205

ing the weights and converting them to 3,000 ft^2 or 1,000 ft^2 depending on whether the material is paper or paperboard. Although basis weight can be measured on any sufficiently accurate scale with samples at least 80 in^2 (8″ × 10″) there are special basis weight scales used in the paper industry that measure BW directly when a 1 ft^2 sample is used. Although the special balance is precise, meaning that it consistently gives the same results, an analytical balance gives more accurate results. For grammage, samples of at least 500 cm^2 (20 cm × 25 cm), are weighed, and the average is converted to m^2.

For grammage or basis weight, sufficient samples to equal at least 5,000 cm^2 (800 in^2) are required, and grammage results should be reported to three significant figures. The test is performed at the standard TAPPI conditions, 23°C ± 1°C (73.4°F ± 2°F) and 50% ± 2.0% RH, as discussed in Section 8.3.2.

Density is defined as the ratio of the mass (M) to the volume (V). Since paper and paperboard are hygroscopic, their density will vary with their moisture content. This is why paper density is always specified at a given moisture content. Density affects almost all properties of paper, and is affected by how much void space is in the paper structure. The apparent density (D) of a paper or paperboard is the weight divided by the volume (V), or the grammage (G) divided by the thickness (t).

$$D = \frac{M}{V} = \frac{G}{t}$$

Different samples of paper or paperboard can have identical basis weight and different thicknesses—or identical thickness and different basis weights—because they differ in density. Typical values of apparent density range from 0.5 g/cm^3 in bulky papers to over 1 g/cm^3 for heavily bonded sheets like glassine. All papers are less dense than cellulose alone (1.5 g/cm^3), which shows that all paper contains some void space (Scott *et al.* 1995).

Following are two examples that show how basis weight, grammage, and density are calculated and how they can be used to determine the cost of paper or paperboard. Table 8.2 gives some typical basis weight and grammage comparisons.

Example 1:

A 12″ × 12″ piece of 5.6 point kraft paper for bags weighs 8.8 grams. What is the basis weight? What is the grammage? What is the apparent density in g/cm^3?

$$BW = \frac{8.8 \text{ g}}{1 \text{ ft}^2} \times \frac{1 \text{ lb}}{454 \text{ g}} \times 3,000 \text{ ft}^2 = 58.15 \text{ lb}$$

$$\text{grammage} = \frac{8.8 \text{ g}}{1 \text{ ft}^2} \times \frac{1 \text{ ft}^2}{0.0929 \text{ m}^2} = 94.7 \text{ g/m}^2$$

$$\text{apparent density} = \frac{94.7 \text{ g/m}^2}{0.0056 \text{ in}} \times \frac{1 \text{ in}}{2.54 \text{ cm}} \times \frac{1 \text{ m}^2}{100^2 \text{ cm}^2} = 0.67 \ \frac{\text{g}}{\text{cm}^3}$$

Or, using Table 8.2

$$58.15 \text{ lb } BW \times 1.627 = 94.6 \text{ g/m}^2$$

$$94.6 \text{ g/m}^2 \times 0.614 = 58.08 \text{ lb } BW$$

The answers are slightly different due to rounding of numbers.

Example 2:

50 lb BW SBS paperboard sheets are 60″ × 72″. The cost of the paperboard is \$1,120/ton. (1) What is the weight of 10,000 sheets? (2) What is the cost of 10,000 sheets?

(1) Weight

$$\text{area} = 10,000 \text{ sheets} \times \frac{(5 \text{ ft})(6 \text{ ft})}{\text{sheet}} = 300,000 \text{ ft}^2$$

$$\text{weight} = 300,000 \text{ ft}^2 \times \frac{50 \text{ lb}}{1,000 \text{ ft}^2} = 15,000 \text{ lb}$$

(2) Cost

$$\text{cost} = 15,000 \text{ lb} \times \frac{\$1,120}{2,000 \text{ lb}} = \$8,400$$

TABLE 8.2. Typical Basis Weights and Grammage.

Material Type	Basis Weight	Grammage
Kraft multiwall bag paper	50 lb/3,000 ft^2	81.4 g/m^2
Butcher wrapping paper	40 lb/3,000 ft^2	65.1 g/m^2
Cereal carton clay-coated recycled board	55–140 lb/1,000 ft^2	268–684 g/m^2
SBS cartons (eg. butter, frozen foods)	45–85 lb/1,000 ft^2	220–415 g/m^2
Kraft linerboard (corrugated containerboard)	42 lb/1,000 ft^2	205 g/m^2
Corrugated medium	26 lb/1000 ft^2	127 g/m^2

Weighing is very accurate. The repeatability of the test is 0.94% and the reproducibility is 2.84%. The test is subject to errors only if the moisture content is too high, if the scale is out of calibration, or the calculation is incorrect. Since paper readily absorbs moisture, it is necessary to test at standard conditions, or to specify the temperature and relative humidity at which the grammage or basis weight measurements were made.

To reduce the influence of moisture on the expression of the density, *specific gravity (SG)* is sometimes used. SG is the ratio of the density of a material to the density of water at standard conditions (1 atm, 4°C). The density of water at these conditions is exactly 1 g/cm^3.

$$SG = \frac{\text{density of material}}{\text{density of water}}$$

For paper, it is common to talk about the SG as based on the oven-dry density. It should be noted that SG has no units. SG can be converted to density by multiplying by the standard density of water. Since this, as stated above, is 1 g/cm^3, simply adding units of g/cm^3 converts SG to density.

The SG of the cell wall of wood fibers, estimated as 1.5 (Haygreen *et al.* 2003) can be used to calculate the percentage of voids in paper or paperboard if the SG is known, using the equation below:

$$\text{Void, \%} = 100 \times \frac{1.5 - SG_{\text{paper}}}{1.5}$$

where 1.5 is the *SG* of the cell wall and SG_{paper} of the paper.

The percentage of voids in a paper with a *SG* of 0.5 is equal to 66.6%. The volume of voids is almost double that of the solid cell wall fibers due to lumens and other cavities in the fiber structure, resulting in a lot of volume in which water, inks, and other substances can be absorbed.

8.2 CALIPER (THICKNESS)

TAPPI T 411: Thickness (caliper) of paper, paperboard, and combined board

ISO 534: Paper and board determination of thickness, density and specific volume

The *caliper* of a material refers to its thickness. Caliper and basis weight are related, since the higher the basis weight, the greater the thickness tends to be. However, the thickness is also affected by the density and degree of compression of the sheet, so there is not a one-to-one correspondence between basis weight and caliper.

In the United States, caliper is usually expressed in thousandths of an inch, or *points* (1 point = 0.001″). So a typical sheet of paperboard with a thickness of 0.020″ would be 20 points. (It should be noted that this terminology is unique to the U.S. paperboard industry. The thickness of plastic, metal, and glass in the United States is also expressed in thousands of an inch, but in those industries are called *mils*: 0.001″ = 1 mil. And in the printing industry, "points" refer to the size of a typeface.)

Outside the United States, caliper is expressed in *microns*, μm, also called micrometers, which is one thousandth of a millimeter. For purposes of conversion between metric and English units, it should be remembered that one point equals 25.4 μm (because there are 1,000 points/in, 25.4 mm/in, and 1,000 μm/mm). But the thickness increments in μm do not correspond exactly to the U.S. point increments, since they are usually expressed in numbers that are rounded to the nearest 5 or 10; for example, some common grades are 340 μm, 380 μm, and 480 μm. Likewise, when converting from microns to points, the value is rounded to the nearest point.

Since most of the examples in this book are based on the point system, the equivalent micron values shown are rounded to the nearest 10, as in the following examples:

Example:

What is the thickness of 20-point paperboard in SI units?

20 point × (25.4 μm/1 point) = 0.508 mm = 508 μm

Practically, this would be specified as 500-μm board

How thick is 380-μm board in points?

380 μm × (1 point/25.4 μm) = 14.96 points

Practically, this would be specified as 15-point board

Thickness is a more pertinent quality for paperboard than for paper. In the United States, paperboard (generally over 10 points, 250 μm) is more commonly referred to by thickness than it is by basis weight, whereas paper is more commonly referred to by weight. In other countries, paperboard is usually specified by grammage (grams/meter^2). The thickness designation also includes the paperboard's clay coating, which makes this measure consistent with the practice of also specifying clay-coated paper by points.

An easy way to remember this difference is that thickness is paperboard's greatest virtue, and therefore thickness is how board is specified. The most common thickness of board used for folding cartons is between 14 and 28 points (360–710 μm).

Thickness affects strength and machinability. It adds bending stiffness to paperboard, which affects how it runs on a cartoning machine. It helps carton walls to resist bulging.

The standard method for determining caliper uses an automatically-operated micrometer with a small circular metal pressure foot that presses the paper against a base plate. The foot is 200 mm^2 (0.31 in^2) in area, the pressure foot lowering speed is 1 mm (0.04 in)/sec and the squeeze pressure is 50 kPa (7.3 psi). The ISO 534 standard specifies 100 kPa, which may give a slightly different result. The standard does not allow the use of a manually operated micrometer, because results can be biased by the skill of the operator.

The sample is placed beneath the foot while it is up, and each time the foot comes down on it, the thickness of the paper is measured and the value is displayed. A series of measurements are made along the cross machine direction of the paper, since there is more variation in thickness across the papermaking machine than there is in the machine direction.

The results are reported as an overall average of 20 or 50 readings, in inches to the nearest 0.04 points (or to the nearest 0.001 μm). The highest and lowest measurements are often also reported.

The test is subject to errors because paper is compressible under relatively low loads, and because surfaces of paper and paperboard are not uniform; measurements of thickness made under different conditions can vary considerably. Other sources of bias are clamp pressure and clamp time. The repeatability of the caliper test is 1.25% and the reproducibility of results between laboratories is 5.5%.

8.3 MOISTURE: MEASUREMENT AND EFFECT

Moisture—too much or too little—can dramatically affect the performance of paper-based packaging materials. Labels can curl when too wet, and paper bags become brittle when they are too dry. Corrugated fiberboard boxes can collapse when the relative humidity rises, and the ability of paperboard cartons to run through a cartoner without jamming depends on their moisture content.

Paper fibers "remember" their role of liquid-filled cells in a tree, and that they were reborn in water during the papermaking process. Paper fibers are hygroscopic, and take on water (sorb) from the atmosphere or lose it (desorb) when the two are not in balance.

This section describes how to measure relative humidity in the air, standard conditions for paper testing, how to measure a paper's moisture content, and its effect on paper properties.

8.3.1 Environmental Relative Humidity Measurement

The relative humidity (RH) is *relative* to the amount of moisture the air can hold at a particular temperature, the saturation vapor pressure. It is the ratio (expressed as a percent) of the actual moisture in the air compared to the moisture that air could hold if it were saturated at the same temperature. For example, an RH of 0% means that the air contains no moisture, and an RH of 100% means that the air is fully saturated.

RH depends on temperature. As temperature goes up, so does the amount of moisture that the air can hold. For example, air with a RH of 90% at 85°F (32°C) has about 3-1/2 times as much moisture as it does at the same RH at a temperature of 50°F (10°C). Similarly, at TAPPI conditions of 50% RH and 73.4°F (23°C), the air contains about the same concentration of moisture as air at 20% RH and 100°F (38°C).

Relative humidity can be measured by using a thermometer and a *hygrometer*. Typically, these are combined in a *psychrometer*, which consists of two thermometers arranged side by side with their bulbs in front of a small fan. One thermometer measures the *dry bulb* temperature of the air. The second thermometer measures what is called a *wet bulb temperature*, because it is covered with a water-soaked material. The wet bulb reads a lower temperature because heat required for evaporating the water on the fabric is drawn out. The drier the air, the more evaporation there is, resulting in a lower wet bulb temperature and a greater difference from the dry bulb reading. The only circumstance that would make the two bulbs show equal temperatures would be if the air was completely saturated (RH 100%), so there is no evaporation.

The difference between the wet and dry bulb readings can be used to calculate the RH. To simplify the relationship, a *hygrometric* table is used. It plots the dry bulb temperature against the difference between the wet and dry temperatures, to give the RH of the air.

8.3.2 Standard TAPPI Conditions for Tests

TAPPI T 402: Standard conditioning and testing atmospheres for paper, board, pulp handsheets, and related products

ASTM D685: Conditioning paper and paper products for testing

ISO 187: Paper, board, and pulps–standard atmosphere for conditioning and testing and procedure for monitoring the atmosphere and conditioning of samples

Since the properties of paper vary so much with RH, standard test conditions have been established. These are known as TAPPI conditions. The standard *TAPPI conditions* are 23°C ± 1°C (73.4°F ± 2°F) and 50% ± 2.0% RH.

Conditioning is a two-step process. Samples should first be preconditioned at 10–35% RH and 22–40°C to correct for the hysteresis effect, as discussed in Section 8.3.3.1. TAPPI T 402 recommends 1 hour for single sheets of paper and 16 hours for sealed boxes and shipping containers; ASTM D685 recommends a minimum of 24 hours.

Next, the samples need to be in the standard TAPPI conditions for long enough so that they can come to equilibrium with the atmosphere, also discussed in Section 8.3.3.1. TAPPI T 402 and ASTM D685 recommend a minimum of 4 hours for paper, 8 hours for paperboard and unsealed boxes, and 16 hours for sealed boxes and shipping containers.

8.3.2.1 Special Atmospheres for Testing

ASTM D4332: Conditioning containers, packages, or packaging components for testing.

ISO ISO 2233: Packaging—Complete, filled transport packages: conditioning for testing

In addition to the TAPPI standard conditions for paper, ASTM and ISO have developed a number of special atmospheres. These are used to reflect more extreme distribution or use conditions. In equatorial countries, the tropical condition is usually used as the standard.

ASTM's special atmospheres include the following:

- Frozen food storage: −18°C (0°F)
- Refrigerated storage: 5°C (4°F), 85% RH
- Temperate high humidity: 20°C (68°F), 90% RH
- Tropical: 40°C (104°F), 85% RH
- Desert: 60°C (140°F), 15% RH

The ISO standard incorporates these, and adds six more intermediate conditions.

8.3.3 Moisture Content Measurement

TAPPI T 412: Moisture in pulp, paper, and paperboard

ASTM D644: Moisture content of paper and paperboard by oven drying

The moisture content of a piece of paper or board is easily measured. A sample is weighed and then dried in an oven at 105°C for 1 hour and allowed to cool in a sealed container. The sample is then reweighed.

Moisture content is expressed differently for paper than it is for wood, as discussed in Chapter 3. Like wood, paper is weighed, dried to a constant weight in an oven, and weighed again. But the paper industry generally expresses the moisture content on the *wet basis* rather than the dry basis, because this is more convenient when they calculate the required pulping chemicals. It is calculated as follows:

$$MC_w = 100\frac{W_m - W_d}{W_m}$$

where MC_w is the percent moisture content, W_m is the weight when moist, and W_d is the weight when dry.

Example:

If a sample of paper weighing 2 grams is dried, and the dry weight is 1.9 grams, what was its original moisture content?

$$MC_w = 100\frac{2\text{ g}-1.9\text{ g}}{2\text{ g}} = 5\%$$

The weighing is usually done in a room with TAPPI conditions, so the time between removing the sample from the sealed container and weighing must be very short, because the paper will begin gaining moisture as soon as it is removed from the container. The test is also subject to errors if the sample is exposed to humid air as it cools, before weighing. The precision is good: repeatability is 0.075% moisture (not % of percent moisture), and reproducibility is 0.93%.

8.3.3.1 Equilibrium Moisture Content

The equilibrium moisture content is the moisture content that a piece of paper will have when it reaches equilibrium in an atmosphere with a given relative humidity and temperature.

Moisture sorption isotherms are developed by measuring the equilibrium moisture content at a constant temperature and varying relative humidity[18].

---

[18]Moisture sorption isotherms can be prepared with the y axis giving moisture content on either a wet or dry basis.

As shown in Figure 8.1, sorption isotherms for paper have three stages. As the humidity rises from 0% to 20%, there is a steep rise in moisture content. From 20% to 65%, the moisture content rise is less steep and steady. Above 70% another sharp rise in moisture content occurs.

Figure 8.1 also shows that the moisture content obtained from drying the paper (desorption) is higher than that obtained from exposing the paper to increasing RH. Studies show that the difference can be as much as 2%, depending on the previous moisture history (Uesaka 2002).

This is why the TAPPI standard for conditioning paper requires for it to be somewhat dried before it is conditioned to 50% RH. This phenomenon, common to many materials and discussed in Chapter 3 as a property of wood, is called *hysteresis*. Repeated drying and wetting retards the paper's ability to absorb moisture.

Hysteresis is attributed to the fact that during the first drying, crystallization occurs in the fiber walls, and subsequent wetting does not succeed in reopening all of the places where the fiber previously held water. By approaching preconditioning from the dry side, test results are more comparable.

Pulps differ in their equilibrium moisture contents. The most important factor in these differences is believed to be the relative amounts of accessible non-crystalline areas in the fibers. Sorption occurs mostly in non-crystalline regions, since the forces involved are not sufficient to disrupt the crystalline

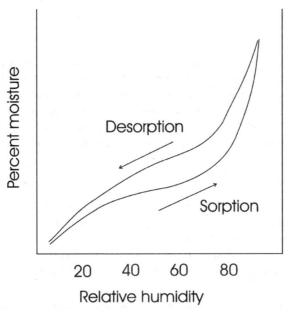

FIGURE 8.1. *Moisture sorption and desorption isotherms for paper.*

regions to any appreciable degree. Thus, the higher the percentage of non-crystalline regions with available hydrogen bonding sites, the greater will be the moisture sorption. Mechanical pulps, therefore, have higher moisture sorption than chemical pulps.

The equilibrium moisture content for paper at the TAPPI standard conditions of 50% RH is about 5%. This differs somewhat for different paper compositions, with the range usually between 4% and 9%, and for paperboards in the range of 6–8%.

8.3.3.2 Effects of Moisture Content on Mechanical and Structural Properties

The TAPPI conditions were chosen to test the properties of paper at its best, in the ideal range of moisture content, which for paper is 3–7% (wet weight basis). If the moisture content in paper falls below 3%, paper becomes brittle. Above 7% moisture, the fibers, and therefore the paper, lose stiffness and the paper becomes soft and mushy. This is the sharply rising region of the moisture isotherm at high RH, shown in Figure 8.1.

Moisture content affects all properties of paper, but it does not affect them equally. This is because changes in paper properties due to changes in moisture are caused by changes in the inter-fiber bonding forces (hydrogen bonds), as well as by internal changes in the fibers. These effects may be complementary or competing.

Figure 8.2 shows the relationship between tensile strength, tear resistance, and stiffness. Wet paper is easier to break; tensile breaking strength decreases with increasing moisture, but stretchiness (extension at break) causes the strength to increase slightly to a maximum at 30–35% RH; higher humidity weakens the fiber-to-fiber bonding. Likewise, stiffness decreases with increasing moisture content because the fibers in wet paper get softer. Paper gains tear resistance and fold endurance at higher moisture content. Tear resistance increases up to 80% RH because the increased stretchiness allows the paper to distribute and absorb the stresses over a larger area before tearing occurs; dry paper is much easier to tear. Folding endurance increases up to 65–75% RH, and then decreases as RH continues to increase, because folding is a complex interplay of fibers and inter-fiber bonds (Uesaka 2002; Scott *et al.* 1995).

Changes in relative humidity also change the dimensions of paper and paperboard. As discussed in Chapter 3, fibers swell when they gain moisture. The amount of dimensional change varies with the type of pulp, extent of refining, amount of internal stress relaxation, built-in tensile stress during paper making, and the fiber orientation.

Since the fibers expand and contract much more in thickness than in length, changes in the CD and thickness are greater than in the MD, and the dimen-

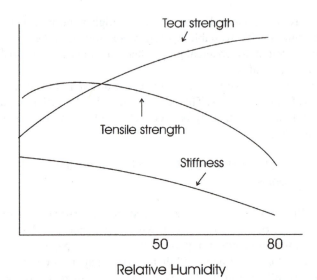

FIGURE 8.2. *The effect of humidity on paper tensile strength, stiffness, and tear resistance.*

sional stability is affected by the degree of fiber orientation in each direction. Paper dried under restraint has a lower ratio of dimensional change to change in moisture content than paper dried freely. The ratio of CD to MD expansion varies from 2 to 10 for most papers, with the average value about 5. Changes in thickness for most grades of paper are typically 1.2–2.1% for each percent change in moisture content when moisture is increasing, and 0.8–1.8% when moisture is decreasing. Also, changes in dimensions are larger in the first sorption/desorption cycle than in subsequent cycles (Uesaka 2002).

Curling and warping is the non-uniform dimensional changes in the three directions of the paper (machine direction, cross direction, and out of plane direction) due to moisture absorption/desorption in the fibers. Curling is exacerbated by differences in the fiber orientation and density on the felt versus the wire side of paper, by coating on one side, and in multi-ply paperboards made using pulp with different properties. Any treatment on one side of the paper or paperboard to improve surface properties may create imbalances in absorption and desorption and therefore differential dimensional changes in the machine, cross, and out of plane direction.

A single-ply paper has a slightly different fiber orientation on the wire side versus the felt side. Coated multi-ply paperboard has even more variation from one side to the other. Labels that have foil or plastic laminated faces always curl away from the stable surface, with the foil or plastic on the inside of the curl, if they gain moisture.

If only one side is wetted, as in the MD test described in Section 8.4.1, the curling effect is even more pronounced. The wet side expands first as its fibers swell, but the paper is restrained by the dry fibers on the other side. This is why labels applied to cylinders are designed with the MD horizontal, to counter the tendency to curl in the CD.

Dimensional instability can also cause problems in multicolor printing. Wet inks and the web tension can cause slight stretching of the paper between color stations, resulting in misaligned and out of register print.

Cyclic changes in humidity, like those experienced from day to night, increase paper's tendency to creep. This factor, significant for corrugated boxes, is discussed in Section 8.4.9.

8.4 PHYSICAL/MECHANICAL PROPERTIES

The physical properties of paper and paperboard vary depending on the source of the fibers, how they are pulped, and the papermaking process.

Paper fibers can be short and smooth (hardwood or recycled paper) or long and strong (softwood). Mechanical pulping shortens them, and chemical pulping keeps the fibers long, changing their shape from straight and tube-like to flexible because the removal of lignin. Pulp with less lignin removed tends to produce stiffer paper. These factors are why mechanically pulped paper has poorer tensile strength and tear resistance than kraft paper.

Strength is also affected by the degree of inter-fiber bonding. Increasing refining (beating) of the fibers tends to increase inter-fiber bonding, but too much refining reduces fiber length and makes the paper easier to tear. Other conclusions that can be drawn are that, at the same fiber length, thin-walled fibers produce greater tensile strength since they conform better at fiber cross-over points, thus producing stronger inter-fiber bonds. At equal fiber length, thicker fibers produce paper with better tear resistance. Fibers with thicker walls form paper sheets with lower density, as they do not collapse as readily. Long fibers with moderate wall thickness provide the most balanced combination of high tensile strength and high tear strength (Peel 1999).

The following sections cover key physical properties.

8.4.1 Machine Direction

TAPPI T 409: Machine direction of paper and paperboard

As discussed in Chapter 7, paper is produced in a continuous web process that results in differing properties along each axis. The forming section of the papermaking machine determines the extent of this difference.

The *machine direction* (*MD*) is the direction in which the paper comes out of the paper-making machine. More fibers are aligned in the MD than in the *cross direction* (*CD*), because the fiber slurry flows in the direction of the forming fabric and the fibers tend to line up in the flow direction. This grain effect is more pronounced for cylinder board, because of the strong directional movement through the vat and around the cylinder than for fourdrinier-produced papers.

The determination of MD is necessary before doing almost any other test, since the direction has so much effect on other properties. Stiffness, tensile strength, and ring crush are higher in the MD. Fold endurance and tearing resistance are greatest across the grain. Brightness and coefficient of friction also differ by direction.

Specification of MD is necessary for most paper-based packaging in order to use the properties to their best advantage. For example, folding cartons for cereal are designed with the stiffer MD horizontal to prevent bulging. Labels for bottles are also designed with the MD horizontal, to counter the label's tendency to curl parallel to the MD with the wet side out when a wet adhesive is used.

There are several standard methods to determine the machine direction of paper and paperboard. They illustrate the basic differences in each direction's stiffness, tensile strength, and reaction to water.

The *axis of curl procedure* moistens one side of the paper or board. Before the water has a chance to soak through, the paper or board will curl around an axis parallel to the MD with the wet side out. This is because the fibers on the wet side swell. The curling must be observed right away, as it will disappear as the moisture penetrates evenly through the paper.

The *drying procedure* likewise depends on the fact that unrestrained paper will dry into a curl with the MD parallel to its axis.

The *bend procedure* compares two thin strips cut from each orientation. The strips are held together at one end and suspended horizontally like a diving board. First one strip is held on top, and then the other one. The one that droops the most is the one cut in the CD. The one with the least deflection is the one cut in the MD, because paper is stiffer in the MD.

The *tensile test procedure* described in Section 8.4.2 can be used because the specimen cut with its length parallel to the MD will have greater tensile strength and less stretch. The burst test procedure is related, since the line of rupture will usually be across the grain.

The *surface fiber procedure* involves viewing the paper with a light at an angle to the paper. A magnifying glass or microscope may be used. The fiber orientation can be observed, usually more easily on the wire side.

The *hand tearing procedure* is based on the fact that paper tends to tear in a more nearly straight line in the MD than in the CD, since the tear can propagate more easily alongside the fibers. In the CD, tearing requires more fiber pull-out. This also gives the CD tear a more ragged, feathery appearance.

Any of these test methods will work for paper or board with a pronounced fiber orientation in the MD, but more balanced materials may have more uncertain results. The standard recommends using more than one procedure when the results are in doubt.

8.4.2 Tensile Strength

TAPPI T 494: Tensile properties of paper and paperboard (using constant rate of elongation apparatus)

ISO 1924: Paper and board: determination of tensile properties

Tensile strength, the ability of paper or board to resist breaking under tension, is one of the most basic strength properties. It is an important factor in package performance. For example, it affects the ability of a bag to support weight and of a peg hole to support a blister card. It affects the ability of paper or board to efficiently run through a printing or converting operation.

The ability of paper or board to resist breaking under tension depends on the fiber length, strength, and bonding. Kraft pulping produces the strongest, longest, purest fibers, giving kraft paper its high tensile strength. Beating/refining tends to improve tensile strength because it increases the inter-fiber bonding. Tensile strength is also related to the internal drying stresses created in the dryer press section of the papermaking machine. Extensible, or freely dried paper, will stretch more before it breaks than will paper that has been pressed under tension.

A tensile testing machine is used to grip the ends of a strip and pull on it at a constant speed until it breaks. The standard sample for paper tests is 25 ± 1 mm (1 in) wide and long enough to span an initial jaw separation of 180 ± 5 mm (7 in). The paper strip must be aligned and the grips must not slip; a test result is rejected if there is slippage or the paper breaks in or adjacent to a grip.

The constant standard crosshead speed is 25 ± 5 mm (1 in)/minute. As the sample is pulled, it first stretches and then breaks. The speed of the test can contribute to bias because in a slowly applied test, paper stretches more before breaking than it does in a test carried out at a higher speed. Typically, 10 samples in each direction are tested.

The tensile testing machine can be attached to a computer or plotter that shows the profile of the material's stress-strain curve as shown in Figure 8.3, from which the following can be derived:

- *Modulus of elasticity (MOE)*, the ratio between stress and strain as the paper stretches during the initial straight line slope of the curve.

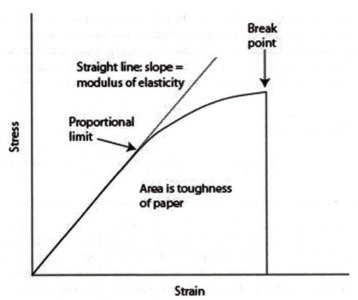

FIGURE 8.3. *Stress-strain curve.*

- *Proportional limit*, beyond which elongation is not directly proportional to stress.[19]
- *Break point*, the peak force and elongation.
- *Tensile energy absorption* (*TEA*), toughness, the area under the curve.

The *breaking force* and *elongation at break* are recorded to three significant figures. Breaking force is reported directly in lb./in of width (this is easy since the sample is only 1 in wide) or kN/m (convert lb/in to kN/m by multiplying by 0.1751 or kN/m to lb/in by multiplying by 5.710).[20] The percentage stretch, also known as *percent elongation*, is calculated by dividing the elongation at break by the initial length and then multiplying by 100%.

Example:

If a 1 in wide strip breaks at 6 lb in the tensile test with an initial jaw separation of 7 in and final jaw separation of 7.1 in, what is its tensile strength? What is its percent elongation?

Tensile strength = 6 lb/in % elongation = $(0.1''/7'') \times 100 = 1.43\%$

---

[19]The stress-strain behavior of paper differs from that of plastics, in that paper usually has no yield point.

[20]In tensile tests for other materials such as plastic and aluminum, the breaking strength is commonly converted to lb/in^2 (by dividing by the sample's cross-sectional area, thickness × width) to draw conclusions about the strength of the material in general rather than that of a particular strip. This is not common in the paper industry which tests to find the property of the sheet itself, rather than of the pulp from which it is made.

TEA, the area under the stress-strain curve is reported in energy units: stress (poind/ft^2) multiply by strain (foot) equal pound per foot (lb/ft) or joules per square meter (J/m^2). TEA is an indicator of the durability of paper when subjected to repeated and/or dynamic stress and strain. Paper with a high TEA is used in multiwall bags that will be subjected to frequent dropping, and TAPPI 494 notes that, for bags, good drop test performance and low failure rates have been found to correlate better with TEA than tensile breaking force.

Tensile strength is greater in the MD because the long fibers strengthen it, and lower in the CD because the fibers are being pulled apart. The percent of elongation at break is greater in the CD because the paper is more elastic in the CD.

Tensile strength results are more precise than elongation results. Tensile strength repeatability is 5% and percent elongation is 9%. Reproducibility (between labs) is 10% for tensile strength and 25% for elongation.

8.4.3 Tear Resistance

TAPPI T 414: Internal tearing resistance of paper (Elmendorf-type method)

Tearing resistance is the ability of paper to resist propagating a tear that has already begun. Tear resistance gives strength to a paper bag after the paper has begun to tear. But the lack of tear resistance can also be an advantage, as a measure of how easy it is to open an easy-opening tear strip.

Tearing resistance depends on the energy absorbed within a small area of the paper at the point of the advancing tear. The energy is used in breaking fibers and fiber-to-fiber bonds, and so it is affected by fiber length, fiber strength, and the amount of bonding between fibers. Tearing resistance is greater in the CD because the tear goes across the fibers. It is lower in the MD, where the tear propagates along the fibers; therefore a more effective tear strip would be designed to run along the MD.

Tear resistance can be measured in an Elmendorf tearing tester. Samples cut from each direction are tested. The sample (or a multiple sheet sample, depending on the thickness of the paper and the sensitivity of the apparatus) is clamped in two places and a slit is cut between them, as shown in Figure 8.4.

The pendulum holding one of the clamps is released and swings away from the other one, registering the minimum force to propagate the tear as a number of *Elmendorf units*. Ten samples 53 mm × 63 ± 0.15 mm (2″ × 2.5″) are tested in each direction. The average tearing force, as well as the minimum and maximum values, are reported for each direction (MD and CD) to three significant figures. The tearing force in grams is equal to 16 times the Elmendorf

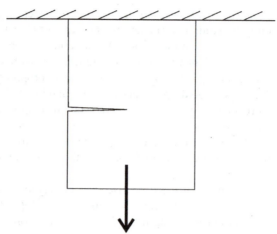

FIGURE 8.4. *Slit sample to propagate tear.*

scale reading. TAPPI T 414 notes a repeatability of 4.2% and reproducibility of 12.5%.

8.4.4 Coefficient of Friction

TAPPI T 815: Coefficient of static friction (slide angle) of packaging and packaging materials (including shipping sack papers, corrugated and solid fiberboard): inclined plane method

The *coefficient of friction* (*COF*) is a measure of the slipperiness of paper and board. The ability of sheets to slide across the surface of each other affects many aspects of packaging and distribution.

Slipperiness causes instability when handling or transporting a stack of boxes, bags, or sheets. Packages sliding off from the top of a stack will not only cause damage, but if a heavy one falls on a worker, injury or death can result. But if package surfaces are not slippery enough, they can jam conveyors, chutes, and erecting/filling equipment, and a high COF can also cause abrasion and scuffing.

The coefficient of friction depends on the roughness or smoothness of the paper surfaces, and is influenced by calendering, coatings, and surface treatments, especially coatings applied over printing. Often, a smoother surface is more slippery than a rough one, because the rough surfaces interlock. But it is difficult to make generalizations without the evidence of test results. Sometimes a smoother interface is less slippery because of the greater contact area than a rough surface. There are coatings, however, that are specifically intended to increase or decrease COF.

In the inclined plane method, a large sample is clamped to the plane, with the surface to be tested face-up. A second sample is affixed to a weight (called the *sled*), face-down; it is slightly larger than the base of the sled. The test standard specifies the weight's pressure, recommending dimensions of 90 mm × 100 mm (3.5 × 4 in) for the surface and a mass of 1,300 g (2.9 lb.), a foam surface, and a means for mounting the sample. The sample-covered sled is placed on the plane such that the faces to be tested are those that will be on the exterior of the package.

After a 10 second dwell time, one end of the plane is raised at a rate of 1.5 ± 0.5°/second until the sled begins to slide, as shown in Figure 8.5. At this time the incline is stopped, and the test is repeated three times on the same samples. The result of the third slide is recorded as the *slide angle*. At least five sets of samples should be tested, and the results of their third slides are recorded.

The tangent of the slide angle is the *static coefficient of friction*. The static coefficient of friction is a measure of the force required to initiate sliding. It can be obtained from a table of trigonometric functions or by using a calculator with these functions. For example, a 40° slide angle is equal to a COF of 0.84.

The MD and CD COF differ as do the values obtained when the test is done with the MD of the two surfaces aligned or perpendicular to each other. TAPPI T 815 specifies that the samples' MD should be perpendicular, to simulate interlocking stacks of boxes or bags. Other orientations can be tested if justified, and the report needs to state the orientation in which the samples were tested.

Since this is a test intended to judge the COF of a package in use, the procedure attempts to overcome the effect from fresh paper surfaces. Therefore, the test is repeated three times on a single sample, but only the results of the third slide are recorded.

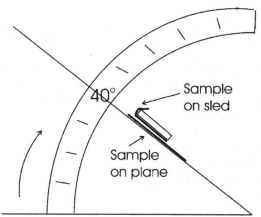

FIGURE 8.5. *Inclined plane coefficient of friction.*

The preliminary dwell time of 10 seconds before the plane is lifted accounts for the minute molecular bonding that occurs between sheets. TAPPI T 815 notes that this occurs in the first few seconds of contact, one reason why the starting COF is greater than the sliding, kinetic COF.

The *kinetic coefficient of friction* is a measure of the force required to keep two surfaces moving past one another at a constant speed. The TAPPI standard does not measure kinetic COF directly, but a related ISO standard does: ISO 15359, Paper and Board—Determination of the Static and Kinetic Coefficients of Friction—Horizontal Plane Method. The kinetic COF can be also estimated by lowering the incline of the plane once the sled starts to move to see how much it can be lowered without stopping the sliding.

The static COF test is moderately precise, with a repeatability of 6% and reproducibility of 15% (TAPPI 2000).

8.4.5 Paperboard Bending Tests

Stiff walls and flexible scores are the key to the success of paperboard cartons. Stiffness gives strength and reduces the propensity of a carton to bulge under the weight of settling contents like cookies or cereal. Stiffness and the quality of a carton affect its ability to run smoothly through a cartoner machine. If the paperboard is not stiff enough or the scores are too hard to bend, cartons will jam and the machine speed will need to be reduced.

8.4.5.1 Bending Stiffness

TAPPI T 489: Bending resistance (stiffness) of paper and paperboard (taber-type tester in basic configuration)

ISO 2493: Paper and board—Determination of Bending Resistance—Part 1: Constant rate of deflection and Part 2: Taber-type tester

Thickness is the primary factor that contributes to a board's bending stiffness (Koran and Kamdem 1989). This is the reason that low-quality recycled pulp is so valuable for paperboard; it gives the bulk to separate the stronger outer layers. Other factors include:

- Tensile strength of the outer layers
- Stiffness, length, and orientation of fibers
- Orientation (MD, CD) of the sheet
- Basis weight, density
- Moisture content
- Coatings

The bending resistance test deflects the board a little like a diving board. The test method determines the bending moment required to deflect the free end of a strip to an angle of 15° (Koran and Kamdem 1990). For thin or flexible papers, the alternative TAPPI T566 configuration is used.

The Taber stiffness test apparatus clamps one end of the 38.1 ± .03 mm × 70 ±1 mm (1.5″ × 2.75″) strip and bends the sample to 15° in each direction by a weighted pendulum. The apparatus reads out in *Taber stiffness units* (*TSU*), a measurement of g × cm force.

The bending moment is calculated as the average of the two readings multiplied by a factor that corresponds to the weight of the pendulum. For example, the factor is 5 for the "500 taber units" weight, 10 for the "1,000 taber units weight," and 20 for the weight marked "2,000 taber units." The test has moderate precision, with repeatability ranging from 3% to 7% and reproducibility from 5% to 11%, depending on the scales used.

An alternative to the Taber test is called the Gurley test (both are named after the men who invented them). TAPPI T 543 (Bending stiffness of paper, Gurley-type tester) is also used for paper. There is also a Taber test for paper, TAPPI T566.

8.4.5.2 Score Quality

Score bend testers are designed to measure the force required to bend paperboard to a 90° angle at a score, in comparison to the force required to bend unscored board.

The score bend test device uses a load cell to measure the maximum force required to bend paperboard to a 90° angle. The force to bend a scored sample is compared to the force needed to bend an unscored one.

The *score bend ratio* of scored/unscored values is multiplied by 100 in order to express the ratio as a percentage. A score that is so poor that the full strength of the unscored specimen is retained would give a score ratio of 100%, while a good crease gives a low number. A ratio of 50% is a common specification.

There is generally a difference in the score bend ratio for the MD versus the CD. Since paperboard has lower elasticity across a fold that is parallel to the MD, scores parallel to the MD require greater penetration than CD scores. Good scores in both directions are necessary for efficient cartoning operations. This test can be used for paperboard or for corrugated board.

Commercial score bend testers often also can be used to measure the residual spring back force when a score is bent, as well as the opening force required to erect paperboard cartons from the flat blank for filling.

There is no standard ASTM, TAPPI, or ISO method for evaluating scores in paperboard. A related standard is T 829: Score quality test, which applies to evaluating scores in corrugated containers; it can be adapted for paperboard.

Another related test is T 423: Folding endurance of paper, which judges the ability of paper to withstand repeated bending, folding, and creasing. It repeatedly flexes a score at 105–125 double folds (360°) per minute until failure. However, this is a test for paper such as that used for maps, and is not suitable for board over 10 points.

8.4.6 Corrugated Board: Edge Crush Test

TAPPI T 811: Edgewise compressive strength of corrugated fiberboard (short column test)

ISO 13821: Corrugated fiberboard—Determination of edgewise crush resistance—Waxed edge method

The *edge crush test* (*ECT*) is the most significant of the corrugated board tests. The remaining parts of this section are all tests of the strength of physical properties of corrugated board and boxes. ECT is used to predict a corrugated box's compression strength.

Most corrugated board in the United States is currently specified on the basis of ECT values (rather than thickness or basis weight). This is discussed further in Chapter 14.

ECT measures the columnar strength of the board, and is also called the *short column* test. It measures the amount of force required to crush a piece of board standing on its edge with its flutes vertical, as shown in Figure 8.6.

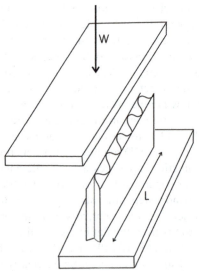

FIGURE 8.6. *Edgewise compression test (ECT).*

The most common form of the ECT test, TAPPI T 811, is performed in a small compression tester that can also be used to perform the ring and edge crust tests. Samples are cut 50.8 ± 0.8 mm (2 in) wide, and the load-bearing edges must be cut squarely with precise, straight parallel edges. The sample height depends on the flute type. For B-flute, the height is 31.8 ± 1.6 mm (1.25 in), for C-flute 38.1 ± 1.6 mm (1.5 in), and 50.8 ± 1.6 mm (2 in) for A flute, double-wall, and triple-wall boards. The corrugation runs in the short (height) direction.

The long edges are dipped in one-quarter inch wax to stiffen and prevent the edges from crinkling or rolling over. This forces the failure to occur in the body of the board (the columns of the flutes) rather than at the edges. The samples need to be conditioned for 2 hours after waxing. The sample is balanced on edge between the compression platens with the help of guide blocks. It is compressed until it buckles. The results are reported in units of pounds-force/inch (kilonewtons/meter) of specimen length:

$$ECT = \frac{W}{L}$$

where

W = load (weight) at which failure occurs
L = length of sample

For example, if 64 pounds pressing on the edge of a 2 in long, 1.5 in tall (flutes vertical) sample of C-flute corrugated board causes it to buckle, what is the ECT?

$$ECT = \frac{W}{L} = \frac{64 \text{ lb}}{2 \text{ in}} = 32 \text{ lb/in}$$

For single-wall board, ECT values range from 20–80 lb/in.

The test is prone to errors if the samples are not cut or prepared properly. If the flutes are crushed when cutting or the edges are not straight and parallel, the ECT value will be reduced. Samples should be cut with a precision sample cutter or some other method that will cut perfectly clean, parallel, and perpendicular edges (Eriksson 1979).

The repeatability of the ECT test is 4% and the reproducibility is 19%.

There are variations of the ECT test in which the samples are not waxed, including ISO 3037. There are two other ECT tests that use taller samples, $2'' \times 2''$, and are designed to better measure compression strength of the combined board. In the T 839 Clamp Method, the sample is placed in a fixture with the ends supported by clamps to ensure that the sample is held and loaded vertically. In the T 838 Neckdown method, parts of the sample are cut away leaving a 1 in neck in the middle, with less bearing capacity than the rest of the sample, where the sample is intended to fail. While all three tests are similar in principle, they do not give the same results.

ECT is a key test because of its relationship to the compression strength of a corrugated fiberboard shipping container, and its relationship to values from the ring crush test, described next.

8.4.6.1 Ring Crush Test

TAPPI T 822: Ring crush test of paperboard

The ring crush test (RCT) is for the containerboard (liners and medium) used to make corrugated board. In this test, a long strip of paper, shaped into a ring, is compressed on its edge.

The cross-machine (CD) values, with the short dimension of the sample parallel to the CD, are of most interest since this is the containerboard's orientation in a stack of boxes and it is also the containerboard's weakest orientation in compression.

The test strips measure 12.7×152.4 mm (0.5 in $\times$ 6 in). Samples are held in a ring shape by a special holder (as shown in Figure 8.7) and a load is applied to the ring until the sample buckles. The apparatus used is the same type of small compression tester used for the ECT and flat crush test (FCT).

The peak crush force is reported as an average of at least 10 samples in kilonewtons per meter to three significant figures. If the length of the sample is the standard 6 in (152.4 mm), the instrument reading in pounds force can be converted to kilonewtons per meter by multiplying by 0.0292. Similarly, an instrument reading in kilograms force can be converted to kN/m by multiplying by 0.0644, and readings in newtons by multiplying by 0.00656. Note that if the sample length differs, these conversion factors do not apply.

Repeatability of the RCT is not too bad, 4%, but the inter-lab reproducibility is only 22.5% for 26 lb medium (17% average for all containerboard tested).

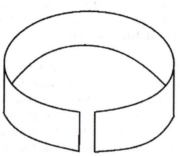

FIGURE 8.7. Ring crush sample.

There is an alternative to the RCT that also tests a strip on its edge in compression: TAPPI T 826: Short span compression strength of containerboard, also known as the STFI (Svenska Traforskininges Institute) test. This test provides a similar measure of a sheet's CD compression strength by clamping a 15 mm (0.59 in)-wide flat strip in two clamps that are 0.7 mm (0.0276 in) apart. The resulting test values are called SCT. Some argue that STFI is a more representative test than RCT because it does not have the cylindrical shape bias. On the other hand, RC samples fail by a combination of buckling and compression, as do real boxes, whereas STFI samples fail only by compression. The results of the two tests cannot be easily correlated, except as they both relate to the basis weight of the board (Urbanik and Frank 2002).

Tests for the "Fluted edge crush of corrugating medium" are TAPPI T 824 (flexible beam method) and T843 (rigid support method).

8.4.6.2 The Relationship of RCT to ECT

Containerboard manufacturers and corrugated board converters use RCT and STFI as a predictor of the ECT of combined board. They have a good incentive to increase the CD RCT, because corrugated board is specified by ECT, and a higher RCT is directly related to higher prices for containerboard.

The CD is significant because most corrugated fiberboard boxes are made with their flutes vertical, so their columns best contribute to stacking strength. But because the corrugated board is combined from rolls, the fibers in the MD are perpendicular to the flutes rather than reinforcing them.

Understanding the relationship between RCT and ECT is strategic for corrugated board converters because it can be used to estimate the effect of additives and orienting more fibers in the CD of containerboard. Orienting fibers in the CD improves the RCT and ECT values. Since the ECT standard was adopted in the 1990s, there have been increased efforts to orient more of the paper fibers in the CD in order to achieve a given ECT value with less fiber.

Generally, the higher the sum of the CD RCT values, the higher is the ECT. In theory, the relationship of ECT to RCT has been proposed as:

$$ECT = \frac{1}{6}[RCT_{L1} + T(RCT_M) + RCT_{L2}]$$

where

L_1 and L_2 = liners
M = medium
T = takeup factor for the flute size

There is disagreement about the strict accuracy of this relationship because of the other variables involved (Maltenfort 1988; Koning 1995; Frank 2007, 2014). For example, in the RCT test, the cylindrical shape contributes relatively more to the strength of lightweight board than to heavier board, so at minimum there should be a correction factor for board weight.

The quality of corrugating and gluing also affect ECT results, so a predictive equation can allow comparisons between two corrugator machines or production runs where the RCT values of the components are known. They also allow a corrugator/box plant to confidently select containerboard to use in production to meet customer ECT specifications without having to test every box order produced. Most box producers have their own version of the equation that is closely guarded, since accurate ECT prediction can provide a competitive advantage.

Some of the problems in corrugating operations that cause low ECT are poor bond strength and crushed flutes. The role of the medium is to support the faces, holding them apart and rigid, so the ECT result is very much affected by the strength of the formed medium and the quality of the glue bond.

8.4.7 Corrugated Board: Burst Test

TAPPI T 810: Bursting strength of corrugated and solid fiberboard.

ISO 2759: Board—Determination of bursting strength

The oldest corrugated board test is the Mullen burst strength test. It measures the ability of a sheet to resist rupture when a small, concentrated ball of pressure (in the form of a hydraulically loaded diaphragm) is pushed through from one side. It is named after its inventor, John W. Mullen.

Burst strength is related to the tensile strength of the liners, and is sometimes preferred as a quality control test by paper mills because it is simple and quick. The sample is clamped between two platens with circular openings. The lower platen has a rubber membrane stretched across the opening that expands until it bursts a hole through the sample. The function of the apparatus is shown in Figure 8.8.

The test is subject to errors if the instrument, diaphragm, and gauges are not properly maintained or if the procedures are not strictly followed. The measurement is sensitive to the clamping pressure. When the pressure is too low, samples will slip and higher readings will result. When it is too high, the flutes will be crushed, resulting in low readings. The test does not strictly measure the board because the result includes the pressure required to stretch the membrane. The measurement is in pounds per square inch.

Burst strength is easy for a corrugated board manufacturer to certify, since it is directly related to the tensile strength of the linerboards. It is largely unaffected by the tensile strength of the medium; because of the slack, the medium

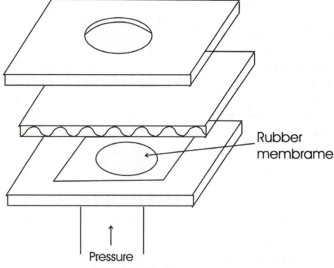

Pressure

FIGURE 8.8. Burst test set-up.

just unfolds as the linerboards stretch. The linerboard tensile strength, in turn, is directly related to the combined weight of facings and the strength and quality of the fiber furnish used. Burst strength has an additional advantage in that it is additive: the burst strength of the linerboards (measured in TAPPI T 807) can be used to predict the burst strength of the board.

The traditional single wall C-flute board made of virgin kraft 42 lb BW linerboard and 26 lb BW medium has a minimum Mullen burst strength of 200 psi. This board would be specified as 200 lb test, 42-26C-42 single wall corrugated fiberboard.

Burst strength was historically considered to be an important indicator of shipping container adequacy, but this is a matter of debate. While it may be an appropriate test for the strength of the elbow region in the sleeve of a coat, it does not necessarily simulate a generalizable form of package damage, and does not reflect a package's overall strength.

Burst strength is most useful as a measure of containment, for predicting internal resistance to the kind of force that occurs when a box is dropped, and the contents rupture the box wall (or something from outside pushes in). The Mullen burst test produces such a force. This is more of an issue in small parcel shipments, where packages are conveyed at high speeds and repeatedly slam into one another with contents that may not be adequately blocked and braced inside.

Furthermore, the burst test is an important test for corrugated fiberboard because of its tradition of use in the transport carriers' requirements for 100 years, discussed in Chapter 14. Burst strength is still the way that some corrugated fiberboard is specified in the United States, and is preferred by some users, especially for small parcel shipments.

TAPPI finds a repeatability of 5.98% and reproducibility of 13.4%. There are also two burst tests for paper, but the test machines differ in the force applied. TAPPI T 403 is for the "Bursting strength of paper," and TAPPI T 807 is "Bursting strength of paperboard and linerboard." There is no bursting test for triple-wall corrugated board, for which the puncture test, described next, is preferred.

8.4.7.1 Puncture Test

TAPPI T 803: Puncture test of containerboard

ISO 3036: Board—Determination of puncture resistance

The puncture test is a more violent form of the burst test. The puncture test measures a board's resistance to penetration by sharp and solid objects, such as the corner of a pallet. It is commonly used in place of the Mullen burst test for heavier grades of corrugated board, like triple-wall, where the Mullen diaphragm will not pop through the thicker boards.

It uses a *Beach test apparatus* with a swinging pendulum that has a special pyramid shaped puncture point that breaks through the corrugated board. Weight is added to the pendulum until it is sufficient to puncture the board. The apparatus, illustrated in Figure 8.9, has a scale that gives the puncture reading.

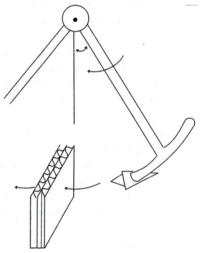

FIGURE 8.9. *Puncture test.*

The results are given by the test apparatus in *Beach units* (named for the inventor of the test), and reported to three significant figures. Beach units can be converted to either *inch-pounds* or *joules*, measures of mechanical energy, using the following:

$$1 \text{ Beach unit} = 0.0299 \text{ joules} = 0.264 \text{ inch-pounds}$$

Although the puncture test is used for specifying triple-wall board in the same way that the Mullen test is used for single or double-wall board, the two tests are not the same, and do not yield comparable units (Mullen results are reported in psi). Both tests stress the board in a similar way, by a force perpendicular to the surface. However, they differ in two significant ways. First, the puncture test uses a concentrated point stress, while the Mullen uses a stress distributed over an area. Second, the puncture test uses a rapid loading, while in the Mullen test the load increases slowly.

Puncture test results are affected most by the tensile and tear strength of the board's linerboards and the stiffness of the combined board. The puncture test results for single-wall board can be as low as 100 Beach units (3 joules) and triple-wall board ranges from 700 to 1,300 or more Beach units. The test has a repeatability of 10% and a reproducibility that ranges from 17% for lighter boards to 29% for the heaviest triplewall. The test is subject to errors if the puncture point is damaged or if the apparatus is not level.

8.4.8 Corrugated Board: Flute and Adhesion Tests

There are two corrugated board tests that focus on the formation of the combined board: flat crush and pin adhesion.

8.4.8.1 Flat Crush Test

TAPPI T 825: Flat crush test of corrugated board (rigid support method)

ISO 3035: Corrugated fiberboard—Determination of flat crush resistance

Flat crush testing (*FCT*) measures the rigidity of flutes. Board with a low FCT value is more subject to crushing in converting and printing equipment. Low FCT resistance can also reduce a box's compression strength, because crushed board is less stiff and easier to bend.

The test is performed in a small compression testing machine like the one used for the RCT and ECT. A disc of corrugated board, preferably with an area of either 5 in² or 10 in² for ease of calculation, is crushed flat, as shown in

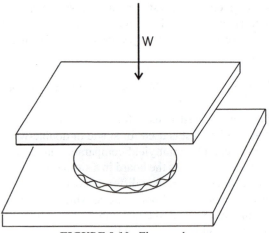

FIGURE 8.10. *Flat crush test.*

Figure 8.10. Failure is defined as the maximum load sustained before complete collapse. The results are reported in pounds/inch². A typical FCT for single wall C-flute is 20 psi.

$$FCT = \frac{\text{force}}{\text{area}}$$

Example:

120 pounds of force on a circular sample of SW C-flute just causes the flutes to collapse. What is the FCT if the sample is 3″ in diameter?

$$FCT = \frac{F}{A}$$

$$F = 120 \text{ lb}$$

$$A = \pi r^2 = \pi \frac{D^2}{4} = \frac{\pi (3 \text{ in})^2}{4}) = 7.069 \text{ in}^2$$

$$FCT = \frac{120 \text{ lb}}{7.069 \text{ in}^2} = 17.0 \text{ psi}$$

The FCT test is generally used only for single-wall board. Double-wall measurements are not meaningful because the middle liner shifts as the load is applied.

Flat crush is another useful quality control test in a corrugator plant. A high value indicates good flute formation and a stiff medium. Low flat crush values can indicate weak medium or flutes that are crushed, leaning, or poorly formed. Too much roll or printing pressure during manufacturing can reduce flat crush resistance.

8.4.8.2 Adhesive Bonding: Pin Adhesion Test

TAPPI T 821: Pin adhesion of corrugated board by selective separation

To realize a combined board's full potential rigidity and strength, the liners and medium must be firmly bonded together. The *pin adhesion test* measures the corrugator's glue bond strength.

The test applies a tensile force perpendicular to the liner and medium to pull them apart. Two matching sets of pins (specific to the flute type) are inserted into the flutes from opposite edges of a sample as shown in Figure 8.11. The 2″ × 6″ board sample (for C-flute) is cut so that the flutes are 2 inches long. The measure is the force required to separate the plies per inch of glue line, so the force is divided by 2 inches. Only the bottom liner separates, so the test needs to be performed in two orientations to test both sets of glue lines.

The pin adhesion test is widely used as a quality test. It should be noted, however, that the stiffness of the liners plays a role in the results, and that there is more than one potential cause, such as inconsistent glue and low strength adhesive, that could give the same results. Furthermore, the results of the test are not necessarily correlated with box performance, because the material is rarely pulled apart in this way during use.

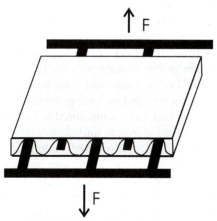

FIGURE 8.11. *Pin adhesion test.*

There is a related test for wet ply separation, TAPPI T 812. It requires submersion in water for 24 hours followed by a visual inspection and brushing a person's thumb across the cut edges to check for adhesion. This can also be followed by a pin adhesion test.

8.4.9 Corrugated Box Compression Tests

Compression strength is a laboratory value for the strength of a box, measured under ideal conditions. Although an empty corrugated box's compression strength is not equal to its realistic stacking strength in use conditions, it provides a starting point for predicting the box's contribution to stacking strength.

Compression testing judges the box materials. There are two types of test methods for compression strength. They can both be performed on either empty or full boxes, although when a box's compression strength (BCT) is given, it generally refers to that of the empty box in a dynamic increasing load test with its flaps sealed.

8.4.9.1 Dynamic Increasing Load Test (BCT)

TAPPI T 804: Compression Test of Fiberboard Shipping Containers

ASTM D642: Standard Test Method for Determining the Compression Resistance of Shipping Containers, Components, and Unit Loads

ISO 12048 Packaging—Complete, Filled Transport Packages— Compression and Stacking Tests Using a Compression Tester

The BCT is directly related to the ECT of the board, but it also judges the quality of the box's construction, including the quality of the scores and slots. These factors are discussed in Chapter 14.

A box is placed between the platens of a compression test unit, and a preload is applied to ensure proper contact with the top platen and to level off. The preload should be 5% of the maximum compression resistance or a value agreed upon by the parties engaged in testing; 50 lb (223 N) is common for single-wall boxes. The applied force is measured as the top platen descends at a rate of 0.5 in/min. The load increases until failure or until a specified force (for example, the weight of a stack) is reached. Failure is usually defined at the point where the box first begins to buckle and lose its resistance.

The top platen can either be fixed or floating. In the *fixed platen* method, the top platen is fixed in a level horizontal position, perfectly parallel to the bottom platen. The fixed platen tests the ability of the stronger parts of the box to resist load. Most of the compression data in the literature relating ECT to BCT relies on the fixed platen test, and TAPPI T804 requires that the fixed platen be used for all peer-refereed published research.

TAPPI T804 and ASTM D642 alternatively allow for a *floating platen* that contacts the entire flat top face of the box, even if that surface is not exactly perpendicular to the base. In contrast to the fixed platen, the floating platen tests for the weakest part of the box (Singh *et al.* 1992). If the box were perfectly square, the two methods would give identical results, but for real boxes the results will differ.

At least five samples should be tested. The flaps are sealed as they would be in distribution, although if this is by glue, TAPPI recommends using a fixture to ensure good adhesion. There is a general recognition that the method of sealing affects the BCT, especially if the box is empty (which causes un-taped inner flaps to fold down into the case and can actually improve BCT values slightly).

The precision of the compression test is pretty good. ASTM gives a reliability of 8.5% and reproducibility of 11.3%, and TAPPI gives a repeatability of 7% and reproducibility of 10.6%. ASTM notes that "these values may reflect the inherent variability of the test specimen as much as the actual variability of the test method and apparatus" (ASTM D642, Note 4).

8.4.9.2 Constant Load/Creep Tests and Cyclic Humidity

ASTM D4577, Standard Test Method for Compression Resistance of a Container under Constant Load

ASTM D7030 Standard Test Method for Short Term Creep Performance of Corrugated Fiberboard Containers Under Constant Load Using a Compression Test Machine

This test is intended to reproduce the situation where packages are stored for a long period. The load is chosen to match the load on a bottom box in the stack or to be a percentage (less than 85%) of the value obtained in the increasing load test, D642. The test apparatus can be either a standard compression tester or a dead load apparatus that guides a weight onto the top of a box or stack. Initial deformation is compared to observations over increasingly long intervals (first 5 minutes, then 10, 30, 60, and 120 minutes). The test continues until failure occurs (the box or stack buckles) or a specified test period is over.

This is a test for creep, a time-dependent test in which the stresses are constant but the strains increase with the duration of the load application. It has been found that the time to failure decreases logarithmically as the dead load approaches the BCT; the time to failure with a dead load of 95% of the BCT is less than 2 minutes, and at 75%, boxes fail within 7 hours (Kellicutt and Landt 1951).

Cyclic humidity, like that experienced in day/night cycles, has been found to accelerate creep failure. Cycles reduce time to failure even more than constant high temperature and humidity.

8.4.11 Cushioning and Insulation Properties

ASTM D1596: Shock absorbing characteristics of package cushioning materials

ASTM D3103: Thermal insulation quality of packages

Although corrugated board is not the usual choice for cushioning or insulation, its structure provides both. The properties of corrugated board are not as predictable as those of plastic foams and the weight is considerably greater. Higher weight increases freight and material costs.

The collapsible flutes are not as resilient as plastic foams, but they are capable of absorbing shocks and of amplifying vibration. Forms and blocks made from built-up layers of corrugated board are used as interior packaging, usually intended more as a blocking or dunnage material to separate the walls of the package from the product, than as a cushion designed according to predictable cushion curves (which trade off cushion density and thickness against weight bearing area). The cushioning characteristics depend on whether the material is used in an edge crush or flat crush orientation. The walls of a shipping container made from corrugated board provide a thin cushion and give surface protection.

Problems in performance as a cushion include fatigue sensitivity due to gradual compaction of the flutes, RH sensitivity, and tendency to create dust. Also, consumers tend to perceive corrugated as a cheap, although environmentally favorable, material. Corrugated cushioning is most often used for light weight relatively non-fragile products.

The insulating properties of corrugated board are about the same as for an equivalent thickness of plastic foam due to the air trapped between the faces (Ramaker 1974). This provides a benefit for shipping refrigerated products because the box alone can protect for a short period. On the other hand, it is a problem for shippers of freshly harvested fruits and vegetables that need to be quickly chilled, and is the reason that most produce boxes have ventilation holes.

8.5 POROSITY: RESISTANCE TO AIR, GREASE, AND WATER VAPOR

Porosity is the resistance to air flow and the penetration of oil and water. It is defined as the fractional void volume of the paper, a measure of the openness of the web that is related to density. Actual measurement of the size and distribution of pores in the paper is difficult and rarely carried out, but air resistance can be measured by the volume of air that can pass through a web of paper.

8.5.1 Air Resistance

TAPPI T 460: Air resistance of paper (Gurley method)

Air resistance is measured and specified when it is needed for a specific reason. For example, air resistance is needed when packages are pneumatically conveyed or erected in packing machinery. Porosity is needed when air evacuation is required, for example, in flour bags.

Specification of air resistance is critical for many medical products that are sterilized inside of a sealed paper package with ethylene oxide gas. The paper must be porous enough to permit ethylene oxide gas to pass through it for the purpose of sterilizing a product. Medical devices in paper pouches are commonly sterilized in this manner, and so the air resistance of the paper is a critical factor in specifications.

The Gurley method uses a special apparatus in which the paper is clamped, and a free-floating cylinder slides inside of a fluid-filled cylinder by gravity, forcing air through the sample. Results are reported as 100 ml/sec, or converted to volume of flow per unit area and unit pressure difference across the sheet. The method's repeatability is 8% and reproducibility is 16%, but it is particularly subject to errors if the samples are incorrectly clamped.

8.5.2 Grease and Water Resistance

Since it is a porosity test, the air resistance test gives an indirect indication of a paper's absorbency to oil and water, and its water vapor transmission. But these can also be measured directly. Even though grease and water resistance can be improved, in most cases, grease and water will ultimately penetrate paper if a long enough time is allowed.

Grease resistance is desirable in packages for food and greasy metal. Grease-resistant paper helps to avoid greasy spots on fast-food containers. It can be measured directly using TAPPI T559 (Grease resistance test for paper and paperboard). In this test, a succession of mixtures of castor oil, toluene, and n-heptane in different proportions are dropped on the surface of the paper. The lowest concentration of oil that causes a stain yields the rating.

Water resistance is desirable since it decreases the strength-reducing effects of moisture, but it can also make paper harder to glue and print, since the glue and ink must be partly absorbed by the paper. Sizing agents, added to the pulp or to the paper surface, as well as wet strength resins, increase water resistance.

Water resistance can be measured directly using a number of standard methods. TAPPI T 433 (Water resistance of sized paper and paperboard (dry indicator method) measures how long it takes for water to pass from one side of paperboard to the other. TAPPI T 441 (Water absorptiveness of sized paper, paperboard and corrugated fiberboard—Cobb test) involves clamping dry paper in a fixture, pouring water over it, letting it soak in for 2 minutes, and then comparing its weight to that of the dry sample. TAPPI T 491 (Water immersion test of paperboard) also evaluates wicking through the board's cut edges.

8.5.3 Water Vapor Transmission

TAPPI T 464: Water vapor transmission rate of paper and paperboard at high temperature and humidity

ASTM E96: Water vapor transmission of materials

ISO 2528: Sheet materials—Determination of water vapour transmission rate—Gravimetric (dish) method

Specially-treated paper has long been used as a water vapor barrier. Before plastics were used, waxed paper (usually waxed glassine) was the primary barrier material, used mostly as the liner for paperboard cartons. Now, plastic film dominates the market for flexible barrier materials, and when paper is used for this purpose, it is usually plastic-coated. The *water vapor transmission rate (WVTR)* can also be improved by calendering, which increases density and surface smoothness.

A standard way to measure the WVTR of a paper, film, or foil is to directly measure the weight of moisture picked up by a *desiccant* inside of a sealed package made from the material. Heatsealed packages like pouches can be tested in this way following ASTM D3079: Water Vapor Transmission of Flexible Heat-sealed Packages for Dried Products.

A desiccant is a material that has a high affinity for water vapor, drawing it from the environment. The most common desiccants are anhydrous calcium chloride and silica gel. Desiccants have another important use in packaging; they are used to control the humidity level in sealed packages of moisture-sensitive products. The desiccant is usually dried in an oven before it is used, to give it the maximum capacity for moisture pick-up.

A sheet material is tested by using it to cover open dishes of desiccant. The material is sealed to the edge of the dish with wax. The dishes are weighed, put into a high humidity chamber, and then weighed again at regular intervals. The standard high temperature, high humidity conditions for this test are 37.8°C 0.6°C (100°F ± 2°F) and 90% 2% RH. If the test is run at different conditions, such as ASTM's tropical conditions of 40°C (104°F), 85% RH, this should be noted in the test report.

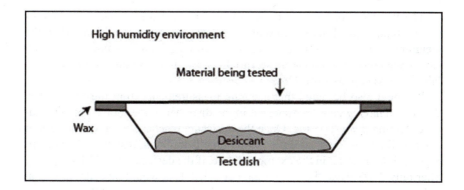

The water vapor transmission rate is calculated as:

$$WVTR = g/m^2 \text{ per day}$$

where

g = weight gain in grams for the time period
m^2 = exposed area of specimen

Example:

If the dish containing desiccant gains 5 grams of water in 5 days, and the exposed surface area of the barrier material is a 3 in diameter disc, what is the WVTR?

$$\text{Area} = \pi r^2 = 3.14159\left(\frac{3 \text{ in}}{2} \times \frac{2.54 \text{ cm}}{\text{in}} \times \frac{\text{m}}{100 \text{ cm}}\right)^2 = 0.00456 \text{ m}^2$$

$$WVTR = \frac{5 \text{ g}}{0.00456 \text{ m}^2 5d} = 219 \ \frac{\text{g}}{\text{m}^2 d}$$

In addition, a material's *permeance* can be calculated. Unlike WVTR, permeance is dependent only on temperature, which affects the vapor pressure, not on relative humidity. It is obtained by dividing the WVTR by the water vapor partial pressure difference across the material. In the dish method, this is equal to the saturation vapor pressure at the test temperature x the RH in the humidity chamber, since the desiccant maintains a water vapor pressure of approximately zero inside the dish. This measure is more commonly applied to plastic barrier materials, is rarely applied to paper, and is not part of the TAPPI standard.

The WVTR *dish test* is subject to large errors if the samples are not correctly prepared or if the samples are exposed to varying conditions when the weighing is done. The repeatability is not too bad, 8.1%, because a single lab can presumably standardize these variables. But the reproducibility (between laboratories) is poor, only 23%.

It should also be noted that accurate measurements from the WVTR test depend on having enough desiccant in the dish that it does not approach saturation during the test period. Once the desiccant reaches saturation, no further water will be absorbed, so if the testing time is too long, results will be meaningless. For instance, in the example above, if the desiccant was at its capacity after only 2-1/2 days, the weight gain at 2-1/2 days would be equal to that at 5 days. The calculation shown above would, then, give a WVTR value twice as large at 2-1/2 days as at 5 days, and neither number would reflect the real barrier ability (or lack thereof) of the material being tested.

What the dish test lacks in accuracy, it makes up for in simplicity. It can be used for paper, film, or foil. More accurate methods, such as using a Mocon Permatran W apparatus, are more commonly used in industry. The related test methods are *TAPPI T557* and *ASTM F1249: Water vapor transmission rate through plastic film and sheeting using a modulated infrared sensor.*

8.6 SURFACE AND OPTICAL PROPERTIES

Brightness, opacity, gloss, smoothness, and color are optical properties of primary economic importance in paper-based packaging. They give paper the ability to contrast dark, light, and colors in print. They also determine whether paper absorbs ink or carries it on the surface.

Bright glossy packages are more effective sales tools, but excellent optical properties increase the cost of the paper. To maximize the value of paper's printing advantage, it is best to specify only those optical properties that are required.

Optical properties are related to the roughness and color of the surface. Coatings and pigments affect gloss, brightness, opacity, and whiteness. Calendering polishes the surface and increases gloss and brightness.

Optical properties are influenced by basis weight and age. Basis weight primarily affects opacity. Aging reduces whiteness and brightness turning some papers yellow, and it causes ink to fade.

Optical properties depend on the interaction of light with paper. They can be divided into the general categories of reflectance, scattering, transmission, and absorption.

- *Reflectance* (or specular gloss) occurs in the top surface of the sheet; a glossy paper has more reflectance. Whiteness and color are generally measured by reflectance.

- *Scattering* is the result of multiple (diffuse) reflections and refractions inside the surface of a sheet of paper. Paper is a mass of intertwined, almost colorless cellulosic fibers, and each time light strikes a fiber, some rays will reflect from the surface and others will pass through, being refracted (bent) as they go. The light makes numerous direction changes, caused by multiple reflections and refractions, scattering light rays in all directions. If a sufficient amount of all wavelengths of light are scattered back in the direction of the viewer, the sheet will appear white.
- *Transmission*, also known as *opacity*, refers to how much light passes through the sheet, rather than being reflected.
- *Light absorption* occurs when light strikes a non-transparent or colored material that absorbs some or all of the light and converts it to heat rather than reflecting it. Paper that absorbs light, rather than reflecting it, lacks brightness.

There are many methods for testing the optical properties of paper. They use optical instruments in order to make unbiased measurements, since the human eye cannot be trusted to make consistent and fine distinctions. The values measured depend on the electromagnetic spectrum of the light source and the photometry (measurement system), as well as on the geometry of the instrument since the paper surface differs in the machine and cross machine direction.

8.6.1 Brightness, Whiteness, and Color

A bright, white printing substrate is necessary for good true color printing. Its contrast gives life to inked images. *Brightness*, *whiteness*, and *color* are the power of a sheet of paper to reflect light. They are commonly measured by shining a light source on a sample from a 45° angle and measuring how much light is reflected (although there are also standards like ISO 2469/2470 and TAPPI T525 that measure the reflectance from a light source straight on).

Color shades are described on the basis of three primary colors: red, yellow, and blue. Color strength refers to how light or dark the shade is. Paper and paperboard color are commonly expressed in three dimensions in the Hunter *L, a,b tristimulus* system. The L value measures color strength, the a indicates a value on the red-green continuum, and the b is a value for yellow or blueness. It is measured on a device known as a tri-stimulus colorimeter, and the relevant test method is ASTM E1455.

Whiteness is related to the color of the sheet. It is actually the equal presence of all colors, because a truly white sheet will reflect all wavelengths of visible light equally. When a sheet appears to be a color like red, that is because the red light is reflected and the other wavelengths are absorbed. TAPPI T 560 is a reflectance test with a light source that includes the entire visible spectrum. It provides values that measure how far red or green the sample is from the perfect white reflecting diffuser.

8.6.1.1 Brightness

TAPPI T 452: Brightness of pulp, paper, and paperboard (directional reflectance at 457 nm)

Brightness is a special papermaker's term that is defined as the amount of blue-white light that a paper reflects. When paper is observed indoors, it has been found that blue light reflectance makes it appear to be whiter. People prefer a blue-white paper; it seems more clean and new because yellow paper is associated with age. Since this measure is more related to what people perceive, it is the most common optical measure that the paper industry uses.

Brightness is measured as the reflectance of light of an effective wavelength of 457 nm, in the blue region of the spectrum. It compares the amount of reflected blue-white light to magnesium oxide, which is the standard for 100% bright. Since this measurement ignores the yellow and red spectrum, two different colors may have the same brightness.

TAPPI T 452 uses a Brightness Meter to provide a directional measure of the light reflected at 45°. Since this geometry is directional, different results will be obtained when paper samples are measured in the machine and cross-machine direction. The geometry eliminates completely the specular gloss effects of the paper surface, and emphasizes the scattering effects. The result is expressed as a percentage of the light reflected, known as *TAPPI Brightness* (or *directional blue reflectance* to avoid confusing it with the brightness of a color), reported to one decimal place.

Brightness of pulps and paper can be increased by chemical purification and/or bleaching to remove residual lignin. It can be improved by fillers such as clay and TiO_2, which also improve opacity and reduce ink strike-through. Fluorescent dyes can also improve brightness. Fluorescence is the process by which ultraviolet light is absorbed and re-emitted at visible wavelengths in the blue part of the spectrum causing the level of reflected light to be higher than that of the incident light at the bluish wavelengths.

Whereas unbleached kraft has a brightness of only 20–25%, fully bleached kraft pulps can be as high as 92%. SBS is commonly sold as *88 Bright*.

The brightness of unbleached mechanical pulp is generally in the range of 53–65%. Unbleached sulfite pulps with a slight red-brown tint are brighter than unbleached kraft. It should be noted that the chemical changes in lignin resulting from chemical pulping cause the lignin to become dark brown in color. Therefore mechanical pulps are much whiter and brighter than unbleached kraft pulps, even though most of the lignin has been removed by the chemical pulping process, but they are more prone to yellowing with age.

The brightness of paper and paperboard is also increased by clay coating. *80 Bright* is a common grade for the clay-coated recycled board in folding cartons. Clay-coated unbleached kraft is commonly sold as 70 Bright. Since the human eye cannot readily detect fine differences in brightness below a level of 85, the L.a.b. value may also be specified.

Other brightness testers and test methods have been developed. ISO 2470 (Paper, board, and pulps—Measurement of diffuse blue reflectance factor), for example, measures using a different geometry, and there is no way to convert the results mathematically to TAPPI T 452 results (Scott *et al.*1995).

8.6.2 Gloss and Smoothness

TAPPI T 480: Specular gloss of paper and paperboard at 75°

TAPPI T 653: Specular gloss of paper and paperboard at 20°

Gloss is a measure of the reflectance of light from the surface of the paper. Gloss is specified when a special package surface is desired. There are degrees of gloss from matte to glare. Gloss and luster connote a pleasing effect, while glare can be blindingly unpleasant. There is a directional effect, based on the angle of reflection. A matte surface diffuses the light in all directions so that the surface appears the same from every angle.

Specular reflectance refers to the measure of light reflected at an angle equal to the angle of incidence. For paper, gloss is generally measured at an angle of 75° from perpendicular (15° from the surface) and is referred to as 75-degree gloss (Figure 8.12). For very high gloss materials such as coated papers and high gloss inks, a steeper angle such as 20° is commonly used. In some cases, different angles of incidence and reflection are used. The angle used must be specified, since it affects gloss values.

The amount of light reflected is compared to a reference standard consisting of a polished black glass surface. It is expressed in terms of the percent reflec-

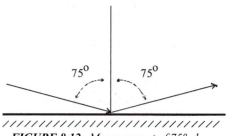

FIGURE 8.12. *Measurement of 75° gloss.*

tance compared to the standard. While the black glass reference is categorized as 100% gloss, the amount of specular light reflected from the glass surface is actually only about 26% of the incident light. Measurements are normally made in both the MD and CD, since the directions of the incident and reflected light lie in a plane perpendicular to the paper surface.

Coatings and calendering are commonly used to increase gloss and smoothness. Fillers improve surface uniformity. Uncoated papers typically have 75° gloss of less than 20%, while the gloss of coated papers is typically between 20% and 80% (Kline 1991). Surface smoothness and uniformity are particularly useful in printing. The smoother the paper, the better is the ink coverage. Uniformity yields sharper print detail.

Surface roughness and gloss are related, although not identical. Generally, rough surfaces are diffuse and smooth surfaces are good specular reflectors. However, some dull coated papers are very smooth, while some glossy papers are relatively rough.

A number of measurements of surface smoothness/roughness are available. Bekk smoothness (TAPPI T 479), with units of sec/10 ml, measures the air flow between the paper surface and a polished glass surface under defined conditions. Bendtsen roughness (ISO 8791) and Sheffield roughness (TAPPI T538 and ISO 8791), with units of ml/min, also depend on air flow measurements. The Parker Print Surf roughness test (TAPPI T555) is a refinement of the Bendtsen test. The Kunz-Lippke (KL) optical smoothness meter measures both smoothness and gloss, based on the width and height of the scattered light intensity curve, using incident light at either a 75° or 20° angle.

8.7 REFERENCES

ASTM (formerly American Society for Testing and Materials, now ASTM International). No date, annually updated. *Annual Book of ASTM Standards.* West Conshohocken, PA: ASTM International.

Chamberlain, D. and M.J. Kirwan. 2013. Paper and Paperboard–Raw Materials. In *Handbook of Paper and Paperboard Packaging Technology,* 2nd ed., edited by Mark J. Kirwan, pp 1–19. Chichester: John Wiley & Sons.

Coffin, D.W. 2005. The Creep Response of Paper. In *Advances in Paper Science and Technology: Transactions of the 13th Fundamental Research Symposium held in Cambridge: September 2005,* Vol. 2, edited by S. J. I'Anson, pp. 651–747. Bury, Lancashire: Pulp and Paper Fundamental Research Society.

Eriksson, L. 1979. A Review of the Edge Crush Test of Corrugated, *Boxboard Containers.* In two parts, 86 (8): 34-36, 38 and 86 (9): 64–67.

Frank, B. 2003. Ring Crush and Short Span Compression Test for Predicting Edgewise Compressive Strength. *TAPPI Journal,* 2 (11) November: 13–16.

Frank, B. 2007. Revisiting Paper Strength Measurements for Estimating Combined Board Strength. *TAPPI Journal,* 6 (9): 10–17.

Frank, B. 2014. Corrugated Box Compression—A Literature Survey. Packaging. *Technology and Science,* 27 (2): 105–128.

Kellicutt K.Q. and E.F. Landt. 1951. Safe Stacking Life of Corrugated Boxes. *Fibre Containers*, 36 (9): 28–38,

Kline, J.E. 1991. *Paper and Paperboard: Manufacturing and Converting Fundamentals.* 2nd ed. San Francisco: Miller Freeman.

Koran, Z. and D.P. Kamdem.1989. The Bending Stiffness of Paperboard. *TAPPI Journal,* 72(6): 175–179.

Koran, Z. and D.P. Kamdem. 1990. Multilayered Paperboard from Kraft Pulp and TMP. *TAPPI Journal,* 73(2) pp. 153–158.

Mark, R.E., C.C. Habeger, J. Borch, and M.B. Lyne. 2002. *Handbook of Physical Testing of Paper.* 2nd ed., New York: Marcel Dekker.

Peel, J.D. 1999. *Paper Science and Paper Manufacture.* Vancouver: Angus Wilde.

Scott, W.E., J.C. Abbott, and S. Trosset. 1995. *Properties of Paper: An Introduction.* 2nd ed, Atlanta: TAPPI.

Singh, S.P., G. Burgess, and M. Langlois. 1992. Compression of Single-Wall Corrugated Shipping Containers Using Fixed and Floating Test Platens. *ASTM Journal of Testing and Evaluation,* 20 (4): 318–320.

TAPPI (Technical Institute for the Pulp and Paper Industries). No date, periodically updated. *TAPPI Standards.* Atlanta: TAPPI. The standards current in 2008-9 are the basis for the descriptions in this chapter.

Uesaka, T. 2002. Dimensional Stability and Environmental Effects on Paper Properties. In *Handbook of Physical Testing of Paper,* edited by Richard E. Mark, Charles C. Habeger, Jens Borch and M. Bruce Lyne, vol 1, 2nd ed., pp. 115–171. New York: Marcel Dekker.

Urbanik, T. and B. Frank. 2002. Ring Crush and Short Span Compression Test for Predicting Edgewise Compression Strength. *TAPPI Journal,* 2(11): 13–16.

8.8 REVIEW QUESTIONS

1. What is the difference between material tests and package performance tests? How does their use differ?

2. What do these acronyms mean: TAPPI, ASTM, and ISO?

3. What standards organization is the most important source of tests for the properties of paper and paperboard?

4. What is the principal difference between ASTM and ISO?

5. What is the difference between repeatability and reproducibility in test standards?

6. What is the difference between accuracy and precision? What is bias?

7. How many square feet is the basis for paper versus paperboard? How many square feet for containerboard (the components of corrugated board)?

8. If an 8″ × 10″ piece of 5.6 point kraft paper for bags weighs 4.4 grams, what is its basis weight? What is its grammage? What is its thickness?

9. If 42 lb. BW corrugated linerboard (containerboard) costs $350/ton, what is the weight of two liners for a box blank with the dimensions of 12″ × 36″? What is the cost for these two liners?

10. In general, what is the thickness borderline that defines the difference between paper and paperboard?

11. How is paperboard caliper (thickness) expressed in U.S. units versus SI metric units?

12. How thick is 24-point board in μm? How thick is 450 μm board in points?

13. What is the range of thickness most common for paperboard?

14. Is it more common to specify paperboard by thickness or basis weight? Which is more common for paper?

15. What is relative humidity "relative" to?

16. What are standard TAPPI temperature and RH conditions, and why are they used?

17. If a piece of paper that weighs 5 grams is dried, and the dried weight is 4.7 grams, what was the original moisture content?

18. Why is wood moisture content calculated on a dry basis and paper on a wet basis?

19. What, in general, is the equilibrium moisture content of paper at 50% RH?

20. What is the ideal range for the moisture content of paper? Below what moisture content does paper become brittle? Above what moisture content does it lose stiffness?

21. Why should a paper-based sample be preconditioned in drier conditions first?

22. What are the temperature and RH for ASTM's standard tropical conditions?

23. How are tensile strength, elongation at break, stiffness, and tear strength affected by an increase in moisture content?

24. Be able to use the following procedures to determine the MD and CD of samples of paper and paperboard: the axis of curl procedure, the bend procedure, the surface fiber procedure, the hand tearing procedure, and the tensile test procedure.

25. Does paper have more of a feathery appearance when torn parallel to the MD or CD? In which direction does a paper wetted on one side curl (which is the axis and which side is out)?

26. If a 1″ wide strip breaks at 13 lbs. in the tensile test, with an initial jaw separation of 7″ and ending at 7.1″, what is its tensile strength? What is its percent elongation?

27. Does paper have more tearing resistance in the MD or CD? In which direction does it have higher tensile strength? In which direction is it stiffer? In which direction does it have the greatest fold endurance?

28. When would a high tear resistance be a benefit? When might it cause problems?

29. If a paper has a slide angle result of 30°, what is its static COF? Is its kinetic COF higher or lower?

30. If static and kinetic coefficients of friction are 0.4 and 0.3 respectively, at what angle does the sample begin to slide?

31. When would a low COF be a benefit? When might it cause problems?

32. Which factor most affects the stiffness of paperboard?

33. What are the names of the instruments and units of measurement for the tearing and bending stiffness tests?

34. What is a score bend ratio and when is it relevant?

35. In which direction do the flutes run in most boxes? Why? In which direction does the MD of the containerboard run in most boxes? Why?

36. If an ECT sample (2″ × 1.5″) buckles at 64 lb., what is its ECT?

37. What purpose is served by waxing the edges of ECT samples?

38. To which property of corrugated fiberboard is RCT related?

39. To what properties of the linerboard is the combined board Mullen burst test most closely related? What is the effect of the medium in the Mullen test?

40. What is meant by "200-lb. test" board?

41. Under what conditions is the Beach Puncture Test used instead of the Mullen Burst Test?

42. Are Beach Puncture Units a measure of force, pressure, energy, or momentum?

43. What characteristics of corrugated board does the flat crush test evaluate?

44. If a 3″ disk of C-flute buckles in the flat crush test at 140 lb., what is its flat crush value?

45. What does the pin adhesion test evaluate?

46. How is the wet ply separation test for corrugated board performed?

47. Give examples of circumstances in which the air resistance is specified.

48. If a dish covered with glassine and containing desiccant gains 6 grams of water in 16 hours, and the exposed surface area of the barrier material is a 3-inch diameter disc, what is the WVTR?

49. Which optical properties are most affected by aging?

50. Which is more commonly specified in packaging: whiteness or brightness? What is the difference?

51. What is the most common brightness value specified for SBS and clay coated paperboard?

52. What is meant by opacity and gloss?

CHAPTER 9

Types of Paper, Paperboard, Laminations, and Adhesives

A variety of types of paper and board are used in packaging applications. This chapter gives a brief introduction.

As discussed in Section 8.1, the distinction between *paper* and *paperboard* is not clear. The term *paper* can be used generically to mean both paper and paperboard. Paperboard is thicker and stiffer than paper, but at the boundaries, whether something is called paper or paperboard is likely to depend as much on its application as on its properties.

Paper can be further subdivided into fine and coarse papers. Fine papers are used for applications such as the printing of books and magazines, and coarser papers for applications such as packaging. Still another categorization is to divide paper and paperboard into printing and writing grades, industrial papers and boards (including packaging, wallpaper, label and filter paper, plaster board, insulating paper, etc.), and sanitary papers (toilet tissue, toweling, etc.). These grades are a reminder that paper has diverse uses in today's world.

This chapter describes the principal materials used in paper and paperboard packaging, like kraft, containerboard, and paperboard, and specialty papers like greaseproof, glassine, cellophane, and molded pulp. It explains how their manufacturing processes and uses differ. The chapter ends with food contact and sustainability considerations. Chapters 11–14 will discuss how these materials are converted into packages ranging from bags to cartons to boxes.

9.1 KRAFT PAPER

The highest volume paper used in packaging is *kraft*, made by the sulfate pulping process, usually from softwood, on a fourdrinier machine. *Natural kraft* is a coarse brown paper; *bleached kraft* is white.

Unbleached natural kraft is the strongest and most economical type of paper. It is used to make shopping bags, multi-wall sacks, crumpled paper void filler, and plain brown wrapping paper. Kraft paperboard is used to make corrugated board, gable-top cartons, frozen food cartons, fiber drums, and cans.

Basis weights for kraft paper typically range from 18 to 80 lb (per 3,000 ft^2, also 30–130 gsm) with the most common weights between 40 and 60 lb (65–100 gsm). Kraft paper can be coated or laminated for barrier and additional strength.

Impact strength and tensile energy absorption of kraft paper can be improved during the drying process. *High performance, extensible* kraft paper is mechanically creped, mainly in the MD, and has 6–7% stretch and higher TEA than does ordinary kraft. *Free dried* kraft is allowed to freely shrink in the CD. Paper dried in these processes is used to improve the impact resistance of multi-wall paper shipping sacks.

Wet-strength kraft paper is made by adding polyamide or polyamine resin and curing it with heat, which allows the paper to retain strength when it is wet and to better regain strength after drying. Wet-strength kraft paper retains about 25–33% of its dry tensile strength when it is saturated with water. The primary use is as an outer ply in multi-wall paper bags that will be exposed to high humidity conditions.

Because of its dark brown color and coarse surface, it is not possible to print fine detail on natural kraft. It can be made more opaque, smooth, and bright with a clay coating.

Bleached and *semi-bleached* kraft are used when good printing is a requirement and the nature of the product requires a package with a clean appearance. Bleached kraft paper is used for flour and sugar bags, envelopes, and labels. Many paper grades, including tissues and printing papers, which formerly were made from bleached sulfite are now made from bleached kraft.

Bleached kraft fibers can be made into many paper grades including printing and fine papers. They are used to make surgical kraft, a high-quality grade used for packaging medical products, discussed in Chapter 11. Kraft is used to make *solid bleached sulfate*, a high quality paperboard described in Section 9.7. Clay-coated bleached kraft is the highest quality (strongest, whitest) paper used for packaging. It is used for overwraps, labels, and the outer ply of some bags.

9.2 GREASEPROOF/RESISTANT PAPERS

There is a great demand for packaging papers that can resist the penetration of oil and grease, especially for products like fast foods, snack foods, baked goods, pet foods, butter, and soap. Although consumers love to eat fatty foods, no one likes to see a greasy package. The need for grease resistance is growing as food processors switch from hydrogenated to unsaturated fats that melt at room temperature.

Because cellulose is hydrophilic, it can be a good substrate for resisting the penetration of hydrophobic liquids like oils. Organic oils do not dissolve or swell the cellulose fibers, nor do they diffuse through fiber walls; they are transported through the pores between the fibers by capillary forces. In *parchment, greaseproof, glassine*, and *grease resistant* papers, these pores are filled. Typical basis weights range from 18–50 lb (30–81 gsm).

9.2.1 Parchment

The process for making *parchment* paper was developed in the 1850s, making it one of the oldest specialty packaging papers. It was invented to look like parchment made from animal skins, and is sometimes called vegetable parchment.

Parchment is produced by passing bleached kraft paper through a sulfuric acid bath that reacts with and partly dissolves the cellulose and causes the fibers to swell, plugging the pores, filling in voids, and increasing the hydrogen bonding. Next, the paper is rinsed with water and the fiber network consolidates, resulting in a pore-free paper. Special finishing processes are used to provide qualities ranging from rough to smooth, brittle to soft, and sticky to releasable.

The process produces a tough, grease-resistant paper that is stronger wet than dry (and is even strong when boiled). Parchment is odorless, tasteless, and has a chemically pure, fiber-free surface that may be coated with silicon to resist adhesion to food. It was first used for wrapping fatty products like butter, but is now used for many kinds of prepared foods.

9.2.2 Greaseproof and Glassine Paper

Greaseproof paper achieves a similar result of plugging the pores from mechanical refining. The bleached kraft pulp is literally beaten to a gelatinous pulp. The prolonged beating fibrillates, breaks, and swells the fibers, resulting in a relatively pore-free sheet that resists penetration by fluids.

Glassine derives its name from its glassy smooth semi-transparent surface. It is sometimes called *glazed greaseproof paper* because it is also produced from highly-refined pulp, and the paper is further subjected to heavy pressure in the presence of steam heat during a supercalendering operation, pinched between a hard cylinder and a softer one. The high pressure and moisture collapse the remaining fiber structure, greatly increase hydrogen bonding, and produce a glossy, nearly transparent sheet with enhanced grease resistance. Besides its grease-resistance, glassine has a high resistance to the passage of air and many essential oil vapors used as food flavoring. It is often waxed. The usual basis weight ranges from 15 to 40 lb (25–65 gsm).

Greaseproof and glassine papers are used for fast foods (like hamburgers and fries), bakery products, and for fluted candy cups. Since they are permeable to water vapor, they can help fried food retain crispness. They may be plasticized, as for soap wrappers, or plastic-coated to further increase their toughness and liquid resistance. They have a reputation for running well on high speed packaging lines.

9.2.3 Grease-Resistant Papers

Grease-resistant papers have less resistance to grease and oil than greaseproof paper, glassine, or parchment. They are designed to provide sufficient resistance to penetration of grease or oil to be adequate for relatively short use times. Often these papers depend more on surface treatments to limit the penetration of oil or grease than on modification of the actual paper structure.

Fluorocarbon additives or coating have been common because they have a low surface energy. Fluorocarbon treatment decreases the wettability of the paper structure, eliminating the tendency for oil to wick through, and gives a non-stick coating. Fluorocarbon-treated paper has been used in multi-wall bags for perishable bakery goods, and in form-fill-seal packages where edge-wicking needs to be prevented. When both water and grease resistance are desired, oleophobic chemicals are also added.

But many of the previously common fluorochemicals are being phased out because they may pose health risks, as degradation products include long-chain perfluorinated chemicals such as perfluorooctanoic acid (PFOA) that have been found to migrate from packaging into foods. This has resulted in new, proprietary treatments that are advertised as "fluorochemical-free."

9.3 WAXED PAPER AND CORRUGATED BOARD

Waxed paper has a paraffin wax coating added to one or both sides of the paper. This can be during in the dry end of the papermaking process, or during printing and converting. Wax provides moisture resistance, and waxed paper was the first flexible water-barrier packaging material. Most kinds of paper and board can be waxed. There are two waxing processes, called *wet* and *dry* waxing.

Wet waxed paper is made by cooling the wax coating quickly, so that it remains on the surface of the paper. This surface layer of wax adds heatsealability and a vapor barrier. Highly-calendered bleached kraft and sulfite papers are usually chosen for wet waxing, and provide a high gloss surface.

Waxed glassine is the most common form. Waxed glassine is made into bags that are easy to open and relatively reclosable by folding. It was formerly used for the pouches inside cereal and cracker and cake mix boxes, and was particularly suited to run on the double-package-making machines that were once common in the cereal industry. Waxed glassine has been replaced in many cases by sealed plastic film bags which are better barriers, but have the disadvantage of being difficult to open.

Dry waxed paper results when heat is used to allow the wax to soak into the paper. The paper does not have a waxy feel, and allows additional surface treatments for special release applications or for lamination. Low-density bleached kraft and sulfite papers are usually chosen for dry waxing. Dry waxed paper has more ability to "breathe" moisture and gases.

Waxed papers are economical, versatile, and safe for food contact (tasteless, odorless, non-toxic, and inert). They are widely used for food service applications like wraps and carryout cartons, and for the liners inside paperboard cartons.

Corrugated board can also be waxed for water resistance, a common practice for shipping containers used for fresh fruits and vegetables as discussed in Chapter 14. The wax can be applied to the containerboards before they are made into combined board, or it can be applied to the combined board using a cascade to saturate it or a *curtain coating* to cover the surface.

One problem with waxing corrugated board is that it makes it hard to recycle the paper—the same feature that makes the waxed board water-resistant makes it resist resuspending the fibers in water during recycling. The Corrugated Packaging Alliance has a voluntary standard for "recyclable wax alternatives" that can be used to certify that the containers are "repulpable and recyclable" and can be marked with a special symbol. As of October 2013, 46 such systems were registered with the Fibre Box Association. These are typically water-based coatings that add water resistance without the use of insoluble wax. Formulations are proprietary, but they generally contain a latex or other polymer as a binder along with one or more minerals as a filler to provide moisture barrier.

9.4 CELLOPHANE AND OTHER CELLULOSE-BASED FILMS

Cellophane (or regenerated cellulose film) is made by extensive chemical treatment of wood that results in a transparent film. The name combines the words *cellulose* and *diaphane*, the French word for "transparent." In the United Kingdom and some other countries, the name is a registered trademark of the Innovia Films Group, but in the United States, the term is generic.

Cellophane was the first transparent flexible packaging material, and was very popular as a shiny wrap for high end candies and cosmetics and a moisture barrier wrap for food products and tobacco.

It is produced from high-purity wood pulp (such as eucalyptus) by dissolving the cellulose fibers in carbon disulfide, then adding sodium hydroxide that converts the solution into viscose. The gelatinous material is ripened for a few days and then cast through a narrow slit orifice onto a casting drum on which it then passes through further stabilizing liquids, mainly sulphuric acid. This regenerates the film by coagulating the viscose solution. After passing through washing baths, the material is plasticized to make it less brittle and more useful by adding ethylene glycol or propylene glycol.

Pure cellophane is extremely permeable, and for most applications it is coated with nitrocellulose (NC) or polyvinylidene chloride (PVdC) to add barrier properties and heatsealability. PVdC is more durable and provides a better barrier to moisture and gas.

Most of the previous applications for cellophane have been replaced by plastic film. Polypropylene (PP) and other plastics have become available as lower cost substitutes. Plastic film production also causes less pollution compared to cellophane's carbon disulfide and other by-products.

But unlike most plastics, cellophane is biodegradable, which has contributed to a recent increase in popularity. It is mostly used in niche markets where its unique characteristics, such as its stiffness and ability to hold a *dead fold*, are useful for applications like hard candy twist wrapping and wrapping gift baskets and homebaked gifts. It is still used to wrap some cigars because of its ability to "breathe." It is easy to tear, cut, and seal (when appropriately coated), has a high level of gloss, dazzling colors, and good machinability.

One of the disadvantages of cellulose is that it does not melt (or dissolve). To get some of the benefits of cellulose without those drawbacks, it is possible to chemically modify the cellulose structure. That is done in making cellophane, but then the added chemical groups are removed again. If we leave the modifying groups as part of the structure, we can obtain cellulose-based (cellulosic) plastics. The most common of these is cellulose acetate. Cellulose acetate is a clear, biodegradable film that consists, as its name indicates, of cellulose with modifying acetate groups. It can be used as a window in cartons. It is clear and resists scratching, age-brittleness, and yellowing. Due to its high breathability, moisture and gas pass through it easily. It competes with PS, PP, and PET films for these applications. There are other cellulosic plastics as well, but since they truly are plastics, further discussion falls outside the scope of this book.

9.5 SPECIALTY WRAPPING PAPERS: BUTCHER, BAKER, TISSUE, AND ACID-FREE PAPER

There are many kinds of specialty papers used in packaging. This section provides a brief listing of some of them.

Dry finish butchers wrap is a well-sized, bleached wrapping paper that has been calendered without the application of surface moisture. It can be made from mechanical and/or chemical pulp. It is used for wrapping meats and has a basis weight of 35–50 lb (57–81 gsm). It is strong and resists the penetration of meat fluids. Similarly, *bloodproof* and *delicatessen paper* are sized and waxed to make them even more resistant to blood. *Freezer paper* remains pliable at low temperatures and easily releases from frozen meat. On the other hand, *blood soaking paper* is a multi-ply tissue used for lining meat and poultry trays to absorb unsightly liquid.

Bakers wrap is a lightweight wrapping paper used for baked goods. It is made from bleached chemical pulp in basis weights of 20–25 lb (33–40 gsm). Most is machine glazed on a Yankee dryer.

Tissue paper is a generic term for any light paper, usually less than 18 lb BW (30 gsm). In packaging it is used for a protective wrapping for a variety of products and as a laminate layer. Tissue papers can be made from any kind of pulp, but the lightest tissues used for teabags are made from long fibers such as Manila hemp. They are generally MF or MG, unglazed, or creped.

Acid-free and *archival paper* and board is used for museums and libraries. Ordinary paper and board made from wood pulp degrades over time, in part due to acid-catalyzed hydrolysis of cellulose, exacerbated by oxidation. It can transfer stains to art that is affixed or wrapped in it and to archival materials. Paperboard or corrugated boxes can deteriorate, losing the ability to contain and protect the materials stored in them. To prevent this, acid-free and archival paper is made with an alkaline sizing (rather than the standard acid sizing) and with calcium carbonate filler as an "alkaline reserve" to neutralize acid groups that may be released from the paper itself. It has a pH between 8 and 10 (U.S. National Archives 1991).

Anti-tarnish paper, used for wrapping coins, silver, and other metals that can tarnish, is made from paper that is relatively free from acids, alkalies, and sulfur compounds that can accelerate corrosion. Copper salts or other corrosion inhibitors, including vapor phase inhibitors, are used to provide additional protection.

Vapor corrosion inhibitor (*VCI*) (also called *volatile corrosion inhibitor*) paper is often used to wrap machinery and weapons to protect them from rust. VCI chemicals can also be applied to corrugated boxes used for metal parts. The chemicals vaporize inside the package over time and form an invisible, thin coating on the metal that resists water vapor. VCI effectiveness increases if the outer package is well-sealed, as that slows down the loss of the chemicals, but is designed to work when simply wrapped loosely around the product.

9.6 NEWSPRINT, BOOK, AND OFFICE PAPERS

Although newsprint, book, and office papers are not generally used directly for packaging, they are indirectly used in paperboard made from recycled paper. They differ from most packaging paper in that they are primarily made from high yield mechanical (groundwood) pulp. Book and office papers are usually bleached.

Newsprint has the highest percentage of mechanical pulp, usually about 95%. Standard newsprint has about a 25–32 lb basis weight (40–51gsm). It has a relatively low brightness (55–65) and low physical strength. It has high oil absorbency, which helps it to accept the oil and carbon newspaper ink.

Magazine and catalog papers are sized and clay-coated to varying degrees to improve the surface for higher quality printing. Magazine paper is generally about 50% mechanical pulp, clay coated, and calendered to provide a bright white surface.

There is a wide variety of commercial papers used for books and in offices. They may be coated, and brightness ranges from 73 to 92. Most office paper has a 20-pound BW, but since this is based on 500 sheets, it has a different conversion factor to 75 gsm. Paper used for photocopying is designed to withstand the heat with a minimum of curl and distortion, with a controlled electrostatic surface resistance to provide a uniform image.

9.7 PAPERBOARDS: RECYCLED AND KRAFT

Paperboard, also known as *boxboard* or *cartonboard*, can be either a single or a multi-layer material. It can be made of more than one type of pulp, and it often incorporates recycled fibers. If it is multi-layer, the inner layers are usually made from lower quality pulp than the outer layers. To improve its appearance and printability, it may have a clay coating or a lining of bleached fibers on one or both surfaces.

Paperboard is made in bending and non-bending grades. Non-bending board is used for set-up boxes, and bending board is used for folding cartons. Paperboard is commonly specified by thickness, rather than basis weight; the most common thicknesses are between 16 and 24 points (also known as 0.016–0.024 in or 410–610 μm).

The lower grades of paperboard are made from recycled pulp on a cylinder machine in multiple layers. Although the terminology is not standardized, names used for different grades include recycled paperboard, chipboard, bending chipboard, lined chipboard, bleached manila, clay-coated paperboard, clay-coated newsboard, white lined recycled, and plain and lined kraft board. (*Lined* refers to a surface coating of bleached fibers.) Clay-coated grades of cylinder board are commonly used to make folding cartons for cereal and crackers.

The premium boxboards are made from kraft sulfate pulp on a fourdrinier machine. The highest grade is *solid bleached sulfate* (*SBS*), also called *bleached kraft*, made from bleached virgin fiber. The all-white board conveys an impression of cleanliness and is used for health and beauty aids and in packaging that is in direct contact with foods, such as cartons for frozen foods, milk, and butter. *Solid unbleached sulfate* (*SUS*) is a strong lightweight board used for beverage carriers, carded packaging, and heavy boxes for detergent.

Paperboard packaging is the subject of Chapters 12 and 13 where the properties of and uses of these types of paperboard are discussed in more depth.

9.8 CONTAINERBOARD, CORRUGATED BOARD, AND SOLID FIBERBOARD

Containerboard is the name for the relatively thin paperboard that is combined to make corrugated board. There are two types of containerboard that are combined in the corrugated structure: linerboard and medium.

Linerboard is the name for the flat facings of corrugated board. It is usually made from softwood kraft pulp to give it strength. Most is formed in the duplex fourdrinier process described in Section 7.1.3, using recycled fibers in at least one layer. Basis weight ranges from 26 to 90 lb (per 1,000 ft^2, 125–440 g/m^2). When white linerboards are desired for higher quality graphics, bleached kraft is commonly used.

The *medium* is the fluted sheet. It can be formed by either the fourdrinier or cylinder process, and is usually a lighter basis weight than the liners. Most common is 26 lb (per 1,000 ft^2, about 125 g/m^2). This is the thinnest board considered to be paperboard for purposes of basis weight, 9 points. The pulp is made from recycled corrugated board and hardwood pulped by the NSSC process, although other materials (including straw) have been used in the past. The main requirement for medium is that it be malleable enough to form flutes that are stiff enough to maintain their structure.

Containerboard has the highest production of any kind of paperboard, since so many products are packed in corrugated fiberboard shipping containers. This material is so universal in the United States that its sales per year correlate directly with the state of the economy. Corrugated board packaging is the subject of Chapters 14 and 15 where the manufacturing process and applications are described.

Solid fiberboard, as its name implies, is a thick sheet of paperboard, formed by gluing two or more plies of paperboard together under pressure. The thickness of solid fiberboard ranges from 35–135 points (0.89–3.43 mm). The most common structure has a thickness of 70–80 points (1.8-2 mm), and is made from four plies of paperboard. The laminated board is printed, diecut, and coated if desired. Boxes made from solid fiberboard are sometimes used as an alternative to corrugated boxes. They are used mostly for reusable boxes, as their initial cost is significantly higher than that of a comparable corrugated box, but they can be used more times. Solid fiberboard is also used for slipsheets, an alternative to pallets.

Kraft paperboard can also be assembled into honeycomb structures, containing a continuous series of hexagonal cells forming the core of a sheet with solid paperboard facings on each side. These very strong structures can be used as interior cushioning materials, pallets, and building construction.

9.9 MOLDED FIBER

Molded fiber, also known as *molded pulp* (or in the United Kingdom, moulded pulp), is the only packaging material in this book that is not converted from a sheet form. The fibers are simply molded to a shape that will hold and protect the product. The molded pulp egg carton is the classic example, but newer forms are more refined and designed for higher-value goods.

Molded pulp packaging has a good environmental reputation for its use of waste. It can be made from any kind of waste paper or other natural cellulosic fibers. The properties of the molded piece depend on the source of pulp. The baled, recovered paper or other pulp is hydropulped in a 99% hot water solution along with additives like color and waterproofing agents. It is molded in one of the processes described in Chapter 13 according to the precision required. It is recyclable, biodegradable, and compostable where facilities are available.

9.10 CLAY-COATED PAPER AND PAPERBOARD

Paper and paperboard are often coated with an aqueous mixture of white clay or other mineral pigments. Clay-coating makes the surface white, bright, and smooth, which is necessary for good color printing. Some clay-coated paper is used for labels and bags, as well as for lamination or in combination with plastic sheets where it provides the printed surface. Most of the paperboard used to make folding cartons and beverage carriers is clay-coated.

Paper or board that is coated with clay on one side or both sides is designated *C1S* and *C2S* (Figure 9.1) respectively. When clay-coated paper is specified,

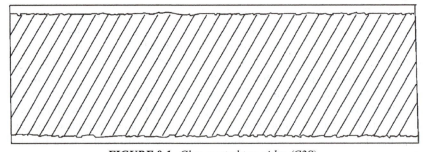

FIGURE 9.1. *Clay-coated two sides (C2S).*

the weight of the clay is included in the basis weight. For example, 39-pound BW C2S clay-coated paper has about 6 lb of clay applied to 33 lb of paper (per 3,000 ft^2), and 50-pound BW clay-coated paper has about 8 lb of clay on 42 lb. BW paper.

9.11 EXTRUSION COATING AND LAMINATION

Paper and paperboard add value to food packaging because of their stiffness, high-quality printability, opacity, and strength. But paper is vulnerable to moisture, so plastic coatings and foil laminations are added to make it resist water, to separate the paper from food, and to make it heat-sealable.

In *extrusion coating*, as described in Section 7.4.2, plastic is melted and extruded through a slit die across the width of a paper web as the reel is unrolled beneath it. PE is the most common coating. PE-coated paper is used to make sachets filled with granular products, like sugar, which are discussed in Chapter 11. PE-coated paperboard is used to make containers for liquids and wet and/or frozen food, as discussed in Chapter 13. The PE coating adds heat-sealability and a water barrier.

Laminations are produced by bonding the paper to a sheet of foil or plastic. Kraft paper/aluminum foil laminations are most common, with the foil adding a good barrier to water vapor and oxygen. While sheets can be glued together with a water-based starch or PVA adhesive (casein and sodium silicate can also be used), extrusion lamination is more common.

In *extrusion lamination*, foil (or another sheet) is joined to the paper with a thin tie layer of molten plastic (usually PE), in a manner similar to that used for extrusion coating. The plastic forms a *tie-layer* film between the two sheets, and joins them together. Laminating a thin sheet of aluminum foil to paper makes the structure an excellent barrier to water vapor and oxygen. An outer and/or inner coating of plastic (usually LDPE or PET) can be added for heatsealability and surface protection. Figure 9.2 shows the extrusion laminating process.

Aluminum foil has been the traditional laminate layer to provide paper with a high barrier to gases, moisture, aromas, and light for drink boxes and composite cans (described in Chapter 13). Paper/foil laminations are used to make pouches for dehydrated powdered mixes for drinks, sauces, and side dishes (discussed in Chapter 11). Aluminum foil is also laminated to the outside of paperboard packages for a decorative effect.

The most common example of a package that combines both extrusion coating and extrusion lamination is the aseptic juice box (discussed in Chapter 13). Each of the six layers of the composite material has a special function (in this order):

• The inner LDPE layer gives a heatsealable watertight inner surface.

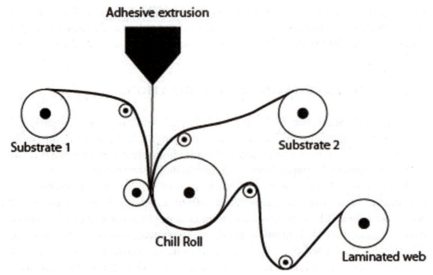

FIGURE 9.2. *Extrusion laminating process.*

- The aluminum foil gives a nearly absolute barrier to light and oxygen.
- LDPE forms a bond between the foil and paper.
- The paper provides the printing surface, tensile strength, and stiffness.
- The outer LDPE layer gives a waterproof outer surface and protects the print.

Specification of these composite structures can be tricky, because there are different conventions for each material. The extruded plastic is specified based on the basis weight of the paper, the number of pounds/3,000 ft^2 or grams/ m^2. But if the plastic is a film sheet laminated to paper, it is specified in terms of thickness in mils (0.001″) or μm. And aluminum foil is specified in inches (e.g., 0.003″) or μm.

9.11.1 Polymers Used in Combination with Paper and Paperboard

Polyethylene (PE) is the most common polymer used with paper. It is inexpensive, heat-sealable, and a good adhesive and barrier to moisture. *Low density polyethylene* (LDPE) is by far the most common, as it is low in cost and provides a good moisture barrier and excellent heat seal properties. It is commonly used, as stated above, in cartons for aseptic packaging of juices and also for coating paperboard milk cartons.

High density polyethylene (HDPE) and *linear low density polyethylene* (LLDPE) are used to a lesser extent. They offer somewhat better moisture barrier, higher melting temperature, and increased strength, but typically at a somewhat higher cost than LDPE.

Polypropylene (PP), another member of the polyolefin family, has similar properties to polyethylene. PP has a higher melting point than PE so it can be used to coat microwaveable paperboard trays.

Ethylene vinyl acetate (EVA) and *ionomers* provide improved heat seal performance that may be needed for dusty or greasy products. Ionomers are related to LDPE, but have better *hot tack*, which is the strength of a fresh, still molten seal, because they contain ionic bonds which provide stronger inter-molecular attractions. This makes them very tough and strong and, very importantly, allows them to seal well even if the seal area has been contaminated with dust or grease.

Higher barrier plastics can also be combined with paper and board. A high-barrier is required for products such as long shelf life dairy products. *Poly-ethylene terephthalate* (PET) (also known as PETE, and frequently referred to simply as polyester) is used as a coating for paperboard intended for dual-ovenable (microwave and conventional oven) applications because of its high melting temperature.

Ethylene vinyl alcohol (EVOH) copolymers, *polyvinylidene chloride* (PVDC) copolymers, and *polyamides* (PA, also known as nylons) can be used for long shelf life products and as barrier adhesives. PET and PA also have applications in frozen food packaging because they can go directly from the freezer to the oven. EVOH is used as the inner adhesive for orange juice cartons, instead of PE, to prevent flavor scalping (explained in Chapter 13) and provide an oxygen barrier because it is more polar than PE. Polymers with polar bonds and high crystallinity tend to be good barriers, especially for non-polar substances such as oxygen.

Other specialty polymers are also used as coatings or adhesives. *EAA, ethylene acrylic acid*, can be used as a heat seal coating layer in pouches, as discussed in Chapter 11. A solvent-based vinyl (PVC) coating may be used for blister board, to facilitate heatsealing to *polyvinyl chloride* (PVC), the primary plastic used for blisters, an application of the general principle that "like sticks to like."

In *cold seal coating*, the sealing surface of the substrate is coated with a slightly sticky water-based rubber latex emulsion that includes an acrylic component. When the two surfaces are pressed together, they form a cohesive bond. The cold seal adhesive is printed onto the paper with a gravure press, only in the sealing areas. The material has either a release lacquer or film to make it easy to unwind in an FFS machine.

Instead of laminating paper to aluminum foil, the same look, and close to the same barrier properties, can be achieved by laminating paper to a metal-lized plastic film. The vacuum metallizing process vaporizes aluminum and deposits a very thin layer on the film. The result is the look of a foil lamination, but at a lower cost. Because the aluminum layer is not 100% continuous at the microscopic level, metallized plastic does not have quite as high barrier properties as a foil lamination, but it is less likely to develop pinholes or disfiguring wrinkles (so its barrier performance after it has been through the distribution process may actually be higher than a lamination). Metallized PET (M-PET) and PP (M-OPP) are used with paper for FFS applications. M-OPP is chosen when a superior moisture barrier is needed, for example for stand-up pouches for cookies or ground coffee. M-PET is chosen when a good oxygen barrier is required, for example, in composite cans, as a substitute for foil.

9.11.2 Metallized Paper, Microwave Susceptors, and Shields

Like plastic film, paper can be vacuum-metallized. Metallized paper can replace aluminum foil or foil-based laminations when the primary goal is to get the appearance of foil. Metallized paper does not generally have good barrier properties as the roughness of the surface prevents the creation of an uninterrupted layer of metal. But metallized paper works well for applications like labels, where the primary goal is the shiny foil appearance and where the paper-based material's stiffness improves the efficiency of the packaging line.

Microwave susceptors, also produced using vacuum or sputter deposition, are used to crisp and brown microwaved foods and popcorn. They are produced by applying a thin aluminum coating (typically a thickness of 3–6 nm) to a heat-resistant surface, such as 12 μm (0.5 mil) biaxially oriented PET film, which is adhesive laminated (typically with PVA) to a temperature-stable substrate like SBS, often referred to as heater board. When the susceptor is placed in a magnetic field, a current flows through the coating, and heat is generated by resistance between the metal particles at their contact points.

In some microwave-cooked meals, it is desirable to shield some parts of the meal from heat. Aluminum foil can be laminated to the paperboard in strategic areas to redirect the microwave energy either towards or away from parts of the meal.

9.12 ADHESIVES USED FOR PAPER-BASED PACKAGING

To go from sheet form to three-dimensional shapes, paper-based packages are often glued. As mentioned, they can also be laminated to other materials. This section describes the process of adhesion and outlines the common adhesives used.

To combine two materials into a single structure, there must be sufficient forces of attraction between them to result in *adhesion*, in the materials sticking to each other with enough strength to enable them to be handled as a single unit. An *adhesive* is a material that assists in bonding two surfaces together by forming reasonably strong bonds between itself and both of the surfaces being joined. Adhesion can be due to either mechanical or chemical forces, or to a combination of both.

9.12.1 Adhesion

In *mechanical bonding*, the force joining two materials is due primarily to mechanical interlocking of the materials (Figure 9.3). Paper-to-paper bonding often relies primarily on mechanical bonding, especially if the paper is porous and rough. The fluid adhesive flows into the paper's surface irregularities and pores, and is solidified by drying. If a force is exerted on the adhesive, the mechanical interlocking that has been produced resists its separation from the paper. In general, the higher the initial solids content of the adhesive, the more quickly the bond will be formed, since less drying will be required to produce an initial level of mechanical entanglement.

In contrast to mechanical bonding, *chemical bonding* depends on attractive forces between the molecules in the two materials being bonded. These forces include van der Waals forces (London dispersion, induction, and dipole forces) and hydrogen bonds.

Hydrogen bonds are particularly important for paper and wood materials, because, as has been discussed, they contain a large amount of –OH groups that can form strong hydrogen bonds. Therefore, to get the best adhesion to paper, an adhesive should also be capable of hydrogen bonding.

Chemical adhesion forces are very sensitive to the distance between the molecules involved, decreasing with the sixth power of increasing distance. When paper is bonded to a smooth non-porous material such as plastic or foil, the adhesion between the adhesive and the plastic or foil depends almost entirely on chemical bonding, as no significant mechanical bonds can be formed.

For either chemical or mechanical bonding to occur, the surfaces must come into close contact. Since solid surfaces, even if they appear smooth, have significant surface irregularities when you examine them at a microscopic level;

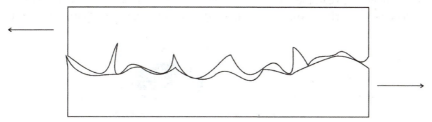

FIGURE 9.3. *Mechanical bonding.*

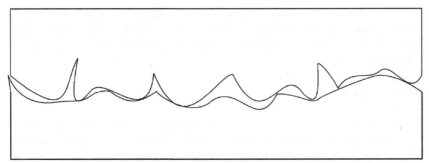

FIGURE 9.4. *Surface roughness diminishes the actual contact area between two solids.*

it is extremely difficult to get two solids into close enough contact to adhere to each other. When the surfaces are pressed together, there is only a small percentage of the surface area where close contact is actually made (Figure 9.4). Therefore, a liquid adhesive that can flow and wet the surface provides the necessary close contact. While the adhesive is liquid, it has little cohesive strength (the internal forces keeping the adhesive itself from breaking apart), and the adhesive must be cured before a strong bond will be formed.

To form a strong bond, either through chemical or mechanical bonding, an adhesive must wet the surface of the substrate. *Wetting* a surface does not necessarily have anything to do with water. It means that the adhesive flows onto the surface, forming intimate contact between the surface material and the adhesive, rather than beading up on top of the surface (Figure 9.5). An adhesive will wet a substrate if the forces of attraction between the substrate and the adhesive are greater than the surface tension of the adhesive.

The viscosity (resistance to flow) of the adhesive is also important. An adhesive with a higher viscosity will take a longer time to flow sufficiently to establish good coverage of the surface than one with a lower viscosity. This will slow down the process of spreading the adhesive on the surface. On the other hand, if the viscosity is very low, more solvent or carrier liquid must be evaporated before the adhesive will have sufficient cohesive strength to establish an initial bond between the substrates, so this will also slow the process down.

The characteristics of the substrate are also important. For paper like kraft

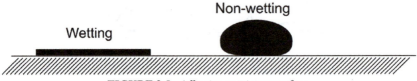

FIGURE 9.5. *Adhesive wetting a surface.*

with a porous but non-absorbent surface, an adhesive with a relatively high viscosity is used in order to prevent it from traveling too far into the pores and being lost from the surface. For absorbent but non-porous recycled paper with a short fiber mat, a lower viscosity adhesive can be used, since it will stay on the surface.

Cohesive strength is the internal bonding strength of a material. A glue-joint or laminated material has *cohesive failure* if an applied stress exceeds the cohesive strength of either of the substrates. For example, if you open a glued carton flap and the carton tears, there has been cohesive failure in the carton material.

Adhesive strength is the force of attraction between two separate materials. Adhesive failure occurs when the applied stress exceeds the adhesive force between the adhesive and the substrate. For example, if you open a glued carton and the adhesive peel cleanly away from one of the surfaces, without any fiber tear, there has been adhesive failure of the joint.

Therefore, in a typical adhesive joint, failure can occur in five locations, as shown in Figure 9.6.

Failure within a substrate (in location 1 or 3 of Figure 9.6) or within the adhesive (location 2 of Figure 9.6) is cohesive failure. Failure at the interface between materials (locations 4 or 5 of Figure 9.6) is adhesive failure.

The adhesive should generally create bonds that are at least as strong as the substrates themselves. Then failure will occur when the stress exceeds the substrate strength; the adhesive joint will not be the weak point in the structure. To achieve this, the adhesive forces and the cohesive strength of the adhesive must all exceed the cohesive strength of the weakest of the two substrates. Then, for example, when we open the flap of a glued-shut corrugated box, the paperboard will tear because the weakest part of the joint is the paper itself.

Sometimes, however, adhesive failure is preferred to cohesive failure. For example, in some resealable packages, it is important for the adhesive to release, so we can open and reclose the package without it tearing. A non-packaging example is U.S. postage stamps: we buy the stamps on a backing sheet,

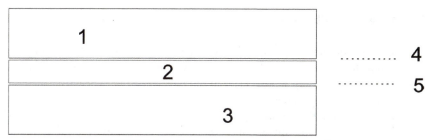

FIGURE 9.6. *Failure in a laminated material.*

and we want to be able to peel them off the sheet so we can put them on the envelope we wish to mail. Proper functioning of the system depends on the "weak link" being the adhesive bond between the adhesive on the back of the stamp and the backing material.

In other cases, we want cohesive failure to occur within the adhesive itself. For example, a temporary adhesive can be used to stabilize a load of cartons. The adhesive is applied to spots on the outside of the carton. When the customer goes to pick up a carton, we do not want the cartons to tear. Instead, the low tensile strength of the adhesive causes cohesive failure within the adhesive itself, so the cartons are easily lifted apart from each other. During distribution, the high shear strength of the adhesive resists side-to-side motion, so the adhesive works well in stabilizing the load of cartons, even though it does not resist the up-and-down stress imposed when a carton is lifted.

The most common test for adhesion is a simple *seal peel* test in which the two surfaces are peeled apart. The presence of *fiber tear* indicates a good bond to the paper, as well as high internal cohesive strength in the adhesive. The absence of fiber tear indicates that the adhesive joint is the weak point in the structure. The absence of fiber tear with adhesive present on both surfaces indicates low internal cohesive strength of the adhesive.

In addition to the strength of the adhesive materials, some other characteristics are important in determining whether a particular adhesive will function properly in a given application.

Tack, or stickiness, is the ability to form an initial bond of measurable strength immediately after the substrates being bonded are brought together. *Setting time* refers to the time required for an adhesive to form an acceptable bond.

Open time refers to the time during which the adhesive remains active without curing, after being applied to one substrate. Also called the *working time*, this is the maximum amount of time that can elapse before bringing together and applying pressure to the two substrates.

Drying time refers to the time (after application) that it takes for adhesive particle coalescence to occur and form the final bond.

9.12.3 Natural Adhesives

The most common natural adhesives used for paper-based packaging are made from starch. Water-based starch adhesives are commonly used for paper-to-paper bonding. Starch is also used in the paper-making process because of its ability to bind with paper.

Starch is a natural vegetable polymer, usually from corn, grains, or potatoes. It is the most economical adhesive. But starch adhesives usually have a low solids content, 20–30%, and therefore a high water content so they require a long drying time. Starch-based adhesives are the most used adhesives for combining corrugated fiberboard, where heat is used to gel and dry the adhesive. They are also used for tube winding, paper laminating, bag bottoms, and some labeling applications.

Dextrin is a partially depolymerized starch with a lower molecular weight. It dissolves more easily in water so it can be used at a higher percentage of solids, 40–50%. These adhesives are tan-brown, and usually have an acid pH. Dextrins have high wet tack and they set faster than starches, but not as fast as synthetics. They are used when a long setting time is not a problem, for set-up boxes and bag tubing.

Borated dextrins have been mixed with borax, to improve tack and stabilize the viscosity. They are white to dark brown in color, very tacky, and moderately fast setting. They have an alkaline pH and fairly good resistance to relative humidity. Borated dextrins are widely used for carton and case sealing, laminating, and tube winding.

Jelly gum adhesives are made from alkaline-treated starch. They have very high tack, excellent internal cohesion, and good humidity resistance. At one time, jelly gums were widely used labeling adhesives, particularly for adhesion to glass and metal containers, but they have been mostly replaced by synthetics.

Casein adhesives, based on milk protein, are used for labels on refillable bottles. They are light in color, stick well to wet bottles, and resist moisture in an ice bucket, but are easy to remove in the caustic solution used for cleaning refillable bottles. Casein adhesives are sometimes used, combined with latex, for laminating of foil.

Gelatin glue, also known as *animal or protein glue*, is rarely now used for packaging, although it was once common. Today, most of it is made from pharmaceutical scrap generated from making gelatin capsules. It is a collagen derivative, tacky, and protein based. Animal glue is used to adhere the paper cover to setup boxes and in tube winding. Since it dissolves in warm water, it presents no problem in repulping.

Latex adhesives are suspensions of rubber in water. While they are expensive and have limited stability, they are very useful for production of self-sealing cases and bags.

9.12.3 Synthetic Adhesives

Water-based *synthetic resin emulsion adhesives*, sometimes called *white glues*, are based on PVA or vinyl acetate copolymers (often ethylene vinyl acetate). Fine particles of the polymer are surrounded by a protective colloid such as polyvinyl alcohol and suspended as an emulsion in water. These emulsions have 50–80% solids, which hastens drying. Common household "Elmer's®️ glue" is a PVA adhesive.

As the water dries, the suspension breaks, and the polymer units combine to form the adhesive film. A tack (or *green*) *bond* forms almost immediately, faster than with dextrin. The green bond is able to hold labels or carton flaps in place during packing and conveying to a palletizer. The bond strength continues to increase as the water evaporates.

Resin emulsion adhesives are economical and are used in all forms of paper packaging. They are used to glue carton joints, box flaps and bottle labels. They can be made to be water resistant or remoistenable. They have non-curl and non-wrinkle properties that are good for litho laminating. They are easy to use and clean up.

Solvent-based adhesives contain a polymer dissolved in an organic solvent. The bond is formed as the solvent evaporates. Environmental regulations require that the solvent be collected or destroyed, and this has fueled movement away from solvent-based systems. They are not generally used with paper-based packaging.

Hot melt adhesives are 100% solids. They are heated to a liquid state, applied while molten, and set by cooling and re-solidifying. Melt temperatures range from a low of 250°F (121°C) to the more common 350°F (177°C). Adequate heat is required for a good bond, and cooling sets the bond very quickly since they do not have a water or solvent vehicle that must be removed.

Hot melts are thermoplastic adhesives with a polymer backbone such as ethylene vinyl acetate. The polymer is diluted with a material like wax to improve melting and reduce cost. Modifying resins are added such as tackifiers and adhesion promoters.

The outstanding property of hot melt adhesive is its ability to bond quickly to any substrate. Hot melts are water-resistant, but they do not resist oil, which can dissolve the bond. Hot melt adhesives are widely used for sealing wrap-around style shipping containers and multi-wall paper bags where quick bonding is essential. They are also used for labels.

However, hot melt adhesives are the most expensive kind, and the equipment for applying them requires a higher degree of maintenance. They can also cause problems for paper recycling, since they cannot be readily removed during repulping and cause "stickies" in recycled paper.

Pressure-sensitive adhesives (*PSA*) are used on pressure-sensitive labels. The surface of a pressure-sensitive adhesive can be considered a high viscosity liquid that provides instant bonding when pressure is applied. PSAs are most often based on acrylics or on rubber/resin blends. They can be applied to the first substrate in the form of solvent-borne solutions, water-borne emulsions, latexes, or hot melts. The adhesive-coated labels are generally placed on a silicone-coated release sheet, from which they are peeled off and applied to the package.

The balance between cohesive strength and adhesive ability is controlled by the type of monomer(s) and the degree of polymerization. Increasing the molecular weight tends to result in better cohesive strength but worse adhesion. Although they are more expensive to purchase than wet glues, PS labels offer many production advantages, as explored in Chapter 11.

9.13 FOOD CONTACT CONSIDERATIONS: MIGRATION OF INDIRECT ADDITIVES

Packaging materials in contact with food must not only be compatible with the food and the preservation method, they must not allow *migration* of harmful substances from the package to the food. To say that "food packaging should be safe" seems obvious, but the details are controversial as the food industry, researchers, and public representatives debate over the risks. Some of the earliest concern was for lead solder in tinplate cans which was clearly toxic, caused cases of lead poisoning, and was ultimately outlawed in most countries.

The government protects consumers from unsafe packaging. In the United States, the Food, Drug and Cosmetic Act (1938) prohibits unsanitary food packages. The Food Additives Amendment (1958) requires additives, even *indirect additives* like compounds that migrate from the package, to be safe. EU Directives have also established test protocols and migration limits. Other countries have similar regulations.

Although most of the current concern is about plastics, there are cases where harmful substances in paper, paperboard, and wood have been found to migrate to food. And although paper and paperboard food packaging is not currently directly regulated in the United States or European Union, like all food contact materials they should meet the requirement to not transfer constituents to the food that would, under "normal and foreseeable conditions of use," endanger human health (EC Framework Regulation No. 1935/2004, article 3).

The contaminants of concern for paper-based packaging are not from the fiber itself, but from bleaching, printing, and recycling. They include dioxins, benzophenone, isopropylthioxanthone, mineral-oil-saturated hydrocarbons, and other contaminates from recycled paper.

Dioxin is the name for a family of polychlorinated dibenzo-p-dioxins and polychlorinated dibenzofurans, some of which are known to be toxic. Dioxins have been found in SBS paperboard pulp bleached with chlorine. In the 1990s, after they were found to migrate from milk cartons into the milk, the World Health Organization recommended a reduction in the amount of acceptable dioxins in food products, and in 2006 regulators in the European Union set maximum levels for dioxins in milk and other food. As a result, suppliers of the SBS used for milk cartons have adopted bleaching processes that avoid generation of dioxins. They have improved the processes of delignifying and washing of pulp, to reduce dioxin precursors prior to bleaching, and replaced elemental chlorine with chlorine dioxide and/or oxygen bleaching compounds.

Fluorochemical additives that were used to impart grease-resistance to packages for oily food like pizza, cookies, and microwave popcorn, were named a "likely carcinogen" by the EPA in 2005. Fluorine-based substances were found to migrate into the very greasy products that they were intended to protect, and cooking in the package, like microwave popcorn, exacerbated the process. As a result, many users switched to safer alternatives.

Benzophenone (BP) and *isopropylthioxanthone* (ITX) are used as photoinitiators for inks, varnishes, and lacquers that are cured with UV light. UV inks dry faster and produce less volatile organic compounds (VOCs) than the more common solvent-based inks used to print paperboard cartons and labels. Photoinitiators act as a wetting agent for pigments and as a reactive solvent and a drying catalyst, but some remains in the paperboard after curing. BP has been found to migrate from printed cartons into foods like chocolate, frozen food, and cereals, even with the print on the exterior, and a plastic film or coating between the board and the food. It has been found to taint the flavor of milk in plastic bottles with paper labels printed with UV-cured ink. ITX has been detected by scientists in drink boxes and the milk contained in them, but unlike BP, has no obvious off-taste. As a result, the printing ink industry is developing new low-migration inks.

Paper and board recycled from post-consumer waste is rarely used for food contact because there are so many potential sources of unhealthy contamination, most of which are associated with the ink from the originally printed paper. *Mineral oil saturated hydrocarbons* (MOSH) come from the ink in newsprint, which is widely recycled to make the paperboard that is, in turn, printed with ink for which MOSH may also be in the solvent. *Diisopropylnaphthalenes* (DIPNs) and *trimethyldiphenylmethanes* (TMDPMs) are from carbonless copy paper or waste paper that has been ink-jet printed.

All of these recycled contaminants have been found to migrate from paperboard cartons to dry foods, like powdered baby milk and pasta, by evaporating from the carton and condensing on the food (see Biedermann *et al.* 2011). Adding a plastic film barrier can slow down the process, but polyolefins like PE and PP may not be sufficient because the permeation occurs too quickly. Oriented polyethylene terephthalate (12 μm) or 15 μm oriented polyamide have been found to be effective barriers towards mineral oil components (Welle *et al.* 2013).

Although there are currently no firm regulatory limits, there is continuing research into the contaminants that can migrate from recycled paperboard and their health effects.

For more information on the subject, an excellent review of the legislative and safety aspects of paper and other types of food packaging has been published by Gordon Robertson (2013).

9.14 PAPER-BASED PACKAGING SUSTAINABILITY ISSUES

As discussed in Chapter 2, wood has inherent sustainability advantages, as it is a renewable resource. Since paper-based packaging comes from this same resource, it has similar sustainability advantages. It is made from a renewable resource, and after it has been used, it is biodegradable and/or easily recyclable (as shown in Chapter 6).

If paper-based packaging is not recycled, it will end up in either a landfill or an incinerator. In those places that have *waste-to-energy incineration,* paper-based materials are an important source for generation of energy. The primary products of incineration of wood and paper-based materials are carbon dioxide and water vapor. However, because paper and wood contain other materials such as extractives, salts, etc., other chemicals can be generated during incineration, including HCl and dioxins. Proper control of incineration conditions and pollution control equipment is essential for clean burning of wood and paper, as it is for other materials.

In a landfill, paper-based materials will eventually biodegrade. All biodegradation processes in landfills are slow, as conditions soon become anaerobic, and landfills are generally managed in a way that minimizes moisture content, in order to protect groundwater from contamination. In dry, anaerobic conditions, microorganisms do not grow very rapidly. When the paper finally does biodegrade, the primary products are methane and carbon dioxide. Methane can be collected as "landfill gas" and used as an energy source. In the United States, if it is not collected, it must be flared, rather than simply allowing it to escape. But some methane does escape even from well-managed landfills, and it is a potent greenhouse gas. The lignin fractions of the paper-based packaging degrade much more slowly than the cellulose and hemicellulose fractions. To the extent that the carbon in the paper-based packaging does not degrade, it sequesters carbon that was once in the air, acting as a greenhouse gas sink.

Paper-based materials that are contaminated with food are not desirable in most paper recycling facilities, as cleaning them is difficult, leading to contamination issues. One option that is increasing in popularity is *composting.* In a composting operation that is designed to handle residential or commercial food waste streams, food-contaminated paper can often be included without significant difficulty, as the cellulose, hemicellulose, and lignin components of the paper are all biodegradable, although in composting, as in landfill, lignin biodegrades at a much slower rate than the other two components. Nonetheless, just as yard waste can be composted, paper can generally be composted if the facilities are designed to be able to handle it.

One of the tools for evaluating the sustainability of products and processes is *life cycle assessment* (*LCA*). LCA is a method for developing an "environmental accounting" for the environmental impacts that are associated with a product or process. Once the goal and scope of the LCA are established along with appropriate boundaries, all the inputs and outputs from the process are quantified (to the extent possible). This life cycle inventory (LCI) is then converted to a set of environmental effects, or impacts, using an impact assessment methodology in a process termed life cycle impact assessment (LCIA). Various impact assessment methodologies have been developed. All are designed to convert the huge list of LCI data into equivalent amounts of "reference substances" in a much smaller number of impact categories. For example, all greenhouse gas emissions are converted into equivalent amounts of carbon dioxide, the reference materials for the "global warming" impact category. Then, in the Interpretation phase of LCA, the information is evaluated, including evaluation of completeness and uncertainty of the information.

A number of LCA studies have addressed various comparisons between paper-based and other packaging systems for a variety of products. The International Standards Organization (ISO) has published guidelines for performing LCAs: ISO 14040 and 14044. One of the key requirements is that studies that are intended to be used to make public "comparative assertions" (saying one choice is environmentally better than another) must be subjected to peer review. Especially in the early days of LCA, different studies have reached radically different conclusions about the relative environmental merits of packaging options, in large part because they used differing data, boundary conditions, and assumptions. While this issue has not disappeared, it has been diminished by the advent of these standards.

A more limited approach to life cycle analysis, carbon footprinting, concentrates on the greenhouse gas profiles of products and systems. There are also ISO standards for carbon footprinting.

There is currently not a consensus on how best to account for the global warming potential of paper-based packaging (or for wood, for that matter). But the relationship to the *carbon cycle* is clear. Carbon dioxide from the air is removed as the tree grows, converted into biomass through the process of photosynthesis. When the components of the tree biodegrade, the contained carbon is returned to the air, generally as carbon dioxide or as methane that is later converted to CO_2. For as long as the carbon remains in solid form, that carbon has been sequestered.

Some approaches to LCA attempt to account for all these time-dependent carbon flows (and do so for a fixed period of time—often 100 years). Others simply regard the whole system as "carbon neutral" except for the influence of methane. Because methane is a much more potent greenhouse gas than carbon dioxide, there is general agreement that methane flows to the atmosphere must be accounted for separately; they cannot just be ignored.

These differences in accounting for carbon flows, as well as other factors, can result in significant differences in the LCA profiles for paper-based packaging materials. In general, it is found that use of recycled paper reduces the overall energy demands for paper-based packaging (see Mourad *et al.* 2008). However, because the manufacturing of paper from biomass resources generally uses a large proportion of biobased energy while recycling tends to use mostly petro-based energy for collection and processing, it is not at all uncommon to find that manufacture of paper with a high recycled content uses more fossil fuel than manufacture of similar virgin paper (see Gilbreath 1996).

Regardless of these findings, most experts agree that the recycling of paper-based materials brings significant environmental benefits. Of course, that does not make recycling the best choice for all materials in all places, but recycling does contribute to maximizing the useful life of these bio-based and biodegradable materials, which with proper management can be produced sustainably.

9.15 REFERENCES

Biedermann, M., Y. Uematsu, and K. Grob. 2011. Mineral Oil Contents in Paper and Board Recycled to Paperboard for Food Packaging, Packaging. *Technology and Science,* 24 (2): 61–73.

Bruno, M.H., Ed. 1995. Label Industry Facts & Guidelines, 2nd ed. Alexandria, VA: Label Printing Industries of America.

Chamberlain, D. and M.J. Krwin. 2013. Paper and Paperboard—Raw Materials, Processing and Properties. In *Handbook of Paper and Paperboard Packaging Technology,* 2nd ed., edited by Mark J. Kirwin, pp 1–50. Chichester: John Wiley & Sons.

Eldred, N.R. 1993. *Package Printing.* Plainview, NY, Jelmar.

Ewender, J., R. Franz, and F. Welle. 2013. Permeation of Mineral Oil Components from Cardboard Packaging Materials through Polymer Films. *Packaging Technology and Science,* 26 (7): 423–434.

Gilbreath, K.R. 1996. Life Cycle Assessment, Energy, and the Environment from a Pulp- and Papermill's Perspective. *Life Cycle Environmental Impact Analysis for Forest Products,* Proceedings, Forest Products Society, pp. 61–106.

Hanlon, J.F., R.J. Kelsey, and H.E. Forcinio. 1998. *Handbook of Package Engineering,* 3rd ed. Lancaster, PA: Technomic.

Jenkins, B. and staff. 2009. Cellophane. In *The Wiley Encyclopedia of Packaging Technology,* 3rd ed., edited by Kit L. Yam, pp. 252–253. New York: John Wiley & Sons.

Kouris, M., Ed. 1996. *Dictionary of Paper.* 5th ed. Atlanta: TAPPI.

Labarre, E.J. 1952. *Dictionary and Encyclopedia of Paper and Paper-making.* 2nd ed, revised from a 1937 edition that included samples, reprinted 1969. Amsterdam: Swets & Zeitlinger.

Mourad, A.L., E.E.C. Garcia, G.B. Vilela, and F. von Zuben. 2008. Environmental Effects from a Recycling Rate Increase of Cardboard of Aseptic Packaging System for Milk Using Life Cycle Approach., *International Journal of Life Cycle Assessment,* 13(2):140–146.

Paine, F.A. 1991. *The Package User's Handbook.* Glasgow: Blackie (published in U.S. by AVI, New York).

Robertson, G.L. 2013. *Food Packaging: Principles and Practice,* 3rd ed. Boca Raton, FL: CRC.

Sikora, M. and staff[CM14]. 2009. Paper. In *The Wiley Encyclopedia of Packaging Technology*, 3rd ed., edited by Kit L. Yam, pp. 908–912. New York: John Wiley & Sons.

Smook, G. 2002. *Handbook for Pulp and Paper Technologists*. Vancouver: Angus Wilde.

Soroka, W. 2002. *Fundamentals of Packaging Technology*. Naperville, IL: Institute of Packaging Professionals.

US National Archives. 1991. Acid-Free Archives Box. http://www.archives.gov/preservation/storage/acid-free-boxes.html

9.16 REVIEW QUESTIONS

1. Why is it difficult to print fine detail on natural kraft paper?

2. What kind of paper is used to make grocery bags, glassine, and corrugated board?

3. For what kind of products is grease-proof paper used? Name two kinds.

4. How does the production of glassine differ from parchment?

5. Which is a better moisture barrier: glassine or waxed paper?

6. Which feels more waxy: wet or dry waxed paper?

7. How is cellophane made?

8. What is the advantage of cellophane over PP?

9. For what purpose is acid-free paper used?

10. What are the differences between recycled and kraft board?

11. What is the highest grade of paperboard?

12. What is meant by C1S and C2S? To what kind of coating does it refer?

13. To what does the word "containerboard" refer? What materials are made from it?

14. How does molded fiber/pulp differ from paper?

15. Which polymer is most common for extrusion-coating paper and board? Why?

16. Explain mechanical adhesion. What is the most common test for paper adhesion? What is meant by "fiber tear" and what is indicated by its absence?

17. What is the difference between an adhesive and a cohesive bond failure?

18. List the following adhesives in order of decreasing solids content: starch, dextrin, hot melt, and PVA emulsion. What advantage is a high solids content?

19. For what purpose is starch adhesive used?

20. What kind of adhesive is Elmers® Glue?

21. What kinds of adhesives can be used for labels?

22. What are some advantages of hot melt adhesive? What are some disadvantages?

23. What happens if the open time in a gluing or laminating process exceeds the range of the adhesive?

24. In the context of food packaging, what is meant by "migration?"

25. What are the sources of dioxins, fluorochemicals, and mineral oil saturated hydrocarbons in food packaging?

26. What are the benefits and drawbacks of waste-to-energy incineration of used packaging?

27. Can paper-based packaging be composted?

28. Use the language of LCA to describe the carbon cycle with respect to paper-based packaging.

CHAPTER 10

Printing on Paper and Paperboard

Before the 1850s, just about the only printing on packages was black letterpress text on labels attached to bottles, or ink stenciled onto wooden barrels and boxes. One hundred and fifty years later, printing is one of the most valuable parts of packaging. Printed words identify the package's contents. They present the facts required by regulation, like net weight, ingredients, nutrition facts, or drug facts.

But package printing is more than words: it is color, excitement, and motivation!

Eye-catching point-of-purchase graphics are effective advertising, enticing consumers at the moment when they are shopping and most receptive. Brand symbols on packages reinforce other advertising images and previous experiences with a product. In the store, graphics can draw attention to a package that is new, contrasting it from others on the shelf. Visual elements convey meaning about the package's contents, emphasizing benefits that make products more useful: mouth-watering photos can make processed food seem to taste better, colors and symbols can emphasize flavors or features.

Colors signify emotions and arouse feelings. In general, red represents vigor, appetite, and energy. Blue is calming, peaceful, and clean. Green is fresh, healthy, and natural. Yellow gives hope and energy. Orange can be warm or edgy. Pastels are gentle, and fluorescent colors are vibrant. Black is sophisticated, white is pure, and touches of gold convey opulence. The meaning of colors can vary by culture, and package color preferences follow other cultural trends.

Printing helps to segment markets. Fun, bright illustrations target children, and easy-to-read fonts aim for their grandparents. A product can be cloaked in sophistication or homespun, high-tech or organic, hedonistic pleasure, earnest religion or in-your-face cynicism—by a simple stroke of the press.

303

The strategic elements of print extend throughout the supply chain. "Smart" graphics like bar codes even tie packages to retailing information systems, from pricing to tracking to inventory control. Even though printing is one of the most expensive components of a package, when it successfully contributes to selling the product, it earns its cost.

Package printing is a significant part of our visual culture. The art and beauty of packaging are an inherently enjoyable part of the shopping experience (Hirschman and Holbrook 1982). The images on packages are an expression of our lives and time (Belk 1988).

The history of modern retailing is illustrated by package printing, and most of it is due to paper-based materials. Paper is the most printable packaging substrate. After all, paper is the material for which most printing processes were developed in the first place. Package printing processes have been adapted from the ones used for publishing books, magazines, newspapers, and for commercial and decorative printing.

Packaging is a big component of the printing industry. About 40% (by value, 25% by volume) of all ink is used for package printing (Eldred 2007). The package printing industry now leads the publication industries in many new technology developments.

This chapter describes the basic package printing processes, inks, and how the design process is managed. It does not deal directly with graphic design, which is a separate artistic subject. It begins with a discussion of text elements, and then shows how images and colors are created.

10.1 TEXT AND OTHER (MOSTLY) BLACK PRINTING

Package printing is a manufacturer's best medium for communicating with its customers and supply chain members. The words on a package are chosen carefully to convey the maximum amount of information in a small space. Words describe the product, provide nutritional or drug facts, net weight, and instructions or recommendations for use. Fonts are chosen for their legibility. Most text and bar codes are simple and printed in black.

10.1.1 Labeling Regulations

Some text on labels is regulated. In the United States, most consumer packages must comply with the *Fair Packaging and Labeling Act*. Food packages must also comply with the *Nutrition Labeling and Education Act*, and over-the-counter drug packages must carry *Drug Facts*. The purpose of this legislation is to protect consumers by giving them useful information about what the package may hide. The government's statement is this:

"Informed consumers are essential to the fair and efficient functioning of a free market economy. Packages and their labels should enable consumers to obtain accurate information as to the quantity of the contents and should facilitate value comparisons. Therefore, it is hereby declared to be the policy of the Congress to assist consumers and manufacturers in reaching these goals in the marketing of consumer goods." (U.S. FDA 1966, Fair Packaging and Labeling Act, section 1451)

The laws are published in the U.S. Code of Federal Regulations. There are many specific provisions that are beyond the scope of this brief discussion.

The Fair Packaging and Labeling Act, FPLA (U.S. FDA 1966) applies to food and other consumer goods. The Food and Drug Administration oversees it for food, and the Federal Trade Commission oversees it for non-food products. It does not apply to all consumer products, but it does apply to products for which a package's net contents cannot be easily ascertained, like the number of Christmas lights on a string, or for products that are purchased by weight, like kitty litter.

There are five elements required by the FPLA, as shown in Figure 10.1. There must be (1) a *principal display panel* (*PDP*) that gives (2) the *product identity* in terms of its commodity name. For example, the PDP for a Cheez-It® carton identifies the contents as "Baked Snack Crackers." There must be a declaration of the package's (3) *net contents* (mass, measure, or numerical count) in both the inch/pound system and metric units. If the pack-

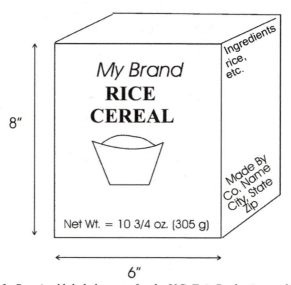

FIGURE 10.1. *Required label elements for the U.S. Fair Packaging and Labeling Act.*

age is labeled by weight, it should be expressed in pounds. If the contents are liquid or linear measure, the package should be labeled with the largest whole unit (quart, foot, etc.). Any remainder should be expressed either in terms of ounces or inches or a decimal fraction of the whole unit (like 0.23 lb.). There is a requirement for the size of the type font, based on the size of the PDP; for example on an 8″ × 6″ PDP, the type must be at least three-sixteenths inch. The net content statement must be parallel to the package's base as it is displayed, in the lower 30% of the PDP. The (4) *name and place of business* of the manufacturer, packer, or distributor must be given, and (5) *ingredients must be listed* in order of decreasing predominance. The company and ingredients are not ordinarily printed on the PDP. The FPLA also specified how big the print needs to be, based on its proportion to the package size (U.S. FDA 1966).

The *Nutrition Labeling and Education Act* (U.S. FDA 1990) requires a *Nutrition Facts* panel on food packages, as shown in Figure 10.2. It can appear on any label panel, and must be large enough to be easy to read. A standard format is specified. The minimum font size is 8 point for nutrient names and 6 point for the values, and "Nutrition Facts" must be the largest boldest words. Values are expressed per serving size, which must be in terms of a common household measure. A package that contains between one and two servings must be listed as a single serving, and if there are between two and five servings, the number is rounded to the nearest 0.5.

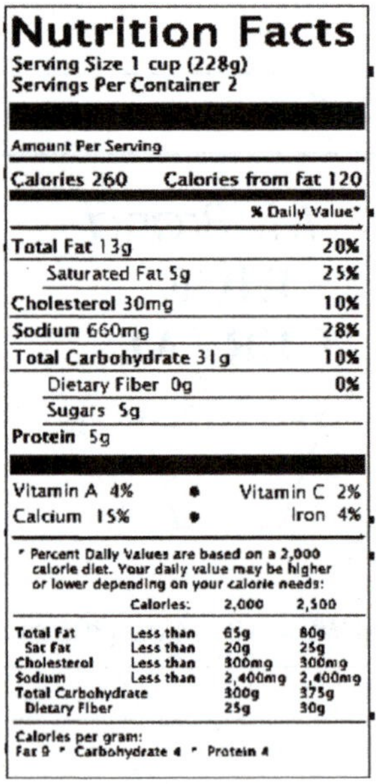

Nutrition Facts

Serving Size 1 cup (228g)
Servings Per Container 2

Amount Per Serving

Calories 260   Calories from fat 120

	% Daily Value*
Total Fat 13g	20%
Saturated Fat 5g	25%
Cholesterol 30mg	10%
Sodium 660mg	28%
Total Carbohydrate 31g	10%
Dietary Fiber 0g	0%
Sugars 5g	
Protein 5g	

Vitamin A 4%	•	Vitamin C	2%
Calcium 15%	•	Iron	4%

* Percent Daily Values are based on a 2,000 calorie diet. Your daily value may be higher or lower depending on your calorie needs:

		Calories:	2,000	2,500
Total Fat	Less than		65g	80g
Sat Fat	Less than		20g	25g
Cholesterol	Less than		300mg	300mg
Sodium	Less than		2,400mg	2,400mg
Total Carbohydrate			300g	375g
Dietary Fiber			25g	30g

Calories per gram:
Fat 9 • Carbohydrate 4 • Protein 4

FIGURE 10.2. Required label elements for the U.S. Nutrition Facts (U.S. FDA 1990).

Drug Facts

Active ingredient (in each tablet) — ***Purpose***
Chlorpheniramine maleate 2 mg .... Antihistamine

Uses temporarily relieves these symptoms due to hay fever or other upper respiratory allergies:
■ sneezing ■ runny nose ■ itchy, watery eyes ■ itchy throat

Warnings
Ask a doctor before use if you have
■ glaucoma ■ a breathing problem such as emphysema or chronic bronchitis
■ trouble urinating due to an enlarged prostate gland

Ask a doctor or pharmacist before use if you are taking tranquilizers or sedatives

When using this product
■ You may get drowsy ■ avoid alcoholic drinks
■ alcohol, sedatives, and tranquilizers may increase drowsiness
■ be careful when driving a motor vehicle or operating machinery
■ excitability may occur, especially in children

If pregnant or breast-feeding, ask a health professional before use.
Keep out of reach of children. In case of overdose, get medical help or contact a Poison Control Center right away.

Directions

adults and children 12 years and over	take 2 tablets every 4 to 6 hours; not more than 12 tablets in 24 hours
children 6 years to under 12 years	take 1 tablet every 4 to 6 hours; not more than 6 tablets in 24 hours
children under 6 years	ask a doctor

Other information store at 20-25° C (68-77° F) ■ protect from excessive moisture

Inactive ingredients D&C yellow no. 10, lactose, magnesium stearate, microcrystalline cellulose, pregelatinized starch

FIGURE 10.3. *Required label elements for U.S. Drug Facts (U.S. FDA 1999).*

The nutrients that must be declared in terms of amount and percentage of daily requirement (based on a 2,000 calorie diet) are: calories, fat, cholesterol, sodium, potassium, total carbohydrates, fiber, protein, and a number of vitamins and minerals. A footnote can be added that also gives the caloric requirement percentages terms of a 2,500 calorie diet. If the product is commonly combined with other ingredients, and directions for that combination are provided, additional columns may present the nutrition content for the food as prepared. Shorter simplified formats are permitted on small packages.

The FDA requires a similar *Drug Facts* label on all over the counter medications (U.S. FDA 1999). The label must include the product's active ingredients, the amount in each dosage unit, the purpose of the medication, uses for the drug, specific warnings and side effects, dosage instructions (when, how, and how often), and inactive ingredients for those with allergies, as shown in Figure 10.3.

Other labeling legislation covers alcohol, tobacco, and hazardous materials. Alcohol and tobacco have warning elements mandated by the U.S. Bureau of Alcohol, Tobacco, and Firearms, and there is some state-mandated information like the distributor's name. Hazardous materials labeling is overseen by the U.S. Department of Transportation, which also has performance standards for packages as described in Chapter 15.

Legislation varies by countries and states. For example, in some countries there is a legal requirement to indicate a food product's shelf life. This is usually expressed in terms of a *sell by* or *use by date*. Such code dating is generally printed on the filling line by an ink jet, laser image, or other code printer.

10.1.2 Other Product Information

The package is the link between the product, the manufacturer, and the consumer. It can inform, entertain, and instruct. In choosing the words and images for a package, successful marketing and advertising strategies focus on the consumer.

Our attention is drawn to pictures before we read words, and if the pictures convey the message, we may not read the words at all. Pictures can show what something looks like (especially for food after it is cooked), they can show how to do something or a sequence of events, and they can show what the product is expected to achieve. Food looks best in photographs, but sometimes drawings, since they are simpler and easier to understand, are better than photos.

Often a consumer buys the promise of a solution, not just a product. In addition to the legally mandated identity statement, a good package shows the consumer how the product may solve his/her problem. For example, a package for laundry detergent that emphasizes color symbolizes the fact that the detergent will preserve colors.

A consumer may make incorrect assumptions, and so information should be easy to understand, and hard to misunderstand. For example, an infamous package failure was the "Wine and Dine" product that included a bottle of cheap cooking wine to be used as an ingredient. The message that consumers got, however, was that the wine was meant to be drunk with the meal, resulting a bad beverage and no repeat customers, because the graphics did not make the message clear (McMath and Forbes 1999).

Information should be appropriate and practical. Package opening and product use instructions should be clear and understandable. Safety warnings should be helpful. Just as graphics and advertising campaigns are tested, informational text elements should be tested with real consumers to make sure they are easy to comprehend.

Aside from the consumer focus, the label represents the manufacturer and brand. The graphic brand depiction is the company's coat of arms, and a great deal of effort is expended to develop and protect it. Package graphics and colors are based on creating a good image for the brand.

10.2.3 Supply Chain Printed Elements

Printed elements on packages can also make a supply chain more efficient and secure. Inventory management in supply chains is growing increasingly efficient, but in order for networks of computers to track the movement of goods through distribution networks, there must be accurate recording each time a package's status changes.

This is commonly done by recording and verifying the *stock keeping unit*

(*SKU*) number, printed on the package, against the computer information that controls the supply chain. The SKU symbolizes a single variation of a product (brand, color, flavor, package size, etc.). For this reason, the SKU number on a shipping container must be easy to read accurately, even in a dark warehouse. The SKU number on consumer packages is in the form of the bar code.

Automatic identification, such as bar codes, reduces the possibility of errors in data entry, and makes it easier to track distribution in real time. For example, by scanning a bar code at check-out, the retail information system can simultaneously look up the price to charge the customer, track inventory, and automatically generate an order when the inventory gets low. For automatic identification to work, however, there needs to be a common symbology that can be applied across the supply chain. For example, retail consumer goods are assigned a bar code number in the United States by the *Uniform Code Council* (in Europe it is the *Article Number Association*). It is estimated that the accurate real-time inventory information made possible by retail bar codes facilitates a 25% reduction in inventory holdings (Hall 1999).

The *Universal Product Code* (*UPC*) and the *European Article Number* (*EAN*) are made up of two sets of five digits; the left-hand set identifies the company (like Kraft) and the right-hand set identifies the specific product (like a 10 oz. bag of mini-marshmallows). There is also a number system digit for the product category and a check sum digit at the end, plus three sets of bars that frame the code, at the beginning, the middle, and the end.

Clean, sharp printing is essential for accurate bar code scanning, because the technology relies on the laser scanner's ability to differentiate between the spaces and the thick and thin lines. The sharpness of lines depends on the printing method.

All printing processes have some image gain, but flexography is worst, due to the halo affect (discussed later in this chapter). For flexo web-fed printing, this is greatest in the printing machine direction, and so the effect is reduced by aligning the bars with the machine direction.

Scanning accuracy also depends on color contrast and the presence of a *quiet zone* that must be kept free of any other printing that the scanner might mistake for a code element. Since the scanner detects reflects red light, the background should be white, yellow, or red, and the bars should be black, dark blue, or dark green. If the bars are printed with red, the scanner will not detect them.

There are bar code symbologies other than UPC and EAN that are used during distribution rather than in retail stores. Typically they have more digits and are standardized within an industry's supply chain. Test methods for bar codes have been published, and most users expect a minimum of 97% first-time readability (LaMoreaux 1995).

There are also two-dimensional matrix codes that can encrypt more infor-

FIGURE 10.4. *Bar code and QR code.*

mation. Originally used for inventory control, the *Quick Response code* (QR code) is the most popular. It is comprised of square dots arranged in a square grid on a white background. The code can be read by an imaging device like an electronic camera that can interpret the image and extract data from the patterns of black and white. Figure 10.4 shows the difference between the one-dimensional UPC code and a two-dimensional QR code.

On-package QR codes are now used for marketing to smart phone users who can scan a code to display text or a website, or connect directly with a producer. This is a powerful marketing tool because of how it directly engages with consumers at the moment that they initiate an inquiry. QR codes can convey information far beyond what can be printed on a package, including a product's source, a manufacturer's backstory, uses, and recipes, and can lead a user to social media sites where they can interact with other users.

Radio frequency identification (RFID) is an emerging automatic identification technology. A programmable computer chip and antenna located on the package communicates with a host computer via radio frequency waves. RFID technology is being first used for palletload and shipping container identification through the supply chain. Although high-priced, there are great expectations for the cost to fall, especially as methods are developed for printing the antenna directly on the package with conductive inks.

Security labeling is growing in this uncertain age. Labels are being used to prevent tampering, or at least make it evident, by incorporating materials that shatter or otherwise react (like changing color) when opened. They are being used to prevent theft, by incorporating *electronic article surveillance* (*EAS*) radio frequency tags that set off an alarm unless deactivated at checkout. Labels can also be used as anti-counterfeiting devices, with an overt hard-to-reproduce element (like a hologram) or with a secret mark known only to the manufacturer (Hall 1999).

Security and product tracking needs are expected to increase, and so-called *smart labels*, and the information systems to enable them, will continue to change to meet the challenges.

10.2 IMAGES AND COLORS

In general, there are two kinds of images: line and continuous tone. *Line* refers to text and line drawings. *Continuous tone* refers to shaded images, like black-and-white photographs or drawings, with a variety of grey tones.

Line images are a straightforward matter of printing the lines and solid dark fields. Line images are inexpensive to produce, but they look flat. Line images have no subtle gradation of tone or color, no shadows, and no highlights.

Continuous tone breathes life into an image. But shaded images cannot be printed on a press in the same way that they are painted by simply using darker shades. A printing press is not capable of making tone gradations; the ink is either printed or it is not. (This is not strictly true because the gravure process is capable of some variation in ink thickness.) Therefore, tone and shading are created with a pattern of dots. In black printing, this is called a *halftone image*.

10.2.1 Halftones

A halftone is an optical illusion. It is a grid of very small black dots of various sizes with equal spacing between their centers. Gradations in tone are represented by larger or smaller dots; a light part of the image has small dots, and a shaded or dark part has large dots. The original image is broken into dots by scanning and digitizing or by directly digitizing a computer image. Originally, this was done by photographing the image through a fine screen, and the number of dots per inch is still referred to as the *screen ruling* or *line screen*.

When viewed from a distance, the pattern creates the illusion of a continuously shaded tone. This is because the human eye has a limited power to resolve small dots at a screen ruling of 133 lines or dots per linear inch (dpi) and higher. This principle can be easily observed in any black-and-white newspaper photo with an ordinary magnifying lens, as shown in Figure 10.5.

10.2.2 Process Color

Colors can be achieved in two ways: spot or process color. *Spot colors* use a different ink for each color, and are used to produce a limited number of flat effects. Spot colors are used for large single-color blocks, line art, and for a brand's special color identification. A spot color is consistent, used for special brand colors (which may be trademarked) like the Kodak gold, Kellogg red, Cheer blue, or Michigan State University green. Spot colors are specified according to the *Pantone Matching System*®.

But most presses can only carry four to seven colors and a separate printing station and plate are needed for each spot color. Inks cannot be mixed on a press in the same way that they are mixed in painting. Therefore a second

FIGURE 10.5. *Halftone dots magnified. Beaumont Tower newspaper photo, reprinted by permission of University Relations, Michigan State University.*

optical illusion is employed: the halftone principle is applied to color to create an entire palette of colors, made from dots of just four primary colors of ink.

This is called *process color*. In order to get a range of colors from a limited number of primary colors, the printer tricks the eye with tiny colored dots of varying sizes that combine to reflect a spectrum of color when viewed from a distance.

The primary process colors of ink are *yellow, magenta* (a pinkish red color), *cyan* (a greenish blue color), and *black* (for shading and contrast), called by the industry *CMYK*.[21] These differ from the *additive* primary colors of mixing pigments or opaque paint (red, yellow, and blue). They are more like watercolors, because they depend on the light reflected through the ink from the bright white substrate below.

Cyan, magenta, and yellow (CMY) inks are known as the *subtractive primary colors*. They are transparent inks that selectively filter (absorb) some wavelengths of light reflected back from the white surface below.

When dots of two primary colors are printed next to or overlap each other, the illusion of an intermediate color is created. Combining dots of yellow and magenta creates the range of colors from yellow to red. Magenta and cyan make the colors from reddish-purple to blue. Yellow and cyan color lemons and limes. A combination of all three theoretically creates black, although in practice the colors are not pure enough to absorb all light, and so the fourth primary color, black, is used in process color printing to enhance contrast.

Like halftones, the size of the dots determines the density and shade; the larger the dot, the stronger is that component of the color. Combining dots of the primary ink colors by interspersing them on the reflective white paper creates the range of colors. The fine dots are printed sequentially in register with each other, like halftones, on equally spaced centers. The dots can only be seen with a magnifying glass, which reveals a surprising world in which almost everything in our graphic world is reduced to the same four primary colors of dots!

The transparent inks and reflected light are the reasons that a bright white substrate is essential, and why the surface brightness of paper and paperboard is usually specified. A colored paper, or even dull white, would alter the colors. In order to get a range of colors and good color reproduction, the substrate should reflect all wavelengths (colors) of light equally. The surface is generally bleached and/or coated with clay or titanium dioxide to increase brightness. For some applications, like transparent plastic or a dark t-shirt, a white layer of ink may even be printed first.

Commercial artists, and especially package designers, take advantage of the process color optical illusion. Process color is used whenever a package is to carry a range of colors, because using the standard colors cuts the cost of changing the press. Most presses have four to six process color stations. Additional stations are used for a spot color, metallic color like gold, or varnish.

It is surprising to realize how pervasive the colors of magenta and cyan are in package, publication, and commercial graphics today. They are the Andy Warhol/Disneyland colors that symbolize modernity itself. This color scheme may be emphasized in many designs because it is already on the press, or it

[21]"K" is used to distinguish black from blue.

may be a solution to reduce variability problems because a pure cyan or magenta are more likely to stay pure throughout a print run. Process color tends to give much of our culture a faint magenta and cyan cast, never looking truly red, blue, or green. Our western eyes and brains seem to like this color cast, finding a magenta-like red and cyan-like blue more pleasing than a more pure red and blue.

A new process color system has been introduced that has brighter, more fluorescent versions of the standard cyan, magenta, and yellow, plus it has a vivid orange and green, with black printed first. Called the *Pantone Hexachrome®*, this system is used to produce brighter and more vibrant colors. The press must have at least seven color stations and could have 11 or more. There are the six colors, plus a darker denser black for text, and provision for two varnishes, a dull and a gloss for a 3-D effect that is less expensive than embossing.

In order to print using process color, the four colors (CMYK) must first be separated. *Color separation* refers to photographing or digitizing the image through four different color filters. The colored negatives (or positives, depending on the process) are converted to halftone dots electronically (or photographically through a screen with fine lines). The colors are recombined by making plates to separately print each of the colors.

It is necessary for the dots of each color to be in the proper places in relationship to the dots of other colors. The dot grids of each of the four process colors are set at a 15° angle to each other in order to avoid a *moiré* effect with unwanted patterns due to one color slightly overlapping another.

The print resolution quality depends on the number of lines per linear inch (lpi); the metric measure is lines per centimeter. The higher the lpi, the better the resolution. But a high resolution image cannot be printed on a rough surface. A coarse screen density is used for rough surfaces like kraft paper and linerboard, as low as 60 lpi. The industry standard for smooth clay-coated paper and board is 133 lpi.

The choice of screen density also depends on the type of printing press. Flexography screens (used for kraft paper) range from 60 to 133 lpi. Lithography and gravure screens begin at 133 lpi and go as high as 300. Screens with more than 150 lpi are used for art reproductions.

The degree of alignment is referred to as *register*. As an example, very poor registration can be seen sometimes in newspaper comic strips, where one color may be way out of register. A print that is *in-register* has perfect alignment. The printed sheet will have a register mark, like a cross (+) printed in the same place by all colors, to verify that each print station is in-register with the others.

10.3 INKS

Inks are dispersions of *pigments* suspended in a liquid, either oil or resin, called the *vehicle*. The pigments and its vehicle are dissolved in a solvent or water with additives to facilitate drying.

10.3.1 Pigments and Vehicle

Pigments give ink its color. They are finely ground (usually organic) colored materials. For example, black ink is made from *carbon black*, and cyan ink is made from *phthalocyanine* (from which it gets its name). In addition to pigments, soluble dyes can be used, but they are not used much in packaging because of their tendency to run, bleed, and fade.

Since batches of pigment can vary slightly in color, the inkmaker carefully blends batches to always achieve the same ink hue. Much color matching is done by computer-assisted spectroscopy, but this analysis is always supplemented by human eyes.

The *Pantone Matching System*® (*PMS*) is used to specify spot colors. PMS provides a common color chart and terminology for describing, selecting, matching, and color control. It has the formulas for over 1,000 colors, based on nine basic colors and eight process and supplementary colors. Its color charts show the printed color on both coated and uncoated material.

The use of standards helps to make package colors match when they are produced by more than one printer. Each type of printing has its own standards maintained by a professional association. Gravure standards are maintained by the Gravure Association of America. Offset standards are referred to as SWOP, Specifications for Web Offset Publications, and have also been adapted to some flexographic printing. GCMI standards for flexo inks used on corrugated board are issued by the Glass Container Institute (formerly the Glass Container Manufacturers' Institute).

The liquid *vehicle* binds the powdery pigments together. The vehicle can be oil or a natural or synthetic resin, which is determined by the printing process. Lithographic and letterpress ink is oil-based, and most gravure, flexographic, and screen print inks are resin-based.

10.3.2 Solvents: To Disperse and Dry

Solvents and/or water dissolve the oils, resins, and additives to disperse the pigment so that it can adhere to the surface of the paper without being absorbed. Solvents affect the ink viscosity, which determines color intensity, press operations, and ink usage rate. The highly volatile solvent is also key to an ink's ability to dry.

Traditional solvents like alcohol and petroleum distillates offer a number of benefits that water does not. They easily dissolve and carry the pigments, oils, and resins. They evaporate quickly without wicking into the paper.

But solvent-based inks can be bad for the environment. They release volatile organic compounds (VOC) into the air as they dry. VOCs are air pollutants; they smell bad and contribute to illness. VOCs are highly flammable and potentially explosive. Government-mandated air pollution control requires that evaporating solvents be recovered and disposed of or destroyed, a costly operation.

There is an increasing effort to reduce this environmental and cost impact by using water- and soy-based inks. Waterbased inks are used for flexography and gravure when printing on paper, but are not practical for lithography (because of its oil/water basis).

Waterbased inks solve many of the environmental problems caused by solvents, but pose technical challenges, especially for paper. Waterbased inks dry more slowly and require more heat to evaporate. They dry partly by being absorbed into the paper, which tends to become distorted by the wetness, resulting in problems with registration. This also requires more ink to be used. Water-based inks behave differently from solvent-based inks on the press because of differences in viscosity and surface tension. But with growing environmental concern, the technical ability to use water and other solvent alternatives like soybean oil will continue to improve.

A key difference between water and solvents is the rate of evaporation. Drying is one of ink's most important jobs because it affects the productivity of the printing operation. Ink is considered to be dry when its viscosity has increased to the point where the substrate can go on to the next operation like stacking, rewinding, or cutting. Most ink dries by a combination of evaporation and absorption.

A higher speed alternative is *energy curing* of inks by exposure to *UV* or *electron beam* (*EB*) radiation. Instead of evaporating/drying, the ink is cured instantly by polymerization that is initiated by UV or EB radiation. The resulting print is hard, glossy, and resists damage. This allows the ink from an earlier station to dry before being over-printed with a different color.

Although they are higher cost than most other inks, energy-curing inks practically eliminate the use of solvents and the emission of VOCs. They reduce printing costs because the process is faster, requires less energy, and eliminates the need for solvent recovery. They have superior gloss and resistance to abrasion, chemicals, and fats. For these reasons, the use is growing rapidly for labels, folding cartons, ice cream cartons, and gable-top cartons.

Energy-cured inks are available for all printing methods but predominate in flexography. Hybrid oil-based inks have both UV-cured and evaporation components, and are used in lithography.

10.4 PRINTING METHODS

The printing industry uses two types of processes for printing on packaging materials: *plate* or *press* processes, and *plateless* processes. Plateless processes are digitally produced; examples are photocopying, digital, laser, and ink jet printing.

The three printing processes used most for packaging are press plate processes: *flexography, offset lithography*, and *rotogravure*. Inked printing plates that press the image onto the paper are used most commercial printing of magazines, newspapers, packaging, and books. The print is applied as the paper passes between the inked *plate* (or *offset*) *cylinder* and the *impression cylinder* that supports the paper as it is printed.

Each of the three plate processes has the image at a different height than the non-image background. The plates for letterpress and flexography are *relief* plates, with the inked images raised from the surface. The images in rotogravure printing are *engraved* in the plate. Offset lithography and screen printing plates have a flat *planographic* surface (Figure 10.6).

Paper-based substrates that are printed include paper labels, bag paper, paperboard, and corrugated board. In Chapters 11–15 where these types of packages are described, reference is made to the printing methods used. Table 10.1 summarizes the primary package forms (including glass, metal, and plastic) and the print processes used for each.

10.4.1 Flexography

Flexography, also called *flexo*, has the broadest range of uses in packaging because it is inexpensive and versatile. It has the ability to print on a wide

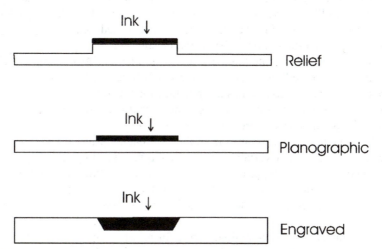

FIGURE 10.6. *Printing plate surfaces: Relief (flexography), planographic (lithography) and engraved (rotogravure).*

TABLE 10.1. Packaging Printing Methods, In Order from Most Often to Least Often Used (adapted from Eldred 2007).

Package	Process
Labels	Litho, flexo, letterpress, gravure, screen, digital
Paper bags	Flexo, gravure
Paperboard folding cartons	Mostly litho, also flexo, gravure, digital
Milk cartons	Flexo
Aseptic juice boxes	Flexo
Composite cans	Gravure
Corrugated board	Flexo, digital
Corrugated preprint	Flexo, litholabel, gravure
Flexible plastic films	Flexo, gravure
Thick plastic sheets, Tags	Letterpress, litho
Glass bottles	Screen
Metal cans/Caps	Litho
Plastic bags	Flexo
Plastic bottles and Tubs	Flexo, screen, hot stamping, container litho, pad printing

range of materials, including plastic film, and its can print any repeat length. It can be used to print any kind of paper, and is commonly used for corrugated fiberboard boxes, multiwall paper bags, milk cartons, pressure-sensitive and wet glue labels, and some specialty packaging like gift wraps and shopping bags.

Flexographic printing was developed by the twentieth century packaging industries. Unlike other printing methods, it is used more for printing packages than for publications. Almost all flexographic ink is used for packaging, mostly on corrugated fiberboard, paper bags, and plastic film.

The resilient relief plate has a raised image like a rubber stamp, which is why it is a good method to print a textured substrate like kraft paper and corrugated board. Flexography has a reputation for low quality print, but newer technology has enabled quality improvements that make it approach the quality of lithography and gravure.

The name *flexography* comes from the fact that the printing plates are flexible. Flexography is a variation of the letterpress printing that began with Gutenberg's movable type, and today letterpress printing also uses flexible plates. Line-art plates have a solid raised surface; halftone and process color plates have a series of relief dots (in a different position on each plate for each color).

Flexography is better than lithography for printing solid fields and line art, but not as good as gravure. It is better than gravure for printing the fine dots for halftone and process color, but not as good as lithography.

Flexo plates are made by exposing a *photopolymer* reactive material to a UV lamp through a photographic negative. The polymer hardens where it is exposed while the non-image area remains fluid and is dissolved away. In the past, flexo plates were made with molded rubber, and although these are still used, they are not dimensionally stable enough for high quality printing. Newer digital computer-to-plate processes laser-etch (*ablate*) a carbon-coated photopolymer sheet using digital information and then the sheet is exposed to the UV light. The best flexo plates are *laser-engraved rubber*, which produces a seamless and highly precise plate.

The flexo plate is mounted on a plate cylinder that is the right size for the package or label *repeat*, without wasting material. The ability to use different size cylinders is a big advantage for package printing, since print repeat lengths can be easily varied (unlike the fixed cylinders in offset lithography that are better suited to publications with standard dimensions). The plates are mounted on sleeves that are easy to put into the press in-register and easy to remove.

The flexo plate picks up ink from a textured ceramic *anilox roll* that meters the ink from a rubber-covered *fountain roll* that has rolled through a reservoir of ink, as shown in Figure 10.7. Sometimes an offset blanket cylinder is also used, similar to the way in which lithography is offset. The ink is transferred to the paper directly as the web passes between the rotating printing plate and impression cylinder.

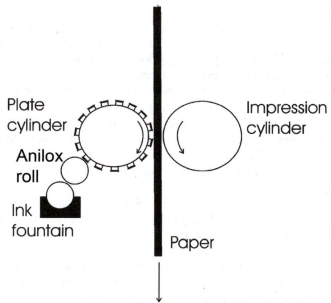

Plate
cylinder

Anilox
roll

Ink
fountain

Impression
cylinder

Paper

FIGURE 10.7. *Flexography has a raised relief plate.*

The anilox roll is the key to flexo printing. It is laser-engraved with millions of tiny cells (100–1,000 per inch) on the surface to hold the ink. The cells are at least five times smaller than the halftone dots in order to get good definition, the finer the better. The development of finer and more consistent anilox rolls during the 1990s has greatly improved the quality of flexo printing.

Flexographic presses are web-fed, and there are three configurations for printing more than one color. *Stack* and *in-line presses* have separate plates, anilox rolls, and impression cylinders for each color, either stacked vertically or in line. In-line presses are used for stiff corrugated board and paperboard, and can be set up in line with other converting operations. The *central impression cylinder* type has just one large impression cylinder, surrounded by plate and anilox pairs for each color, and is used for flexible substrates like plastic film but rarely for paper. Press size ranges from wide (44″, 110 cm) used to print folding cartons, mid-web (22″–44″, 60–110 cm) for linerboard and flexible packaging, and narrow-web (4″–10″, 10–25 cm) used for labels, tickets, and tags.

Since it is web-fed, a flexo printing press can be in line with other converting operations like extruders and laminators (on the inbound side) and (on the outbound side) coating, diecutting, slitting, gluing, folding, and windowing. Quick-drying flexo inks make it possible to print, score, slot, fold, and glue all on the same line, which is how corrugated boxes are converted in a sheet-fed flexo folder-gluer machine, described in Chapter 15. Milk cartons are also printed and converted on a single line.

Most flexographic ink is water-based. The thin inks have a tendency to soak into paper and make colors look a little flat. Flexo has traditionally used quick-drying inks, but it increasingly is using UV-cured inks with higher solids content that cover better and dry even more quickly.

Because of the soft plates and thin inks, careful control of the printing pressure is essential. This is the source of many quality problems. If the pressure is too light, skips or misses leave white spots in solid color blocks. If the pressure is too high, it forces ink to the edges of each letter and squeezes it beyond, causing the dots in the image to spread, called gain. This has led to using screens with low resolution, usually between 60 and 133 lpi, to avoid image distortion.

In flexography, a slight *halo* ridge of ink is always raised at the edge of lines, and the print area immediately adjacent to the edge may be lighter since the ink that forms the ridge was forced from this area. Figure 10.8 shows the halo effect. It also compares how the edges of flexographic lines differ from gravure and lithography. Gravure can be recognized by the saw-tooth edges of its image. Lithography has the sharpest, cleanest lines. These effects are easy to see under magnification, especially in a bar code.

Unlike litho printing, it is easy to change colors on a flexo press. This makes

Flexo   Litho   Gravure

FIGURE 10.8. *Differences in the edges of line print: flexography's halo effect, gravure's saw-tooth edges, and lithography's smooth edges.*

the option of using a special spot color more viable. Automatic wash down and color change systems are employed.

Flexography is the simplest and least expensive of the printing processes. It can print at high speeds. It can print on uneven surfaces, on most materials, and in dirty, dusty, and/or humid conditions.

Flexo is responsible for 70% of all package printing in the United States and 50% in the world, and the share is growing because the technical capability and quality are continually improving (Eldred 2007). It is the only printing method that can be used directly on corrugated board, and also dominates in preprinted linerboard, P-S labels, and paper bags. Flexography shares the plastic film market with rotogravure, and the paperboard market with (primarily) lithography. It has been taking market share from both by approaching the quality of lithography and the speed of rotogravure, at a lower cost than either one.

As the newest of the plate printing processes, flexography is changing and improving most rapidly. The quality improvements are due to better plates, inks, presses, color separations, color standards, and press operator training.

10.4.2 Offset Lithography

Offset lithography, also called *offset* and *litho*, is used for printing most paperboard folding cartons and some labels. It is also widely used for publication and commercial printing, including newspapers, books, greeting cards, business forms, and direct mail advertising. It is the most paper-specific of all the printing methods.

Lithography means literally "stone writing," because it originally used limestone plates.[22] It employs a printing plate that is smooth—neither raised nor engraved. On its flat *planographic* surface, the image area is separated

---

[22]*Lith* is the Greek word for stone, as used in *monolith* and *Neolithic*.

from the non-image area chemically. The image area attracts ink and the non-image area repels it.

Lithography is a printing method with an artistic history. Since its development in 1796, it was used first for illustrations. Color was added in 1837 to create *chromolithography*, and later it was applied to reproducing photographs. The process makes sharp clear images and is economical.

Lithography is based on the principle that grease and water do not mix. The plates are made from aluminum, coated with a photosensitive material that is exposed to the image, and developed with a chemical that differentiates the image from non-image areas in terms of grease- and water-resistance. The image area of the flat printing plate is grease-receptive and therefore repels water (hydrophobic). The non-image area repels the oil-based ink but holds water.

The litho printing plate is rotated first in contact against a wet roller that covers the non-image area with a thin film of water (with some additives). Next the plate rotates against inked rollers, picking up the oil-based ink onto the ink-receptive image area, but repelling it from the non-image area because of the film of water.

The process is called *offset*, because the inked image is next transferred to a rubber blanket-covered cylinder that in turn transfers it to the paper passing between the rubber blanket and impression cylinder (Figure 10.9). Although other printing methods can be offset, lithography is so frequently printed using the offset principle that it is sometimes called just *offset*. The flexible offset blanket makes it easier to print on rough or uneven surfaces, and also makes it possible to print aluminum cans.

The soft offset blanket makes a clear impression on a wide variety of paper surfaces. It is a good method for printing paperboard made from recycled stock because it tolerates some variation in paper thickness. It conforms and squeezes into rough or uneven surfaces and is also used for printing sheets of metal for cans and microflute corrugated board (but not ordinary corrugated board, which is too bumpy and dusty).

The print has sharp edges since the image and non-image areas are at the same level on the plate surface. Offset lithography can be recognized by sharp lines and the lack of discontinuities at the edge of the print, as shown in Figure 10.8.

Only solvent based inks can be used because water-based inks would wet the non-image area. Offset inks are a paste consistency, stiffer, and more tacky than gravure or flexo inks. But they produce a thinner film of ink on the paper because the stiff ink lays on the surface of the paper rather than soaking in, and so it makes finer dots and sharper images. But because the ink film is thin, color density needs to be higher than for other printing processes.

Litho's common quality problems are related to the ink. The sticky ink has a tendency to *pick* fibers and dust from the surface of the paper and leave lint on

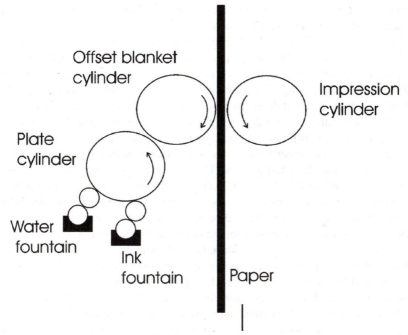

FIGURE 10.9. *Offset lithography.*

the blanket, interfering with the ink transfer and creating unwanted patterns in solid fields. Dust and dirt on the blanket reduce quality and necessitate frequent cleaning that reduces productivity. When the plate prints a heavy image, the rolls may not completely replenish the ink on the next revolution, developing *ghosts* in solid areas.

Offset lithography has more color variability than any other printing process because of the ink/water interface. Small droplets of water in the ink may cause some fuzziness, and small patterns may be seen in the printed areas. Bubbles or streaks appear if there is excess water, and if the ink has emulsified in the water, color appears in the non-image areas.

Lithography presses are either *sheet fed* or *web fed*. The feeder choice depends on the substrate, the size of the imprints and the length of the run. Different inks are used for each.

Sheet-feeding is still the most common for lithography. Folding carton printing has traditionally been sheet-fed because the image size varies so much from one job to another, a large sheet (44″–63″ wide) can accommodate various sizes, shapes laid out together on a single plate with less waste than a web-fed process, and the stiffness of paperboard is well-suited to feeding by sheets. So-called *combination labels* are also printed in sheets, with many labels on the same sheet. Sheet-fed offset is economical for short to medium length runs since the plates are relatively low cost

Most inks used for sheet-fed lithography dry slowly, in 2–4 hours, although printers usually allow 24 hours to dry. Unlike flexo or gravure inks, litho inks are not dried between print stations, so the degree of dot pattern separation is critical. The slow-drying ink can cause print to *set off* onto the back of the next sheet in a stack before it dries. To protect the ink, many lithographers coat the printed sheet with a varnish that is UV-cured as the sheet leaves the press.

Narrow *web-fed* litho presses are common for printing books, and are gaining popularity for longer runs of cartons because of their speed, about five times faster than sheet-fed presses. The economic crossover between sheet-fed and web-fed offset is between 20,000 and 50,000 impressions for process-color (Eldred 2007). Web-fed presses have the added advantage of the ability to be set up in line with coating, diecutting, laminating, and gluing, whereas sheet-fed presses are separate from the other carton converting operations.

To achieve high speeds, the ink used in web-fed presses is less viscous than that used in sheet-fed presses, and is dried by evaporating a solvent. To do so, the freshly printed paper passes through a hot dryer and is then chilled, a process that must be carefully controlled to avoid brittleness or shrinkage. The solvent contains VOCs that must not be discharged into the atmosphere, and so these are collected and recycled or incinerated. UV or EB ink curing systems can be formulated for web presses to avoid the heat/chill cycle, but the cost for the ink and curing is higher.

A major disadvantage of web-fed offset is that some presses have a *fixed cutoff*, meaning that they can have only one size of plate. This limits them to printing packages of a single size, and marketers may not want standard-sized packages. Fixed cutoff presses are used for printing standard half-gallon ice cream cartons and citrus juice cartons. *Variable cutoff* web presses use different sized cylinders, but it is costly to change sizes because all of the plate and blanket cylinders must be changed; these are used for printing folding cartons that are produced in large volumes.

The advantages of lithography are in production and quality. The plates are the lowest cost of all of the processes, and they are easy to produce. Litho is economically practical for small to large print runs, from a thousand to more than a million. The press is made ready more quickly and easily than for any other printing method. Lithographic printing is a very competitive, low-price business that serves many commercial purposes.

Lithography gives sharp images with excellent detail. It holds highlight and shadow halftone dots well. The edges of lines are straight and sharp. Plates can be made with 200 lpi and finer screens, but 133–150 lpi are more common in packaging. However, it does not produce large solid fields as well as flexography or gravure.

But lithography requires an expensive substrate. The force needed to split the ink between the paper substrate and the offset blanket can pick bits from the surface, and so paper or board must be clean with high surface strength. Clay coating is common. To properly feed into the press, sheets must be stiff and strong.

There are also disadvantages related to production. Offset litho is the most complicated printing method because of the complexity of the inking system. Setting up a press to bring the process into ink/water balance is time consuming and wasteful. Sheet-fed litho is a slow operation and the inks dry slowly. Water-based inks, which would be safer and cleaner, are impractical. Web offset wastes a great deal of substrate while the press is being set up and in production.

Of all the printing processes, offset lithography is the most paper-specific. For folding cartons and paper labels with good sharp images, lithography is still the state of the art.

Most plastic films are not stiff enough for litho sheet feeding, and are too flexible for litho web presses that tend to tug on the substrate. The exception is thick (11–15 mils) HDPE and PS plastic sheets, used to make horticultural tags that are commonly printed using offset lithography.

10.4.3 Rotogravure

The rotogravure process is also called *gravure*, *roto*, or *intaglio* because it uses an engraved metal cylinder. A gravure press can only print on smooth surfaces like fine (or coated) paper and plastic films. It is widely used for magazines, newspaper supplements, mail-order catalogs, linoleum, and plastic laminates. Paper currency and postage stamps are gravure-printed.

Gravure is best known for long runs, high quality print, and expensive plates. The high cost of gravure plates–several times that of other processes— limits its use to long-run jobs that spread the cost over millions of impressions. Plate-making lead times are longer too. Gravure printing is web-fed and is the highest speed printing method.

Gravure is limited to very high volume packages, with the expectation of using the plate again for future runs in which the image does not change. Examples are detergent and cigarette cartons, high volume beverage carriers, long runs of labels (like those used for beer bottles), and spiral-wound labels on composite cans. Such applications are rare since most consumer packages have graphics that change frequently with promotions.

The name *gravure* refers to the engraved plates. The gravure image areas are etched onto a copper (or chromium) plated cylinder, and the non-image area remains smooth. The etching is composed of tiny pits that collect the ink. The plates can be made by chemical etching, electromechanical engraving, or direct digital laser engraving. The ability to vary the depth and area of the tiny engraved pits gives gravure the ability to vary the thickness of the ink film, giving more tonal range to an illustration. Heavy ink applications can produce exceptionally bright glossy colors.

The image cylinder rotates directly in the ink, and the excess ink is wiped from the flat surface with a flexible steel *doctor blade*. The ink remaining in the image's tiny cells is then transferred directly to the paper as it passes between the plate and impression cylinders (Figure 10.10). Gravure ink is mostly solvent-based with low viscosity (similar to flexo ink) that enables it to be drawn from the engraved pits onto the paper. Heavily pigmented metallic inks can be used.

The ink dries almost instantly as the solvent evaporates with the help of hot air dryers. This makes it a good printing method for plastic, which does not absorb ink like paper does. The industry is moving towards more environmentally friendly solvents than the toluene and xylene of the past. A range of more environmentally friendly solvents have been developed, using alcohols, esters, and ketones. Solvent recovery systems are used to reduce the pollution caused by evaporated solvents. Water-based inks predominate.

Because of the fast drying, each color printed by gravure in a series is printed on a dry surface, which improves color quality. Up to 10 press stations are used. It is the fastest of the printing processes and (aside from the high cost of the plate) the lowest cost printing operation.

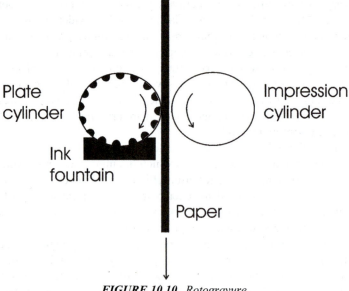

FIGURE 10.10. *Rotogravure.*

Of the three printing methods, gravure gives the best color consistency, partly because of high contact pressure. But it requires a very smooth surface; skips and missing color halftone dots are characteristic on a rough surface. To overcome this tendency, an *electrostatic assist* charge is induced on the impression cylinder to pull the ink from the gravure cells onto the paper. *Offset gravure*, in which the cylinder prints first on an offset blanket, is used for paperboard with an insufficiently smooth surface.

Gravure has excellent print quality that cannot be matched for long and repeated production runs. The amount of ink delivered is fixed by the engraving and so is less subject to operational variation. Gravure produces exceptionally fine halftones, accurate skintones, and uniform solid fields. Its lpi resolution equals lithography. Since the printing plate is not attached to the cylinder, there is no seam line, and so gravure can be used for continuous patterns like gift wrapping and wall paper.

Gravure's primary quality problem is its characteristic ragged or *sawtooth* edge of lines, as shown in Figure 10.8, because of the engraving's dot-like pit pattern. This causes legibility problems for very small type and poor resolution of fine curved lines. Gravure requires a smooth surface such as plastic film or clay-coated paper or board. It cannot be used to print metal, rigid plastic, or rough paper and board.

10.4.4 Digital Printing

Digital printing in general refers to all non-contact printing methods. The plateless processes include electrostatic copiers and printers, and electronic printing such as ink-jet, thermal transfer, and laser (and its cousin, ion deposition). These are used for printing relatively small quantities of copies compared to most packaging applications. Most have some customizable element such as in insurance policies, personalized billing, direct mail, etc.

Among these, *ink jet printing* is used most in packaging, primarily for variable information. Since ink jet printing can be postponed and applied to the package on the filling line, it is ideal for printing code dates and other production-specific identification for tracking. It is also used for printing specific SKU information on shipping containers that are used for a number of different SKUs, and for mailing labels on direct mail merchandise.

Conventional ink jet printing is only one color. The quality of most ink jet printing is poor (but usually legible) because the ink jets print in relatively crude dots. For this reason bar codes are not usually printed by ink jets.

A new generation of higher quality, wide-format, six-color ink jet technology is now being used for corrugated board. It is being used primarily to print point of purchase displays, which are made in short runs.

On the other hand, the *electrostatic digital printing press*, first introduced by Indigo in 1993, is based on photocopier technology. It uses a single printing drum, and some processes have an offset cylinder to print up to seven colors.

In digital printing, an electrostatically-charged drum is exposed by an imaging system in such a way as to attract a resin-based dry or liquid toner to the image area by an electrical charge. The laser imager creates an image in a different color with each revolution and transfers each ink color separation from the image cylinder directly to the paper or onto an offset blanket that in turn, bonds the ink to the paper with no drying. Four to seven colors are typical.

Digital printing has no traditional set-up costs or plates. Printing is direct from computer or artwork. The process is slow and expensive for long runs, but is economical for short runs. When combined with a laser scanner, the whole process is digital. Digital printing is slower than conventional printing, but speeds are increasing. It is still relatively rare, but installations and applications are growing. The quality approaches that of offset lithography.

By moving away from the fixed printing press, digital printing signals the dawn of a new era in package printing. Digital printing, combined with digital prepress, dramatically shortens the time it takes to plan and print a job. It offers the ability to localize operations and customize graphics.

The first widespread packaging use for digital printing has been for labels. They can be produced in sheets (cut-and-stack) or on rolls. It is optimal for short and medium runs. Offset digital production has a higher quality than does traditional offset, and enables innovations using special materials and hybrid processing or relief screen print.

Digital printing is especially suited for graphics that change frequently or warrant postponement. An example is in the drug industry, where regulatory changes can occur unexpectedly; digital printing makes it easier to postpone label printing to reduce risk of the labels becoming obsolete. Another example is custom-printed wine bottle labels that commemorate a special occasion.

Mass customization and personalization offer opportunities for new marketing strategies. For example Coca-Cola® used digital printing for labels in its worldwide "Share a Coke With" marketing campaign for over 750 million packs to 35 nations in Europe (Smithers Pira 2014).

As just-in-time deliveries grow, there is more need for packages to be printed in small quantities on demand, and digital printing can even be distributed to several local presses from a central location. This is called the *distribute and print* model, in contrast to the *print and distribute* model instituted by Gutenberg.

As target markets become increasingly segmented, digital printing can serve the increasing desire by manufacturers to move towards *mass customization*, economically producing for niche markets. It even offers the ability to personalize package graphics, producing for a series of markets of one. For example, an on-line seller of shampoo offers it in a package with the consumer's name featured on the label. Digital printing, capable of changing graphics from one imprint to the next, is well positioned to develop with these trends.

10.4.5 Other Printing Methods

Letterpress is the oldest form of printing, deriving from variations on Gutenberg's original press. Letterpress was once used to print all books, newspapers, magazines, and labels, but now web offset and flexographic printing are also used for publications, especially those with illustrations.

The letterpress process itself has evolved to a flexible plate, like flexography but harder. It still prints the best text. Letterpress is still used for label printing where the sharpness of the letters and quality of the halftone color are advantageous, and specially designed narrow-web rotary letterpresses have been developed for labels. Embossing and die cutting are derived from and done on letterpresses.

Screen printing has limited applications in paper-based packaging. It uses a screen made from fine metal or plastic mesh (originally silk, hence the term *silk screening*) that has the non-image area masked off. The screen is placed against the substrate and a wiper blade pushes an ink puddle across the screen. Ink is deposited through the porous image area. Since it transfers a heavy coating of ink, it is used for some labels where brilliant, fluorescent, or thick pastel colors are desired. The heavy coating is also useful for printing metal cans, glass bottles, and molded plastic. Screen printing is best for solid color or line art. Halftone screens are limited to a relatively coarse 85 lpi. Thick layers of varnish are applied by screen printing, as are electronic printed circuits. The screen-printing process is relatively slow and rare for paper-based packaging.

Besides screen printing, there are a number of other methods for printing directly on metal, glass, and (especially) rigid plastic containers, eliminating the need for a label. These include *hot stamping, container offset printing, transfer print,* and *pad printing.*

Holograms are made from micro-embossed plastic films and are not printed onto paper. However, they can be applied to paper, especially paperboard, to give a 3-D effect.

It is becoming more common to use more than one printing process on a package. For example, flexo printers may apply a coating or variable information like price on a package that was first printed by rotogravure or offset lithography. Ink jet printers can add unique information like lot codes to any kind of preprinting. Hybrid presses typically use three or four printing processes on one in-line machine.

10.5 PREPARATION FOR PRINTING AND QUALITY CONTROL

There are usually four roles involved in graphic production:

- The product manufacturer's marketing department commissions the work.
- The graphic designer develops the art.
- The pre-press operation converts the art to plates.
- The printer/converter runs the press and other converting operations.

These roles may be taken by four different firms, or shared by the manufacturer and the converter with the manufacturer's design department doing most of the artwork and the printer doing the pre-press.

The various players work together, with the product manufacturer approving each stage in design and production. There is an increasing trend to integrate and streamline the design process by the use of computer links. Linking is the logical next step in the already largely electronically controlled process for preparing plates and presses for printing.

Once the product manufacturer has approved the design, the convertor prepares for printing. A comprehensive *key line* layout is developed, which accurately represents the size and spacing of print elements. The *prepress operations* include preparing the text and pictures for printing by separating colors, assembling them into the configuration to be printed, proofing the form, and preparing the plates. The prepress operation is critical for achieving the right colors and registration, considering slight variations in a press as well as the nature of the substrate.

Most of the pre-press work is done with the use of a computer, in a process called *computer to plate*, which streamlines platemaking and makes it easy to translate the same design to different shapes and substrates. Text has traditionally been handled differently from illustrations, but now *imagesetters* set both type and halftone illustrations. Text is composed on a keyboard, and illustrations are either scanned or digitally photographed in a process that electronically generates halftone dots and color separations. The areas of the package that will be glued or hidden beneath flaps are not printed. The image is assembled electronically and reproduced and fit together if multiple impressions are to be made from a single plate. The illustrations and type are digitally set by exposing a photosensitive film or plates to a laser source. Even the mounting of the plates is electronically controlled so that they will print in register with one another.

There are several types of *proofs* that are generated throughout the process. Proofs are a means of communication among the printer, the designer, the separator, and the product manufacturer. They are used to make proofreading corrections, verify color separations, to get a manufacturer's sign-off, and to instruct the press crews.

Digital proofs can be generated on screen and/or printed with a digital color printer. Since the electronic image on a screen (generated by red, blue, and green light) can appear much different from the final package, a hard copy provides a closer approximation. Even digital proofs will not perfectly match the production print because digital printers use only CMYK inks instead of the spot colors used on the press, they use lines instead of dots, the digital process does not reproduce some press characteristics like dot gain, and the production print is affected by stretch and absorbency of paper moving through a wet press.

Nevertheless, digital proofs are accepted by many customers so that problems can be identified early. The further the process goes, the more expensive it is to make corrections.

The traditional pre-press color proof was the *overlay proof,* or *color key,* with the image printed in each of the four process colors on transparent sheets. When the four sheets are overlayed, the proof represents the color on the final print. *Integral* or *single-sheet proofs* are prepared by photographically exposing a sheet to each of the four colors and developing one color before adding another.

The final step in prepress operations is making the plate. The most common process is to expose a photosensitive material to light and to develop the image. Some variation of this process is used for every method except for gravure plates, which are mechanically or laser engraved by electronic control from a digitized image. The computer-aided design is coordinated with manufacturing process (CAD/CAM) for cutting and finishing packages like paperboard cartons. The *position proof* shows the layout for the job and helps to verify copy.

Once the plates are made and approved, the job is ready to set up. This is called the *press makeready* process. The plates or plate cylinders are placed into position, the press is inked, the paper substrate is placed or threaded into position, and everything is adjusted to achieve the proper color and registration.

The press makeready process depends on the type of press and the type of job. One of the trickiest parts is compensating for *dot gain,* which occurs in every printing process but is a greater problem in flexography, and is the chief cause of color variation in process-color printing. Each printing press differs slightly, and the makeready process depends on the skill of the operator and his/her understanding of the press.

There is continuing effort to shorten makeready times, in order to improve printing productivity. For short runs, the makeready may take longer than the actual printing. It also includes cleaning up from the previous job, and to reduce this time jobs with the same ink are run in sequence.

10.6 PRINT QUALITY CONTROL AND TESTING

The most basic quality control during printing involves visually scanning the print and comparing it to a standard to verify sharpness and registration. For sheets, this is a simple matter of looking at them as they come off the press. For web-printing, a special vision inspection system is used. An operator may use a simple stroboscope that stops the image in the lens as the web whizzes by, or an automated system that uses video and computer technology to give a warning when register or color go out of specification.

Printed symbols and *control targets* are generally printed in a place on the web where they will not be visible on the package, in a trim margin or under a sealed flap. *Registration marks* are usually in the form of a cross on each plate; when all printing plates are aligned, the cross will appear solid black. Color bars and gray scales enable the operator to measure and judge the colors and halftones against a standard.

Devices are used to provide objective measurements of the print. The optical density of the ink is measured by *densitometers* that check for coverage of every ink color. Color can be quantified by colorimeters and spectrophotometers. *Colorimeters* record colors as the eye sees them, in terms of red, green and blue. *Spectrophotometers* measure colors in narrow bands. Color measurement is tricky, and there are a number of variables that can affect the result, including the light source and angle, procedural errors, variation between instruments (the colorimeter varies more than the spectrophotometer), and calibration (Myers 2002).

It is common for the printer and the product manufacturer to agree on the acceptable print quality during the first print run. Samples are saved, showing the quality and color tolerance ranges for reference.

Printing is paper-based packaging's strategic advantage. This chapter presents a brief overview of the elements of printing. The next chapters in this book explore specific paper-based packaging forms: labels, bags, paperboard cartons, and corrugated fiberboard boxes. They discuss the role of printing and why the types of printing presses differ for each.

10.7 REFERENCES

Belk, R.W. 1988. Possessions and the Extended Self, *Journal of Consumer Research*, v. 15, September, pp. 139–68.

Danger, E.P. 1987. *Selecting Colour for Packaging*. Aldershot, UK: Gower Technical Press.

Demetrician, R.. 1996. *Label and Graphic Design*. Plainview, NY: Jelmar.

Eldred, N.R. 2007. *Package Printing*, 2nd ed. Pittsburgh: PIA/GATFPress

Fleming, P.D. 2003. Printing, in *Encyclopedia of Forest Science,* ed. Jeffrey Burley, v. 2, pp 678–686. Oxford: Elsevier Academic Press.

Garber, L.L., R.R. Burke, and J.M. Jones. 2000. *The Role of Package Color in Consumer Purchase Consideration and Choice.* Report No. 00-104, Cambridge, MA: Marketing Science Institute.

Giles, G., Ed. 2000. *Design and Technology of Packaging Decoration for the Consumer Market.* Sheffield, UK: Sheffield Academic Press (and Boca Raton, FL: CRC Press).

Hirschman, E. and M. Holbrook. 1982. Hedonic Consumption: Emerging Concepts, Methods and Propositions, *Journal of Marketing*, v. 46, Summer, pp. 92–101.

Holkham, T. 1995. *Label Writing and Planning: A Guide to Good Customer Communication.* London: Blackie Academic and Professional.

LaMoreaux, R.D. 1995. *Bar Codes and Other Automatic Identification Systems.* Leatherhead, Surrey: Pira International.

McCaughey, D.G. 1995. *Graphic Design for Corrugated Packaging.* Plainview, NY: Jelmar Publishing.

McMath, R.M. and T. Forbes. 1999 *What Were They Thinking?* New York: Times Business.

Meggs, P.B. 1998. *A History of Graphic Design*, 3d ed. NY: John Wiley & Sons.

Mohan, A.M. 2004. Pepsi POPs with Digital Ink-jet Printing, *Packaging Digest*, January, 40–42.

Myers, B.L. 2002. Packaging Color Control Procedures; How to Optimize Physical Instrumental and Colorimetric Standards, *Flexo*, v. 27, no. 12, 22–27.

The Pocket Pal. 2000. 18th ed. Michael H. Bruno. Memphis, TN: International Paper Co.

Rauch Guide to the US Ink Industry, 2000–1. 2002. 4th ed. Manchester Center, VT: Impact Marketing Consultants Inc.

Smithers Pira. 2014. *The Future of Digital Printing to 2018.* Leatherhead, UK: Smithers Pira.

Soroka, W. 2009. *Fundamentals of Packaging Technology,* 4th ed. Herndon, VA: Institute of Packaging Professionals.

Steele, P. 2000. *The Printing Process in Design and Technology of Packaging Decoration for the Consumer Market,* ed. Geoff Giles, pp. 15–24. Sheffield, UK: Sheffield Academic Press (and Boca Raton, FL: CRC Press).

U.S. FDA (Food and Drug Administration). Annual. *The U.S. Code of Federal Regulations, Title 21, Parts 101 and 102.* The Fair Packaging and Labeling Act, 1966, Nutrition Labeling and Education Act, 1990. Format and Content Requirements for Over-the-counter (OTC) Drug Product Labeling, 1999. Washington, DC: U.S. Government Printing Office. http://www.gpo.gov/fdsys/browse/collectionCfr.action?collectionCode=CFR&bread=true

10.8 REVIEW QUESTIONS

1. What are the five elements required by the U.S. Fair Packaging and Labeling Act?

2. What is an SKU and what does its number signify?

3. Why is clean, sharp printing essential for accurate bar code scanning?

4. What do the two 5-digit numbers of the UPC code represent? What is the purpose of the quiet zone?

5. What three colors are recognized as a bar by a bar code scanner? What three colors are recognized as a space? What is the problem with using red ink to print a bar code?

6. Give an example of a 2-D bar code and how it is used.

7. What kind of ink would be needed to print the antennas for necessary for RFID?

8. What is a halftone and how does it differ from line art?

9. What is the difference between a spot color and a process color?

10. What are the four primary colors of process color printing and how are they combined to make other colors?

11. Why is a bright white surface necessary for good color printing?

12. How are the four process colors separated?

13. If process colors require only four ink presses, and most presses have more stations, what may the other print stations be used for?

14. What is meant by a printing screen density of 133 lpi? Is this fine or coarse? Which printing process has the coarsest screen density?

15. What are the three main ingredients in ink? What is the purpose of each?

16. What is the Pantone Matching System®?

17. What is the problem with solvent-based inks? What are the alternatives? What are some problems with water-based inks?

18. What is the advantage of energy-cured inks, compared to inks that dry from evaporation?

19. Why are the three main printing processes called "lithography," "flexography," and "gravure?"

20. What is an anilox roll, and what is its purpose?

21. Which printing process produces a halo around lines and has the most problem with gain? Why? What problems can this cause for barcode scannability?

22. Which printing method produces microscopic saw-tooth edges on lines? Which has the smoothest lines?

23. What is the difference between the plate cylinder and the impression cylinder?

24. Which printing process has developed with the twentieth century printing industry, and has the broadest range of uses in packaging including flexible plastic films and paper-based packaging?

25. Which printing process is most commonly used for paper bags and corrugated board?

26. Which printing process is best known for reproducing illustrations?

27. How does an offset lithography press work? What does the word "offset" mean?

28. What are the following and which process uses them: offset blanket cylinder, anilox roll, impression cylinder, central impression cylinder?

29. Why can't water-based inks be used for offset lithography?

30. Which printing process is most commonly used for paperboard folding cartons? Why?

31. What is the difference between a web-fed and a sheet-fed process? Which type of packages are lithographically printed in a sheet-fed process?

32. Which printing process is used only for very long repeated runs of cartons and labels?

33. What are some advantages and disadvantages of gravure printing?

34. Which printing process is best for fine lines? Which is best for large solid fields?

35. Which printing process is best for very short runs, and can be customized from imprint to imprint? Give an example.

36. What are some of the uses for proofs during the preprint process?

37. Why is a digitally printed proof better than viewing it on a computer screen? What are the limitations of a digitally printed proof?

38. What is a printed registration mark, and how is it used?

39. What are some of the devices used for checking the quality of print?

CHAPTER 11

Labels, Bags, and Other Flexible Paper-based Packaging

Paper labels, bags, and wraps have a long history of packaging success. It is easy to forget, in light of the current growth of flexible plastic films, that paper was the original manufactured flexible packaging material.

But paper lacks heatsealability and barrier properties, which can be added by a plastic coating or lamination, enabling the substrate to be used in form-fill-seal operations. Paper adds stiffness and superior printability to plastic/paper substrates. Paper's printing advantage and strength, combined with the strength, barrier, and heatsealability of plastics, continue to make paper a useful element in flexible packaging.

The paper in coated and laminated structures is specified by basis weight (lb/3,000 ft^2 or g/m^2) and by the kind of paper (e.g., kraft, bleached or not, coated or not and with what, recycled content, etc.). Plastic coatings are also specified in terms of their contribution to basis weight (lb/3,000 ft^2 or g/m^2). Film and foil laminations are specified in terms of thickness.

The companies that make paper-based packages are called converters, which means they begin with rolls of paper (or sheets of paperboard) that may have been made by another company, and coat or laminate, print, and convert it into something else, like labels, bags, cartons, or boxes.

11.1 LABELS

Labeling was one of the first uses for paper in commercially-produced packaging. Long before other forms of paper packaging were invented, paper labels were used to identify bales of cloth, glass bottles and jars, wooden fruit crates, cigars, and metal cans.

Paper is still the most popular material used for labels because it offers many advantages. It is inexpensive and available everywhere. It is easy to print and easy to glue. It can be colored or coated, shaped, or square.

The right label design gives a brand its face. It enables consumers to recognize a familiar brand or to notice something new. Labels offer differentiation options for relatively standard shapes of containers like cans and bottles, using various print styles, illustrations, textures, shapes, and positions. Labels on wine bottles, for example, provide "the only sensory clues about what lies within:"

> *A successful label beckons from a distance, then invites personal discovery the closer you get. Like revealing clues in a detective novel, the message on a bottle must engage the reader, skillfully maintain the suspense, and be compelling enough to lead you to the mystery inside (Caldewey and House 2003).*

Labels can be glued to almost anything. Paper labels work best on surfaces that are flat or cylindrical and smooth. The adhesive method employed depends on the properties of the label, the surface to which it will be adhered, and the labeling process.

Labels are typically adhered in one of two ways: with *wet glue* (water-based or hot-melt) applied during the labeling process or with *pressure-sensitive adhesive* that has been pre-applied by the label manufacturer. There are also various labeling processes, from manual to highly automated, which will be discussed after the following sections about how labels are made.

11.1.1 Label Production

Anyone can make a label; all that is needed is a substrate, printing method, and glue. At the most basic level a scrap of paper, hand-script, and a bottle of household glue can be used to label almost anything.

11.1.1.1 Paper Label Substrates

Almost any kind of printable paper can be used to make labels, so the choice depends on the effect or properties desired. Labels can be glossy or dull, rough or smooth, embossed or foil-embellished. Labels used in the beverage industry usually require wet-strength. Labels that will be exposed to a damp environment for a long period may incorporate an anti-fungicide to prevent mold growth.

Most paper used for labels is made from bleached chemical pulp and is calendered to have a white, smooth surface. It may be *MF*, *MG* with high gloss on one side or clay coated on the side to be printed, referred to as C1S Label.

The paper may be supercalendered if resolution higher than 110 lpi is desired. A smooth surface on the backside makes the label easier to glue. Basis weights range from 35–60 lb/3,000 ft² (57–100 gsm). Other papers may be used for special effects such as colored, embossed, metallized, or antique looks. A typical label paper is 60-pound, C1S Label with a brightness of 80.

The properties of the paper affect label printing and appearance. Each surface accepts print differently. Good press runability depends on the paper's stiffness, tear resistance, moisture content, ink absorption, and pick strength. The print quality depends on the paper's brightness, smoothness, opacity, and gloss.

Paper can be laminated to foil for a reflective high-gloss look, used to make labels for premium bottled beers. Foil-laminated paper is the most expensive label substrate; the foil layer is about 0.00025″ to 0.00035″ with a paper backing basis weight of 40–50 lb A very smooth paper must be used to achieve a smooth foil surface.

Another way to achieve a shiny metallic appearance is by metallization, which deposits an aluminum vapor on the surface of the paper in a vacuum chamber. Metallization offers the advantage of being less expensive (the layer is one-onehundred and fiftieth of the thickness of a foil layer) and avoids the problems with foil printing of dead fold and curling. Since aluminum is difficult to print, a lacquer, shellac, or water-based acrylic coating must be applied before printing.

A thin film of plastic like PE and PP can be laminated to a C1S label paper, providing labels with additional wet strength. Plastic-coated labels are used on bottles for water and juice, and are applied using hot-melt adhesives.

Pressure-sensitive (P-S) labels consist of three layers: the label face material, the pressure-sensitive adhesive, and the release-backing material. The adhesive is applied between the back of the label and the release surface. When the backing is peeled away, it releases the adhesive and leaves it on the label.

P-S label stock is available in a wider variety of materials and adhesives than wet-glue labels; one-third of P-S labels are made from plastic film (TMLI 2013). It is produced in rolls by specialty converters, many of whom are members of the Tag and Label Manufacturers Institute, or TMLI (http://tlmi.com). Pressure-sensitive labels are two to three times more expensive than wet-glue labels.

The P-S release-backing may be a super-calendered unbleached kraft paper or glassine that has been coated with silicone or other release coating, or a coated plastic film. It is usually thinner than the label itself, with a lower basis weight, as low as 30 lb/3,000 ft² (50 g/m²). It must have a high enough tensile strength to withstand the intermittent, jerking motion as the labels are dispensed in a labeling machine.

11.1.1.2 Label Printing and Finishing

There is a wide range of graphic effects possible for labels. Printing can range from simple, one-color typeface to photographic representations highlighted with gold. The ink usually is formulated to be rub-resistant, and/or a protective lacquer is coated over it.

All major printing processes described in Chapter 10 can be used to produce labels, but offset lithography and flexography are the most popular.

Sheet-fed offset lithographic presses are used for most wet-glue labels. Lithography is best able to produce the soft halftones in a four-color process that is used for high quality labels for bottles of food, wine, and liquor.

Most offset label printers offer a *combination label* service in which labels for several brands and/or customers are printed on the same sheet, up to 100 labels/sheet (depending on the size of the press and the labels), typically in run length of 30,000 sheets. If there is more than one color, the sheets are fed through multiple print stations before cutting. These so-called *cut-and-stack labels* offer the advantage of supplying small orders in volumes of 30,000 (Eldred 2007). The FDA does not allow this practice for pharmaceutical labels to avoid the possibility of mixing them up.

Wet-glue labels can be *guillotine-cut* with straight edges, or *die-cut* for rounded corners and more complex shapes. For straight cutting, a stack of up to 1,000 sheets is positioned, clamped, and the cut is made with a guillotine knife; subsequent cuts are based on programmed gauges. Die-cutting, flatbed, or rotary, yields slightly greater accuracy than guillotine-cutting. For foil-laminated labels, die-cutting work-hardens the edges so that there is less of a tendency to curl. For short runs, digital die-cutting (sometimes as a companion to digital printing) is available. Die-cutting is popular because differentiating labels by shape is becoming more prevalent as a marketing tool, especially now that label application equipment is better able to handle various shapes.

Roll-fed printing is used for P-S labels. Presses have multiple print stations and incorporate die-cutting, waste-stripping, and rewinding in one sequential operation. They can be used in combination with other processes, like screen printing, hot-stamping, or UV varnish for high-value applications like cosmetics or wine bottle labels. Flexography is the most widely process for printing P-S labels, used for over 90% in the United States and 40–50% in Europe (Fairley 2013).

Although mainly used for black printing, there is a trend to flexographic presses that can print five to six colors or more. Flexo label presses are web-fed and narrow, between 6 and 16 inches. They can have print stations in line or a common impression drum, and use water-based, UV-cured ink. The P-S labels are die-cut in a roll-fed operation that slices through the label stock without cutting, impressing, or weakening the release-backing web. The integrity of the web is a critical quality for its ability to carry through the subsequent labeling process. In the case of shaped labels, the excess face material is stripped away as the roll is rewound. The industry standard roll has a diameter of 12 inches on a 3 inch core, with one-eighth inch gap between labels and one-sixteenth inch strip along the edges. For high speed operations, roll diameter may be as large as 24 inches (Henry 2012).

Letterpresses using hard photopolymer plates have more controllable resolution than flexography. In Europe, rotary letterpress is commonly used for P-S labels. In the United States, rotary letterpress is used primarily for high definition pharmaceutical labels. In Japan, flatbed letterpress predominates for P-S labels.

Gravure is used for long runs of high quality (wet-glue) beer labels, in the three to four million range. It is sheet-fed with up to 67,000 sheets/run. Body and neck labels can be run together on the same sheet in so-called *combo* runs. Gravure is declining for label production however, because there are fewer mass marketed products produced in such high volumes in identical packages. Gravure is still used to produce labels for the highest volume beers, but now there are more product variations like light beers and specialty beers, with few of them in large enough volume to justify the high cost for gravure plates.

Screen printing offers the ability to add texture and depth. It can be printed in combination with letterpress, flexo, or litho to achieve special effects. Since it lays down a thicker film of ink, it can add visual impact with fluorescent inks, on dark substrates, or add tactile impact, including for braille. It can be a flat or rotary operation.

Digital printing is increasingly used for short-run, multicolor labels in either sheet or roll form. These have a high added value with up to seven colors (extended color range, including opaque white) on special substrates (including metallized or textured), cold or hot foil and varnish.

Digital printing offers the ability to customize labels for individual customers or small markets. It can produce short-run labels for test markets. It can provide special brand protection, anti-counterfeiting features, or serial numbering.

The cost of digital printing has steadily decreased as the quality has improved. The label industry was the first packaging industry to take advantage of digital printing. While the early digital presses were cost-effective only for very short runs (up to 10,000 labels), newer presses can compete with higher-volume plate processes for up to 100,000 labels.

For packages printed by conventional press methods, variable information can be printed in a separate operation. Variable information includes lot or date codes, price/weight information, mailing labels, or serialization (which is increasingly used in the pharmaceutical industry as an anti-counterfeiting measure). Variable information printing methods include ink-jet, thermal transfer, and laser (and its cousin, ion deposition).

After printing, a final press station applies a coating or varnish to increase gloss, make the print appear more saturated, and increase resistance to scuffing and smudging. *Press varnish* and *UV curable coatings* are a type of clear ink. *Press coating* is water-based and applied to sheets to help prevent ink set-off (transfer of ink to adjacent surfaces) in a stack. A coating or laminated plastic layer can also be added off-press in a separate operation.

Touches of gold (or silver) are imparted with bronzing, stamping, or ink. The most economical method is to print with a gold bronze ink made from a clear vehicle and a pigment of fine metal flakes called bronze powder. The result is poorest with lithography because the ink film is so thin; flexography is better; and gravure and screen print give the best results because they deposit the thickest layer of ink. A problem with gold ink is scuffing, and varnish that prevents scuffing reduces its brilliance.

Gold bronzing applies bronze powder to areas that have been coated with a special adhesive called gold size. *Gold stamping* transfers coated foil to the label with heat and pressure. Gold bronzing and stamping, usually done off-press, are slow and expensive, and are used primarily on labels for luxury products like wine and cosmetics.

Embossing can give texture and attractive dimensionality. *Pebbling* is an all-over pattern created by an engraved press cylinder; some common patterns are eggshell and linen. Pebbling also increases flexibility by breaking, stretching, and separating the fibers in the paper, making labels easier to apply to a curved surface. Spot embossing runs the paper through two engraved platens, one convex and one concave, which sets an impression in it. Plate embossing creates luxurious effects but is slow, expensive, and labor intensive.

11.1.2 Wet-Glue Labeling

Despite fierce competition from pressure-sensitive and plastic labels, pasted paper labels are still the best solution in many cases, especially for high-volume operations like labeling beverage bottles or food cans, operating at speeds of 60–100,000 labels/hour. They are inexpensive and so is the labeling process. Wet-glue labels represent 41% of the North American label market, although the percentage is steadily declining (TLMI 2013).

Bottles may have a combination of face, neck, and back labels. Can labels usually wrap around, and may be glued only at the overlap joint (which may be left uncoated during printing to better adhere to the glue).

The choice of glue depends on the surface of the container or object to which the label will be applied. It also depends on the conditions of filling (for example, wet bottles in a brewery), use (for example, wine bottles in an ice bucket), and whether it is desirable to make the label easy to remove.

Most wet glues are water-based synthetic emulsions like PVA or dextrin/

starch, although hot-melt adhesive is gaining wider use. Dextrin/starch-based adhesives are used in paper-to-paper labeling. Hot-melt is used for applications like can labeling where the bond needs to be faster and is concentrated in a small area.

PVA is used for paper labels on glass bottles; this adhesive is similar to white household (like Elmer's®) glue. The setting time for water-based adhesives is determined by how quickly the water can evaporate or be absorbed into the label paper. At least one surface needs to be absorbent to form a strong bond. PVA and other synthetic emulsions set faster than dextrin/starch adhesives because of the ability of the polymer particles to draw together to form a continuous film, whereas dextrins have a tendency to crystalize. Since water-based adhesives dissolve and release when wet (like in an ice bucket), additives are used to provide longevity.

Hot-melt adhesive sets quickly as it cools, and has a higher initial tack than water-based adhesives, so it is a good choice for high-speed operations like wrap-around labels for cans. Hot-melt adhesive is not suitable for wet bottles.

For refillable glass bottles, casein adhesive is used because it sticks well to wet bottles and resists moisture, but is easy to remove in the caustic solution used for cleaning. Refillable bottles are rare in the United States, but are more common in other countries in Europe and South America, and in Canada and Mexico.

The back of the label is usually left uncoated for better penetration of the glue. The water absorbing capacity of the back is essential because the water in the adhesive must be temporarily absorbed by the label, but the adhesive should not penetrate to the front of the label. When a wrap-around label is used, like those used for cans, the overlap strip on the face may likewise be left unvarnished to improve adhesion.

Labels that will be attached by water-based glue to cylindrical objects like bottles and cans are oriented so that the MD goes around the cylinder, as shown in Figure 11.1. This compensates for the tendency of paper that is wetted on one side to curl in the CD (described in Chapter 8). This orientation prevents the label from curling off (called *flagging*) immediately after wet glue is applied to one side.

Wet glue labels today are almost all applied by high-speed machines, but they can also be applied manually. Manual application is common in home-canning operations, as well as in small-scale production and in less-developed countries. This is a good reminder that a simple pot of PVA glue (such as white household glue) applied by paintbrush can be an adequate labeling operation for small scale production.

Labeling machine performance depends on properties of the label like basis weight and density, flexibility/stiffness, tensile strength, absorbency and sizing (which affects backside water absorption), wet strength, and the amount of

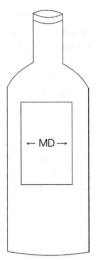

FIGURE 11.1. *The proper MD for a label attached by water-based glue.*

dimensional variation. Labeling machines can either stand alone or be placed in line with other operations like filling and closing.

Labeling performance also depends on how flat and dry the labels are, since humidity affects stability and can cause them to curl, which can cause machines to jam. Labels are always shipped flat and in water-resistant wrappers, and they may even be preconditioned to 70–80% RH at 75–80°F, a slightly higher moisture level than TAPPI conditions, which increases their flexibility.

When the face of a label is laminated or coated with plastic or foil that is not affected by moisture, there is even more tendency to curl. They curl toward the non-paper surface when the moisture content of the paper is high, and away from that surface when the moisture content is low.

Wet-glue labelers for jars and bottles feed the label from a stack in a magazine. The label is picked up with a transfer mechanism, such as suction or compressed air, and adhesive is applied to the whole back surface of the label or in stripes or spots by a glue-application roller. The container is moved into position and the label is wiped or pressed onto it by pads, brushes, and rollers.

Wrap-around labels are glued to cans by first applying the adhesive to the leading edge of the label, which is fed from a magazine or cut from a roll. The can picks up its own label and it is rolled on. Adhesive is applied to the trailing edge of the label that is then pressed against the can and the label overlap.

Wet glue labeling machines are of heavy duty construction, and have the highest speeds of all labeling systems. *Straight line labelers* have line speeds of up to 400 containers per minute, and rotary labelers, like those used for beer bottles, run up to 3,000 containers per minute.

11.1.3 Pressure-Sensitive Labeling

The use of *P-S*, also called *self-adhesive*, labels has grown to almost half (48%) of the labels in North America, despite their higher cost compared to wet glue.

P-S labels have production advantages over wet glue. They are able to adhere to almost any substrate. They simplify the labeling operation, shorten the set-up time, and eliminate the need for clean-up. However, there are the environmental disadvantages related to the release-backing waste and problems with recycling packages that have been labeled with permanent P-S adhesives.

P-S adhesives are based on acrylics or rubber/resin blends. P-S adhesives are formulated to be either removable or permanent (or at least permanent enough that the label is defaced before it can be removed). Removable adhesives are used for short-term applications, such as point-of-sale pricing or promotion that is designed to be removed before the product is used. Usually, removable adhesives are preferred, to facilitate package recycling and reuse or when the label is placed directly on a product like fruit.

P-S labels can be applied by hand, or in semi-automated or high speed systems, but the speed is generally slower than for wet glue. The speed at which the container is conveyed through the labeler must be carefully synchronized with the speed of label application, in order to avoid wrinkles or misplacement.

P-S machinery has four stations. First, the *label unwind* pulls the web from the roll; the orientation of the label, called the *unwind position*, must be specified in order for the labels to be dispensed right-side up. Second, the peeler plate transfers the label by sharply bending back the carrier web over a stripper plate, peeling the backing from the label. Third, the label is then *blown, rolled,* or *tamped* onto the container and pressed into place by rollers, brushes, air jets, or pads. Fourth, the web is pulled by a *powered nip roll* that has intermittent motion, and the backing is rewound for disposal. The machine is controlled by a PLC (programmable logic controller) that governs timing and label positioning through the use of photosensors (Henry 2012).

The fastest machines, capable of labeling 600–1,000 containers per minute using a multi-station applicator cannot achieve the top speed of wet glue machines. But they are versatile, capable of applying labels onto various shapes and surfaces, and are about half the cost of comparable wet-glue machines.

11.1.4 Label Tests

In addition to paper-industry tests, some of which are discussed in Chapter 8, and printing/ink tests, some of which are listed in Chapter 10, the label industry has developed a number of tests specifically to evaluate adhesion and other properties unique to labels. Two trade associations representing P-S la-

bel manufacturers have developed test methods: TLMI and the comparable
European FINAT (originally the Féderation Internationale des Fabricants et
Transformateurs d'Adhésifs et Thermocollants sur Papiers et Autres Supports).

Most of the tests are for P-S labels. Table 11.1 lists some of the most common test methods. *Peel adhesion and tack tests* use a tensile tester to peel off
the label from a substrate or a test fixture at various speeds, measuring initial
tack and tack after 24 hours. The wettability test evaluates the label's ability to
receive wet glue. *Die cutting tests* evaluate how clean a P-S label's facestock
is cut without slicing or weakening the release-backing web, both of which are
essential for machine-application. The *printed/applied label tests* evaluate how

TABLE 11.1. Selected Label Test Methods.

Container Type	% Recovered
Peel Adhesion	TLMI 1: 180° Peel Adhesion—Face Stock from Substrate TLMI2: 180[CM3] Peel Adhesion—From Release Liner TLMI 3: Ramp Velocity High Speed Release FTM 1: Peel Adhesion (180°) at 300 mm/min FTM 2: Peel Adhesion (90°) at 300 mm/min FTM 3: Low Speed Release Force FTM 4: High Speed Release Force
Tack	ASTM D6195: Standard Test Methods for Loop Tack TLMI 4: Tensile Tester Loop Tack Test TLMI 4: Static Sheer Test TLMI 6: Thickness of a P-S Label TLMI 7: Coating Weight Test
Wettability	ASTM D714: Surface Wettability of Paper (Angle-of-Contact)
Die Cutting	TLMI 8: Die Cut Burst Ratio Test TLMI 9: Ticker Test
Printed, Applied Label Tests	TLMI 12: Cross-Hatch Tape Test FTA 1: Heat Resistance FTA 2: Crinkle or Flexibility FTA 3: Odor and Taint FTA 4: Lamination Bond Strength FTA 5: Water Resistance FTM 26: Wash-Off Labels FTM 27: Ink-Rub for UV Printed Labels ASTM D4060: Abrasion Resistance of Coatings (Taber) ASTM D5264: Abrasion Resistance of Print (Sutherland)
Finishing	TLMI 22: Splicing PSA Coated Label Stock TLMI 23: Unwind Roll Chart

well labels survive scratching, abrasion, rubbing, water, and heat. And the *finishing standards* give a common terminology for how P-S labels are presented on a roll. A comprehensive collection of TLMI, FTM (FINAT Test Methods 2009), and applicable ASTM test methods is available from TLMI (2010).

11.1.5 "Smart" and Active Labels

A new generation of labels goes far beyond identification and branding. Smart and active labels are primarily used to control the safety and quality of fresh food and products likely to be counterfeited.

Smart, or *intelligent, labels* can be printed with symbols, sensors, or indicators to communicate product information. They range from the use of automatic identification to indicators of temperature change. *Automatic identification* like printed bar codes or radio frequency identification (RFID) can be used to track and trace a product through distribution when linked to the appropriate software systems. *RFID* tags include a small, integrated-circuit chip and a metal antenna. QR codes can even enable internet-connectivity.

Temperature change can be indicated by thermochromatic ink which changes color with temperature, a melting-point material like wax, or a chemically reactive material. They are formulated to melt or change color at a given temperature. They can indicate whether a package of perishable food has been subject to temperature abuse, or can indicate when a product (like wine) has reached the best temperature for consumption.

Food freshness indicators go beyond temperature to monitor an aspect of the food that contributes to ripeness or decay. Concepts are either based on a color change of the tag due to the presence of microbial spoilage or based on conducting polymers or fiber optics. For example, indicators for meat and fish are sensitive to pH change, CO_2, diacetyl, amines, ammonia, and hydrogen sulfide, and are an indicator for fruit ripeness responds to aldehydes and ketones.

Active labels go one step further to actually interact with food to extend shelf life. Examples are oxygen scavengers that prevent rancid oxidation of fried chips and ethylene scavengers to slow the ripening of fruit.

There is a great deal of research in smart inks and active packaging. Some are seeking anti-bacterial solutions and indicators of microbial activity. Paper has been researched as a substrate for printed electronics, power source, and memory-storage. Labels can also provide sensors to deter theft, provide evidence of potential tampering, or features to reveal counterfeiting. Reviews are provided by Mills (2009), Scully (2009), and Yam *et al.* (2009).

Smart and active features are usually applied as labels rather than being directly printed on a packaging substrate due to limitations in the package converting process. The P-S labels are usually paper, and when placed inside a food package there may be a plastic coating to prevent direct food contact.

11.1.6 Substitutes for Paper Labels

A theme that recurs throughout this book is the substitution of plastic for traditional paper uses. Although paper is still the major label material, the use of plastic labels is increasing. Reasons include the great improvements in the quality of printing on plastic film, the growth of rigid plastic containers, and new formats like shrink-, stretch- and in-mold labeling. Plastic labels have, in the past, been more expensive and lower quality than paper labels, but improvements in technology and competitive pricing has enhanced the appeal of plastic labels.

The growth of plastic containers has been accompanied by an increase in the use of plastic labels. Their flexibility better matches the flexibility of plastic bottles and other forms, whereas paper labels often suffer from looking "tacked on" to molded plastic. On the other hand, plastic labels are more vulnerable to wrinkling, blistering, and lifted or curled edges. The increased use of rigid plastic containers, many of which have replaced glass jars and metal cans, has brought with it a diversity of labeling materials and techniques.

Plastic labels are capable of some unique effects. They can be made to look like no label at all by using transparent film, which simulates direct printing (especially used with gold ink), or to masquerade as paper. These are popular for glass and clear PET bottles with high end graphic appeal.

Plastic labels can be pressure-sensitive, wrap-around and glued; shrink, stretch, or applied in the bottle-molding process. P-S labels are made from PE, PP, OPP, PVC, and PET. OPP wrap-around labels are used on bottled water. Shrink- and stretch- sleeve labels represent 7.3% of the market and in-mold labels are up to 2.2% (TMLI 2013). Shrink- and stretch-sleeves offer the advantage of a 360° decoration; since they can conform to the contours of various plastic bottles, even squeezable shapes, which are problematic for paper labels. Colored transparent shrink labels can also offer a new look to glass bottles making it possible to vary colors without the prohibitive expense of changing colors in a glass factory. Shrink labels are tricky to design because the graphics need to be pre-distorted, so that they have the right appearance after shrinking.

Another substitute for paper labels is the increase in direct on-package printing for rigid plastic bottles, tubs, and canisters, as well as metal cans and glass bottles. For example, the highest volume cans and bottles of soft drinks and beer are not labeled; they are almost all directly printed lithographically because it is economical for long runs. Other methods for direct printing include *heat transfer, hot stamping, container offset printing, pad printing*, and *screen printing* (Giles 2000).

11.2 PAPER-BASED FORM-FILL-SEAL PACKAGING

Sealed flexible packaging is the fastest-growing sector of the packaging market. The technology has advanced to the point where almost every kind of food can be packed in a flexible pouch with properly chosen barrier properties. Most of these are formed, filled, and sealed in a single sequential operation, called *form-fill-seal* (*FFS*).

FFS bags, pouches, and flow-packs are formed from roll stock and heat-sealed by the packer-filler. They can utilize a wide variety of papers, films, and plastic film laminations and coatings. Although the largest percentage of flexible packaging is made from plastic film, 25% is made from plastic-coated paper or a paper-foil lamination (Impact Marketing Consultants 2006).

Most paper-based FFS packaging is used for dry, powdered food products like dehydrated mixes for soup, sauce, pudding, and drinks. Some are simple pillow pouches that are minimally printed and packed in colorful cartons; others stand on their own and are directly printed with colorful graphics. FFS is used primarily for smaller packages, from sachet to cake mix size, but has also been applied to industrial bags, discussed later in this chapter.

An outer layer of paper in a FFS structure solves three problems that plague all-plastic film pouches: it is easier to print, adds stiffness, and is easier to tear open. Stiff paper-based flexible packages can present a higher-quality appearance than do "flimsy" plastic-film pouches, and the print quality can be better too, giving them a colorful and appetizing shelf-presence without the need of a carton. Paper's easy tearability satisfies the growing demand for easy-to-open packaging and represents a significant opportunity for the paper-based flexible packaging industry.

FFS is a low unit-cost operation, compared to using pre-made bags (discussed later, in Section 11.4.4). Material costs are lower, less labor is required, and the packing and filling is all done in-line. But more skill and investment are required, since the product manufacturer takes total responsibility for the quality of the forming operation, which requires more expertise than simply filling a pre-made bag. Compared to filling and sealing pre-converted bags, FFS requires higher cost equipment, provides less versatility in package sizes, and requires a long-term commitment to a package form.

11.2.1 Paper-Based FFS Materials for Food

The most common paper used in flexible packaging for food is coated or laminated bleached kraft. It has a clean, bright surface with good graphic effect and is suitable for food applications. For pouches with low-end graphics, the paper is simply MG. For higher-end graphics, C1S paper is used.

The paper rollstock is first printed and then coated and/or laminated. The C1S paper for high-end graphics is typically printed by rotogravure in long and repeated print runs, or by lithography for short runs. A lacquer coating is over-printed to protect the surface. UV- or EB-cured coating and inks can be used to improve gloss, heat resistance, and rub-resistance.

The MG paper for pouches with low-end graphics is printed using flexography, or in some cases the pouches are simply ink-jet coded as they are filled. These are generally packed into a more highly decorated folding carton.

One of the most common structures is paper laminated to an inner aluminum foil layer by PE extrusion or an adhesive. Foil provides an excellent barrier for powdered food products that are sensitive to oxygen and/or moisture.

Increasingly, metalized plastic film is replacing such foil laminations. Although foil is a better barrier, it can develop flex cracks and pinholes during the FFS process, and handling the pouches can make them look crinkled and shopworn. Metalized film is less expensive and produces a substrate better able to take abuse. M-OPP is chosen when a superior moisture barrier is needed, and M-PET is chosen when a good oxygen barrier is required.

The film or foil side of the substrate is coated with PE or a PE-copolymer to provide a sealing layer inside the package. Common copolymer sealants include LLDPE, EAA, and ionomers. Copolymers are used when seal requirements are more demanding, such as dusty products that can cling to the sealing area, liquids, and heavy products.

There are also some small sachets, used for products like salt and sugar, which have no foil and are simply made from bleached kraft, because these products do not require a moisture barrier.

For some applications, an alternative to heat-sealing is the *cold seal*. When the two pressure-sensitive adhesive-covered surfaces are pressed together, they form a cohesive bond. Cold seals are used for some foods like chocolate bars that cannot tolerate the heat of sealing, and for bandage strips. However, cold-sealing is more typically used for all-plastic film made from PP and M-PET, and is used for candy and snack bars. Because of its presence on the food-contact side of a wrapper, cold seal materials are strictly regulated.

Paper-based FFS materials are specified from the outside to the inside, in units that differ for each type of material in the structure, since in the United States, each material has a different specification tradition. The paper layer is specified by its basis weight. The coating weight is also based on 3,000 ft^2. Aluminum foil and laminated films are specified in terms of thickness, but the foil is expressed in terms of inches, and plastic laminated film in terms of gauge (which is the same as mils, 0.001 in). Table 11.2 shows several structures, how they are specified, and their applications.

11.2.2 Sterilizable Packages for Medical Products

Medical single-use supplies like bandages, surgical gloves, syringes, and sutures are packaged in *surgical kraft* FFS strip-packs, pouches, or plastic trays

**TABLE 11.2. Examples of Paper-Based FFS Materials
(Used with Permission from American Packaging Corporation).**

Material Specification	Application
OL/Print/ 40# C1S Bleached Kraft /22# LDPE	Dry drink mix
25# Natural Kraft/6# LDPE/1 mil Barrier Sealant Film	Gelatin mix (in carton)
Print/ 20# Bleached MG Kraft /7# PE/0.000285 Foil/10# LDPE	Dry drink mix
OL/Print/ 30# C1S Bleached Kraft /7# PE/0.000285 Foil/15# LDPE	Dry drink mix
20# Bleached MG Kraft /6# PE/60ga Met OPP/10# PE	Dry seasoning mix (in carton)
20# Bleached MG Kraft /7# PE/45g Met OPP/10# mLLDPE	Stuffing mix (in carton)
OL/Print/ 25# C1S Bleached Kraft /7# LDPE/0.000275 Foil/7.5# EAA Coating	Tea
25# Bleached MG Kraft / 7.5# LDPE / 0.0005" F / 12# Ionomer	Effervescent denture cleaner

C1S = Clay-coated one side; MG = Machine-glazed; OL = Overlacquer Met = Metalized; mLLDPE = Metallocene linear low density polyethylene; EAA = Ethylene Acrylic Acid

with paper lidding. Paper wrapping, bags, and lidding are also used for medical devices and implants, and for medical devices that are sterilized in a hospital for reuse.

These products are sterilized after packing, so the package needs to be strong, clean, and porous. High-strength bleached virgin kraft pulp is used for most grades, with only FDA-approved additives that do not support microorganisms. Some include anti-microbial additives. The fibers are highly beaten/refined and the paper is smoothly calendered so that there are no loose fibers that could contaminate the product, and some grades are much like glassine. Basis weights range from 20–90 lb (33–148 g/m^2), but 40–60 lb (64–100 gsm) is most common.

Surgical kraft is also called *sterilization paper* because it must be porous to Ethylene Oxide (EtO) gas or steam, while still preventing the ingress of bacteria. If the packages are to be steam-sterilized, the paper needs sufficient wet-strength. Packages for medical devices can also be irradiated, a sterilization method more common for the non-porous plastic film packages that increasingly predominate in this category.

For higher strength, the paper can be reinforced with latex or other resins. Such binders improve wet-strength, peelability, and dimensional stability. But too much resin or a plastic coating will interfere with porosity and sterilization. Medical paper-making is a highly specialized market niche, and there is a range of proprietary formulations.

A heat- or cold-sealing material is typically coated onto the paper in a pattern, leaving uncoated areas to allow for steam or gas penetration, or it is applied to the seal area only. The seal must be strong enough survive heat, steam, or radiation, but must still allow the pouch to be peeled open easily.

For steam and EtO sterilization, the filled and sealed package is flushed with gas or steam in a vacuum chamber. The vacuum is held for a period of time, and then the gas or steam is evacuated. This process stresses the seals, especially since the fresh heatseals may still be warm. If the package is to be sterilized by steam, it must withstand the hot moist air, at temperatures up to 270°F (132°C), in which case PP is generally chosen as the sealant. Steam sterilization is usually done on a small scale in hospitals for heat-resistant devices. EtO sterilization is used for single-use devices produced in high-volume operations, gassing pouched products in their shipping containers and even on pallets. EtO sterilization is performed at a lower temperature, 100–140°F (27–60°C) at 50% RH, and so a PE or EVA heatseal is sufficient (Yambrach 2009).

Medical packages need to present the product in a sterile manner, whether they are opened by a nurse during the stress of a surgical environment or by a mother applying a bandage to her child's cut. The most common easy-opening method is peeling open the seal. The grip area can be increased with a chevron or open corner on the seal. The peel and tear strength of paper is lower than plastic or foil, so paper-based packages are easier to open. But paper has the disadvantage of generating particulates when it is torn or peeled, which is not permitted in a surgical environment.

Medical packaging is highly regulated in order to protect patients and to assist health-care workers. There are ISO, FDA, and EU mandates for material qualification, testing, and validation. There are standards for pH, chlorine, and sulphate content, strength, microbial penetration, and air permeability. There are a number of ASTM tests for microbial barrier, porosity, particulates, package integrity, and ease of opening. An excellent review can be found in Bix and de la Fuente (2009).

Paper was the earliest flexible medical packaging material, but its use has been supplanted largely by plastics, especially Tyvek®, a polyethylene material made with spun fibers that is also permeable to sterilization gas and steam. Plastic film is used for products that are sterilized by radiation. Compared to paper, plastic provides better puncture-resistance. It can be made in a more uniform thickness, density, and porosity, and is not vulnerable to moisture or microorganisms.

But paper is less expensive than plastic. It is stiffer and easy to run on machinery. It is easier to clearly print with vital identification and instructions. And it is easier to open and recycle.

11.2.3 Form-Fill-Seal Specifications and Operations

Most paper-based pouches are packed in horizontal form-fill-seal (HFFS) operations. The material is folded or sealed on the bottom, and the machine positions the pre-printed substrate correctly by sensing registration marks. The pouch sides are formed, filled, and the top is then sealed. Filler types for powders include pistons, augers, vibratory, volumetric, gravimetric, and count.

HFFS machines are available in three different configurations depending on whether the packages are cut apart before or after filling, and whether the pouch is upright (vertical) or horizontal when as it is packed. The most basic "workhorse" is the vertical pouch form/cut/fill/seal intermittent motion machine, which favors the stiffness of paper substrates, with speeds ranging from 50–120/min. Similar machines with continuous and/or rotary motion are more complex and expensive, but faster, producing 300–500 pouches/min. Machines that cut apart the pouches after filling are even faster (and more expensive) producing up to 1,300/min. Machines for which the product is fed horizontally, like cookies and bars, are called *flow wrappers*, rather than pouches, at speeds up to 300/min (Bardsley 2009). (FFS machines are classified as horizontal or vertical depending on the direction the pouch material travels through the machine, not the filling direction.)

There are several variations on the FFS theme. Vertical form-fill-seal (VFFS) systems can be used, but are more common for plastic film. Pouches can be formed as pillow-packs or with a square bottom or gussets. A flexible lidding material can be sealed onto a thermoformed tray. Flexible materials can be wrapped and glued around a mandrel or product, such as roll-wrapped candy or cookies. And teabags are made from a lightweight tissue with long fibers in its pulp, also on a FFS machine.

11.3 WRAPS AND VOID FILLER

"Brown paper packages tied up with strings" used to hold all kinds of things. In other parts of the world, wrapping is much more common than it is in the United States. For example, British wine merchants still wrap bottles in tissue rather than bagging them. In Japan, wrapping is a common skill, and merchants elaborately wrap items for their customers.

LaBarre, in 1952, presented an extensive discussion of wrapping papers used at the time. He noted that:

> *An extraordinary diversity of papers is used for wrapping purposes, from the vulgar straw paper—scarcely ever seen in the UK—to luxurious glassines and parchments costing seven or eight times as much; from the thinnest tissues to rope brown, frequently stronger than cloth of the same weight.*

After discussing the salient properties to be considered when choosing wrapping papers (ranging from sizing and stiffness to handle and "look-through"), he provides a list of wrappings that are named for their origin (like kraft or straw), for their use or purpose (like butter, sugar, and cartridge) or for their appearance (browns, blues, or glassine).

In Japan, the concept of wrapping is called *tsutsumi*. Tsutsumi plays a central role in a variety of spiritual and cultural aspects of Japanese life. Wrapping gently conceals items, gifts, and architectural space. There are special terms and rules for wrapping certain items, the style varying according to the occasion. "In Japan it is said that giving a gift is like wrapping one's heart. Just as one helps a friend into a coat carefully and courteously, a gift should be wrapped tenderly and conscientiously," (Ekiguchi 1985). One of the most common papers used in tsutsumi is *washi*, a softly lustrous handmade paper made in a variety of weights and types. Figure 11.2 shows a typical Japanese wrapping technique.

However, there is little wrapping used in the United States today because of the easy availability of fabricated bags, cartons, and boxes. As a result, we

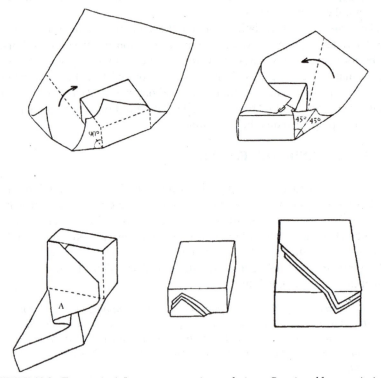

FIGURE 11.2. *Tsutsumi: A Japanese wrapping technique. Reprinted by permission of Kodansha International Ltd, from Ekiguchi (1985),* Creative Ideas from Japan, *drawings by Eiko Ikeda, 21–22.*

may have lost the art of paper wrapping. Even gift wrapping is being replaced by gift bags, removing the human touch from this personal exchange. It is an unfortunate loss of mystery, skill, and grace.

Most food wrapping uses have switched to plastic shrink or stretch film. For example, meats and fish are now wrapped with PE or PVC film in trays by the grocer. Most clear sparkly wrapping is now OPP that has replaced the glossy look of cellophane. Even roofing shingles, which were long bundled in kraft paper wrappers, are now bundled in plastic LDPE film.

Another use of paper is as a dunnage material to fill voids in small parcel packages. The stiffness of the paper, when crumpled, provides a relatively good void-filler for lightweight items. Kraft paper provides a good, stiff cushion, compared to newsprint that provides a softer surface. There are even machines that simultaneously crumple multiple kraft sheets for high volume use. A related material is cellulose wadding, which is a multi-layered tissue material that is softer than newsprint.

11.4 BAGS AND SACKS

Bags (also called sacks) are one of the oldest forms of packaging that is still commonly used. This is because humans have always found that bags are an efficient way to pack many kinds of products, or even an assortment of products. Bags, throughout history, have been made from whatever kind of flexible material has been plentiful: leaves, skins, fabric, paper, or plastic.

Paper bags were once used more widely than they are today. The paper grocery bag, once ubiquitous in the United States, has been largely replaced by less expensive plastic bags. Paper shopping bags are used now mainly by high-prestige stores and have high-end graphics. Plastics are competing with, or at least reinforcing, even rugged multiwall shipping sacks.

Paper bags use most of the kraft paper produced, aside from corrugated containerboard. Grocery and retail bags use about 43% of all kraft paper and multiwall bags and shipping sacks use 38%; the other 19% is used for wrapping and other converting (DeKing 2003).

Paper bags are inexpensive, flexible, and rugged. They use a minimum amount of material to contain a product. They are lightweight packages for heavy products. The bag only needs to provide tensile strength. Since the product inside provides the stacking strength, bags are almost infinitely stackable as long as they have a high coefficient of friction. By comparison, the walls of rigid boxes or cartons for heavy powdered products require a great deal more packaging material to provide stacking strength, since there is always some compressible headspace in a carton.

Since paper is so easy to print, a wide range of graphic effects is possible. The stiffness of paper makes a good display surface.

Paper bags are simple to fill and close. With no plastic liner, the porous paper lets entrapped air escape easily, which is a special problem associated with packing dry granular and powdered products in airtight packages

Paper bags have several drawbacks. They can allow powders to sift through the folds and closures. They suffer from breaks and punctures, and retailers complain about leakage. Consumers complain about the lack of reclosability, especially for products like dog food or fertilizer that are used over a period of time.

Most paper bags have more than one ply. There are three bag categories, defined by their uses, but they have similar materials and constructions. Two are used by consumers: shopping bags and retail bags containing products like flour, sugar, cookies, and pet food. The third type, multiwall shipping sacks, are primarily used by industrial and institutional customers.

11.4.1 Paper, Film, and Foil Plies

Paper for most bags needs to be strong. This is why most of the paper used in traditional brown paper bags is natural kraft. Some special types of kraft paper have been developed for the bag industry; following is a partial list, along with the acronyms used to describe them in bag specifications.

- Natural kraft (NK)
- Wet strength natural kraft (WSNK)
- Extensible kraft (EK)
- Bleached kraft (BK)
- Clay-coated bleached kraft (CCBK)

In addition, kraft paper can be coated with silicone to provide release properties for tacky products. It can be coated with a plastic like PE for moisture resistance.

The outer surface or ply of a bag usually has a special treatment to facilitate printing and to protect during distribution. Bleaching and clay coating give a nice bright smooth surface that accepts print well. Extra calendering or a lacquer finish can add rub resistance, but can also add slipperiness. There may be a special treatment to increase the friction coefficient, because slippery bags are easily damaged and can be dangerous. Extensible kraft may be used to improve impact resistance, compared to ordinary natural kraft. A wet strength treatment, which causes the paper to retain about 25–33% of its dry tensile strength, may be added to the outer kraft ply (WSNK) and is designated as such by a faint colored stripe.

Multiwall bags can be lined with plastic film to act as a water vapor, grease, or odor barrier. A plastic ply can add strength and may be heatsealed. The following plastics are common barrier materials used in bags:

- Moisture barrier: HDPE, LDPE, LLDPE, OPP
- Grease barrier: HDPE, PP, PVDC, Nylon
- Moisture, odor, and grease: PET, PVDC

For the highest barrier properties, a lamination of paper/PE/foil can be used. The barrier ply may require some perforation to vent entrapped air.

Paper bags are also biodegradable, which can be enhanced by the choice of materials (Mondi Industrial Bags 2013). Biodegradability is increasingly of interest for packaging materials.

Greaseproof or glassine paper can be used as a grease barrier, to prevent oil from staining the outer ply. Grease resistance is desirable in pet food bags because pet food is often packed warm and has a high fat content, and a plastic liner may not be suitable because it traps moisture that can cause mold.

11.4.2 Multiwall Paper Bag Converting

The process for making a multiwall paper bag is deceptively simple: the outer ply of paper is printed, and then the plies are combined and glued together. But the quality of a bag depends on every little detail being done perfectly in a process that involves a number of variables.

The outer ply of paper is first printed, usually by flexography, although it can be printed in any manner. Rotogravure is used for very long runs with the highest quality requirements. Lithography is rare.

The printed ply is married to its fellow plies, and the back seam is glued on a machine called a *tuber.* Since the paper is fed in from rolls, the paper's machine direction runs from the top to the bottom of the bag. The tuber makes the *longitudinal seam* along the back of the bag, gluing each ply separately to itself and gluing the plies to each other. The tuber then cuts the bags apart. The plies can each be cut separately in a *stepped end* bag, so that each ply can be independently glued, or they can be *flush cut*. The basic tube construction and dimensional designations are shown in Figure 11.3.

Bag quality depends on control over the tubing operation. The plies need to be nested together precisely within each other so that they work together. An easy test of ply nesting is to cut a one-inch strip like a belt from around the bag, staple the plies together, and compare how each strip is slightly longer from the inner plies to the outside one.

Glue is needed not only to secure the lineal seam, but also to hold the plies together as they move through the machine. But too much glue holding the plies together can weaken a bag in impact, because it constrains all plies to the weakest one.

Gussets are the reverse folds in the sides of most bags. They are always present in square-bottom bags and can be built into other styles. They help to

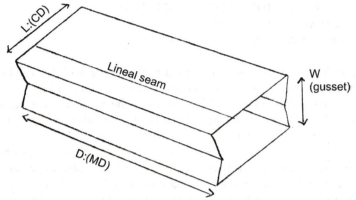

FIGURE 11.3. *Bag tube construction.*

give a uniform three dimensional thickness to a bag, and to make them sit more brick-like on a palletload. *Tube styles* that have no gussets suffer from pointed corners that can cause stacking and damage problems.

The bottom end of the bag is then made in a machine called a *bottomer*, which may or may not be in line with the tuber. This is also known as the *factory closure*, since it is made at the bag factory. It may be sewn or glued into a flat folded or squared bottom.

The open mouth of the bag is also called the *field closure* because it will be closed by the company that fills the bag (in the field). This can also either be sewn or glued, and the method is used to define the names of several bag styles described in the following two sections.

11.4.3 Paper Shipping Sack Styles

Paper shipping sacks, also known as *multiwall bags*, are larger and heavier than most consumer bags, although they are used for some products purchased by consumers, such as pet and lawn/garden care products sold in sizes over 25 lb But most multiwall bags are used for free-flowing granular products sold to industrial or institutional customers. Examples include resin pellets sold to plastics factories, flour and sugar sold to food processors, fertilizers and pesticides sold to farms, and cement, sand, and mortar used in construction.

Most paper shipping sacks are made with a multiwall construction of three to six plies. Each ply has a basis weight of 40–60 lb (per 3,000 ft^2, 65–100 g/m^2). The multiwall construction is stronger than a single layer with the same total basis weight/ thickness because the plies work together, each contributing its individual stretch and tensile strength. The energy of an impact can be more easily absorbed by the multiple plies and some punctures can be deflected by breaking one or two plies, leaving the inner liner intact to carry the product through to its destination. Multiple plies are more flexible than one thick, stiff ply, making the bags easier to make, fill, and handle.

Although the names for multiwall paper bags styles sound very similar, they are in fact very descriptive, and the variations are few. The names refer to how the bag is filled and closed. The terminology has to some extent been standardized by the *Paper Shipping Sack Manufacturers' Association* (pssma.com), and since their customers close one end, this distinction is useful.

The two primary categories are *open mouth* and *valve* bags. An *open mouth* bag can be opened wide to accept filling. Open mouth field closures are sewn or glued or a pre-applied hotmelt adhesive is activated. There are four standard open mouth styles.

The *sewn open mouth (SOM)* style (Figure 11.4) is sewn on both ends. This closure method, inherited from fabric bags, is inexpensive and simple. The field closure is sewn automatically or by use of a hand-guided sewing machine suspended from above, and usually a reinforcing tape is applied either over or under the stitching. The factory closure is usually also sewn.

Since the sewing is usually a single-thread chainstitch, pulling the thread in just the right place will quickly rip out the stitching, which should make this an easy-to-open bag. The "right place" is on the straight stitch side without

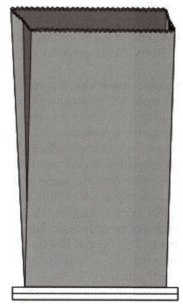

FIGURE 11.4. *Sewn open mouth (SOM) gusseted bag style. Reprinted with permission from Smurfit-Stone Container.*

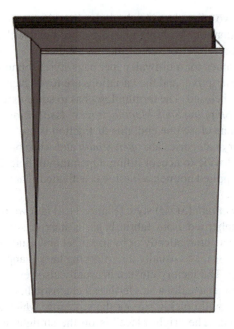

FIGURE 11.5. *Pinch bottom open mouth (PBOM) bag style. Reprinted with permission from Smurfit-Stone Container.*

the chain, beginning at the end characterized by the chain stitch's loop, but (besides being hard to explain) this is generally only known by frequent users.

The sewn closure has some drawbacks. The sewing holes can allow powdery product to sift out or allow insects and moisture to creep in, and the perforated line of stitching adds a weakness in the otherwise strong paper. Taping over the stitching can reduce these problems. SOM bags are used mostly for sugar, flour, rice, seeds, beans, and other granular agricultural products that are low cost and not prone to infestation or moisture damage.

The *pinch bottom open mouth (PBOM)* bag is sealed by hot melt adhesive at both ends, making a nice clean edge (see Figure 11.5). The ends are *stepped*, with each ply cut to a slightly different length so that all plies can be glued. The stepped-end construction makes the end secure. The factory closure is made by applying hot adhesive and folding the end like the flap of an envelope, pressing so that every ply is adhered. In the same machine, the stepped ends of the open mouth are coated with adhesive, but this is cooled without adhering to anything. When the bag is filled, the adhesive is activated with heat and sealed in the closing machine that folds over the field-closure end and seals it (also like the flap of an envelope).

PBOM bags offer more protection than other paper bag styles because they are siftproof and the closure is strong. The positive closure minimizes infestation, and a plastic liner can lend specific barrier and strength properties. The flat front and printable ends present a superior product display in a retail store. PBOM bags are used for pet food, chemicals, agricultural products, and some hazardous chemicals. They are one of the strongest bag styles, preferred by the U.S. government for its export food aid (blended and fortified grain) shipped in one of the most demanding distribution systems in the world.

Pasted bottom bags usually have a plastic liner, which may be independently heatsealed to achieve an even more positive seal. A common reason for desiring a better seal is to protect an extremely moisture-sensitive or insect-prone product. It also slightly strengthens the end seal.

But if the powdered or granular product is conveyed through the filling equipment by the use of air, a sealed liner can cause a number of problems. Air-filled bags waste space, are unstable during handling and storage, and can bounce violently in transit. The usual solution to this problem is to add tiny air vent holes in the liner to allow the air to escape. Depending on the moisture sensitivity of the product, these may just be small holes punched in the gussets of the liner, or there may be a special one-way vent strip added that forces the air through a more convoluted path.

There are two pasted-bottom open mouth styles with flat, square bottoms.

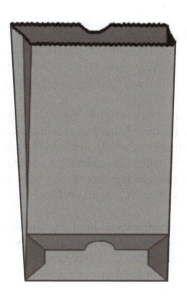

FIGURE 11.6. *Self opening square (SOS) and satchel bottom bag styles. Reprinted with permission from Smurfit-Stone Container.*

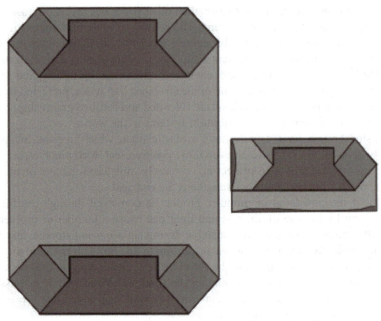

FIGURE 11.7. *Pasted valve sack and tuck-in sleeve. Reprinted with permission from Smurfit-Stone Container.*

The *pasted open mouth satchel-bottom sack* (*POM*) has no gusset on the top, and the pasted open mouth *self-opening square* (*SOS*) has a square bottom (actually rectangular) with a gusset extending to the top (see Figure 11.6). SOS bags are also the style of shopping bags, also called *self-opening style* because they can be shaken open with one hand.

A square bottom adds a number of advantages over a pillow shape. It helps the bag stand up for filling. Square-bottom bags form better stacks when their brick-like ends are oriented towards the perimeter of a palletload. SOS bags are used for many industrial products when the palletization benefit is needed. This style is also used as a *baler bag* that may bundle together several bags of (for example) potatoes, petfood, kitty litter, or flour. It is used, in Japan, as a baler for toilet paper.

Valve style bags are filled through an open corner spout built into the bag instead of an open mouth. Both ends are either sewn or glued into a satchel bottom. Both ends of a valve bag are sealed at the bag factory, leaving a small unsealed area with an inner flap or tuck-in sleeve that functions as a self-closing check valve after filling, as shown in Figure 11.7.

Valve bags usually do not have liners and are used for dusty powdered products. During filling, the product is blown into the valve opening and the bag acts as a filter to trap all of the product and dust but permit the air to evacuate. Filled valve bags are space-efficient and palletize well since the square ends create a brick-like shape. They can be made with only two paper plies, and so are less expensive than open mouth multiwall bags, but are not used for moisture-sensitive products. Common uses are for cement, corn sweetener, and carbon black.

There are a number of special features that can be built into a paper shipping sack. Easy-opening features can be fashioned by burying a rip string in a closure. Anti-skid coatings are sometimes applied to raise the coefficient of friction and prevent accidents from bags slipping off of a stack. Barrier materials (usually LDPE or PP) can help protect from puncture and moisture. A separately sealed inner plastic bag that breaks away from the paper plies can be designed for products that require special containment. For example bags for dried milk that must be emptied in a clean room are packed in a bag that is designed to have its dirty outer paper plies stripped away prior to entering the room.

11.4.4 Consumer Bags

Most of the paper bags for consumer products contain less than 25 lb The most common forms are shopping/grocery bags, flour and sugar bags, cookie and snack bags, and bags for pet food and lawn/garden products.

Pre-formed (as opposed to FFS) consumer bags bear a great similarity to multiwall shipping sacks. The styles are similar (with the same names) and the technology is the same. For example, the common paper shopping bag is named for its self-opening style (SOS).

A version of the SOS style is used for many products like rice, salt, sugar, pet food, charcoal, and garden chemicals. They have a maximum of three walls, no more than 130 lb total basis weight. Unlike shipping sacks, they usually have a bleached and calendered or clay-coated outer ply for better graphics. Sometimes the inner plies are also bleached to give a pure appearance for food contact, or they may be colored, as in the case of using blue plies inside sugar or flour bags to make the product appear whiter. The top can be rolled twice and closed with a hot melt adhesive or sewn.

The advantages of the SOS style for consumer products are that the square bottom makes it stand up straight for display and presents an acceptable display panel when laying down. It is easy for a consumer to use. The disadvantage is that it can allow sifting.

Converted consumer bags can also be made in the satchel bottom style (with no gussets, used for sugar or flour), and the PBOM style, sometimes used for cookies. Microwave popcorn bags are a form of PBOM, with a microwave susceptor incorporated in the inner ply. Although the sewn open mouth style is rarely used for consumer products, it is used for burnable barbecue bags for charcoal, which have wax-coated tape over the stitching to assist burning.

11.4.5 Specifications and Bagging Operations

Bags are specified by buyers in term of the plies of paper, dimensions, and style. The style is designated by its name as described in the previous two sections, such as SOS and PBOM.

Plies are specified from inside to outside. For example, the following describes the material used in a typical bag used in the U.S. Government Food Aid Program: *3.0 mil LLDPE/50 NK/50 NK/60 WSNK*. This means that there is a 3 mil (0.003 in) linear low density polyethylene liner inside the bag. The outer ply is 60 lb (3,000 ft^2) basis weight sheet of wet strength natural kraft. The two middle plies are 50 lb BW natural kraft.

Like FFS substrates, a laminated and/or coated paper ply is specified in terms of its basis weight and the coating weight is also based on 3,000 ft^2. Aluminum foil and laminated films are specified in terms of thickness, but the foil is expressed in terms of inches, and plastic film in terms of gauge (which is the same as mils, 0.001 in). Therefore, a laminated ply that is specified as 45# PE/.00035AL Foil/8#PE/50#NK is made from a 50-lb BW kraft, extrusion-laminated with 8 lb/3,000 ft^2 of PE to aluminum foil that is 0.00035 in thick and coated inside with 45 lb/3,000 ft^2 of PE.

The number and basis weight of plies is chosen based on the strength required. The outside surface is selected on the basis of printing requirements, rub-resistance, and coefficient of friction (COF). For the highest quality printing, a bright clay-coated bleached kraft sheet is used. The wire side of the paper generally goes on the inside with the felt side out for a smoother print surface. Rub-resistance and COF can be modified with coatings or varnish, and performance can be specified in standard tests discussed in Chapter 8. If an inner plastic ply is intended to be a water or oxygen barrier, the barrier characteristics of the film sheet may also be specified.

Dimensions are described in terms of the flat, unfilled bag, to the top of its open mouth (note that this is not the dimension when the end is closed). As with other package forms, the dimensions are always listed in terms of L × W × D, but the designation for bags is a little confusing since the perspective is different. To remember which is the L, W, and D, one must imagine the bag standing up on one end, as shown in Figure 11.8. Dimensions are specified by:

Face Width (*L*) × Gusset Width (*W*) × Top-to-bottom Length (*D*)

The choice of filling machine and the associated bag style depends on the nature of the product's flow characteristics and particle size. There is a wide

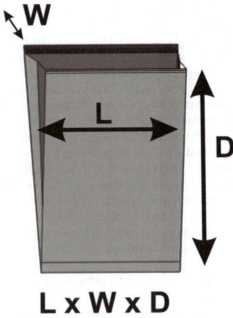

FIGURE 11.8. *Bag dimension specification. Reprinted with permission from Smurfit-Stone Container.*

range of speeds and levels of automation possible, from manually shoveling and sewing open-mouth sacks, to high-speed fully automatic rotary fillers. Free-flowing granular products, like resin pellets and grain, are conveyed and weighed by gravity or belt feeders. Fine powders like flour are first *fluidized* by aeration so that they can flow by gravity or pressure. Products that do not flow freely, like fine pigments or coarse flakes, are conveyed by vibration or screw *impeller*. Weighing occurs either as the bag is being filled or the product is pre-weighed in a hopper before filling the bag. Empty bags are placed, and connected to the filling spout, either manually or automatically. Valve bags are faster to fill than open-mouth bags because they do not need to be closed (although some do have tucked or heatsealed valves). Open-mouth bags are sewn or sealed. The filled bag may be vibrated or flattened to settle the shape for stable and safe stacking (Mondi Industrial Bags 2013).

11.4.6 Size and Cost Estimation

Bag dimensions are chosen based on the product density and weight, the geometry of the bag style, standard tube widths, freeboard (headspace), and pallet pattern.

A bag becomes a different dimension when it is full than when it is empty. Although this may sound obvious, it is easy to overlook. The relationship of the flat bags' dimensions to those of filled bags varies by bag style. Open mouth bags become considerably shorter, by the size of the closure and one-half of the gusset width. Bags without gussets and pasted valve bags become narrower in the face width dimension. They all (obviously) become thicker.

Multiwall bag manufacturers have standard dimensions of paper widths and tuber settings that determine the range of tube dimensions. Within that range, the face and gusset width and length for a given net weight of bag are chosen based on the product density and the intended net weight.

However, a paper bag is rarely truly "full." There is always a certain amount of *freeboard* (a ship term for the space between the waterline and the deck). This headspace in a bag facilitates handling and improves brick-like palletization. Bags that are over-filled take on more of the shape of footballs, making unstable palletloads. Powders that entrap air require an initially larger bag. On the other hand, bags that are too empty use too much material and are awkward to handle.

Dimensions are also chosen based on obtaining an efficient pallet pattern. Palletizing dimensions and characteristics vary by style; for example, bags without gussets are wide at their ends and use more pallet footprint than do square bottom bags.

Estimation of required dimensions is complex. It requires a formula in between that used for the volume of a flat envelope and that of a squared-off box, with variables to allow for freeboard, end styles, entrapped air, standard tuber widths, and palletization alternatives.

To simplify the estimation, the Paper Shipping Sack Manufacturers Association has developed calculations and tables (which can be found at pssma.com) that recommend bag sizes based on experience and the product volume, calculated as:

$$V = \frac{\text{mass of product}}{\text{bulk density of product}}$$

For example, Table 11.3 gives recommended dimensions for two products, one of which has twice the volume of the other. It is important to remember that dimensions are specified for flat bags, from the bottom closure to the open top.

But even the manufacturers' guides remind in capital letters, "ALWAYS VERIFY ESTIMATED BAG SIZE BY FILLING SAMPLE BAGS WITH THE CUSTOMER'S PRODUCT." Given the geometrical changes when bags are filled, it is also a good idea to build at least a pallet layer with filled sample to be sure that they properly fit on the outside, too.

TABLE 11.3. Example: Bag Volume Versus Bag Size in Inches (Smurfit-Stone Container, no date).

Product	Product Volume	SOM	PBOM
Corn starch	2.02 ft³	17 × 5 × 39	18 × 4 × 39
Plastic resins	1.01 ft³	15 × 3-1/2 × 29	14-1/2 × 4 × 29

There are three easy steps to estimate the cost of a bag, similar to the cost estimation method given for paperboard cartons in Chapter 12 and corrugated boxes in Chapter 15.

1. Determine the amount of paper (basis weight × dimensions × number of plies) and other materials used to make the bag.
2. Find the market cost of the materials and multiply it by the number of pounds in the blank. Current prices for kraft paper can be found in *Pulp and Paper Week*, and commodity plastic prices are published in *Plastics News*.
3. Double the cost of material, using this as an estimator for the conversion cost of making the bag. Research has shown that the cost of a finished kraft bag, including basic 1–2 color flexographic printing, is generally double the cost of the material. Printing more colors or gravure printing would add an extra cost.

Example:

A multiwall bag has three plies of 60-lb kraft paper and a 3.5 mil LDPE liner, each ply measuring 10 ft². If the transacted price for kraft paper is about $1,000/ton (*Pulp and Paper Week*) and LDPE resin is $1/lb (*Plastics News*), estimate the cost of the converted bag.

1. The weight of three plies of paper = 3 (60 lb/3,000 ft) × 10 ft² = 0.6 lb

   Plastic liner weight = 10 ft² × (0.0035 in × 1 ft/12 in) × 0.92 g/cm³

   × (2.54 cm/in × 12 in/ft)³ × lb/454 g = 0.1674 lb

   (the typical density of LDPE is 0.92 g/cm³)

2. (0.6 lb × $1,000/2,000 lb) + (0.1674 lb × $1/lb) = $0.3 + $016.74 = 46.74¢ for materials

   It is interesting to note that the plastic sheet costs almost as much as two of the three paper plies.
3. Double for converting estimate: 46.74¢ × 2 = 93.48¢ per bag

Of course, bag prices depend on other factors, like volume discounts and competitive conditions. But an estimation method like this can be used to judge whether a supplier's price quote is fair. It can also be used to estimate the effect of changing bag size or materials. For example, if the bag in the example above was made from two plies of 60-lb kraft paper and a 3-mil liner, the estimated cost would be reduced to 73.4¢ per bag.

11.4.7 Substitutes for Paper Bags

The history of bags is a history of progressive substitution, from animal skins to plastic. In paper bags, plastic layers have increasingly added water and gas barrier properties and heatsealability. Plastics substitutions with no paper at all have taken two forms: woven sewn shipping sacks and single-sheet plastic film bags.

Woven PP (WPP) shipping sacks are used for flour, rice, and corn. Most of them are made from extruded flat PP yarns (about 10 strands per inch). They are woven in the form of a tube or sheet and sewn closed, resulting in a very strong bag. The fabric can be laminated to a film sheet for added barrier properties (although pinholes form) or to give a glossy printable surface, or to paper for reinforcement.

The introduction of photographic-quality color printed WPP has been successful for large dog food bags, because it is stronger than paper. Although more costly than the equivalent multiwall paper bag, WPP can reduce the cost of torn/leaking pet food bags, a top complaint of grocers. However, WPP does not seal well, and sewing creates holes and therefore a vulnerability to insects, so many are simply taped closed. This is a technology problem that is in the process of being solved.

Woven PP is also used to make *bulk bags* (also known as *flexible intermediate bulk containers*), which hold approximately 1 ton of free-flowing product, and these have replaced multiwall bags for some industrial uses.

The most notable plastic bags are the HDPE and LDPE grocery bags, which have replaced most paper grocery and shopping bags. Baggers can be choosers when the so-called *tee shirt* bag (because its shape resembles one) is so much less expensive, and uses less material than the equivalent paper bag.

Improved printing capability, zipper reclosability, and handles have driven some users to plastic bags. Pet food and lawn/garden product manufacturers have switched to plastic bags as the graphics capability has improved, and special features like handles and reclosable zippers have been incorporated. The easy heatsealability of plastic film, compared to paper, has created opportunities for the form-fill-seal format to be adopted for larger bags, up to 100 lb, by high-volume users in the food and chemical industries.

Flour and sugar processors, tired of complaints about leaking and sifting, have developed plastic micro-perforation technology to the point where sealed plastic bags can evacuate air under compression, a major limitation to using plastic bags for powdered products in the past.

The U.S. multiwall bag industry output peaked in 1999 (Impact Marketing Consultants Inc. 2006). Output has been falling steadily as paper bags have been increasingly replaced by substitutes like plastic bags, intermediate bulk containers, and bulk shipping.

Little growth is predicted for the U.S. kraft paper bag industry. The industry has consolidated, kraft paper and bag plants have closed, and there is little margin for innovation. Most consumers do not buy large "bulk" package sizes today; most households have no need for packages larger than a pound or two—even for sugar and flour. There is less home cooking, and more desire for convenience, easy dispensing, secure reclosability, and stable storage.

But paper bags are still a low cost alternative for many products, especially when the paper is combined with plastics. Multiwall bags are more rugged than single-ply plastic bags. They also have the advantage of being porous to air, allowing it to escape from sealed bags that are filled with powdered product. Furthermore, many current users of paper bags in the agricultural, chemical, and pet food industries, are reluctant to switch package forms, since it would require a change in packing and filling equipment. But if and when they do decide to change to a plastic bag or other package form, the change has a tendency to be widespread. The paper bag industry has much at stake in finding its role in the future.

11.4.8 Tests to Evaluate Bags

In many logistical systems, bags are handled roughly. They are tossed, dropped, slid, and stacked. A bag that is not palletized or containerized, shipped *break bulk* through the ports of the world as is U.S. Government Food Aid, will be handled over 20 times by longshoremen and warehouse workers. Even if bags are shipped palletized or containerized, they are handled before and after they are assembled in unit loads. Therefore, it is common to evaluate bags based on their impact performance.

Tests for bag impact resistance are similar to those used for boxes and other shipping containers:

- ASTM D5276: Drop Test of Loaded Containers by Free Fall
- ISO 7965: Packaging–Sacks–Drop Test
- ASTM D5487: Simulated Drop of Loaded Containers by Shock Machines

Either free-fall or shock machine tests can be used to specify a minimum drop height requirement. They can be used to judge the effect of repeated impacts from the same drop height. Or a progressive ("elevator") test that increases the drop height to failure can be used to compare alternative constructions. Different orientations like *butt drop, edge drop*, and *side drop* can be compared.

The free fall drop test is simple but variable and messy. It usually employs a trap door release mechanism to reduce the tendency of the bag to rotate as it is released. This tendency adds to the variability of results. If the bag is to be tested to failure, the free fall test can be messy.

On the other hand, shock machine impacts are more repeatable and tidy. The angle of impact is more controllable, and the plastic programmers generate a short (less than 3 ms) repeatable shock pulse. This reduces the variability and so fewer samples are required. In the shock machine test, the bag can be loosely shrouded by a large plastic bag that will contain the product if/when the bag breaks (which if used in a free-fall test would alter the physics of the fall). Therefore the shock machine test is more likely to be used as a progressive test to failure. The shock machine test increments, however, are measured in terms of velocity change (ΔV), not drop height, because the characteristics of machines vary. This test method is used by the U.S. government to specify the required strength of bags in its international food aid programs (Twede, *et al.* 1990).

Impact testing to failure reveals that paper bags have some built-in weaknesses. A butt drop best evaluates the strength of the paper in a bag in its weakest drop orientation, because it stresses the paper in its weaker cross-machine direction at its narrowest point, the girth of the bag. It is clear that the MD in paper bags is chosen for production efficiency, and not for impact strength.

Butt drop tests can also betray a weak longitudinal seam. They reveal which kinds of reinforcement, such as plastic plies, can withstand impacts that the paper plies alone cannot. Butt drops demonstrate that some ways of gluing the plies together can weaken and others strengthen a bag. For example, if a tough LLDPE liner is used, gluing the liner to the paper in the body of the bag makes it fail along with the paper ply; if the liner is loose and only attached at the ends, it is often still able to continue to contain the product after the paper plies fail.

Side-drop tests are used to judge the integrity of a bag's sealed ends. For example, they show that sewn bags tend to rip open along a perforated sewing line. On the other hand, flat drops cause relatively little damage, because all areas are stressed equally (and bags are designed to lay flat).

Impact performance is based, in part, on each ply's tensile energy absorption (TEA). TEA is a measure of its toughness, related to the paper's tensile strength and percentage of elongation at break, the area under the stress-strain curve.

Impact strength also depends on the relationship between the plies. If their modulus of elasticity differs, and/or if they are not nested properly, the plies with the least room to stretch will have the greatest effect on failure height. The way that the plies are glued together also affects performance.

Besides tensile, elongation, and TEA, paper properties that contribute to bag performance are thickness, burst strength, tear strength, coefficient of friction, wet strength, porosity, and moisture content. Tests for these properties were reviewed previously in Chapter 8. There are also ISO standards for paper sack description and measurement (ISO 6591), specification (ISO 8351), and dimensional tolerances (ISO 8367).

11.5 REFERENCES

Alexander Watson Associates. 2013. *Labeling Markets: North American Market Study & Sourcebook 2013*. Chicago: Alexander Watson Associates.

Bardsley, R.F. 2009. Form/Fill/Seal, Horizontal. In *The Wiley Encyclopedia of Packaging Technology*, 3rd ed., edited by Kit L. Yam, pp. 540–543. New York: John Wiley & Sons.

Bix, L. and J. de la Fuente. 2009. Medical Device Packaging. In *The Wiley Encyclopedia of Packaging Technology*, 3rd ed., edited by Kit L. Yam, pp. 713–727. New York: John Wiley & Sons.

Caldewey, J. and C. House. 2003. *Icon: Art of the Wine Label: A Collection of Work.* San Francisco: Wine Appreciation Guild.

Czerniawski, B. 1991. Requirements for Single-Use Sterilization Papers and Paper Packaging Used for Medical Devices, *Packaging Technology and Science,* 4 (4) 213–223.

DeKing, N. 2003. "Kraft paper: U.S. Decline of Bag and Sack Papers Continues," *Pulp and Paper.* 77 (11), 7.

Demetrician, R. 1996. *Label and Graphic Design.* Plainview, NY: Jelmar.

Dunn, T.J. 2013. Multilayer Flexible Packaging. In *The Wiley Encyclopedia of Packaging Technology*, 3rd ed., edited by Kit L. Yam, pp. 799–806. New York: John Wiley & Sons.

Ekiguchi, K. 1985. *Gift Wrapping: Creative Ideas from Japan.* Tokyo and New York: Kodansha International.

Eldred, N.R. 2007. *Package Printing,* 2nd ed. Pittsburgh: PIA/GATFPress

FINAT. 2009. *FINAT Technical Handbook, Test Methods,* 8th ed. The Hague, Netherlands: FINAT

Fairley, M. 2013. Paper Labels. In *Handbook of Paper and Paperboard Packaging Technology,* 2nd ed., edited by Mark J. Kirwan, pp 125-168. Chichester: John Wiley & Sons.

Fowle, J. and M.J. Kirwan. 2013. Paper-based Flexible Packaging. In *Handbook of Paper and Paperboard Packaging Technology,* 2nd ed., edited by Mark J. Kirwan, pp. 91–123. Chichester: John Wiley & Sons.

Giles, G., Ed. 2000. *Design and Technology of Packaging Decoration for the Consumer Market.* Boca Raton, FL: CRC Press.

Hall, I.H. 1999. *Labels and Labelling: A Literature Review.* 2nd ed. Leatherhead, Surrey, UK: Pira.

Hanlon, J.F., R.J. Kelsey, and H.E. Forcinio. 1998. *Handbook of Package Engineering,* 3rd ed. Lancaster, PA: Technomic.

Henry, J.R. 2012. *Packaging Machinery Handbook: The Complete Guide to Automated Packaging Machinery, Including Line Design.* Published by the author, johnhenry@changeover.com

Labarre, E.J. 1952. *Dictionary and Encyclopedia of Paper and Paper-making.* Amsterdam: Swets & Zeitlinger. Revised from a 1937 edition that included samples.

Marotta, C. 1997. Medical Packaging. In *The Wiley Encyclopedia of Packaging Technology,* 2nd ed., edited by Aaron L. Brody and Ken Marsh, pp. 610–614. New York: John Wiley & Sons.

Mills, A. 2009. Intelligent Inks in Packaging. In *The Wiley Encyclopedia of Packaging Technology,* 3rd ed., edited by Kit L. Yam, pp. 598-605. New York: John Wiley & Sons.

Mondi Industrial Bags. 2013. Multiwall Paper Sacks. In *Handbook of Paper and Paperboard Packaging Technology,* 2nd ed., edited by Mark J. Kirwan, pp. 217–251. Chichester: John Wiley & Sons.

Opie, R. 1987. *The Art of the Label: Designs of the Times.* Secaucus, NJ: Chartwell Books.

Paper Shipping Sack Manufacturers'Association. 1991. *Reference Guide for the Paper Shipping Sack Industry.* Tarrytown, NY: PSSMA.

Plastics News. Published weekly: http://www.plasticsnews.com/

PPI Pulp & Paper Week. Published weekly by RISI. http://www.risiinfo.com/

Scully, A. 2009. Active Packaging. In *The Wiley Encyclopedia of Packaging Technology,* 3rd ed., edited by Kit L. Yam, pp. 2–9. New York: John Wiley & Sons.

Smurfit-Stone Container. No date. *A Reference Guide to Multiwall Terminology and Bag Sizing.* Cantonement, FL: Smurfit-Stone Container.

Stienecker, U. 2008. *The Technology of Multiwall Paper Sack Machinery.* Germany: Windmöller & Hölscher.

TLMI. 2010. *A Manual of Recommended Standard Test Methods for Pressure Sensitive Labels.* Gloucester, MA: The Tag and Label Manufacturers Institute.

TLMI. 2013. *The North American Label Study 2013,* 7th ed. Prepared by the Institute for Trends Research for the Tag and Label Manufacturing Institute. Gloucester, MA: The Tag and Label Manufacturers Institute.

Twede, D., R.Clarke, and J. Tait. 2000. Packaging Postponement: A Global Packaging Strategy, *Packaging Technology and Science,* 13 (3), 105–115.

Twede, D., B. Harte, J.W. Goff, and S.P. Miteff. 1990. "Breaking Bags: A Performance Specification Philosophy Based on Damage," *Journal of Packaging Technology,* 4 (6), 17–21.

Werblow, S. and M. Noah. Labels and Labeling Machinery. . In *The Wiley Encyclopedia of Packaging Technology,* 3rd ed., edited by Kit L. Yam, pp. 633–639. New York: John Wiley & Sons.

Yam, K., P. Takhistov, and J. Miltz. 2009. Intelligent Packaging. In *The Wiley Encyclopedia of Packaging Technology,* 3rd ed., edited by Kit L. Yam, pp. 605–616. New York: John Wiley & Sons.

Yambrach, F.. 2009. Package-Integrity in Sterile Disposable Healthcare Products. In *The Wiley Encyclopedia of Packaging Technology,* 3rd ed., edited by Kit L. Yam, pp. 851–859. New York: John Wiley & Sons.

11.6 REVIEW QUESTIONS

1. What is meant, in packaging terminology, by the word "coverter?"

2. What are the two basic methods for affixing paper labels? What are the advantages of each?

3. What kind of printing and glue are most common for wet-glue paper labels? Do you have this kind of glue and printed paper in your house/apartment/dormitory?

4. What is most common printing method for P-S labels?

5. What is the acronym for labels that are self-adhesive? What is the acronym for labels that are clay-coated on one side?

6. Which kind of label is almost always applied by a high-speed machine? Which kind is used for manual application and in slower machines?

7. What factors may cause labels not to lie flat in a labeling machine? What problem can this cause?

8. In which direction does a bottle label's MD run? Why?

9. List some tests for labels and indicate whether they are for wet glue, P-S, or both.

10. What is meant by a "smart" label? What is an active label? Give examples.

11. Why are paper labels used less for plastic packages than for glass and metal? What are some substitutes for paper labels?

12. What are the primary uses for paper-based FFS packages?

13. What are the advantages of including paper in a FFS substrate, compared to an all-plastic film?

14. What is surgical kraft, and how are paper-based pouches for single-use medical products sterilized?

15. Why are we, in the United States, losing the art of wrapping?

16. What are the benefits of bags versus rigid packages like boxes? What are the drawbacks? How high can bags be stacked?

17. What is the primary kind of paper used to make bags? Why?

18. What are the two machines used to make paper bags?

19. What kind of printing is most common for paper bags?

20. Which is stronger: a single ply kraft bag or a multiwall sack with the same total basis weight? Why?

21. Which direction does the paper's machine direction run in a paper bag? Why?

22. What are gussets? What benefits do they add (versus simple tube styles)?

23. What is the primary advantage of a SOM bag?

24. What advantage does a PBOM bag offer over other styles?

25. Why is air evacuation such an important issue for bags? How is air evacuation accomplished?

26. To what does the bag name SOS refer? What are the advantages of an SOS bag?

27. How is a valve bag filled? For what kinds of products is a valve bag best?

28. What are some kinds of products that are packed in multiwall bags? What are some examples of industrial and institutional uses?

29. What are some kinds of products packed in consumer bags?

30. In what order are the plies of a multiwall bag specified?

31. In what order are the dimensions of a bag conventionally specified? Is the D dimension measured with the bag filled or empty?

32. Upon what factors are bag dimensions chosen? What is the most important rule for verifying bag size?

33. Estimate the cost of a multiwall bag with the following construction:

    3.0 mil LDPE/50 NK/50 NK/60 NK,
    each ply measuring 10 ft^2,
    if the transacted price for kraft paper is about $1,000/ton (*Pulp and Paper Week*) and LDPE resin is $1/lb (Plastics News)
    The typical density of LDPE is 0.92 g/cm^3.

34. What are some substitutes for paper bags?

35. What kind of test best evaluates a bag's paper strength? In what drop orientation are most paper bags weakest in an impact test? Why? Which impact orientation puts most stress on a bag's ends?

CHAPTER 12

Paperboard: Folding Cartons, Set-up Boxes, and Beverage Carriers

Paperboard is the most popular rigid material used for consumer packaging. Its primary virtues are flat stiffness, suitability for conversion (easy to cut, crease, fold, and glue), sharp, well defined folds, and excellent printability. Paperboard provides functional properties such as strength and stiffness at relatively low cost. Cartons dress their contents in a high-quality, three-dimensional billboard.

Three good sources of information about paperboard packaging are:

- The Paperboard Packaging Council (PPC), www.ppcnet.org, an association of the leading suppliers to the U.S. paperboard industry, including the manufacturers of folding cartons, cylindrical containers, rigid set-up boxes, and carded packaging.
- The Paperboard Packaging Alliance, www.paperboardpackaging.org, a joint initiative of the American Forest & Paper Association and the Paperboard Packaging Council, which sponsors an annual student design competition.
- The European Carton Makers Association, www.ecma.org, the European equivalent of PPC, publishes the ECMA Code of Folding Carton Design Styles.
- Other countries or regions may have similar packaging trade groups.

These organizations analyze evolving packaging requirements and market tends, and promote the benefits of paperboard packaging and products. They have developed a common language and standards for the diversity of paperboard packaging forms to facilitate clear communication at all levels of design, manufacture, and use.

Paperboard packaging includes a wide range of forms. This chapter covers the types of paperboard in more depth than in Chapter 9, and reviews the basics of folding cartons: design factors, how they are made, standard styles,

375

specifications, filling operations, and cost estimation. It also covers two related package forms: beverage carriers and set-up boxes.

12.1 PAPERBOARD TYPES AND THEIR PROPERTIES

There are five general types of paperboard, although there are many variations depending on the fiber sources and number of layers. Two grades (bending and non-bending) are made from recycled paper, mainly ONP and OCC, providing a useful market for the increasing amount of wastepaper that is collected for recycling. Two grades (bleached and unbleached) are made from virgin kraft, and one grade (folding boxboard) is made from groundwood.

Paperboard (also known as *boxboard* or *cartonboard*) is nearly always a multi-layer material. The multi-layer structure is a necessity of the manufacturing process, but enables the production of materials with better (and more economical) functional and aesthetic properties and better folding properties. It is usually coated with clay or other mineral pigment on one or both sides to provide a bright, white printing surface. Thickness can range from 10 to 40 points (250–1,020 μm), but the most common thickness is between 14 and 28 points (360–710 μm).

Paperboard can be made on a cylinder or fourdrinier machine, or a combination of both. The cylinder has traditionally produced most of the board that incorporates recycled content. The fourdrinier machine has been traditionally used mostly for board made from kraft (sulfate) fibers. However, the fourdrinier machine is increasingly being used to form multi-ply boards from various kinds of fibers, including secondary and mechanical pulp, using multiple headboxes or formers.

The uses and characteristics of the principal types of paperboard are shown in Table 12.1. The paperboard industry has not standardized the names of these boards, and so sometimes the same type of board can go by different names.

12.1.1 Recycled Cylinder Machine Boards

The cylinder process makes a thick board by consolidating two or more web plies into a single sheet. It uses relatively low-cost, low-grade waste like newspapers and old corrugated containers. Since paperboard is generally specified by thickness rather than weight, there is an advantage for manufacturers to use bulky, low cost materials.

Recycled board is a largely unstandardized material with widely varying properties depending on what it is made from. Manufacturers combine fiber types in the various plies to achieve stiffness and strength inside the sheet and smoothness on the surface. The color of recycled board ranges between grey

TABLE 12.1. The Principal Types of Paperboard.

Board Names/Acronyms	Pulp	Forming Method	Typical Uses	Caliper, BW, Grammage	Characteristics
Uncoated Recycled Boxboard (UCRB) Plain Chipboard	100% recycled fiber	Cylinder	Set-up boxes, partitions		Low grade recycled fibers Poor bending and printing Tan or gray, low cost
Clay-Coated Recycled Board (CRB) Clay-coated News Bending Chip Board White Lined Chipboard Duplex, Grayback Cartonboard	100% recycled fiber, may have white top ply to improve printing, usually clay-coated	Mostly cylinder, some fourdrinier	Folding cartons for cereals, crackers, dry processed food, and detergent; shoe boxes	14–40 pt 360–1020 µm 55–140 lb./msf 270–680 gsm	Low grade recycled fibers Good bending quality 5–9 layers, Can be custom combined Lowest cost carton grade Tan or gray but clay coated or white lined for smooth bright print surface
Solid Bleached Sulfate (SBS) Solid Bleached Board (SBB) Food Board	Bleached, kraft (sulfate), at least 80% virgin fibers, usually clay-coated	Fourdrinier	Premium grade, high end packaging: foods, drinks, pharmaceuticals, cigarettes, cosmetics, health care. PE-coated for wet strength, especially for frozen food	10–24 pt 250–610 µm 40–85 lb./msf 195–415 gsm	Excellent bending qualities Excellent printing surface White sheet throughout Water/moisture resistant Food contact 1 ply in United States 1–4 plies in Europe
Solid Unbleached Sulfate (SUS®) Coated Unbleached Kraft (CUK) Coated Natural Kraft (CNK®) Solid Unbleached Board (SUB) Clay-Coated Kraft Back (CCKB)	Unbleached kraft (sulfate), at least 80% virgin fibers, clay coated	Fourdrinier	Heavy duty packaging; beverage carriers, hardware, powdered soaps, and detergents. May be clay-coated for white surface, PE-coated for wet strength	14–28 pts 360–710 µm	Tough, strong, tear resistant Water/moisture resistant Excellent printing surface Good wet strength 2–3 plies
Folding Boxboard (FBB) Tri-plex Clay-Coated Paperboard Manila Board Bleached Manila	Bleached chemical top, mechanical center, bleached or semi-bleached chemical back	Fourdrinier	High quality folding cartons used for food, confectionery, cigarettes, cosmetics, health and beauty aids More common in Europe	14–28 pts 360–710 µm 50–80 lb./msf 245–390 gsm	Smooth sheet for printing Food contact

Adapted from Paperboard Packaging Council (2000), MeadWestvaco (2002), and Paperboard Packaging Alliance (www.paperboardpackaging.org).

and greyish tan, depending on the source of the secondary fiber pulp. The ink from white newspaper is the source of the grey color, and corrugated board content is the source of the tan.

Recycled board is made in *bending* and *non-bending* grades.

Non-bending chipboard is used to make set-up boxes which have cut scores and taped edges instead of folded scores. These are described later, in Section 12.4. It is also used as a backing stiffener for pads of note paper. Non-bending recycled board accounts for about 9% of the world paperboard supply (Paperboard Packaging Alliance 2003).

Bending chipboard, which is usually has a white surface, is also known as *clay-coated recycled board* (*CRB*), *clay-coated news, greyback, white-lined chipboard* (*WLC*), *recycled cartonboard*, or simply *recycled paperboard*. This is the most economical coated paperboard grade, widely used for folding cartons. As the "bending" designation indicates, it must endure a single fold/score/bend to 180° without breaking the surface or separating the plies (other than at the core of the fold line). It usually has 5–9 plies made from different types of recycled fibers.

The layers of recycled fibers in CRB are combined to take advantage of the benefits of softwood and hardwood properties. Mixed waste recycled fiber is used in the center to provide thickness in a relatively low density paper stock, but it lacks strength, bending ability, and a smooth surface. It can be lined with a higher grade of short fibers, like mechanical hardwood pulp or recycled newspaper to make a smooth surface in preparation for clay coating and printing. A layer of longer softwood fibers, like recycled kraft, may be used to add strength and resist cracking when it is scored (creased). A board is said to be *filled* when it has a base of lower quality inner pulp layers, covered by higher quality fibers which facilitate printing, coating, or gluing.

Since a white surface is best for printing, the recycled board used for folding cartons is coated with clay (or other white mineral pigment), white-lined, or both. Clay-coated on one side is called *C1S*, and coating on both sides is *C2S*. Likewise, a board lined with a surface layer of white fibers on a single side is called *SWL*, and white-lined on both (double) sides is abbreviated *DWL*. Clay-coated recycled board accounts for about 53% of the world's paperboard, and 40% of the paperboard used in the United States (Paperboard Packaging Alliance 2003).

European board mills widely use the German standard DIN 19303 which defines board grades using two letters and a number. For example, High bulk SWL grades are coded GD1 to GD3 (according to bulk) and DWL grades GT1. The terms "Duplex" and "Triplex" are used to describe SWL and DWL grades respectively.

Recycled paperboard has a distinct sustainability and marketing advantage in being able to claim recycled content as well as recyclability. The Recycled

Paperboard Alliance has its own "RPA-100%" trademark which indicates that a paperboard is made from 100% recycled fibers.

Recycled board is not normally used for direct food contact because of the possibility of contamination by the recycled material. A primary source of migration in many cases is from mineral oil in the ink of the recycled paper (Biedermann *et al.* 2011). In the United States, soy-based inks are typically used rather than mineral oil based inks, so this is much less of a problem (DesRoberts 2012). A plastic bag inner liner is commonly used inside cereal and cracker cartons to provide a barrier to contaminant migration (as well as a barrier to oxygen and water vapor). Polyethylene terephthalate (PET) or polyamide (PA) are more effective barriers to mineral oils than polyolefins like PE or PP (Ewender *et al.* 2013).

Cartons for free-flowing products are designed with the machine direction (MD) horizontal to resist panel bulging from internal pressure. This is most critical for cylinder board, because of its proportionally lower stiffness in the CD. The dimensions can also affect the tendency to bulge; thin cartons with large front and back panels tend to bulge more than stocky square ones. A horizontal MD also reduces warping and shrinking due to humidity changes, since most dimensional changes are in the cross machine direction. Although a vertical machine direction would be stronger in compression, this consideration is outweighed by the need to prevent distortion, as shown in Figure 12.1.

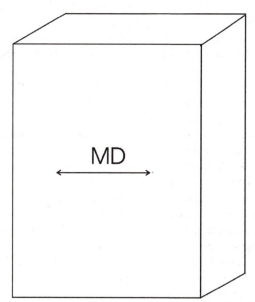

FIGURE 12.1. *The proper machine direction for a cylinder board folding carton.*

12.1.2 Fourdrinier Machine Boards

The fourdrinier machine makes a higher grade of paperboard than the cylinder machine, especially because most of its grades use a high proportion of virgin kraft fibers. The thinnest boards are produced on fourdrinier machines, which can make multi-ply board by the use of secondary headboxes or formers.

Solid bleached kraft (*SBS*) and *coated unbleached kraft* (*CUK*) are made on the fourdrinier machine. The kraft boards are strong for their weight and have good bending qualities. They are the best boards for retaining reasonable strength under humid and damp conditions. They are made from at least 80% virgin kraft fibers. Some hardwood (for smoothness) and/or softwood (for strength) may be combined. They can be clay-coated for a good printing surface. A board is called solid when it is made of the same material throughout, as contrasted with a board where a combination of pulps are used.

The premium boxboard is *solid bleached sulfate* (*SBS*) (*GZ in the DIN 19303 system*). It is white throughout. It is valued for its cleanliness, high quality printability, and smoothness. A clay coating can give an even brighter appearance. Sometimes called *bleached kraft* and *foodboard*, SBS is more expensive than other paperboards.

The clean, odorless, sanitary SBS is valued for foods that come into direct contact with it, such as frozen foods, ice cream, milk, and butter. Cartons made from SBS are widely used for food, drugs, perfumes, cigarettes, and cosmetics. SBS accounts for 16% of the world's paperboard supply, but is more popular in the United States (35% of U.S. supply) than in Europe (where it is only 0.5%, where recycled and mechanically pulped boards predominate and unbleached kraft is preferred where a premium grade is needed) (Paperboard Packaging Alliance 2003).

An LDPE coating makes SBS water-resistant on the coated face(s), for wet and frozen food, and heatsealable. A coating of PET adds the ability to contain food while it heats in an oven.

Due to its foldability, SBS performs well in high-speed machines. It can be used in lighter caliper than recycled board to give the same performance. The thickness can range from 10 to 24 points (250–610 μm), with most applications between 14 and 20 points (360–510 μm).

Coated and *uncoated unbleached kraft* is also called *solid unbleached sulfate* (*SUS*, a trademark of Graphic Packaging International), *coated natural kraft* (*CNK*, a trademark of MeadWestvaco), and GN according to DIN 19303. This is the strongest paperboard used for heavy products such as cans and bottles of beverages, premium frozen foods, powdered detergent, kitty litter, electronics, sporting goods, tools, and auto parts. It may incorporate a wet-strength additive which, combined with long softwood kraft fiber, provides excellent strength when wet, making it the top choice for 12-pack beverage carriers.

A thin top ply of hardwood fibers can be used to give a smooth surface, and a clay coating is used for brightness. Thickness ranges from 14 to 28 points (360–710 μm). CUK accounts for about 20% of the U.S. paperboard supply (Paperboard Packaging Alliance 2003).

Folding boxboard, manila board, also called *bleached manilla*, and simply *clay-coated paperboard*, is made on the fourdrinier machine from groundwood fibers. It typically has bleached chemical pulp on the surfaces and mechanical pulp in the center, which creates a board with high stiffness. Since the sheet tends to yellow when exposed to sunlight, a clay or mineral pigment coating is applied (C1S) to improve its whiteness, smoothness, and brightness. Folding boxboard accounts for 22% of the world's paperboard supply (Paperboard Packaging Alliance 2003). It is more popular in Europe, where it is a substitute for SBS, especially in food packaging because of its virgin fibers and consistent purity (Kirwan 2013).

There is some recycled board made by the fourdrinier process, and this is growing, as multi-ply forming technology has improved. For example, for board below 20 points, a fourdrinier machine with three headboxes can lay down a top layer of white recycled paper pulp, a middle layer of recycled corrugated pulp, and an inside layer of recycled newspaper pulp. A fourdrinier machine has higher speeds that reduce cost. A fourdrinier can even be combined in series with cylinder machines to add a smoother top ply to recycled board.

12.1.3 Paperboard Stiffness and Compression

Most of the problems of folding carton performance occur during transportation, stacking, and storage in warehouses. Puncture resistance, foldability, delamination, tear, stiffness, and compression strength contribute to the durability of cartons.

The prediction of compression strength and stiffness of folding cartons from the intrinsic properties of fibers and the folding carton design style has been a challenge in paper physics for decades. Such information is paramount for the selection of the appropriate paperboard basis weight, especially in this era of sustainability and light-weighting, without compromising the protection function of the packaging.

The stiffness of a paperboard is defined as the bending moment divided by the radius of curvature generated when a load is applied at one end and expressed as (Koran and Kamdem 1989):

$$S = \frac{MOE \times t^3}{12}$$

where S is the stiffness in pounds-inch (or other units such as N·m), MOE is the modulus of elasticity in pounds/in^2 (or other units such as N/m^2), and t is the thickness of the paperboard in inches (or other units such as m).

This equation shows that an increase in MOE and/or thickness will result in an increase in stiffness. For a given pulp with fixed MOE, the best way to improve stiffness is to increase the thickness. Knowing that thickness and density are related through the grammage, therefore an increase of thickness without a change in density corresponds to an increase of grammage:

$$\text{thickness} = \frac{\text{grammage}}{\text{density}}$$

Of course, this also results in increased cost. Thickness can be increased without increasing grammage if density is decreased. However, low density corresponds to low tensile strength and low MOE. Therefore, it is not simple to optimize stiffness (Koran and Kamdem 1990).

The stackability of cartons is related to the stiffness/rigidity of the paperboard to reduce the out-of-plane bulging of the carton under stress in compression. Estimates have been reported (Peel 1999) relating the carton compression strength (P_{carton}) and the compression strength of the carton panels (P_{panel}). The equation below provides an estimated strength for a folding carton having four panels, if the P values are expressed in N.

$$P_{carton} \cong 3.8\, P_{panel} + 1.2$$

Furthermore, the compression strength of a panel (P_{panel}) is related to short span compression strength of the paperboard (F_c) and the bending stiffness of the paperboard (S_b), as shown in the equation below. F_c is obtained by testing the compression strength in the direction of the loading using a very short (0.7 mm long) sample to avoid buckling.

$$P_{panel} = 2\pi \times \sqrt{F_c \times S_b}$$

The bending stiffness S_b is the geometric mean of the bending stiffness in the MD and CD of the panel:

$$S_b = \sqrt{S_{b(MD)} \times S_{b(CD)}}$$

The equations above can be used to estimate the compression strength and stiffness of paperboard cartons. Other factors to consider are the effect of environmental factors such as storage time, age, and handling.

12.2 FOLDING CARTONS

Folding cartons (sometimes called boxes, as in *cereal boxes*) are valued for their low cost, good strength, and excellent advertising graphics. These advertisements appear in the very best places to motivate consumers to buy and remember a brand—in retail stores at the moment of purchase and in consumers' homes as they anticipate and consume the product.

12.2.1 Folding Carton Design

Carton design is a trade-off between factors throughout the supply chain and consumer perceptions. While this is typical of all package design, cartons have some unique requirements and opportunities for design innovation.

Containment and protection requirements depend on the nature of the product: whether it is free-flowing or solid, and whether a barrier is required. There is a limit to the strength of a folding carton, and generally the content's weight does not exceed a couple of pounds or a kilogram. Puncture resistance, stiffness, and tensile strength determine a carton's performance in distribution. Wet or greasy products require special treatments or coating. Refrigerated or frozen foods require strong water-resistant board. The printing surface must be the appropriate color, brightness, smoothness, and gloss.

The weight of the contents determines the thickness of board required, because thickness is related to the board's stiffness (if the board is homogenous, stiffness is proportional the cube of thickness, as shown in Section 12.1.3) and therefore to its ability to resist bulging. As a rule of thumb, a carton made from clay-coated recycled board (CRB) for light contents (up to 0.5 lb, 230 g) is made from board that is 15–18 points thick (380–460 μm). Board for a 1-lb (450 g) load is 20–24 points (510–610 μm). Board for a 2 lb (900 g) of contents is 28–32 points (710–810 μm). And for contents that weigh over 2 lb (900 g), a microflute corrugated board (E, F, or N flute) may be considered. Cartons can be used for heavier products if they are solid and load-bearing. The ratio of the required paperboard thickness to the content weight depends on the type of fiber in the paperboard. Kraft fiber is stronger than recycled fiber, so for most routine carton uses, SBS can be specified at least 2 points (50 μm) thinner than recycled board. Unbleached kraft board is even stronger; a 12-can soft drink package weighing 10 lb (4.5 kg) is routinely packed in an 18-point (460 μm) CUK beverage carrier (Soroka 2009).

The dimensions depend on the product, including how big the package needs to be to be filled and how to contain the contents before settling. Its proportions may be typical for the product category, or they may be chosen to deliberately disrupt the category, like packing breakfast cereal in a "milk carton."

Cartons are designed to move smoothly and minimize cost throughout the supply chain. The shape and layout of blanks is carefully planned to minimize scrap. Since they can be shipped flat (or folded flat), they occupy less space in transit than other rigid packaging forms like cans and bottles. The design may need to run on a high-speed cartoner with the right coefficient of friction, stiffness, and consistency. The shape needs to be chosen to maximize cube utilization on palletloads.

Most importantly, cartons are designed for a marketing strategy. They identify the brand and are expected to motivate consumers to buy the product. Colorful folding cartons can conceal and yet endow any processed food product (no matter how desiccated or frozen) with a delicious photographed appearance. Lively hologram cartoon characters can reach out in greeting from a delicious-looking bowl of cereal. A brilliant-colored carton of non-descript detergent promises to make the wash brilliant. Or cartons can reveal an attractive product through a window framed with enhancing graphics.

Folding cartons are the source of marketing success for many of our best-known brands. Birdseye®, Cheerios®, Kodak®, and Tide®—their names alone conjure up images of their distinctive carton graphics. Certain shapes are symbolic of a product category, like the cereal carton and hinged-lid cigarette pack.

Cartons can be the basis for a differentiation strategy that sells the same basic product in surface and graphic presentations that appeal to various target markets. For example, a carton printed black with a glossy varnish and gold embellishment can target the luxury adult market, versus crayon-like colorful printing on a matte white surface for the family market, versus a natural kraft surface printed with green soy-based ink for the ecological market.

The trade-off between design factors can be illustrated with an example of breakfast cereal cartons. A new compact (shorter and wider) carton was designed in 2009 to use 8% less paperboard, to fit more easily into consumers' pantries, and to make more efficient use of transport and retail shelf space. But the new carton profile never moved beyond the test market phase when the company learned that consumers perceived it to contain less cereal because it was not the iconic cereal-carton shape.

In the end, consumers' perception and experience with the carton is the most important design factor. Clean, sealed cartons can provide security and tamper-evidence. Easy-to-open and reclose ends improve the consumption experience. Even the fill level affects the consumer's perception of a trustworthy brand. Cartons can be designed to open and dispense products in unique and innovative ways to support a product differentiation strategy.

Cartons can satisfy psychological and sensory needs too; for example, a tastefully-printed tissue box that fits in the home décor can add to the owner's sense of self-esteem. Braille can be embossed for blind consumers (and is mandatory for pharmaceutical packages in Europe). A scratch-and-sniff ink can stimulate hunger and/or preview the product scent. Cartons can be entertaining: they can conceal prizes, present games, or their box tops can be redeemed for fun premiums. A window can frame a product with advertising. They have space for drug facts, nutrition facts, and fun facts. And since the consumer is at the end of the value chain, a carton design can even ease the guilt of consumption because it is easy to recycle.

Cartons are designed in partnership between the converter and the packer-filler, as the converter aims to fulfill its customer's requirements. There may also be a third-party design firm involved in graphic or other creative design work.

12.2.2 Folding Carton Printing and Converting

The folding carton converter prints the paperboard, and then die-cuts and scores (creases) the carton *blanks*. For some carton styles, the converter also glues the joint.

Integrated converters (those who make their own paperboard) predominate in folding cartons, supplying 80% of the U.S. market in 2013, an increase of 10% over the previous 15 years (Paperboard Packaging Council 2013). The industry has become more concentrated in recent years. In 2005, the leading five U.S. companies represented 53% of sales, compared to 44% in 2000, 30% in 1996, and 25% in 1989 (Impact Marketing Consultants 2006). Between 2005 and 2013, two of the five had further consolidated, leaving over half the industry sales to just four companies. The bulk of the supply market is highly commoditized, and so suppliers need to provide good service and quality from a highly efficient operation to be competitive.

The converting process starts with rolls of paperboard which are cut into sheets before printing if a sheet-fed process is used. The sheet size is usually cut for a specific job and the blanks are laid out on the sheet, fitted together to minimize scrap. Processing of large sheets, 44–63 in (112–160 cm) wide, is common in the folding carton industry. The stiff nature of paperboard makes it stable enough to be fed by large sheets into printing and cutting operations.

For some special-purpose boards, a coating or lamination step precedes or follows the printing process. Some coating operations only suit reel-fed conversion and so would precede printing for downstream sheet conversion, or be integrated into a reel-fed printing process. An extrusion coating of molten plastic, such as PE, PP, PET, or PS can be flowed onto the surface of the board if the carton requires water resistance or heat-sealability. For example, paperboard for drink boxes is coated with PE on the inside for water resistance and heatsealability, and on the outside for water resistance and surface protection. PET is used for most dual-ovenable (microwave and conventional oven) paperboard applications. Plastic coated board usually must be corona or flame treated to enable the printing ink to adhere. Acrylic coating is common to improve ink adhesion. For heatsealability and/or water-resistance, paperboard can also be coated with wax, silicone, or a heatsealable varnish.

To further improve the surface, a layer of paper, plastic, metallized plastic, or foil can be laminated to paperboard, using waterbased adhesive or a thermoplastic extrusion. A sheet of high quality printing paper, pre-printed paper, or reverse-printed plastic film can be added to enhance graphics. A layer of plastic, like PE, PP, or PET can add strength, shine, and a water or gas barrier. A layer of foil or metallized film can add barrier and sparkle. A sheet of grease-resistant paper on the inside of a carton can prevent oil staining and is widely used for quality chocolate confectionary as the smooth surface does not damage the chocolate surface. Or a second sheet of paperboard can be added to increase stiffness.

12.2.2.1 Printing

Although any printing process can be used on paperboard, as described in Chapter 10, *offset lithography* is by far the most common because of its high quality and low cost. It gives a nice, clean interface between the image and non-image areas, and the plates are inexpensive.

Paperboard is the only packaging material for which lithography is the dominant printing method, in part because of the traditional sheet-fed nature of carton converting. Sheet-fed offset lithography can accommodate a wide variation in sheet and blank size, and is economical for short to medium length runs since the plates are low cost. The stiffness of paperboard is well-suited to the sheet-fed process.

Six-color presses are usual for cartons: four stations for the process colors (cyan, yellow, magenta, and black), one for spot color (such as a company logo) and one for varnish or a water-based coating. But some carton presses can run up to 12 colors. The VOCs in some lithographic inks have been found to have a persistent odor, which can impart an unpleasant taste to chocolates or other foods. However, there has been a large shift to UV-cured inks and varnishes for food cartons.

Narrower web-fed litho presses are gaining popularity for longer runs because of their speed (about five times as fast as sheet-fed presses) and the ability to use in-line die-cutters. Only one process is required, compared to sheet-fed presses which produce a stack of printed sheets that require a separate process for cutting and scoring. Web offset is used for cartons of standard sizes, such as half gallon ice cream cartons, multi-color juice cartons, high volume cereal cartons, and beverage 12-packs.

There is also a trend towards printing paperboard with web-fed flexography. Flexography has been traditionally used only for lower quality cartons, but the quality has improved to compete with the quality of main-stream lithography.

It has the added advantage of using water-based inks which minimizes the VOCs. Flexography is used for milk cartons, drink boxes and some over-the-counter cartons for drugs.

Higher quality *rotogravure* printing is only used where long and frequent runs can justify the high cost of the etched plate. Gravure printing is generally used for detergent, cigarette, and beer can cartons, where volume is high and packages rarely carry promotional changes. For premium brands, gravure is sometimes chosen for its ability to maintain consistent brand-critical colors over long runs.

Regardless of the printing method, the ink must be dry or cured before the carton-making process can proceed. If the printing and cutting are web-fed and in-line, the intermediate drying/curing operation is critical.

Following printing, a coating is often applied to provide gloss, smoothness, the desired coefficient of friction, smudge and rub resistance, and/or resistance to water, gas, oil, or chemicals. Coatings can be based on water, solvent, or oil, or can be radiation-cured. Coatings are applied by a station on the printing press or by an off-line unit. The most common coating is varnish: water-based which must be dried, or UV types which are cured by exposure to UV light. Other types are wax-resin hot melt or plastic extrusion coating to improve barrier and gloss (Paperboard Packaging Alliance 2003).

12.2.2.2 Cutting and Scoring

The next step is *die-cutting* and *scoring*. Scores are also called creases.[23] Blanks are stamped out using a press with a *die*. Traditionally, the die is made from plywood that has thin, metal blade-like *rules* mounted upright in laser-cuts in the wood. The scores (creases) are made by blunt-edged steel *rules*, whereas the cutting rules are sharp. Figure 12.2 shows the difference. The die presses the paperboard against a counter-base plate, called a *make-ready matrix,* which has grooves into which the scoring rules press.

Dies are generally reused more times than printing plates, since the carton shapes do not change as often as the graphics. The cutting and scoring operation is either sheet-fed or rotary.

For sheets that have been printed by sheet-fed offset lithography, a *flatbed die* is mounted in the upper platen of a press. Mechanical grippers guide each sheet into the press, which uses an intermittent motion to cut and score the entire sheet in a single stroke. This is also called *crush cutting.* Figure 12.2 shows a typical flatbed die.

A counter-plate die, made from hard phenolic plastic or hardened ground-

---

[23]In some parts of the world, the word "score" refers to a cut in the surface layer of the paperboard which is used to facilitate bending (albeit not so attractive and strong as a crease) or an easy-opening strip.

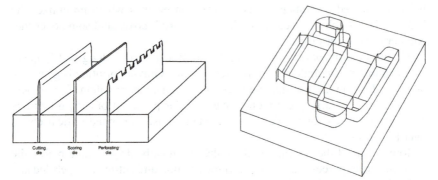

FIGURE 12.2. *Steel rules for cutting and scoring, flatbed die. Reprinted with permission of John Wiley & Sons, from Roth and Wybenga,* The Packaging Designer's Book of Patterns *(2000), 7 & 8.*

flat steel, is mounted in the lower platen. The groove width is usually 1-1/2 times the thickness of the paperboard plus the width of the scoring rule, with slightly narrower scores parallel to the MD of the board (Kirwan 2013).

Score quality depends on the alignment of the rule and make-ready groove, and the condition of the make-ready which deteriorates over time. Small, compressible pieces of rubber or cork are placed alongside the cutting blades to hold the paperboard sheet in place as it is cut and to facilitate the efficient removal of scrap as the press completes its cycle. A second set of grippers then pulls the cut and scored sheet from the press as the next sheet is pulled in.

A carton-maker will have a limited range of machine sizes which it can process. Part of the carton designer's skill is being able to fit blanks in a nesting/interlocking pattern in the die and printing press so as to maximize the use of the machine size and minimize wasted scrap (and have a horizontal MD).

Rotary die-cutting is used for web-fed material, usually in-line with a printing press. In a rotary die, rules are inserted into a solid metal cylinder or a curved die board that is attached to a cylinder which presses against a backing cylinder. The cylinders rotate, drawing the paperboard through the nip, cutting and creasing the blank. Web-fed machines have a fixed maximum width, and the *repeat length* is determined by the circumference of the cylinders.

Rotary die-cutting is fast, but the dies are expensive. It is best suited for longer runs and frequently repeated orders, where design changes from run to run are minimal. High volume milk cartons are an example of the kind of carton that is best suited to rotary die-cutting.

Traditional rotary crush-cut dies must be resharpened after about 1.25 million cartons. For shorter runs, up to 800,000 impressions, chemically etched plate dies have been developed. For higher volume, *pressure* or *shear cutting* dies are used, in which the two cylinders have slightly offset metal strips that cut the paperboard, similar to scissors; shear-cut dies can last for 10 million impressions before resharpening. Shear-cut, wrap-around dies can also be made by chemical etching which yields up to 6 million impressions (Kirwan 2013).

In modern high speed, high volume presses, the carton blanks are automatically separated from the trim waste and blanks are mechanically stacked. In smaller, older flatbed presses, the trim is removed in a separate operation; the cutting blades have small fiber bridge *nicks* which cause the sheet to stay intact long enough for the stack of sheets to be moved to a second station, where they are hand-stripped by knocking off the trim around the edge with a mallet or an air chisel. These nicks detract from the quality appearance of a carton, and are usually confined to areas where they are not easily noticed.

The quality of the scores affects the quality of a carton. The score makes the flexible hinge in the stiff board, but the exterior surface should remain smooth and uncracked. The appearance of a carton on the shelf depends on crisp scores to keep the walls flat. The blank's performance on the eventual carton erecting and filling equipment is influenced by the quality of the score and how easy it is to bend in comparison to the stiffness of the board.

Scores can be evaluated in the *score bend test*, discussed in Chapter 8. For most applications, the score needs to reduce the board's initial bending stiffness by about 50%.

The blunt-edged scoring rule pressing the board into the counter-plate groove produces a ridge in the paperboard, a controlled line of weakness along which the board will later fold without cracking the top ply. The board delaminates along the fold-line to produce the inward-bulging ridge. The depth and width of the ridge depend on the type and thickness of the paperboard as well as the width and height of the scoring rule.

The raised score ridge goes on the inside of the finished carton, as shown in Figure 12.3, forming a clean hinge. This ridge is an indication that the inner layers have been partially delaminated. Reducing the bonding strength between the plies makes the score flexible. The importance of this can be illustrated by how easy it is to bend a phone book as long as the individual pages are free to slide over one another; if the pages are stuck together, it becomes very stiff. The inner layers are forced away from the edge in order to give a smooth bend on the outside without rupturing the surface. Having the ridge on the inside reduces the stretch on the outside and also makes the box more stable and strong.

Especially for cylinder board, since it has a significant difference in stiffness in the MD and CD, the scores have different factors to overcome in each direction. A good score along the MD depends on depth in order to overcome the board's elasticity. A score crossing the MD depends on width in order to give adequate space to accommodate the fibers folding into the ridge. The width of the raised ridge is generally 3-1/2–4 times the caliper of the board.

Besides cutting and scoring, the die-cutting process can add perforation,

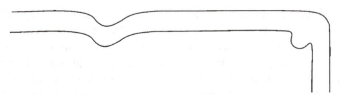

FIGURE 12.3. *Score detail.*

embossing, debossing, or foil. A perforation rule is serrated, as shown in Figure 12.2. A tear strip can be created by slightly offset cut-scores on both sides of the board. Functionality of opening and closing features depend on the quality of perforations and cuts. For embossing or debossing, the paper is pressed by a hot die or slug against an inverted shape to create a three-dimensional surface. Embossing can emphasize a design element, like a coat of arms or brand name, or can add braille. Hot foil stamping may be a separate function or integrated with embossing, and is popular for adding gold or silver highlights.

12.2.2.3 Finishing and Gluing

The operations that follow printing, cutting, and scoring depend on the carton design and how it will be erected and filled.

Off-line treatments can add marketing benefits. Windows made from stiff clear PS, PP, PET, PVC, or cellulose acetate can be glued to the reverse side of the carton blank. A window can make an attractive product visible to consumers and prevent the need for them to open the package to inspect a product. A soft, perforated LDPE window is often added to tissue boxes, and features like a serrated metal edge for cutting wrap can be added to improve customer use and dispensing. Premiums and coupons can be added for a promotion. The blanks can be wax-coated for barrier, gloss, or water-resistance. Microwave susceptor or shield films (discussed in Chapter 9) can be added to improve browning and differential heating of food.

Cartons that will be glued later on the packer/filler's production line have no further processing and are shipped as flat blanks.

But many tube-style carton blanks are *pre-glued* by the converter in a machine called a *straight line gluer.* This stage allows the convertor to add maximum value to their product, and the packer filler to have the simplest packaging machinery (the next stage is the erection of the carton body and so would be uneconomic for shipment). As the blanks are fed in, the *non-working* scores are first *pre-broken*, bent past 90°, and then returned to a flat state to make them easier to erect on the eventual cartoning line. Next, glue is applied to the long seam, called the *glue flap*. The blank is formed into a flat *knocked down* (*KD*) tube by folding the two *working scores* inward 180° and pressing the glue flap and right side panel together in the center to form the glued joint. Straight-line gluers can run at up to 200,000 cartons/hour. This process is shown in Figure 12.4.

The most common glue is dextrin or PVA glue when it will have enough time to set before the cartons will be erected. Hot melt adhesive is used when a quicker bond is needed or the surface is not suitable for PVA; for example, some highly sized and clay coated board and laminated boards can be difficult to bond with water-based adhesive because the liquid will not penetrate. Printing and varnish also reduce adhesion, and often printers leave unprinted the areas to be glued in order to overcome this problem.

Heatsealing is common for plastic-coated boards used for milk, juice, and frozen foods to provide a sift-proof, leak-proof seal. Even if the board is not coated, a thermoplastic adhesive can be pre-applied to the blanks in the glue areas for the purpose of later heatsealing.

Modified straight-line gluers can add plastic tubes for bag-in-box cartons or a second, double wall. For quick erect (crash erect) cartons, trays, and more complex shapes than simple tubes, a *right angle gluer* is used, which turns the partially glued blank 90° to fold and apply glue in the other direction. It is possible to add additional panels that incorporate platforms to display and locate products.

The glue bond at the joint should be stronger than the internal paperboard bonds. Glue bond can be measured and specified by the amount of time it takes to reach a minimum peeling force. Or it can be measured by the amount of force required to peel apart a fully set bond, along with a measure of the amount of *fiber tear* that occurs when the bond is broken. In any glue bond test, one or the other of the substrates should delaminate rather than the glue simply letting go.

The glue joint needs to be straight and clean. Skewing or displacement of the flap and glue *squeeze-out* can reduce packing line efficiency and the appearance of the filled carton.

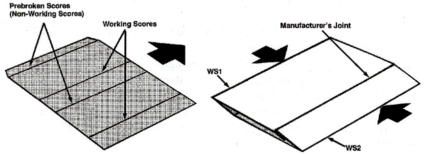

FIGURE 12.4. *Straight line tube gluing operation. Reprinted with permission of Paperboard Packaging Council, copyright 2000.*

The glued cartons are counted, packed in corrugated boxes, and palletized. They may be shrink- or stretch-wrapped to protect from moisture and keep them clean. They are usually packed on edge, to preserve their flat shape (avoiding the banana- or s-shape that results from the extra thickness of the glue flaps in a stack) and to prevent compression of the scores. They are loosely packed, with a little extra space in the box (to preserve the so-called *bounce* or *fluff* in the scores).

12.2.3 Standard Folding Carton Styles

There are two standard categories of folding cartons: *tube style* and *tray style*, as described in the following two sections.

The standard styles are not noted so much for their creative structure as they are for their flat display panels which can feature creative graphics. Their flatness and billboard surface is their greatest virtue. Furthermore, the standard styles fit into standard end-loading cartoning machines, making them easy to erect and fill. The standards are maintained by organizations like the Paperboard Packaging Council and European Carton Makers Association, and are used in most computer-aided design software.

12.2.3.1 Tube Style Cartons: Sealed and Tucked Flaps

Tube style cartons are the most commonly used style. They present a full frontal principal display panel. They have a glue joint along one back edge in the depth dimension, which is glued either by the carton converter, as is usually the case for tuck-end cartons, or on the filling line, as is the case for most glued-end cereal cartons. The tube style allows pre-glued blanks to be shipped folded flat. Tube style cartons are filled through the top or bottom, called the *ends*, in an *end-loading cartoner* (discussed in Section 12.2.6).

The difference in the standard tube styles is the configuration of the ends. The standard styles are designed to minimize the use of paperboard and reduce scrap.

The *partial overlap seal end* (*POSE*) style (also known as *economy flaps*), shown in Figure 12.5, is the classic carton for breakfast cereal, cake mix, and crackers. The box top panels overlap just enough to permit gluing. When there is a lock tab on the top, this is cut from the bottom closure panel of the adjacent carton blank in the die to minimize scrap.

The body glue joint is formed first. The contents are inserted, and then the ends are sealed. The *dust flaps* are folded closed first, followed by folding and gluing the *major flaps*.

POSE cartons are used primarily for flowable, dry products like powders and crackers. Such contents settle after filling, so cartons need to be large

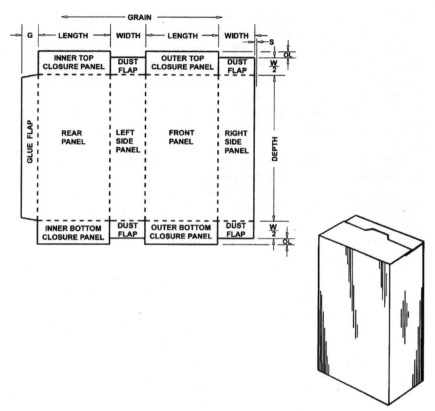

FIGURE 12.5. *Partial overlap seal Eed (POSE) carton and blank. Reprinted with permission of Paperboard Packaging Council, copyright 2000.*

enough to accommodate the initially filled product while being small enough to not seem underfilled by the time they reach the consumer.

There are a couple of variations on the POSE carton, named for the degree of flap overlap on the carton ends. The *full overlap seal end* (*FOSE*) prevents sifting of fine powdery products. Sifting is further prevented when the ends are *double glued* with the minor flaps glued to the inner closure panel which is then glued to the outer panel. If cartons are collated on their sides, the full overlap ends can provide enhanced compression strength, sometimes allowing the use of a lighter board grade. The *economy overlap seal end* (EOSE) with the least overlap, uses the least material.

In a second POSE variation, the style is tipped on its back with the front panel up, for products like facial tissues and cookies in trays. In this orientation, the board's machine direction is vertical, so it can contribute to compression strength. For products which do not flow or settle, there is less tendency for the walls to bulge, so there is no benefit from a horizontal MD.

Tuck end cartons, as the name implies, have ends that are tucked rather than glued. Their *major tuck flaps* lock into the ends, atop *dust flaps*, and are able to be securely reclosed after initial opening. Tuck end cartons are used for a wide range of solid products, from bottles of perfume and spark plugs, to toothpaste tubes and underwear. They typically have a pre-glued joint and are packed automatically or semi-automatically on a cartoner that does not require a glue station.

There are a number of different standard tuck end styles. The difference is in the position on the blank of the tuck flaps.

The *standard reverse tuck (SRT)* shown in Figure 12.6, is the most economical to produce. The tuck flaps are on opposite panels. As Figure 12.7 shows, when a number of SRT cartons are laid out on a sheet of paperboard, the geometry is such that the blanks interlock like jigsaw puzzle pieces when they are die-cut, minimizing scrap.

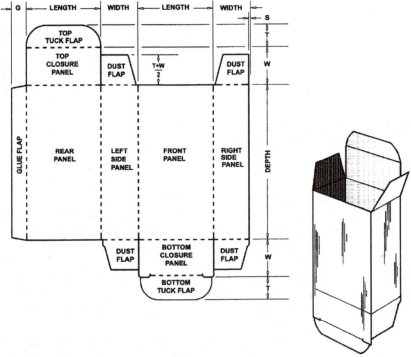

FIGURE 12.6. *Standard reverse tuck (SRT). Reprinted with permission of Paperboard Packaging Council, copyright 2000.*

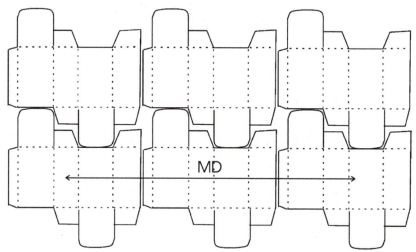

FIGURE 12.7. *Standard reverse tuck (SRT) blanks laid out, interlocking, on a sheet.*

But since one flap tucks in the reverse direction from the other, the SRT carton does not present a clean frame for the front panel. The top front edge has a gap since the top flap tucks from the rear to the front. The FRT and SST styles, shown in Figure 12.8, overcome this problem.

The *French reverse tuck* (*FRT*) is preferred by the cosmetics industry. Its top flap tucks from the front to the rear, giving the top of the principal display panel a more clean and finished appearance. Since the bottom flap is attached to the back (reverse) panel, there is not a similarly clean edge on the bottom, but this shape also has the interlocking die-cut blank advantage.

The *standard straight tuck* (*SST*) has the cleanest look of all, with both of its flaps attached to the front panel, the blank resembling an airplane shape. It is particularly well suited for products with a large window in the front display panel, because it prevents interference between the tuck and the window film material. This look is sometimes preferred for marketing purposes, but it is the most expensive style, because the flap placement will not permit interlocking blanks, which generates more waste that is trimmed away during die-cutting.

If the carton is meant to be opened and reclosed several times, a plain *friction lock* tuck flap will provide a neater appearance over the carton's life. But if a single entry is desired, two *slit locks* cut at the edges of the tuck's score provides a more positive and secure closure, although it has a tendency to tear when the package is opened.

There are other variations of the basic tube style, based on the design of the ends. For example, two styles that are normally erected and filled manually, used in small scale production or retail operations, are shown in Figure 12.9.

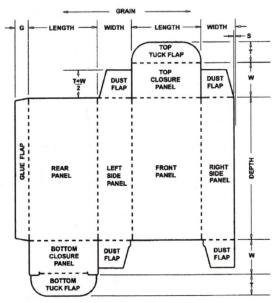

French Reverse Tuck (FRT)

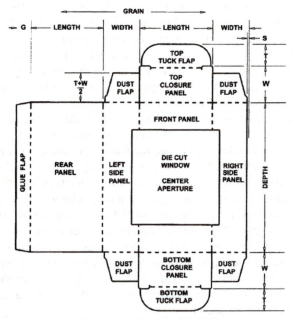

Standard Straight Tuck (SST)

FIGURE 12.8. French reverse tuck (FRT) and standard straight tuck (SST) cartons. Reprinted with permission of Paperboard Packaging Council, copyright 2000.

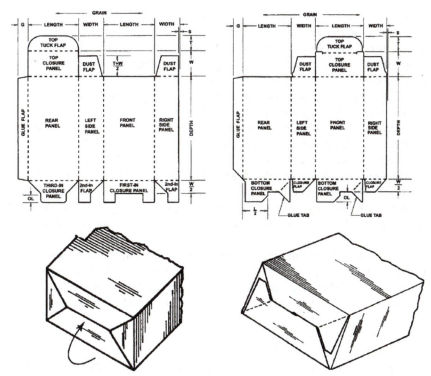

FIGURE 12.9. *Snap lock and automatic bottom cartons are erected manually. Reprinted with permission of Paperboard Packaging Council, copyright 2000.*

The *snap lock* (or crash lock) is tucked into place, and the *automatic bottom* has a pre-glued bottom closure panel that has been folded flat. These bottom styles provide good resistance to accidental opening by heavy contents and so are widely used for large premium bottled products such as spirits.

Gable-top milk cartons and brick packs for juice can also be considered tube-style cartons, but since these liquid-tight sanitary cartons have unique considerations, they are discussed in Chapter 13.

12.2.3.2 Tray Style Cartons

Tray style cartons are valued for their presentation effect. From bakery cartons to gift boxes to display cartons, the open top of the tray style showcases what is inside of it. The tray serves up its contents to the consumer on a solid base with the walls and sometimes a lid extending from the base. Many trays have lids, hinges, and even windows to protect but also to enhance the experience of opening the package and unveiling the contents.

They also called *top loaded cartons*, and there is a diversity of styles, based on how they are erected and filled. Tray style cartons are usually shipped flat and are usually not glued until they are filled. However, some styles have clever pre-glued corners that still permit them to be shipped folded flat. Others have tapered walls so that they can be pre-erected but nest when empty; examples are paper cups and fast food cartons. The paperboard machine direction usually runs in the long dimension of the bottom panel.

Some trays are erected and loaded manually. The corners are secured either by tucking locking tabs or by pre-gluing. They can have tuck-flap lids like a pizza box (which is a tray-style although it is usually made from corrugated board). Other examples include trays used for fast food take-out and fresh bakery goods like pies and donuts. Department store telescoping style gift boxes are made from two trays, the lid slightly larger than the base. The *Beers* (*Biers*) tray and *cake lock box* shown in Figure 12.10 are good examples of the tray style.

Other tray-style cartons are erected and loaded in automatic or semi-automatic machines. Examples of automatically erected trays include those used for factory-made baked goods, checkout counter candy, and frozen foods. The 4- and 6-Corner Brightwood trays shown in Figure 12.11 are common styles.

12.2.3.3 Some Creative Carton Shapes

There are many other carton designs possible. The standard styles are popular primarily because they are easy to make and fill with standard equipment and make efficient use of the sheet from which they are converted. But paperboard can be used to make many non-standard container designs, limited only by what can be folded into a 3-dimensional shape, with the help of cuts and scores, from a thick two-dimensional board.

Clever designs for sleeves, folders, and origami-like shapes ranging from pyramids to dodecahedrons, with pedestals, flanges, and curves have been developed. Competitions sponsored by the paperboard industry highlight creative designs. Resources like *The Packaging Designer's Book of Patterns* (Roth and Wybenga 2013), *Packaging Prototypes* (Denison and Cawthray 1999), and the *ECMA Code of Folding Carton Design Styles* (2009) offer hundreds of designs, along with drawings of the blanks and computer-aided design tools. For example, the two examples shown in Figure 12.12 have curved scores.

For a new design to be useful it must be capable of being made and filled. Often a clever design will require hand-filling, which is adequate for a small scale user. There are plenty of medium-scale specialty manufacturers, especially makers of crafts and prepared food, who can get a distinctive marketing value from using specially shaped cartons, and can justify the extra cost of hand-filling.

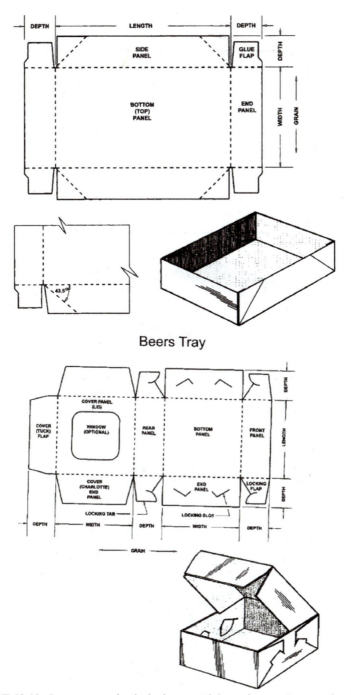

Beers Tray

FIGURE 12.10. _Beers tray and cake lock tray with hinged cover. Reprinted with permission of Paperboard Packaging Council, copyright 2000._

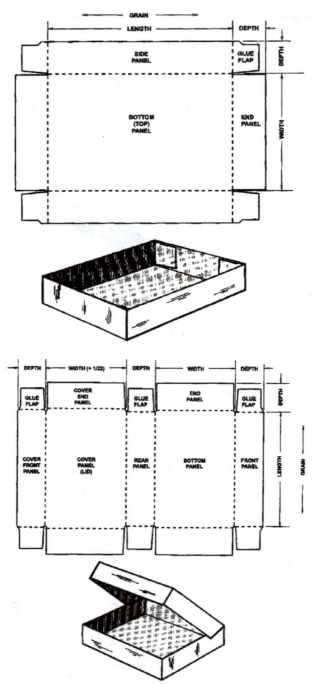

FIGURE 12.11. *4-Corner and 6-corner Brightwood trays. Reprinted with permission of Paperboard Packaging Council, copyright 2000.*

400

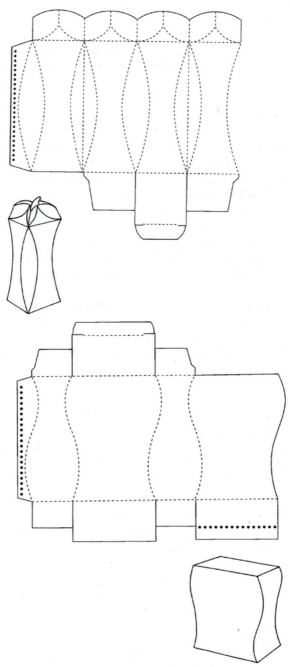

FIGURE 12.12. *Carton shapes with curved scores "allow for a beautiful interplay of shadows." Reprinted with permission of John Wiley & Sons, from Roth and Wybenga,* The Packaging Designer's Book of Patterns *(2013), 219, 228.*

But to be used on a large scale, cartoning must be mechanized. The following three sections describe the buyer's concerns of specification, cost, and cartoning operations.

12.2.4 Carton Specifications

Cartons are specified by buyers in terms of paperboard materials and properties, blank layout, and printing.

In the United States, paperboard is specified in points (thickness) more commonly than basis weight, although both may be used. In other countries it is more commonly specified in grammage (basis weight in grams per square meter). The composition of the board depends on what the paper mill can make, and most board manufacturers identify the different multi-ply boards by their own name or number, which is included in the specification. When a particular property like extra stiffness or coefficient of friction is required, it is specified as a special case.

The coating for each side is specified. For white lined and/or coated board, the brightness is specified; a common grade for CRB is 80 Bright, and for SBS is 88 Bright. The method of printing is specified. The converter and customer work together to develop artwork and designs that optimize manufacturing operations, cost, and marketing benefit.

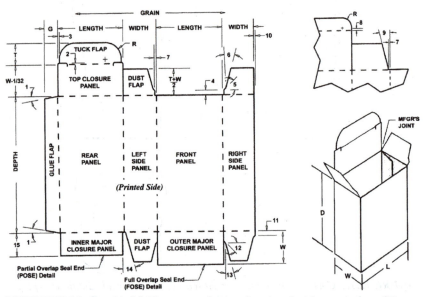

FIGURE 12.13. *Standard folding carton layout. Reprinted with permission of Paperboard Packaging Council, copyright 2000.*

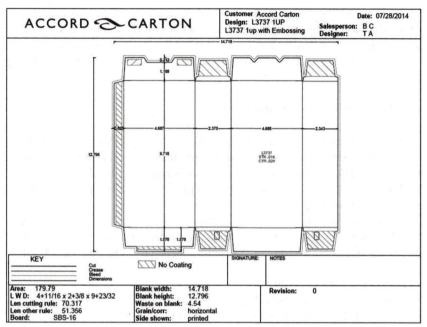

FIGURE 12.14. *Carton specification. Reprinted with permission of Accord Carton, 2014.*

Folding carton dimensions are expressed as length × width × depth, but these are based on the blank layout, not carton orientation, which can be a little confusing. Despite the orientation of the finished carton on the shelf, they are defined as if the carton is upright, as shown in Figure 12.13. The length and width are dimensions of the open ends of the carton. Depth is the distance between the ends.

The standard way to represent a tube-style carton in a specification layout drawing is with the printed side up and the glue flap on the left. The layout shows the cuts as solid lines and the scores as dotted lines. Figure 12.13 shows the standard nomenclature and layout drawing of a carton in a specification, with all of the relevant dimensions and angles. The layout will also show areas of flaps where the carton is not to be printed or coated, in order to facilitate better glue bonding. Figure 12.14 is an actual specification for a POSE carton made from SBS.

All specified dimensions are measured between score centers. Since the score occupies space, a scoring allowance is figured into the layout and the design of the die by adding an amount equal to the thickness of the board to each panel. This differs from corrugated fiberboard box specifications which designate the inside dimensions and specify the score allowance separately.

The Folding Paper Box Association of America developed a voluntary set of standard tolerances for folding cartons. Dimensional tolerances are limited to 1/64" in any single dimension and ± 1/32" for any combination of dimensions. Caliper tolerance is ± 5% or ± 0.0001", whichever is greater (Trefry 1997).

12.2.5 Carton Cost Estimation

The pricing of paperboard cartons is relatively competitive. Although there are many factors such as volume discounts that affect the price of cartons, there is an easy way to estimate the cost. It is based on the size of the blank, the basis weight, and *transacted boxboard* prices. It also includes an estimated up-charge for printing.

The paper trade press *PPI Pulp & Paper Week* publishes a reference standard for current prices of various kinds of paperboard. Representative 2014 prices are listed below, from most to least expensive, dollars/ton (2,000 lb):

- SBS (16-point) ~ $1,120
- Coated, unbleached kraft (20-point) ~ $1,080
- Clay coated recycled board CRB (20-point) ~ $1,000
- Uncoated recycled board (20-point) ~ $675

The published prices are used for reference and tracking trends. Prices vary in different parts of the United States and other countries due to supply and demand.

There are five steps to estimate the cost of a carton.

1. Determine the size of the blank, including all scrap. This is the amount of board that will be used to make the carton. The easiest way to determine it is to draw the blank.

   For example, if a cereal carton (POSE style) measures 6.5" × 2.5" × 10" with 1/2" end flap overlap and 1/2" glue joint, its blank is shown in Figure 12.15.

   $10" + 2 [(2.5"/2) + 0.25"] = 13"$ and $(2 \times 2.5") + (2 \times 6.5") + 0.5" = 18.5"$
   $13" \times 18.5" = 240.5$ in^2, which equals 1.67 ft^2 (1 ft^2/144 ft^2)

2. From the blank size and basis weight, determine the amount of paperboard material in the package.

   If the example has a basis weight of 74 lb (a typical BW of 18 point CRB, approximately 360 g/m^2), there are 1.67 ft^2 × 74 lb/1,000 ft^2 = 0.124 lb (56 g).

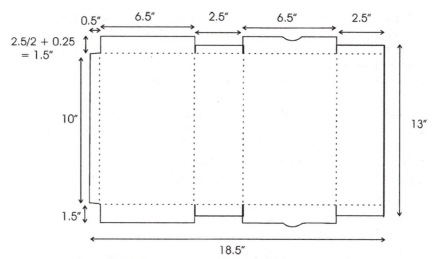

FIGURE 12.15. *Cost estimation example blank.*

3. Find the market cost of the paperboard material and multiply it by the number of pounds.

PPI Pulp & Paper Week gives a 2014 market price of $1,000/ ton (1 ton = 2,000 lb) for 20-point CRB, and it is assumed that 18-point CRB would have about the same price.

$1,000/2,000 lb × 0.124 lb = $0.062

4. Double the cost of material, using this as an estimator for the conversion cost of paperboard to an unprinted carton. Research has shown that the cost of a finished carton is generally double the cost of the material.

$0.062 × 2 = $0.124 estimated cost of unprinted carton

5. Add a printing cost factor. This is less accurate than the estimate of material cost because it also varies based on the set-up time, run length, and printing process. Printing cost is quoted in terms of the amount of ink to cover 1,000 in^2, and so the blank's dimension in square inches is used. Assume that one color costs 5¢/1,000 in^2 and 4-color costs 15¢/1,000 in^2.

If one color, 240.5 in^2 × $0.05/1,000 in^2 = $0.012 for printing alone

Carton cost + printing cost = $0.124 + $0.012 = $0.136 estimated printed carton cost.

If 4 color, 240.5 in^2 × $0.15/1000 in^2 = $0.036 for printing alone

Carton cost + printing cost = $0.124 + $0.036 = $0.16 estimated printed carton cost.

If the carton in the example above was made from the same basis weight of SBS, at a market price of $1,120/ton, the carton would cost $0.174955. While 1.5¢ may not seem like much difference for an individual carton, for a 10,000-carton order it is a difference of $150. This is a second reason, beyond its strength, that SBS is usually specified thinner than CRB for an equivalent use.

Other cost factors could be added for plastic coating or any other special treatments. Usually plastic coating for different polymers is specified in terms of weight per unit of coverage, like pounds/ream.

Here is another example: Estimate the cost of an SRT 5″ × 2″ × 6″ with 4-color printing, 2″ glue and tuck flaps, 80-bright 60-lb BW CRB.

1. Blank size

   0.5 + 5 + 2 + 5 + 2 = 14.5″

   0.5 + 2 + 6 + 2 + 0.5 = 11″

   $(11″ × 14.5″)/144 = 1.1076$ ft^2 (159 in^2 or 0.1029 m^2) blank dimensions; note that nesting/interlocking would reduce this by $(2.5 × 14.5)/144 = 0.25$ ft^2

2. Weight: 1.1076 ft^2 × 60lb/1,000 ft^2 = 0.066456 lb (30 g)

3. Material cost: 0.066456 lb × $1,000/2,000 lb = $0.033228

4. Add converting cost: $0.033228 × 2 = $0.066456

   note that nesting/interlocking reduces this to $0.051456, savings of $150/10,000

5. Print cost: 159 in^2 × 15¢/1,000 in^2 = $0.02385

   Printed carton cost: $0.066456 + $0.02385 = $0.09

An estimation method like this can be used to judge whether a supplier's price quote is fair. It can also be used to estimate the effect of changing carton size, material, thickness/basis weight, and number of colors printed. It is important to reiterate, however, that prices also depend on set-up and transaction costs, so volume discounts are usually offered.

12.2.6 Cartoning Operations

Carton packing operations range from manual to fully automatic, at speeds typically from 10 to 1,000 cartons/minute. Semi-automatic, operator-assisted equipment can run up to 120 cartons/minute. Fully automatic equipment ranges from 60 to 600 cartons/minute. Specialized machines that run even faster have been developed for cigarettes; flip-top cartons are filled at 400–700 cartons/minute, and new machines can run up to 1,000 cartons/minute (Kirwan 2013; Corning 1997).

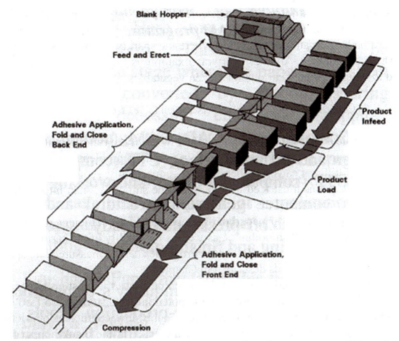

Blank Hopper

Feed and Erect

Adhesive Application,
Fold and Close
Back End

Product
Infeed

Product
Load

Adhesive Application,
Fold and Close
Front End

Compression

FIGURE 12.16. *End-loading cartoner. Reprinted with permission of Paperboard Packaging Council, copyright 2000.*

In a typical *end-loading cartoner*, shown in Figure 12.16, tube-style cartons are fed from the blank hopper, and erected (if pre-glued) or folded and glued into the tube shape. If they are glued, a hot-melt adhesive is used because of its quick drying time. Heatsealing, rather than glue, can be used if the paperboard (or just the glue line) has been pre-coated with a thermoplastic, like PE.

The contents are inserted, and the flaps are tucked closed or glued. End-loading cartons may be filled horizontally (lying down) or vertically (standing up with the bottom end closed first), depending on the product. Cereal and cracker cartons are usually loaded with pre-filled and sealed bags, although some manufacturers still use the traditional *double package maker* that forms the bag and carton together around a mandrel, fills the bag, and then seals it and the carton. When the product is free-flowing and the carton is filled vertically, the machine may include check weighing and "top-up" stations to ensure accurate weight.

Top loading tray-style cartons are erected by plunging a form into the flat blank to form the floor of the tray. The walls pop up around it and are glued or locked into place.

As a final operation, some cartons are overwrapped with a coated PP film or cellophane that is folded and heatsealed. Typical applications are for cigarettes, chocolates, and tea where the overwrap provides a moisture, oxygen, and aroma barrier, and for cosmetics where the sparkly overwrap is valued for its shiny appearance. Another overwrap example is the coated, printed paper used to wrap unprinted cartons for frozen vegetables.

12.2.6.1 Carton Properties Affect Runability and Efficiency

The physical properties of a carton affect its ability to run smoothly through the machine that erects, fills, and seals it—without problems like jamming, breaking, spilling, or causing other defects—consistently, economically, and at high speed. Cartons must feed into the machine smoothly, snap into shape to accept their contents, and the ends must firmly close.

Runability depends on paperboard properties like coefficient of friction, stiffness, porosity, and smoothness, as well as the quality of the carton scores, flatness, and gluing (Hine 1999). These properties can be measured and specified using the test methods described in Chapter 8.

The feeding and erecting part of the machine cause the most problems because of the stresses on the scores and the potential for distortion. To feed into the machine smoothly, cartons need a reasonably low coefficient of friction that permits them to separate easily from the stack in the blank hopper (too slippery may cause problems at a later stage). If a pneumatic extractor is used, the paperboard needs sufficient air resistance and smoothness. The stiffness of the board and the bendability of the scores should be consistent so that the force needed to erect the box is consistent.

The stiffer the paperboard, the better it runs through the machine. Stiff walls give the machine something to hold on to, as grippers, knife blades, or pneumatic pressure manipulate the carton. Stiffness also helps to square the carton as it is filled. Therefore, there can be a trade-off between the desire to reduce the board cost/thickness and the desire to run the machinery at an economical high speed. Decorative windows and cut-outs can reduce panel stiffness and are known to require special attention.

Cartoning machinery is particularly sensitive to the quality of the score and how easy it is to bend. In order to run efficiently, there must be good, deep scoring that makes flaps fold crisply at the score lines. Tucking flaps is a multi-step operation—slitting to make room for the tuck, closing the dust flaps, breaking the tuck flap scores, and ploughing the tuck into the end—and every score must flex into formation.

An insufficient score can make the knocked-down carton resist opening and require slowing down the cartoning machine. It can also result in flaps that spring open and panels that bow. Too hard a crease will cause the board to crack. Since the scoring rules in a die wear unevenly, or may be replaced, cartons from one position may have different score bend ratios than those from another position.

Furthermore, scores have a shelf life. The flattened, non-working scores of pre-glued KD cartons were pre-broken in the conversion process, and fresh scores will easily spring into the box shape. But over time, the non-working scores will set back into the flattened position and become harder to open. For this reason, most pre-glued cartons are shipped loosely packed on edge, rather than lying flat in a way that would compress the open scores, and are not stored for more than 3 months before erecting. Machines differ considerably in how they erect the body of the carton and some are far more sensitive to storage time than others.

Gluing and flatness also affect runability. If pre-glued, the glue joint must be secure enough to withstand erecting and the alignment must be square. If blanks are warped, they will jam in the machine. Paperboard is vulnerable to warping due to temperature and humidity changes, because it is made from layers of different pulps and the printed surface and non-printed surfaces have different porosities. For this reason, palletloads of unerected cartons are kept wrapped, and care is exercised when moving them from one condition to another that exposes them to changing temperature or humidity conditions.

12.2.7 Substitutes for Folding Cartons

Bright displays of folding cartons were largely responsible for creating the twentieth century supermarket shopping experience. The ability to prolong the shelf life of a processed food by incorporating a plastic or paper barrier liner was also key to their success. Folding cartons still predominate in many categories, especially dry and frozen foods.

Until the 1990s, there were few competing substitute package forms in the "dry product" categories. Some cereal and cookie makers used plastic or paper bags, but these did not have the graphic impact on the shelf. That was before the stand-up pouch.

The *stand-up all-plastic pouch* represents a serious contender for some carton applications. It has the advantage of high-quality shiny printing and an upright silhouette, presenting a prominent display panel. It does not need a separate barrier liner, and may have zipper reclosability. It is usually less expensive. The primary disadvantages are that it lacks the flat billboard effect of the paperboard and cannot be filled or closed on a company's existing cartoning equipment.

A second substitute is a plastic folding carton made from PVC, PET or PP. Such clear cartons, partially printed inside or out, can present a striking showcase for an attractive product. These are popular for gifts, toys, and for bottles of cosmetics or fragrance. They tend to not run well on high speed lines.

Another substitute for folding cartons is . . . nothing. There is an environmental trend away from using cartons as a *secondary package* to contain a *primary package*. For products like toothpaste in tubes and cosmetics in bottles, the primary purpose of an outer carton has been for shelf presence. But concerns about sustainability, cost, and solid waste have driven many consumers, lawmakers, and producers to rethink the use of cartons for this purpose, opting rather for a single, easily displayed primary package.

But when it comes to sustainability, paperboard cartons generally win hands-down over the plastic and composite alternatives. They provide a useful market for waste paper, and are, themselves, recycled in most communities, giving them a political advantage over the plastic substitutes.

12.3 BEVERAGE CARRIERS

Beverage carriers are closely related to folding cartons, but they are made from the strongest paperboard: CUK. Carrying heavy beverage cans or bottles is one of paperboard's most physically challenging jobs.

The purpose of beverage carriers is to group 6, 12, or 24 cans or bottles of soft drinks or beer into an easy-to-carry package for retail sale. Multi-pack units are an important sales tool, providing a convenient package for consumers, and a means for the manufacturer and retailer to increase sales. Although shrink-wrap and plastic ring carriers have largely replaced paperboard for 6-packs, paperboard cartons prevail for larger groupings.

Strong CUK (aka CNK® and SUS®) with 70–80% virgin fiber is used because it is the strongest paperboard with the best wet strength. The carriers need to resist moisture since beverages are moved in and out of refrigeration and condensation often forms on the board. The failure of a beverage carrier can be dangerous, especially when carrying glass bottles that can shatter explosively under carbonation pressure when dropped. A dozen soft drink cans, weighing 10 lb (4.5 kg), are commonly packed in 18 point CUK.

The tensile strength and tear strength of the board, seals, and handle are of prime importance. This is why beverage carriers are made with a high percentage of virgin unbleached kraft fiber treated to have high tear strength when wet. Beverage carriers are usually tested under wet conditions and a *handle jerk test* is common.

The clay coating enhances the brown kraft board surface to prepare it for some of the most creative graphics in all of packaging. Beverage carriers have more dramatic point-of-purchase display potential than other cartons because they are such a high-volume product. The in-store billboard effect can be extended in a large display to form an integrated graphic, each carton a brick in the wall of color.

Beverage carriers are a powerful weapon in the "war" to win consumers of beer and soft drinks. Beyond a tie-in to an advertising campaign's colors, graphics, and messages, the carrier offers the potential to actually stimulate thirst. Special effects, like foil or reverse-printed film laminations, paradoxically make the package look cold and wet, at the same time as they actually help to protect the paperboard from weakening due to exposure to moisture and temperature changes. Beyond the store, they are literally walking billboards and status symbols, as beverages are carried home and to parties.

There are two paperboard carrier styles used for cans: fully enclosed sleeve packs and can wraps. *Can wraps*, more popular in Europe and Asia, are only partially enclosed.

Fully enclosed sleeve wraps (Figure 12.17) are more popular in the United States. Similar to tube-style cartons, they are made from pre-glued sleeves that are mechanically erected and filled with collated groups of cans. Then the ends are sealed.

There are two paperboard styles used for bottles: basket carriers and bottle wraps. Since glass bottles are heavier than cans, the count is generally fewer: 4-, 6-, or 8- packs. Outside the United States, the bottle wrap is commonly used for non-refillable bottles.

The *basket-style* was originally designed for returnable bottles. It does not fully enclose or seal in the bottles, so the empties can be replaced in it to be taken back to the store. Basket-style carriers are die-cut and pre-glued. Even though the use of refillable/returnable bottles has declined, the basket remains the most popular style for bottles in the United States.

Beverage carriers are packed on specialized equipment, most of which is sold by the manufacturers of the paperboard. Basket-style equipment is the

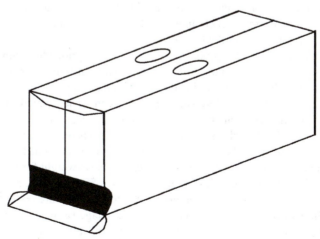

FIGURE 12.17. *Beverage carrier.*

slowest, about 180 baskets/minute. Machines for fully enclosed wraps run at about 200/minute, and wrap-around machines are the fastest, at about 355/minute (Paperboard Packaging Alliance 2002).

12.4 RIGID, SET-UP BOXES

Set-up boxes, also called *rigid boxes*, were the earliest form of paperboard boxes. They differ from folding cartons in that they are made from *non-bending* board. They are two pieces, a box and a separate lid. They are widely used for jewelry, shoes, chocolates, board games, jigsaw puzzles, software, and greeting cards.

Compared to folding cartons, set-up boxes have the disadvantage that they cannot be knocked down flat for shipment. The factories that produce them are therefore usually located near to the customers, and in some cases the box-maker even contracts to pack the customer's product. Another disadvantage is that they are slower to produce and more costly than folding cartons.

But set-up boxes have the great advantage of being able to convey a higher quality, luxurious presentation and have greater durability for long term use. They are made in a thicker grade of paperboard compared to folding cartons. They are stronger and have a premium image, which is why they are used for gifts and jewelry. Set-up boxes can even be made in non-rectangular shapes, such as heart-shaped boxes for Valentine's Day candy, or circular hatboxes. They are decorated with a glued-on wrapping sheet, so there is a wider possible range of graphic effects possible compared to printed paperboard.

Set-up boxes are preferred when the product is consumed over time or is stored in the box for an extended period of time. The separate lid is sturdy and much easier to open and reclose, compared to a folding carton's flaps. This is why they are used for products like board games and greeting cards that are consumed and/or stored in the box over time.

The most common set-up box material is thick, nonbending chipboard or newsboard, lined with a mixture of recycled, white groundwood or chemical pulp (*white lined*) to make the inside surface clean, and to give the outside good pasting qualities. A white sheet of paper can be laminated to it (called *book-lined*) if an even cleaner look inside the box is desired.

Non-bending boxboard is specified in terms of basis weight, but in the United States this is measured differently than for folding paperboard for historical reasons, and it is somewhat counterintuitive. The basis weight for set-up boxes is expressed as the number of 26″ × 40″ sheets in a 50-lb bundle. For example, if there are 40 sheets, the basis weight is 40 (note that there are no units). Other parts of the world are more rational, using grammage (grams/meter^2). Thicknesses range from 16 to 60 points (410–1,520 μm), although they can go as high as 98 points (2,500 μm).

Set-up boxes can be made manually or on high-volume production lines that synchronize the operations, depending on the size of the order. First, the blank is cut and scored on a machine called a *scorer*. This is usually through a straight-line operation with circular knives, or on a flat-press die-cutter if the order is large. The scores differ from those on folding cartons; since the board will not bend, it is sliced partway through to make the score. For a box and lid, the scorers must be set twice, since the lid is larger by one-eighth to three-eighths inch (3.2–9.5 mm). Next, the corners are notched out.

The walls of each blank are then folded up and its corners are formed with glued-on stays, seven-eighths- inch (22 mm) strips of kraft paper, in a machine, not surprisingly, called a *stayer*. A quad stayer, that folds and simultaneously stays all four corners, is used for high volume orders.

Last, the outer covering is applied. The paper covering, with cuts at the corners to facilitate clean wrapping, is coated with sticky, protein-based gelatin adhesive that forms a good mechanical bond. PVA glue is used for surfaces that are coated, laminated, or otherwise resist adhesion. The gelatin adhesive yields a cleaner surface and is easier to run on machinery. The glue is dissolved in hot water to make it set faster and minimize warp in the board. The tray or lid is placed (*spotted*) in the center of the glue-covered wrap manually or by

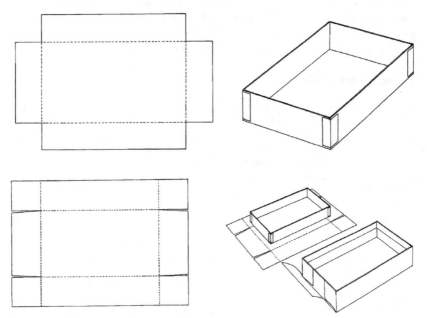

FIGURE 12.18. *Set-up box blank and assembly. Reprinted with permission of John Wiley & Sons, from Roth and Wybenga,* The Packaging Designer's Book of Patterns *(2013), 425–426.*

machine, and then a plunger forces the wrap around the tray. At the same time, rollers and brushes smooth the paper along the sides and tuck it around the edges. Figure 12.18 shows the blank and assembly process.

Since the lid and tray are wrapped with a sheet of printed paper, fabric, foil, or plastic, there is a wide range of print and surface finishes possible, compared to printed paperboard. For example, in the case of jigsaw puzzles, a photographic-quality printed paper wrap, that covers the top and sides of the lid and partially secures the bottom tray, is added after the box is filled.

There are variations on the style that include the *shell* and *slide* which is well-known as a matchbox, boxes with a hinged lid, and magazine/folio slipcases. Accessories like metal hinges and clasps can be added.

To showcase cards, stationery, or other attractive products, a clear rigid PVC or PET lid can be used in combination with a rigid paper-wrapped tray. Such lids are usually sized to fit inside the tray and to make the presentation tight and well-framed.

Set-up boxes cost more than a comparable paperboard carton because the technology is slower and the board used is heavier. More converting steps and higher-skilled employees are required. Production volume varies by season. But since the level of automation and set-up costs are low, at low volumes the cost is competitive. The industry is small and has consolidated; the number of set-up box plants operating in the United States declined to 122 in 2002 from over 260 in 1982 (Impact Marketing Consultants 2006).

12.5 REFERENCES

Biedermann, M., Y. Uematsu, and K. Grob. 2011. Mineral Oil Contents in Paper and Board Recycled to Paperboard for Food Packaging, *Packaging Technology and Science,* 24 (2): 61–73.

Chamberlain, D. and M.J. Kirwan. 2013. Paper and Paperboard—Raw Materials, in *Handbook of Paper and Paperboard Packaging Technology,* 2nd ed., edited by Mark J. Kirwan, pp 1-19, 263-312, 353–384. Chichester: John Wiley & Sons.

Corning, A. 1997. Machinery, Cartoning, End-Load, in *The Wiley Encyclopedia of Packaging Technology,* 2nd ed., edited by Aaron L Brody and Kenneth S. Marsh, pp. 580–588. New York: John Wiley & Sons.

Denison, E. and R. Cawthray. 1999. *Packaging Prototypes.* Switzerland: Rotovision; distributed in U.S. by Watson-Guptill Publications, NY.

DesRoberts, E. 2012. Mineral oil: a need for more data. *Packaging World,* Nov. 2, 2012. http://www.packworld.com/sustainability/material-health/mineral-oil-need-more-data, accessed 6/26/14.

Eldred, N.R. 2007. *Package Printing,* 2nd ed. Pittsburgh: PIA/GATFPress.

European Carton Manufacturers Association. 2009. *ECMA Code of Folding Carton Design Styles.* The Hague, Netherlands: EMCA (electronic resource).

Ewender, J., R. Franz, and F. Welle. 2013. Permeation of Mineral Oil Components from Cardboard Packaging Materials through Polymer Films, *Packaging Technology and Science,* 26 (7): 423–434.

Hanlon, J.F., R.J. Kelsey, and H.E. Forcinio. 1998. *Handbook of Package Engineering,* 3rd ed. Lancaster, PA: Technomic.

Hine, D. 1999. *Cartons and Cartoning.* Leatherhead, UK: Pira International.

Impact Marketing Consultants. 2006. *The Marketing Guide to the U.S. Packaging Industry.* Manchester Center, VT: Impact Marketing Consultants.

Jukes, M. 2013. Rigid Boxes. In *Handbook of Paper and Paperboard Packaging Technology,* 2nd ed., edited by Mark J. Kirwin, pp 253–263. Chichester: John Wiley & Sons.

Kirwan, Mark J. 2013. Paper and Paperboard—Raw Materials (with Daven Chamberlain), Processing and Properties; and Folding Cartons, in *Handbook of Paper and Paperboard Packaging Technology,* 2nd ed., edited by Mark J. Kirwan, pp 1–19, 265-312. Chichester: John Wiley & Sons.

Koran, Z. and D.P. Kamdem.1989. The Bending Stiffness of Paperboard. *TAPPI Journal,* 72(6): 175–179.

Koran, Z. and D.P. Kamdem. 1990. Multilayered Paperboard from Kraft Pulp and TMP. *TAPPI Journal,* 73(2) pp. 153–158.

Kuhr, C.R. 2009. Cartoning Machinery, Top-Load, in *The Wiley Encyclopedia of Packaging Technology,* 3rd ed., edited by Kit L. Yam, pp. 228–241. New York: John Wiley & Sons.

Lynch, L. and J. Anderson. 2009. Boxes, Rigid, in *The Wiley Encyclopedia of Packaging Technology,* 3rd ed., edited by Kit L. Yam, pp 170–173. New York: John Wiley & Sons.

MeadWestvaco. 2002. *Paperboard Knowledge Seminar.* Phenix City, AL: MeadWestvaco.

Obolewicz, P. 2009. Cartons, Folding, in *The Wiley Encyclopedia of Packaging Technology,* 3rd ed., edited by Kit L. Yam, pp. 234–241. New York: John Wiley & Sons.

Orta, O. 2009. Carriers, Beverage, in *The Wiley Encyclopedia of Packaging Technology,* 3rd ed., edited by Kit L. Yam, pp. 225–238. New York: John Wiley & Sons.

Paperboard Packaging Alliance. www.paperboardpackaging.org, accessed 3/30/2013.

Paperboard Packaging Alliance. 2002 *Paperboard Packaging Syllabus*, prepared by John Latham. PPA.

Paperboard Packaging Council. www.ppcnet.org, accessed 3/30/2013.

Paperboard Packaging Council. 2000. *Ideas and Innovation: A Handbook for Designers, Converters and Buyers of Paperboard Packaging.* Edited by Richard DePaul. Alexandria VA: Paperboard Packaging Council.

Peel, J.D. 1999. *Paper Science and Paper Manufacture.* B Vancouver, Canada: Angus Wide Publications.

PPI Pulp & Paper Week. Published weekly by RISI. http://www.risiinfo.com/

Roth, L. 1990. *Packaging Design.* New York: Van Nostrand Reinhold.

Roth, L. and G.L. Wybenga. 2013. *The Packaging Designer's Book of Patterns,* 4th ed. New York: John Wiley and Sons.

Savolainen, A. 1998. *Paper and Paperboard Converting.* Helsinki: Fapet Oy.

Soroka, W. 2009. *Fundamentals of Packaging Technology,* 4th ed. Herndon, VA: Institute of Packaging Professionals.

Trefry, A. 1997. Carton Terminology Brief, in *The Wiley Encyclopedia of Packaging Technology,* 2nd ed., edited by Aaron L. Brody and Kenneth S. Marsh, pp. 176–181. New York: John Wiley & Sons.

12.6 REVIEW QUESTIONS

1. What are the three primary virtues of paperboard packages?

2. Is most recycled multi-ply board made on a fourdrinier or cylinder machine?

3. How does the color of recycled paperboard vary by its pulp source?

4. What is the difference between bending and non-bending chipboard and how do their uses differ?

5. Why is recycled board not used for food contact?

6. Which way should the paperboard's MD run on a folding carton? Why?

7. Why is SBS favored for food contact?

8. What type of board is typically used for frozen food and ice cream containers? What type is used for cereal and cracker cartons? What type is used for 12-can beverage carriers?

9. In general, how thick should recycled board be for a carton containing 1 lb (450 g) of product?

10. Discuss the factors to consider when designing a folding carton.

11. Propose a good use for a carton printed with scratch-and-sniff ink.

12. For what kind of products would a carton window give a marketing advantage?

13. What is meant by an "integrated converter?"

14. Which printing process is most commonly used for paperboard folding cartons? Is this sheet-fed or web-fed?

15. When is a flatbed die used, versus when a rotary die is used?

16. Should the score's raised ridge be on the inside or the outside of a carton?

17. Why is it necessary for a score to delaminate paperboard plies?

18. What is meant by pre-broken scores? What is meant by knocked down? What is the difference between working scores and non-working scores?

19. When is PVA used to glue folding cartons? When is hotmelt adhesive used, and why?

20. What differentiates the various tube styles of folding cartons?

21. Draw blanks of the following tube style cartons: POSE, SST, SRT, and FRT.

22. Give some examples of the kind of products for which the POSE carton is most popular.

23. Which tuck style carton style is most economical to produce? Why?

24. Why is a FRT style carton preferred by the perfume industry?

25. Why is a SST style carton the most expensive tucked style?

26. What is the difference between *friction lock* and *slit lock* tuck flaps?

27. What are some uses for tray-style cartons?

28. What is the difference between a Beers tray and a Brightwood tray?

29. Is paperboard specified in the United States by weight or thickness, and in what units?

30. How are the dimensions measured for specifications? Is there a separate score allowance? Is this the same for corrugated fiberboard boxes?

31. What is the standard way to represent a tube-style carton in a layout drawing in terms of which side (printed or unprinted) is up and the position of the glue flap?

32. Lay out the blank for a POSE carton of dimensions $6'' \times 2.5'' \times 10''$ with $1/2''$ flap overlap and $1/2''$ glue flap.

33. Estimate the cost of the POSE carton in the previous question. Four-color printing, 80-bright 60 lb BW clay-coated recycled board (assume the market price is $1,000/ton).

34. Label the dimensions on an unlabeled carton blank from the dimensions of the box it corresponds to, and vice versa.

35. Why are KD folded and glued tube-style cartons shipped on edge?

36. What properties of a carton affect its runability on a cartoner? Explain why.

37. What are some alternatives to folding cartons for dry products?

38. Why do beverage carriers have more dramatic display potential than most other paperboard packages?

39. From what kind of paperboard are beverage carriers made? Why?

40. What is a "jerk test" and why is it used?

41. How does a set-up box differ from a folding carton?

42. What are some common uses for set-up boxes?

43. What are the advantages and disadvantages of set-up boxes, compared to folding cartons?

CHAPTER 13

Liquid-Filled Cartons and Other Specialized Paperboard Forms

Folding cartons and beverage carriers make up almost 64% of all paperboard packaging in the United States (aside from corrugated board). Sanitary food containers are the second largest percentage, almost 20%. This chapter presents paperboard packages for food applications, like liquid-filled cartons and freezer-to-oven plates, as well as fiber cans and drums, carded and solid fiberboard packaging, and molded pulp. Figure 13.1 illustrates the relative usage of these package forms (Impact Marketing Consultants 2006).

The differential advantage of paperboard for food packaging is its inexpensive stiffness, strength, and printability, compared to alternative materials like plastics, metal, and glass. The properties of the paper or paperboard are tailored to protect the food product from the environment, but also from indirect contamination by the package surface, as discussed previously in Chapter 9.

13.1 LIQUID-FILLED CARTONS

The packaging of liquid is paperboard's hardest job, considering the fact that paper is absorbent and porous. Add to that the challenge of sanitation for food contact, and it becomes obvious that milk cartons and juice boxes are among the paperboard industry's highest achievements.

An extrusion-coating with plastic (or, in the past, wax), gives paperboard the ability to be made heat sealable, and to contain liquid. The two primary forms, described in the following two sections, are gabletop cartons and aseptic cartons, but other patented shapes can be found in the market.

Liquid-filled cartons are made from SBS because it is strong and clean. They are extrusion-coated with plastic, usually LDPE, for water-resistance and heat sealing, and may also have a laminated foil layer. The board's pulp contains more internal sizing than that used for folding cartons, to ensure that the

419

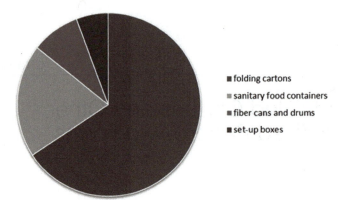

FIGURE 13.1. *Paperboard packaging U.S. usage.*

cut edges of the board do not readily absorb water. They are printed by flexography (or gravure if the runs are long) before coating.

There are several advantages of using paperboard cartons for liquids, compared to cans or bottles. They can be shipped to the filler knocked down (KD) or in roll format. They give a good substrate for colorful graphics. They can be made to be relatively strong and stiff. Their squareness minimizes the cubic space in transit and display. And they are inexpensive.

Because they are made from coated or laminated paper, aseptic and gable-top cartons have a disadvantage because they are not widely recycled. Often there are too few of them in the waste stream to make it economical to collect, sort, and transport them to the few specialized mills that are able to repulp and separate the high-quality kraft fiber from the plastic and foil. Regardless, most life cycle assessments have found that beverage cartons score better than cans or bottles on measures of climate change, cumulated energy demand/fossil resource consumption, and acidification (von Falkenstein *et al.* 2010).

By far, most liquid-filled cartons are for beverages, especially dairy products. They are the overwhelming choice for milk and for yogurt drinks that are popular in Europe. They are used for other pumpable food products like cooking oil, soups, baby food and pet food. They have been used in food service for condiments and fountain syrups and for soap dispenser refills. This category of liquid cartons does not include bag-in-box, in which a separate bag contains the liquid, usually inside a corrugated box.

13.2.1 Gable-Top Cartons

Today's gable-top cartons are made from SBS coated with LDPE. The SBS for a typical quart or liter milk carton has a basis weight of 195–210 lb/1,000 ft^2 (950–1025 gsm) with an LDPE extrusion coating of 0.0005 in (12.7 μm) on the outside, and a thicker coating of 0.001 in (25.4 μm) on the inside. The bigger the carton, the thicker are the paperboard and the coatings (Lisiecki 2009).

When a better gas barrier is required, a layer of aluminum foil coated with ionomer for heat sealability is laminated to the SBS, or a plastic like EVOH or SiOx-coated PET is extrusion coated to the inside surface.

The material and filling machine are *tied goods*, tied to each other by patents and technology, and so one company's cartons will not run through another company's machine. The leading producer is Elopak, a successor to Ex-Cell-O and the original Pure-Pak® licenses.

The typical blank layout is shown in Figure 13.2. At the dairy or other filling location, a form-fill-seal machine erects the cartons from flat blanks. The bottom folding, sealing, and profile are critical to prevent leakage. The top seal is formed by induction sealing.

The clever geometrically-designed top seal allows it to be tight, yet able to be opened by peeling apart the seal, pressing back the two wings and then compressing them to form a spout. Unfortunately they are not always very easy to open (the challenge being the need for a very secure seal up until that point), and the fiber left from tearing apart the seal is in the path of the milk or other food.

Alternative closures have been developed to make gable-top cartons easier to open and reclose. The most successful have been threaded spouts and caps, made from PP. These are applied by the carton manufacturer. While they add

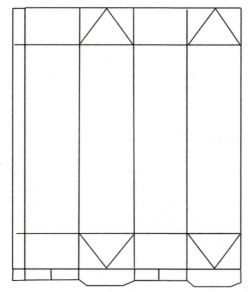

FIGURE 13.2. *Gable-top carton blank.*

convenience, they detract from recyclability because they must be separated from the paperboard for repulping.

Gable-top cartons were once the primary package for milk in the United States. Milk cartons symbolized cleanliness and sanitation. They offered many production and distribution advantages over reusable glass bottles, not the least of which were the abilities to be shipped into the dairy KD and to guard the milk from UV light. But gable-top cartons are used less for milk in the United States today, except in small "school lunch" sizes, due to competition from PE blow-molded bottles, which have the additional advantage of being shipped as resin pellets and blow-molded at the dairy.

The gable-top form has moved on to dominate the market for pasteurized fruit juice, due to the advantage of screw-top closures and the improved quality of flexographic printing that can now provide more vivid graphics. For example, manufacturers of pasteurized orange juice can present a fresher image by picturing a juicy orange on a carton than by revealing the actual juice through a transparent bottle.

The gable-top carton for single-strength (pasteurized, refrigerated) orange juice is a good example of the need for a careful choice of plastic coating and how a food can interact with a package's contact surface. The PE coating used for milk could not be used for orange juice, because it was found to absorb essential oils that carry important flavor compounds from the orange juice. Also called *sorption* and *flavor scalping*, it made the juice taste dull. The solution was an interior EVOH coating; EVOH is a much more polar molecule than PE, and is thus less attractive to the limonene and other flavor compounds. This so-called *j-board* is then overcoated with a very thin coat of PE, because EVOH cannot be heatsealed. This construction allowed a whole new line of single strength products into the market place and underscores the importance of packaging material selection in product acceptance.[24]

Gable-top cartons have been used for other wet and dry products ranging from concentrated refill soap to granola to soup.

13.2.2 Aseptic Cartons

Aseptic cartons, also known as *juice* or *drink boxes*, have a different story. Like gable-top cartons, aseptic cartons can be used for fresh liquids that are kept refrigerated. But they are most often used for *shelf-stable* drinks that have been sterilized. They compete primarily with canned juices and milk, although they can be used for any liquid with minimal particulates.

---

[24]The sorption of limonene by plastics has been the subject of extensive research. Its hydrophobicity allows easy sorption by some synthetic polymeric materials, especially non-polar polyolefins like PE (Kwapong and Hotchkiss 1987; Fayoux *et al.* 1997). For example, it has even been found that incorporating an active packaging material can improve juice flavor by removing bitterness (Soares and Hotchkiss 1998).

Aseptic cartons were invented in tandem with the UHT milk sterilization process. The aseptic packaging concept is that sterilized milk or juice is filled into a sterile package in a sterile environment. The juice or milk is sterilized in an *ultra high temperature* (*UHT*) process for a short time, which results in less flavor and nutrient loss (and, for milk, a less caramelized taste) than canning. Whereas canned baby formula is retorted in the can at 250°F (121°C) for 40 minutes, aseptic milk is flash-heated for only 2–4 seconds at 284°F (140°C). The sterilized milk or juice is cooled and conveyed through sterile pipes to the form-fill-seal (FFS) packaging machine.

In the patented Tetra Pak process, shown in Figure 13.3, the multi-layer paperboard is sterilized at the same time as the liquid, as the roll unreels through a sterilizing hydrogen peroxide (H_2O_2) bath and passes into an aseptic filling zone. The sterilized material is dried and then formed into a tube in the vertical FFS operation. The longitudinal heat seal is made with the help of a *longitudinal seal strip* that was applied before sterilization, comprised of PET or EVOH, depending on the O_2 sensitivity of the product, coated with LDPE or LLDPE. The longitudinal strip covers the seam, prevents contact between the inside and outside of the carton, and protects the edges of the foil and paperboard layers from contact with the product that could cause swelling and corrosion.

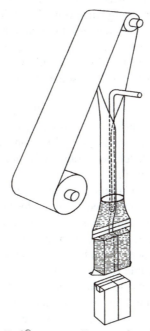

FIGURE 13.3. *The Tetra Brik® process. Reprinted with permission of Tetra Pak, copyright 2004.*

Induction sealing jaws (that heat the foil layer and efficiently heat the adjacent seal layer, avoiding the lag of heating through the paperboard layer) form the top of one carton and the bottom of the next one simultaneously, as the liquid fills the tube, sealing right through the liquid and leaving no headspace in the carton. The reason that Tetra Pak cartons cannot be used for products with particulates is that particles interfere with the seal and are difficult to sterilize.

Lastly, the top and bottom are folded square and glued, and if required, a wrapped straw is glued to the back. The cartons are usually shipped in shrink-wrapped bundles, rather than full corrugated boxes, to minimize cost and potential dynamic response to in-transit vibration.

Like gable-top cartons, Tetra Pak paperboard is a tied good, and will only run on the Tetra Pak machines. Each layer has a special function. The SBS paperboard (186 gsm, 38.13 msf or heavier) gives the stiffness, strength, and printed surface. Unbleached kraft board can also be used, but the bright white SBS is preferred for its graphic potential and clean, sanitary appearance. The board is flexographically printed before extrusion coating and lamination.

Tetra Pak cartons are made from three materials: SBS, PE, and usually aluminum foil, with the PE used as adhesive and coating. The printed SBS is extrusion-laminated with a PE adhesive (20 gsm, 4.1 msf) to a thin aluminum foil (6.3 μm, 0.00025″) that acts as barrier to light and oxygen. Inside the package, a thin layer of polyethylene-co-methacrylic acid (6 gsm, 1.23 msf) ensures good adhesion between the aluminum and the thick extrusion coating (29 gsm, 6 msf) of PE that provides the heat sealable barrier layer. The outside surface has a thinner coat of PE (12 gsm, 2.5 msf) to protect the ink and enable the heat sealing of corners to form the squared-off ends and the sealing of the longitudinal seam.

The first aseptic cartons weren't boxes at all, but were a tetrahedron shape, the source of Tetra Pak's name. The compact, rectangular Tetra Brik® shape has the advantage of minimizing cube utilization during distribution. In recent years Tetra Brik has developed a range of more interesting aseptic carton shapes to enhance shelf appeal.

Other paperboard manufacturers offer different patented designs of aseptic cartons. For example, the Combibloc® process does not use a vertical FFS process, but fills prefabricated cartons in a horizontal FFS machine, using a hot peroxide fog for sterilization. Since the cartons are erected, filled, and then sealed, they do have some headspace, which builds in a stacking weakness, since air in the headspace is compressible (but the liquid is not). However, the seal is not made through the product and so liquids containing particulates can be filled.

On the other hand, sometimes a headspace is deliberately added to make the carton easier to open or to give space for shaking to disperse pulp or particulates before consumption. In this case, steam or an inert gas such as N_2 is added.

The most basic drink box is intended to be a single serving with a wrapped straw attached. The straw can be used to puncture through an easy-open hole in the paperboard, covered with a pull tab (as in the case of kids' drinks), or the top is cut with scissors (as in the case of tomato puree). Both of these can be messy, resulting in some mothers calling them "juice cannons." A number of reclosable alternative closures, including pull tabs and plastic threaded caps, have been introduced to overcome problems that make the basic box hard to open and easy to spill.

Other shapes have been designed, like the eight-faceted Tetra Prisma®. Aseptic cartons aren't just for drinks, and uses have expanded to alcoholic beverages, non-particulate soups and purees.

As an alternative to aseptic filling there are even retorted forms: Tetra Re-cart® and SIG Combibloc's Combisafe® are substitutes for cans. The retortable laminates include SBS, aluminum foil, and PP with heat sealability adequate to withstand retorting at a temperature of 130°C (266°F). They are suitable for a wide range of foods, including chunks, like vegetables, fruits, sauces, soups, and pet food, since they are not sealed through the product. Shelf-life is comparable to canned foods, and no refrigeration is required.

Compared to cans and bottles, with which they compete, aseptic cartons offer many production, marketing, and logistics advantages. The paperboard material is relatively low cost. It is much less expensive to deliver to the juice factory or dairy, because it comes on rolls, as opposed to buying pre-assembled cans. The ability to aseptically process the liquid improves its taste, compared to retorting it in the can. The FFS process has slower production speeds than canning, but a smaller scale economy is better suited to many applications. No labeling operation is required. Brick packs occupy less space in distribution as well as in landfills. They offer benefits in foodservice, including their lighter weight, ease of opening without a can opener or sharp metal edges, and the ease of crushing to minimize space in garbage containers.

Aseptic juice boxes have faced political opposition from environmental groups who object to the multi-layer material on the basis that is difficult to separate in recycling the packages. The state of Maine even went so far as to ban the sale of juice boxes in 1990. The ban was lifted four years later after intensive industry efforts to increase the recycling rate, as well as extensive public relations campaign to extoll the environmental virtues of using less material and energy compared to cans or bottles. The cartons can be hydrapulped to separate the high-quality fiber, or the whole structure can be ground to make filler for mixed plastic lumber substitutes.

Despite the limited recycling rate, aseptic cartons have been a great success in human sustainability. The Institute of Food Technologists has called aseptic processing "the most significant food science innovation of the last 50 years" because it uses a minimal amount of resources to do one of the most important packaging jobs. Shelf stable products are distributed at ambient temperature and so avoid the environmental impact associated with chill-chain distribution. The shelf-stable aseptic package is best known for its beneficent success in providing affordable sanitary liquid milk to people all over the world, particularly to children in developing countries where there is no refrigeration to distribute fresh milk safely.

13.2 CYLINDRICAL SHAPES: COMPOSITE CANS, FIBER TUBES, AND DRUMS

A cylinder is the strongest shape. Although thickness contributes most to the stiffness of a sheet of paperboard, the stiffness of a cylinder owes even more to its shape. There are three basic types of cylindrical shaped paperboard packages: composite cans, fiber tubes, and fiber drums.

13.2.1 Composite Cans

Composite cans (also called *fiber* or *fibre cans*) are used for products ranging from frozen orange juice to nut snacks. Several brands have built their reputations on the basis of composite cans, including Pringles® snack chips, Comet® cleanser and Pillsbury's Pop-N-Fresh® dough. They are also a good option for products sold in vending machines because they easily roll.

Composite cans have a good strength-to-weight ratio. They are not as strong as thick-walled metal cans but they are less expensive and lighter weight. They can have attractive built-in graphics without the need for a label. There are three kinds of composite cans, named for their forming processes: spiral-wound, convolute, and linear draw.

13.2.1.1 Spiral-Wound Cans

Spiral winding is the most common, least expensive process. Spiral wound cans are the most common type, used for many products, such as frozen orange juice, shortening, peanut butter, unbaked dough, snack chips, and nuts.

In the spiral winding process, narrow-width rolls of paperboard 2–10 in wide (depending on the size of the finished can) are first coated with adhesive and then fed in sequence at an angle to a cylindrical *mandrel*, which temporarily supports the structure as it is formed. They wind around it to form a continuous tube (see Figure 13.4). As the layers are fed in sequentially, their seams are offset from each other to interlock the plies together into a strong structure.

The can liner ply forms the barrier, the can body has one or more layers of paperboard, and the last ply wound is the label. Each ply is glued to the next

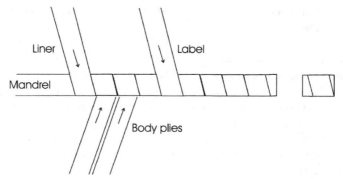

FIGURE 13.4. *Spiral wound composite can forming.*

one. The dimensions of the mandrel determine the can's diameter. The cans are cut to length at the end of the mandrel.

Spiral-winding is a faster, continuous process, compared to the convolute and drawn processes, which are discussed next. But it is most common, and the "materials and options" discussion that follows applies especially to spiral-wound cans.

13.2.1.2 Convolute, Single-Wrap, and Linear-Drawn Cans

Convolute composite cans take more time to make and are more expensive than spiral-wound because they are made in a discontinuous process. Each layer is glued on top of another as cut blanks are fed in sequence straight onto a revolving mandrel. The mandrel does not have to be cylindrical, and so this is the process used to make tubes in other shapes, such as those used for edge protection in appliance and furniture packages. The same kind of glues and ply materials are used as in spiral winding. Once the structure is completed, the can is removed and trimmed.

There are some patented *single-wrap* cans that are made on such a scale that the cans can be assembled in the filler's factory, which reduces inbound transportation costs because the containers are KD. For example, single-wrap convolute containers with a heatsealed seam and sealed paperboard bottom end are converted in-plant for dried fruit, oatmeal, cereal, bread crumbs, snacks, and ice cream.

Linear forming (also known as linear drawing) brings four webs together and folds them around a mandrel with a longitudinal overlap seam.

13.2.1.3 Composite Can Materials and Options

Since the typical composite can is made from four kinds of material, the liner, paperboard body, label, and can ends, the barrier, strength, graphics, and dispensing can be tailored to the use.

Plies of paperboard body form the structure. The body plies must provide enough strength to resist stacking and impact forces in distribution. Recycled paperboard is the most commonly used body material. It is less expensive than unbleached kraft, which is used when additional strength is required. Recycled paperboard, in this case, is acceptable to use for food, because the can's liner provides a barrier between it and the food. The edge of each ply is *skived*, sliced on an angle, so there is less extra thickness where it overlaps itself. If there are more than one, the plies' seams are offset from each other to give greater strength.

It is common for spiral-wound cans to have two body plies, but they can have more, up to a dozen, if additional strength is required. Snack chips and frozen orange juice cans have just one body ply, but compensate by having thicker board and more secure seams. Various adhesives, most water-based, are used to meet specific needs. Silicate adhesives are inexpensive and stiff when dry. Dextrin adhesives are used when less water-resistance is needed. PVA and PE are also common.

The inner liner ply of a spiral-wound can is kraft paper, either laminated to aluminum foil, coated with plastic, or both. The choice of liner depends on the product protection requirements and compatibility. The long sides of the liner ply are double-seamed to prevent wicking in what is called an *Anaconda seal*.

Some products, like coffee, snacks, shortening, powdered beverages, and refrigerated dough, require a barrier to moisture and/or oxygen. A can-liner for salty snacks needs to resist abrasion. Aluminum foil provides the best barrier, but it is the most expensive, and is not used where less expensive metalized PET (M-PET), will suffice. The foil or M-PET is coated (inside the can) with a layer of PP, PE, or ionomer,

Other products only need a liner to prevent leakage. For products with medium barrier requirements, the liner is coated with one or two plastics, PET, ionomer, or OPP plus PE. For low barrier requirements, like frozen juice, the paper is simply coated with HDPE or LLDPE. Table 13.1 shows the range of options.

The outer label ply is the can's billboard. The most common materials are kraft paper, aluminum foil, or a lamination of both. Foil is used to add shine, a barrier, or when the surface needs extra protection. For example, foil and kraft paper are used for refrigerated dough and cleanser because of the wet environments. Bleached paper is less expensive, commonly used for snacks.

Since the outer layer of most spiral-wound composite cans is pre-printed, gravure or flexography is used, depending on the length of the printing run. (The outer ply of convolute cans, which are sheet-fed, is printed with lithography.) A plastic coating is usually applied over the print to provide protection

**TABLE 13.1. Composite Can Liner Ply Structures
(Used by permission from Sonoco, 2013).**

Common Liner Structure	Barrier Properties	Typical Uses
Kraft/foil/polypropylene	High	Hermetic foods (nuts, snacks, shortening, peanut butter, infant formula, etc.), solvent and latex based paints
Kraft/foil/ionomer	High	Hermetic foods
Kraft/foil/HDPE	High	Solid shortening, powdered beverages, snacks
Kraft/MPET/ionomer	High	Hermetic snack foods
Kraft/PE/foil/HDPE	High	Refrigerated dough
Kraft/foil/vinyl coating	Medium	Refrigerated dough, adhesives, caluk, powdered beverages, non-hermetic food
Kraft/PE/PET/ionomer	Medium	Pet food, paper bottom cans
Kraft/LDPE/PP	Medium	Solid shortening
Kraft/LLDPE/HDPE (white)	Low	Hard-to-hold frozen concentrate
Kraft/ionomer/HDPE (white)	Low	Hard-to-hold frozen concentrate
Kraft/HDPE (white)	Low	Frozen citrus concentrate
Kraft/HDPE	Low	Motor oil

and a barrier to moisture. The label ply adds graphics and nutrition information without the need to add a label in a separate operation.

Alternatively, cans can be labeled after forming or after filling, similar to metal cans, with a paper or plastic label. This makes it possible for a small business, private-label brand, or contract packer to use cans that are standard stock sizes.

The units in specifications can be a little confusing because of the combination of materials (and the traditions from each of their industries). The paperboard plies are specified in terms of basis weight per 1,000 ft^2 or grammage and thickness in points, inches, or μm. The paper liner and label are specified in terms of basis weight per 3,000 ft^2 and the plastic coatings on them are likewise specified in terms of number of pounds per 3,000 ft.^{2} or grams/meter^2. Foil is usually specified in terms of thickness in inches or μm.

For example, a typical can specification for frozen orange juice shows a single body ply made from 21-point (81.9-lb BW) recycled board with a water finish to improve adhesion. It has a liner ply of 20-lb machine-glazed natural kraft coated with 10 lb of white HDPE.[25] The label is 35-lb bleached paper with a solvent-based catalytic coating.

[25]The HDPE is too thin for flavor scalping (compared to the heatsealed gable-top carton) and the frozen juice is less susceptible than the liquid juice.

In the United States, composite can dimensions are specified the same way as for metal cans: the external diameter of the can end and the height are each specified by three-digit numbers designating whole inches and one-sixteenth fractions of an inch. Composite can sizes range from 1″ to 7″ (25–178 mm) in diameter and from 1″ to 13″ (25–330 mm) tall. There are several popular standard sizes; the common peanut can's diameter is 4-1/16″ (103 mm). Thus, a 401 × 602 can measures 4-1/16″ in diameter and 6-2/16″ tall. In Europe, the diameter, expressed in mm, is based on the internal diameter of the tube; popular European diameters (and their U.S. equivalents in parentheses) are: 65 (211), 73 (300), 84 (307), 93.7 (313), 99 (401), and 155 (603) (Henderson 2013).

The bottom can end (and sometimes both) is usually tin-free steel (TFS); in the United States, can ends are specified like steel cans, in terms of lb/basebox. A basebox is equal to 32,360 in^2 (an old measurement that derives from a standard unit of 112 sheets that are 14″ × 20″). Commonly used weights range from 55 to 107 lb/base box (1.2–2.3 kg/m^2). The ends may be coated (with tinplate, vinyl, epoxy, or phenolic) for additional protection. Steel ends are the most secure, capable of withstanding vacuum or nitrogen flush. For example, the typical frozen orange juice can has one seamed TFS end that measures 210.5, an easy-open end that is slightly larger, 211 (65 mm), to accommodate the easy-open strip; and both ends are specified as 65- lb (1.4 kg/m^2), with a non-PBA coating.

The ends are crimped on with a double seam similar to that used for steel cans. The double seam prevents leakage. But this seam line is the weakest part, and too tight a crimp for an impact transferred from an adjacent can in a box may break the paperboard or rip it out of the seam. For additional strength, a sealing compound is applied. Composite cans with metal ends are typically opened with a can opener (which can be problematic if the paperboard walls are wet) or a tear strip wedged in the seam for one end.

Similar paperboard cans for rolled oats and salt have always had at least one paperboard end, since these products do not need a barrier and the net weight is low.

Other kinds of easy-open ends are made from aluminum or flexible membrane, used on only the top end. The typical aluminum end has an easy-open scored ring-pull panel.

Flexible membranes are made from plastic film, or plastic-coated foil or paper. A peelable, flexible membrane can be sealed to a TFS ring; it can even have a one-way release valve to allow freshly roasted coffee to off-gas in the package. Snack chip cans have the simplest closure of all: the top of the can is flanged and curled outward, giving a heat sealable surface to which the peelable, metalized seal can be attached.

Cans with easy-open ends usually include a reclosable PE or PP snap cap.

More complex dispenser ends include the following:

- Simple sifter used for household cleanser
- Spout for salt
- Rotary dial for powders
- Syringe-like nozzle and plunger ends on a caulk tube.

But the *self-opening can*, used for refrigerated dough, is the most brilliant application of this package form. It exploits the spiral windings in a careful balance of ply strength and seam placement. Once the tough, natural kraft, outer ply is stripped off, the single spiral-wound laminated body ply splits open along its joint, because the liner joint is strategically positioned in the same place, due to pressure from the dough's yeasty expansion.

Fiber cans offer great value providing an inexpensive package. They dress up recycled board in label-quality graphics and a clean food-contact surface barrier. They provide strength and structure through their material and shape.

For dry foods like snacks and powdered beverages, fiber cans seal so well that the contents can be nitrogen-flushed during filling to remove oxygen in order to retard rancidity. Although they could be made strong enough to hold a vacuum, they are rarely vacuum packed because of the high cost of the extra material that would be needed. They have been successfully used for semi-viscous and/or oily products like shortening and peanut butter. The shape, which is used to best advantage for circular-shaped snacks and pastry, has potential uses for other circular and cylindrical products, from balls to plumbing fixtures.

Composite cans are not recycled in some communities because the materials cannot be safely separated. Many consumers are unsure about how to recycle them and some material recovery facilities (MRFs) will not accept them. This prejudice focuses on the can's inability to be recycled for its fiber.

But, unlike juice boxes, composite cans can be recycled in the steel can material stream, because about half the weight of an average composite can is its steel end(s). They can be sorted by magnet. Under the high temperature of steel recycling, the fiber and plastic are burned off and even serve as a (small) extra energy source. Therefore, the package is considered by the industry to be fully recoverable (Sonoco 2013).

13.2.2 Fiber Tubes

Fiber tubes have found uses besides just cans. Strong, spiral wound tubes with more than two plies, up to a dozen, are used as mailing tubes or industrial coil cores. Single-ply tubes are used as cores for wrapping paper, toilet paper, and paper towels.

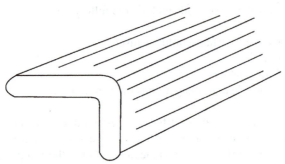

FIGURE 13.5.   *Paper tube form for edge protection.*

Convolute tubes are often shaped to make forms to protect the edges of products like tables and appliances in boxes or shrink-wrap, as shown in Figure 13.5. They are also put in edges of boxes to improve box compression strength.

13.2.3 Fiber Drums

Fiber drums are large, industrial versions of the fiber can. They can range from 2 to 60 gallons (10–270 L) and are capable of holding up to 550 lb (250 kg) of dry or semi-liquid products. The two most typical capacities are 55 lb (25 kg) and 110 lb (50 kg).

They are used for a variety of foods, chemicals, and adhesives shipped in semi-bulk quantities to manufacturing plants. They are used mostly for powders, granular materials, and very viscous liquids, like adhesives, that practically solidify in the drum. They are used for food ingredients, chemicals, and electrical wire.

Fiber drums are lighter weight and lower cost than the comparable steel drums, and they compete with plastic drums for non-hazardous (and some slightly hazardous solid) materials, but are not allowed for any liquid hazardous materials.

Fiber drums are made in the convolute process, spooling paperboard from a roll to wind around the mandrel. Adhesive is applied as the paperboard is wrapped until the right number of plies has been built up. Sodium silicate (water glass) is the most common adhesive because it is strong and inexpensive. PVA adhesive is used if more water resistance is required.

The number of plies can range from 4 for a lightweight drum to 11 for heavier loads. The thickness of each sheet is typically 12 points (305 μm), but a heavier board can be used to reduce the number of plies. Recycled board or kraft can be used; kraft imparts greater strength. The typical wall thickness is 60–200 points (1,520–5,080 μm).

A barrier ply can be incorporated on the inside, or sandwiched between plies, of PE, PET, PVdC, PA, and/or Al foil. Sometimes a loose PE bag liner is used, tied off at the top, for added water vapor barrier. The outer ply generally has additional sizing or waterproofing; it may be the only kraft ply, to add toughness and strength.

Closures range from a telescoping paperboard structure to plastic or steel ends that are locked on by a ring around the edge. In the United States, steel and PE ends predominate; in the United Kingdom most are plastic. Steel is used for the highest performance applications.

Although most are cylindrical, there are also fiber drums with a more square footprint with rounded corners, which better utilize cube in transportation and storage. There are also several heavy duty corrugated fiberboard structures that have been designed to compete with fiber drums, notably octagon and hexagon shapes.

13.3 CARDED PACKAGING

Carded packaging is used for small products like lipstick, school supplies, toys, and hardware. The card provides the principal display panel and substrate for other graphics, and a blister or skin of plastic encapsulates and displays the item. Mounting products on display cards serves two important retailing functions: they display the product and prevent pilferage.

Carded packaging frames showcase and advertise the product. It gives details about special features. The pegboard hole makes it easy for the retailer to hang the card on wall displays. When combined on a wall, a family of carded packages creates an attention-getting point-of-purchase (POP) display.

The second function is to discourage pilferage. Items that are small—like batteries, cosmetics, hand tools, pens, and toys—present a temptation for theft in self-service stores. The card makes the product large enough so that it is more difficult to conceal.

Although there are other kinds of packages (folding cartons, for example) that could fulfill these same functions, there are production and display advantages for carded packaging. The materials and packing process are inexpensive. The process is small enough in scale to facilitate easy change-overs, so blisters are commonly used for a family of related products (like cosmetics or hardware fasteners).

Carded packaging offers easy change-overs and versatility for products that are sold in a variety of promotional formats. This is often used as part of a *postponement strategy* that separates manufacturing from the packaging operation, waiting to package until the product is nearer to the market or until a promotion is launched or demand can be predicted. For example, products like disposable razors and pens can be sold in one-packs, multi-packs, and back-to-school packs. A postponement strategy would ship the pens and razors in bulk to a location near to the market, and pack them to order when promotions arise. Postponing promotional packaging until demand is more certain reduces inventory and transportation cost (Twede *et al.* 2000).

Because carded packaging adds substantial cubic volume to a product, post-ponement can dramatically reduce transportation cost, because the products are shipped in more compact bulk packages for most of their supply chain. Packaging postponement can be an especially useful strategy for global mar-keting, given the long distances and various presentation formats.

The next two sections describe skin and blister packages. There are also clamshell packages made from two thermoformed blisters, sometimes con-joined and hinged, in which the card plays no role at all in the structure, but is inserted only for its information and advertising content.

13.3.1 Blister Packaging

Blister packages are the most common form of carded packaging, and can be recognized by the pre-formed blister in which the product is relatively loosely encased. They have two components: the pre-formed blister and the paperboard card, as shown in Figure 13.6. The thickness of the blister and board depend on the weight of the product.

Most blisters are made from PVC which is easy to thermoform, is rigid, and heat seals well to properly coated paperboard. Blisters can also be made from PET, which is a little more expensive than PVC but more environmentally friendly, or OPS (oriented polystyrene), which has poorer impact resistance. Blisters are commonly 5–7 mils (0.005–0.007 in, 127–178 μm).

The paperboard can be SBS or recycled board, usually clay-coated. Thick-ness ranges from 14 to 30 points (356–762 μm), but 18 = 24 points (457–610 μm) is most common. Two layers of board glued together may be used for heavy products, particularly to reinforce the pegboard hole. The machine di-rection of the paperboard is usually vertical, to use its tensile strength to best advantage and to resist tearing across the card around the peg-hole.

Thermoformed plastic blister

Heat-sealed blister flange

Product

Heat-sealed blister flange

Heat-seal coated printed paperboard

FIGURE 13.6. *Blister package.*

A coating is applied to the board to make it heat sealable and to add gloss and abrasion resistance. The coating must be compatible with the card's clay-coated surface and the blister, and provide good hot tack to support the still-warm package as it is ejected from the sealing unit. The most common coatings are solvent-based vinyl because of their superior gloss and sealability to PVC.

The process for filling and assembling blister packages is simple. Blisters are loaded into the cavity of the heat sealing unit. A product is loaded into each blister, and they are covered with a card. Heat is applied to the back of the card until its heat sealable coating fuses to the edges of the blister.

The process can be automated, semi-automated, or manual. Its adaptability makes it ideal for small lot jobs. For example, a number of promotional variations can be produced for a single product with almost no changeover cost. The low level of technology, and in some cases high degree of easy manual labor, makes this a common packaging operation that is outsourced to contract packaging businesses, including sheltered workshops that employ disabled workers.

The pegboard hole plays an important role in displaying carded packages. It needs to be strong enough to permit repeated removal and replacement on pegs without tearing. Extra reinforcement may be added. There are several different shapes of pegboard holes.

Carded packages are notoriously difficult to open, and this is the most common complaint against them. The fused bond between plastic and paperboard needs to stay strong throughout distribution, but the problem is that it remains strong once the consumer takes it home. Opening difficulty does not serve the customer well, causes anger and frustration (wrap rage), and can even cause injuries if a tool is misused.

Some designs solve the opening problem by adding perforations to the card so that the item can be pushed out through it. Other solutions involve leaving parts of the blister edge unsealed so that it can be more easily peeled off. Besides the ordinary heatsealed blister, there are sliding blisters in which the blister covers the whole card and wraps around its edges, forming a groove into which the card can be inserted.

13.3.2 Skin Packaging

The plastic skin, in *skin packaging* is molded directly over the product. Since the process does not use pre-formed blisters, it can be adapted to any shape of item. It is popular for short runs of products like hand tools, because a single roll of film can be used for an entire product line.

The film literally forms a skin over the item as it lies on the card of paperboard or corrugated board. The film is heated and then draped over the item as a vacuum is pulled beneath the board. The film is sucked onto the item, encapsulating it against the board that has been coated with a heat-sealable primer.

Skin packaging films are usually LDPE or an ionomer. LDPE is economical, and a thick grade is used (7.5–15 mil, 191–381μm), along with corrugated board, for large heavy items like tools and other industrial products. But LDPE is not as clear, tough, or as easily formed as the ionomer that is used for finer detail, smaller, or sharper items (an application that was formerly filled by PVC, now largely discontinued). LDPE is less expensive but requires a longer heating time than ionomer, which can be more cost-effective in terms of productivity. The film is usually corona-treated on one side to make it adhere better to the board.

Skin packaging board is usually heavier than blister package board because of the tendency to warp under the high heat. Corrugated fiberboard is used for heavy products. For lighter-weight products SBS (20–28 pts, 510–710 μm) or white-lined recycled board (28–54 pts, 710–1370 μm) is used.

The board must be porous so that a vacuum can be pulled through it, and sometimes the board is even perforated throughout. Clay coating is not generally used because it reduces porosity, and if it is used, it must be even more heavily perforated, which limits the graphic capability. An aqueous EVA-based primer improves the board's heat sealability by soaking into the board fibers and porous inks to maximize the mechanical and chemical adhesion between the film, ink, and board. The ink needs to be heat-resistant and compatible with the primer.

Skin-packaging equipment ranges from inexpensive manual models, which run at speeds of 2–3 cycles/minute, to automated machines that can run up to 12 cycles/minute. High-speed machines have automated card infeed, product placement, quick change-over, adjustable vacuum, and hydraulic die-cutter. The number of packages produced, however, is higher since a number of products are usually packed on a single board. The standard size platforms range from the smallest 18″ × 24″ (450 × 600 mm), which yields 10–12 packages, to the largest size of 36″ × 96″ (900 × 2,430 mm).

Radiant heat is used to soften the film that is then lowered onto the carded product as the vacuum is applied. There are two types of vacuum systems. The turbine vacuum is the least expensive type, used for manual models. The positive displacement vacuum, with a pump and surge tank that can pull the air from between the film and board rapidly, is used in higher-speed machines. Both types of vacuum may be used together, for high-profile and odd-shaped products, and for board with poor porosity. The packages are cut apart, and hang holes added, with a rotary or hydraulic die cutter.

Skin packaging is popular for tools and hardware. New machines even have magnetic plates, so that steel parts like fasteners and ball bearings can be reliably placed on the board.

13.4 MOLDED FORMS

While most paper-based packaging starts with flat sheets that are folded, two forms are molded: pressed trays and molded pulp. Pressing is limited to simple relatively flat geometries; molded pulp can have much more complex shapes.

13.4.1 Pressed, Ovenable Paperboard Trays

Plates and trays for ready-to-eat meals can be made by pressing and deep-drawing a sheet of paperboard. The normal depth is about 1 inch (25 mm), and a second drawing stage can extend it to 2 inches (50 mm). SBS is used for its strength, cleanliness, and printability.

The trays, bowls, and plates can be made waterproof, with an ability to withstand freezing to cooking temperatures of 400°F (204°C) for up to an hour, and the ability to perform through up to eight freeze/thaw cycles (during distribution and in the home) by the addition of an extrusion-coating of PET. For products to be baked in the tray, a coating of TPX, 4-methyl-pentane-1 copolymer, is used because it can withstand higher temperatures and prevents sticking to caramelized sugar. Other plastics, like LDPE or PP are only used for applications where the cooking heat is less than 215° F (102°C) for LDPE and 260°F (127°C) for PP, to prevent the plastic from melting into the food (Huss 2009).

The blank is softened with moisture to a level of 8–11%, and then drawn into a heated shaping cavity by a plunger die. A circular shape, or at least rounded corners, is essential to make the geometry work. Since the paperboard does not stretch sufficiently, this type of forming produces creases in the sidewalls. The limit for how deep paperboard can be drawn depends, in part, on how large the piece is. When expressed as the forming height/base diameter, a limit of 0.5 has been proposed, but in industrial applications, 0.3 is usually not exceeded because above that the quality is poor. Research shows that the process can be modified, by controlling the blank holder force and optimizing temperature and moisture, to achieve a limit of 0.63 (Hauptmann and Majschak 2011).

Ovenable paperboard trays are a good choice for many ready-prepared meals since the food processor can cook in the tray that goes directly into the freezer until it is reheated or cooked by the consumer in a conventional or microwave oven. This streamlines food production and eliminates the need for cooking and serving dishes. Trays can be lidded with heat-sealed plastic film (like biaxially oriented PET) or coated paperboard, and a metalized microwave susceptor can be incorporated in key locations to induce local browning. The clay-coated, printed paperboard looks nice when served on the table, and the plate form looks at home for dinner.

Ovenable paperboard (clay-coated SBS, extrusion coated with PET) can also be formed into cartons, including tapered style with attached lid. Likewise, molded pulp trays and plates can be coated with PET to withstand oven temperatures up to 400°F (204°C).

13.4.2 Molded Fiber

Molded fiber, also known as *molded pulp* (or in the United Kingdom, *moulded pulp* and *moulded fibre*) is formed directly from a slurry of fibers to its final three-dimensional shape. As in papermaking, the porous, shaped mold is a fine mesh screen through which the water is pulled from the slurry with the help of a vacuum. With the mold submerged in the slurry vat, the fibers are drawn onto the screen, forming a layer of fibers in the shape of the mold.

Molds are made from aluminum, bronze, or plastic. New developments in molds include fused deposit manufacturing (FDM), which eliminates the use of screens, and multi-axis machines for forming the molds, which reduce the amount of labor required for complex shapes.

Most molding machines are fitted with multiple molds to increase productivity and provide the ability to produce a variety of items. One of the economic design parameters is fitting several forms in the footprint of the mold to minimizing waste; like folding cartons and labels, careful planning of dimensions can enable more to be produced in a single pass.

Molded fiber packaging is made almost exclusively from recycled waste paper. The properties are related to the fiber mix. In Europe and North America, the kraft fibers from OCC are used to produce stronger items; short ONP fibers are used to produce a softer, more resilient material; and pulp made from used office paper is bleached to form very smooth, sanitary forms. By blending fibers that have been subject to different types of processing and refining, the strength can be enhanced. The fiber blend affects the drainage, internal bond strength, shrinkage, and biodegradability. In Asia, waste paper is used, along with bagasse, bamboo, palm, and reeds.

Molded fiber packaging products, because of their three-dimensional capabilities, can be used to make a variety of packaging forms to provide for support, spacing, organizing, and cushioning. It is frequently used as an alternative to EPS because of its environmental benefits and sustainability.

There are four principal markets for molded fiber packaging: dunnage/protective packaging, food, horticulture, and disposable vessels. Dunnage/protective packaging is used for manufactured goods, like electronic equipment, furniture, appliances, assembly parts, etc. Cushioning properties are more difficult to predict than with EPS foam, as they depend on the fiber mix as well as the geometry (Ma *et al.* 2004).

Food-related packaging examples include plates, trays, berry punnets, takeout clamshells, and egg packaging. There are some unique benefits for the horticultural industry because of its biodegradability and compostability. For example, plant pots made from a peat moss slurry with fertilizer additives can be buried with the plant, and seeds can be added to molded pulp, encouraging consumers to plant their used packages. As molding technology improves, new markets are emerging, like bag-in-shell bottles and jars. One of the most unusual applications is urinal bottles and other hospital disposables.

The International Molded Fiber Association (IMFA, www.imfa.org) represents the industry to promote its use and provide technical and other information. They have developed the taxonomy of forms discussed in the following sections. There are three categories of molded fiber processes: plain, transfer, and precision molding and a fourth category for special treatments.

13.4.2.1 Plain Molding

In plain molding (IMFA Type 1) the screened mold descends into the vat filled with the hot, watery slurry of fibers, usually a mixture of OCC and ONP, with a fiber consistency of 0.8–2%. While in the vat, a vacuum is applied, sucking the slurry into the porous screened mold, removing most of the water and depositing fiber on the screen. The mold is then lifted from the vat and the fiber mat is blown off from the mold onto a conveyor-fed oven, or placed in a stationary oven, and the still-soggy part is free-dried at 400° F (205°C) for 6–12 minutes depending on the size and thickness. In some cases the parts are dried outdoors. Drying bonds the fibers together to form a rigid material.

Type 1 produces the thickest walls, three-sixteenths inch to one-half inch (4.8–12.7 mm). It is sometimes called *slush molding*. The single-pass mold makes a product that is very rough on one side, and moderately smooth on the other. The mold depth ranges from 1″ to 20″ (25.4–508 mm), with typical size ranges from 3″ × 3″ × 1″ to 30″ × 40″ × 20″ (7.6 × 7.6 × 2.5 cm to 76 × 102 × 51 cm). These thick-wall parts are typically used as packaging to support and protect heavy items such as vehicle parts, wine bottles, furniture edges and corners, and plants.

13.4.2.2 Transfer Molding

Transfer molding (IMFA Type 2) is the most common process. It is widely used to make egg cartons and filler flats, beverage cup carriers, hospital disposables, fruit punnets, and protective packaging for electronic equipment and appliances. The raw material is usually ONP. If needed, additives are used to impart moisture-resistance or aid in the production process. Type 2 forms have a smoother appearance than Type 1.

Type 2 items are formed on a multi-stage automated machine. Neither the slurry nor the molds are heated. Two wire molds are used for each part. The fibers are first sucked onto the forming mold. Then the second mold mates with the forming mold and uses a vacuum to transfer the wet form onto a conveyor through the hot oven, 400°F (205°C), for drying. The typical forming cycle can vary from a few seconds to one minute.

The matched-mold transfer process presses the fiber mat, imparting a relatively smooth surface to both sides (although one side is slightly rougher). It adds definition and dimensional stability, but limits the wall thickness to between 0.07″ and 0.13″ (1.78 to 3.3 mm). Product sizes are typically 4″ × 4″ × 3″ to 15″ × 20″ × 4″ (10 × 10 × 7.6 cm to 25 × 25 × 20 cm) with most food services items three-quarters inch to 2 inches (2–5 cm) deep.

Transfer molding is used to make endcaps and trays for electronic equipment and for other small household and hardware items, as shown in Figure 13.7. The molded parts block, brace, and/or separate items in a corrugated box, and can serve as a lower-cost substitute for molded expanded polystyrene (EPS) cushioning. Molded fiber trays can be used to arrange items for ease of packing.

The design parameters are a little different than for EPS, due to different properties and manufacturing processes. Molded fiber is heavier than EPS, which can result in higher shipping cost, so shapes must be designed for stiffness and to minimize material usage. Protectiveness depends on the thickness, cavity shape, and ribs which must be designed to support the product as well as support the walls of the box in which it is packed (and if the form does not support the walls, the corrugated board grade may need to be increased). The rough side may be in or out, depending on the protection and appearance required. Dynamic and compression performance testing of filled packages is required to ensure that protection is not compromised. Advantages over EPS include the ability to nest to minimize the space required to ship them to the point of use (generally three

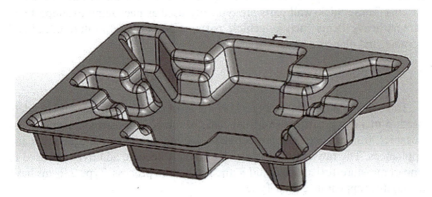

FIGURE 13.7. *Molded fiber tray. Used by permission from the International Molded Fiber Association.*

or more forms occupy the same cube as one EPS cushion), as well as the environmental benefits: the use of recycled fiber, recyclability, biodegradability, and economy. Molded pulp packaging can impart a sustainable, high-touch organic image to high-tech consumer electronic products.

13.4.2.3 Precision Molding

In precision molding (IMFA Type 3) the final drying is done in a heated mold, rather than an oven. The molded product is captured in matched, heated, vacuum-assisted forming molds where it is pressed and dried in a multi-stage process. The result can resemble the appearance of thermoformed plastic. In contrast to plain molding, precision-molded parts are fully dry when they leave the mold due to the heat applied to the matching mold surfaces. Machines can be rotary or reciprocating action. Precision molding is more costly than plain molding, because it takes more time and energy, and the machinery is more complex.

Precision molded parts are denser, smoother, and more exactly dimensioned and contoured than the free-dried forms. The process is used for items where smoothness, appearance and/or accurate positioning is desired, such as dinner plates, insert/trays with fine detail or compartments for small items. Walls are usually about 0.05″ (1.3 mm) thick, although some Asian processes produce walls as thin as 0.032″ (0.8 mm) Cavities are normally no deeper than 3″ (7.6 cm).

The raw material varies with the location of the manufacturing facility; but most common are recycled paper, bagasse, reed, wheat straw, and palm fiber. Food service products like plates, bowls, and trays are made from virgin fiber or specially treated recycled paper that conforms to FDA food contact requirements, to avoid contaminants commonly found in recycled pulp. Very smooth trays for home medical products like thermometers are made from recycled office paper.

A hybrid precision/transfer molding process is used to produce molded fiber bottles and jars. Bag-in-shell packaging has been developed for granular products like protein powder and liquid products like soap or even milk. The shell, made from 70% OCC and 30% ONP, encloses an inner LDPE pouch with an integrated fitment and locking collar using less than half the plastic of an equivalent plastic jug or canister. The pulp and plastic are easily separated for recycling, or the pulp can be composted. The molded jars offer a marketing advantage by presenting an environmentally friendly image (Mohan 2013).

13.4.2.4 Special Molded Effects

IMFA Type 4 is the category for special fiber blends, additives, and items with secondary treatments. Most manufacturers use no additives, or limited environmentally-acceptable additives, in order to emphasize molded fiber's sustainability advantage

But additives can be used to modify properties. Colloidal rosin and wax emulsions provide water repellency. Fluorocarbon chemicals provide grease and oil resistance. Dye, bleach, flame retardant, fertilizer, starch, or wet-strength resins can be added. Thermosetting polymers can even be added to make non-disposable items like automobile door liners and trunk wells

Secondary treatments include printing, embossing, die-cutting, hot pressing, wet pressing, plastic coating, and lamination. For example, egg cartons can be smoothed and embossed with a hot after-press that enhances printability. Like pressed plates, but thicker and stiffer, molded plates and trays can be laminated with a PET film to make them dual-ovenable, capable of withstanding a conventional oven temperature of 400°F (204°C), while still retaining stiffness and structural characteristics.

The four IMFA categories greatly simplify the types of molded fiber products and manufacturing processes. There are hundreds of different molding machines and processes worldwide using varying molding techniques and fibers, producing a wide variety of molded fiber products, from almost paper-thin plates, to large molded fiber tubs and pallets. In China for example, single station machines are used to produce millions of molded fiber products one at a time. This is in contrast to a large rotary machine in the United States producing millions of products 50 at a time. Although the basic process of forming a 3D structure using natural fiber and water appears simple, the variations in equipment and materials present intriguing potential for future advancements and packaging design.

13.5 REFERENCES

Cullen Packaging. 2013. Moulded Pulp Packaging. In *Handbook of Paper and Paperboard Packaging Technology,* 2nd ed., edited by Mark J. Kirwin, pp 385–392. Chichester: John Wiley & Sons.

Eubanks, M.B. 2009. Cans, Composite. In *The Wiley Encyclopedia of Packaging Technology,* 3rd ed., edited by Kit L. Yam, pp. 195–199, New York: John Wiley & Sons.

Fayoux, S.C., A.M. Seuvre, and A.J. Voilley. 1997. Aroma Transfers in and Through Plastic Packagings: Orange Juice and d-Limonene. A Review. Part I: Orange Juice Aroma Sorption and Part II: Overall Sorption Mechanisms and Parameters—a Literature Survey, *Packaging Technology and Science,* vol 10, no 2: 69–82, and no 3: 145–160.

Fibrestar Drums Ltd. 2013. Fibre Drums. In *Handbook of Paper and Paperboard Packaging Technology,* 2nd ed., edited by Mark J. Kirwan, pp 205-216. Chichester: John Wiley & Sons.

Gordon, G.A. 2009. Drums, Fiber. In The Wiley Encyclopedia of Packaging Technology, 3rd ed., edited by Kit L. Yam, pp 368–373. New York: John Wiley & Sons.Hanlon, Joseph F., Robert J. Kelsey and Hallie E. Forcinio. 1998. *Handbook of Package Engineering,* 3rd ed. Lancaster, PA: Technomic.

Carded Packaging. In *The Wiley Encyclopedia of Packaging Technology,* 2nd ed., edited by Aaron L. Brody and Kenneth S. Marsh, pages 161–166. New York: John Wiley & Sons.

Hauptmann, M. and J.P. Majschak. 2011, New Quality Level of Packaging Components from Paperboard through Technology Improvement in 3D Forming. *Packaging Technology and Science,* 24 (7): 419–432.

Henderson, C.R. 2013. Composite Cans. In *Handbook of Paper and Paperboard Packaging Technology,* 2nd ed., edited by Mark J. Kirwan, pp 183–204. Chichester: John Wiley & Sons.

Huss, G.J. 2009. Microwavable Packaging and Dual-Ovenable Materials. In *The Wiley Encyclopedia of Packaging Technology,* 3rd ed., edited by Kit L. Yam, pp 756–759. New York: John Wiley & Sons.

Impact Marketing Consultants. 2006. *The Marketing Guide to the U.S. Packaging Industry.* Manchester Center, VT: Impact Marketing Consultants.

International Molded Fiber Association. www.imfa.org

Kirwan, M.J. 2013. Paperboard-based Liquid Packaging. In *Handbook of Paper and Paperboard Packaging Technology,* 2nd ed., edited by Mark J. Kirwan, pp 353–384. Chichester: John Wiley & Sons.

Kwapong, O.Y and J.H. Hotchkiss. 1987. Comparative Sorption of Aromatic Flavors by Plastics used in Food Packaging, *Journal of Food Science* 52: 761–763, 785

Lisiecki, R.E. 2009. Cartons, Gabletop. In *The Wiley Encyclopedia of Packaging Technology,* 3rd ed., edited by Kit L. Yam, pp 241–243. New York: John Wiley & Sons.

Ma, X., A.K. Soh, and B. Wang. 2004. A Design Database for Moulded Pulp Packaging Structure, *Packaging Technology and Science,* 17 (4): 193–204.

Mohan, A.M.. 2013. Molded-pulp Bottle is Perfect for Protein Powder Line, *Packaging World,* vol. 20, no. 2, 30–33.

Paperboard Packaging Alliance. 2002 *Paperboard Packaging Syllabus,* prepared by John Latham. Washington, DC: PPA.

Robertson, G.L. 2013. *Food Packaging: Principles and Practice,* 3rd ed. Boca Raton, FL: CRC.

Soares, N.F.F. and J.H. Hotchkiss. 1998. Bitterness Reduction in Grapefruit Juice Through Active Packaging. *Packaging Technology and Science,* vol 11, no. 1, 9–18.

Sonoco. 2013. E-mail correspondence from Laura Rowell to Diana Twede, May 10, 2013.

Soroka, W. 2009. *Fundamentals of Packaging Technology,* 4th ed. Herndon, VA: Institute of Packaging Professionals.

Spottiswode, B. 2009. Skin Packaging. In *The Wiley Encyclopedia of Packaging Technology,* 3rd ed., edited by Kit L. Yam, pp 1111–1115. New York: John Wiley & Sons.

Twede, D., R. Clarke, and J. Tait. 2000. Packaging Postponement: A Global Packaging Strategy, *Packaging Technology and Science,* vol 13, no 3, 105–115.

von Falkenstein, E., F. Wellenreuther, and A. Detzel. 2010. LCA Studies Comparing Beverage Cartons and Alternative Packaging: Can Overall Conclusions be Drawn? *International Journal of Life Cycle Assessment,* Springer-Verlag 15:938–945.

Waldman, E.H. 2009. Pulp, Molded. In *The Wiley Encyclopedia of Packaging Technology,* 3rd ed., edited by Kit L. Yam, pp. 1044–1046. New York: John Wiley & Sons.

13.6 REVIEW QUESTIONS

1. What kind of paperboard is used to make sanitary food packages? Which plastics are laminated or extrusion coated onto paperboard to make it water-tight?

2. What is meant by the fact that milk carton and drink box materials and their machines are "tied goods?"

3. What is the name of the Swedish company that developed the drink box?

4. Explain the aseptic packaging concept.

5. From what three materials are most aseptic cartons made? What is the purpose of each? How is the paperboard sterilized before it is filled?

6. Compared to cans and bottles for beverages, what are the advantages of aseptic cartons?

7. Under what conditions are juice boxes recyclable?

8. By what printing process are most aseptic cartons, milk cartons, and composite cans printed?

9. Which contributes more to the strength of a paper tube: thickness or shape?

10. Which manufacturing process is most common for composite can bodies?

11. What is a mandrel? Is it used for making convolute or spiral wound cans?

12. How are composite can dimensions specified in the United States?

13. Are composite cans recyclable? In which material stream?

14. What are some uses for fiber drums?

15. How can postponing the packaging of pens on blister cards, as part of a global marketing strategy, reduce transportation costs?

16. What is the difference between a blister package and a skin package?

17. What plastic is most commonly used to form the blister in blister packaging? What other plastics are used?

18. What plastic film is most often used for skin packaging? Why are clay-coated boards not used in skin packaging, and what disadvantage arises from this?

19. What is the most common consumer complaint about carded packaging?

20. What kind of fiber is used to make most molded pulp packaging in the United States and Europe? What other kinds of fiber can be used?

21. What are the principal markets for molded fiber packaging? Give examples.

CHAPTER 14

Corrugated Fiberboard

Corrugated fiberboard is arguably the most important packaging material used today. It has the highest volume of any single packaging material in the world. It is also called *corrugated board* (and sometimes just *corrugated* in informal industry jargon).

The name *cardboard* is also used, but mostly by the public. The paper and packaging industries do not use the term "cardboard" because it is too general and vague a term that is also used to designate much lighter paperboard grades including those used for greeting cards.

Almost two-thirds (63%) of the value and 78% of the tonnage of all paperboard packaging used in the United States is in the form of corrugated fiberboard shipping containers. The remainder are the cartons and other packages discussed in Chapters 12 and 13 (Impact Marketing Consultants 2006, U.S. EPA 2014). In 2011, the value of the U.S. corrugated fiberboard production was estimated by the industry to be $26.1 billion, a lower number than the census figures given in Chapter 1 in part because of reselling (FBA 2012).

Corrugated fiberboard shipping containers, also called *corrugated boxes* or *corrugated cases*, are used for almost all goods shipped in the United States. For this reason, the economics of the industry are closely linked to the performance of the overall economy. This is not the case in some other countries that may have different transport packaging practices or where a higher percentage of shipping containers are made from other materials. For example, in European retail products packaged in strong primary packs (such as steel cans) are collated using shrink-wrap only, although a strong trend to retail-ready collations favors the use of corrugated.

The annual U.S. production is almost 359 billion square feet, over 33 billion square meters (FBA 2012). This is enough to cover the entire state of Maryland. The containerboard materials, called *linerboard* and *medium*, comprise the highest volume single grade of paper made. It is also the most highly recycled packaging material.

There are three basic operations involved in making corrugated fiberboard boxes:

- Making the containerboard materials (liner and medium)
- Combining the liner and medium in a corrugator to make corrugated board
- Converting: printing, scoring, slotting, and gluing the boxes.
- These operations are often done in separate factories.

Containerboard is predominantly made by large integrated paper-producing companies. It is a commodity that they either convert themselves, trade with one another, or sell on the open market. Transacted prices are reported in *PPI Pulp and Paper Week*, the same publication that reports other paper and paperboard market prices.

The large *integrated* producers do it all—they make the containerboard, combine it into corrugated board, and cut out and print the boxes. At least half of their paper mill production is used to supply their own converting factories. In the United States, about 75% of corrugated box shipments are from vertically integrated producers (FBA 2012). Some also own forests, although this is becoming less common. They are big, with names like International Paper, Rock-Tenn, and Packaging Corporation of America. In 2005, the top five corrugated board and box suppliers in the United States controlled 62% of an industry that is growing increasingly concentrated, compared to 43% in 1996, 36% in 1989, and 29% in 1984 (Impact Marketing Consultants 2006). The integrated suppliers also sell containerboard and combined board on the open market.

Independent converters buy their raw materials—containerboard or combined board—from the integrated suppliers. The independent converters combine the board, make the boxes, or both. They compete fiercely with each other and with the integrated suppliers. This generally has the effect of neutralizing price collusion attempts on the part of the integrated suppliers (Goldberg *et al.* 1986). The independent converters have their own association, *The Association of Independent Corrugated Converters* (www.aiccbox.org).

There are three types of factories for combining and converting the board from paper. A *corrugator plant*, also called a *sheet feeder*, makes combined board exclusively to supply sheets to *box plants*, which are also called sheet plants (which can be a little confusing). A plant that does both is called a *corrugator/box plant*.

The corrugated fiberboard industry is represented in the U.S. by the *Fibre Box Association* (*FBA*) (www.fibrebox.org) and in Europe by FEFCO (The European Corrugated Packaging Association www.fefco.org). They sponsor programs and services responsive to the unique needs of each membership segment (independent sheet plants, independent corrugators, and integrated companies) and monitor and report on statistics, technology, safety/health, and the environment/recycling.

This chapter shows how the board is combined and discusses the properties and grades of corrugated fiberboard. Chapter 15 will describe how the board is converted into boxes, show common box styles, and provide an introduction to shipping container performance testing.

14.1 COMBINING CORRUGATED BOARD

The *corrugator* is a huge machine several hundred feet long. It corrugates the *medium* and glues it to the *liners*. Speeds can exceed 1,000 feet/minute (270 m/minute). Figure 14.1 shows the process in which the three rolls of paper and adhesive are combined.

14.1.1 Preconditioning, Preheating, and Corrugating

Most of the corrugating operation focuses on the medium. It bears the mechanical stress of corrugation and is the part to which the glue is applied. In the corrugating process, the liners are required only to accept the union with the medium.

First, the medium is heated and moisturized to soften it. It passes over a steam-filled drum called a *preconditioner* and/or through a steam shower. The fibers in the medium become soft and pliable due to the heat and moisture. If the medium is too dry it will not form properly or may fracture or break. The high-pressure steam gives off heat to the paper as it condenses; steam is used because it is an easily generated, distributed, and controlled source of uniform heat and moisture. Variations in the heating and moisturizing of the medium would cause sheet instability and distortions such as warp which are highly undesirable.

The medium is then fed between the nip of two long metal rolls with alternating ridges and grooves shaped like the flutes. The grooves mesh together like gears, forming the flutes in between them. Because of the wavy shape, it takes nearly 150 lineal feet of flat paper to make 100 ft of C-flute corrugated medium. The exact amount depends on the flute profile. The ratio of the length of uncorrugated paper to the length of the corrugated medium is called the *takeup* factor, and ranges from 1.15 to 1.54, depending on the size of the flutes.

At the same time, the liners are preheated to prepare them for bonding, adding energy to begin the gelatinization of the adhesive. Both liners are heated to balance the moisture content between the two; if one liner has a higher initial moisture content, it will shrink more when it dries and cause the board to warp.

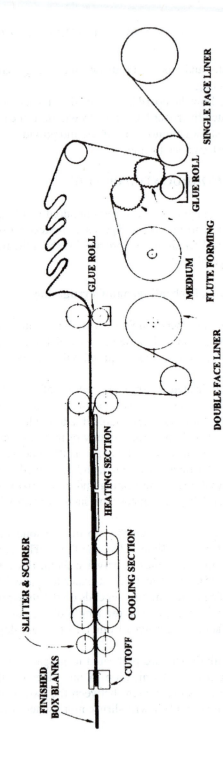

FINISHED
BOX BLANKS

SLITTER & SCORER

GLUE ROLL

CUTOFF

COOLING SECTION

HEATING SECTION

DOUBLE FACE LINER

MEDIUM

FLUTE FORMING

GLUE ROLL

SINGLE FACE LINER

FIGURE 14.1. Double-face corrugator machine. Reprinted with permission of Angus Wilde Publications and Gary A. Smook, copyright 2002.

450

14.1.2 Single-Facing Operation

Next, adhesive is applied to the peaks on one side of the corrugated medium. As the fluted medium exits the nip of the corrugating rolls, the flute tips pass through a film of adhesive carried on an applicator roll. Each flute tip picks up a narrow bead of adhesive.

The starch-based adhesive first wets the surface of the medium and is partly absorbed as the starch adheres to the cellulose in the paper. A pressure roll secures the bond between the liner and the glue-tipped corrugated medium to form the single-face web. The web remains in the nip of the pressure roll and bottom corrugating roll for about 2 milliseconds. In that short time, the adhesive must create enough adhesion, using pressure and heat, to hold the bond until it is cured. The strength of the bond depends on the adhesive formulation, how it is applied, and temperature. The heat gelatinizes the starch, setting the glue.

The face that is applied first in the single-facing operation can be easily distinguished from the second face because it has more pronounced glue lines due to the greater pressure used. Because of these lines, this face is usually designed to be inside the box. Too much pressure in the single-facing operation can cause the medium to be embedded into the liner, resulting in a loss of caliper (thickness). It can cut into the liner, reducing the board strength. To overcome these problems, newer corrugators replace the pressure roller with a belt to increase the amount of time for bonding and decrease the pressure.

Heat is then used to dry the board and cure the bond. The web is accumulated in large festooned loops, giving it an extra 20–30 seconds to cure before it is joined to the second face. The festooning also provides a buffer between the single facer and double facer sections of the machine, allowing machine changes and reel changes without machine stops.

14.1.3 Double-Facer and Dry End Slitting and Scoring Processes

Next, the *double-facer* applies glue to the exposed flute tips on the single-face web by a *doctor roll* from an *applicator roll*. Then, the single-face web is joined to the second liner.

The pressure applied in the double-facer is less than that in the single-facer. This results in the second face having a more uniform appearance with less-pronounced glue lines. It will be the side that is ultimately printed and will be on the outside of the box. When a pre-printed liner (a liner that is printed before the board is assembled) is used, it is also the second face applied.

Pressure rolls and heat from steam chests are employed to finish curing the glue. The corrugator can have additional equipment for adding wax, color, or other treatments. A second corrugated medium and third liner are added (as a further single-faced material) for double-wall board, and triple-wall has yet another single-faced ply.

Once it is cooled, the web is slit to the desired width and cut into sheets. The sheets are ejected and accumulated in a stack at the end. If the board is to be used to make RSCs and other slotted containers, generally the board is slit on the corrugator to the width needed for the height dimension of the box blanks, and two scores are made parallel to the machine direction, across the flutes. These long scores will form the top and bottom edges of the RSC. The board is cut to the length of the box blank or slightly longer to be trimmed later.

It is important to note the reason why flute direction runs perpendicular to the machine direction of the paper: this is necessary for the mechanics of the production process, since the paper runs through the paper-making machine in the same direction that it does on the corrugator. This is not the best combination for adding compression strength to the combined board, and this is the reason why linerboard manufacturers orient as many fibers as they can in the cross-machine direction.

14.1.4 Coatings and Surface Treatments

Most functional coatings are applied on the corrugator. But they can also be applied to the liner and/or medium alone, to the combined board, at the box plant or on the flexo folder-gluer, depending on the function required and the coating's properties.

Some of the most common purposes are to improve the board's resistance to moisture and oil. Water is the biggest enemy of corrugated board. Moisture weakens the paper, the board structure and, to a lesser extent, the box joints and closures.

For products such as fresh fruit and vegetables where there is a high potential for water damage, the paper can be coated and/or impregnated with wax or plastic, and the board can be combined with water-resistant adhesive.

Wax is the most popular water-resistance treatment. It is widely used on boxes for fresh fruit, vegetables, and some meat intended for further processing. Fresh produce is shipped cold and sometimes wet (certain types are even iced), and the moisture needs to stay in the product. Wax provides a good barrier and prevents wicking. It can be impregnated or coated.

Wax impregnation of the containerboard during the corrugating operation imparts a dry wax treatment. The penetration of wax must be carefully controlled so as not to inhibit corrugating and gluing; at most 18% wax can be added in this way. The wax soaks into the board and is not visible, although it often darkens the paper by filling void spaces. It can be applied to one or all components of the board; for example, some double-wall board used for fresh fruit and vegetables has a wax treatment on only the inner liner and first medium.

Curtain coating is applied to flat blanks by passing the sheet horizontally through a thin continuous curtain of polymer modified wax, which forms a film over the entire surface of the board. Wax saturation, or cascading, is applied to a box or blank passing vertically through multiple cascades of paraffin based wax with the objective of saturating the combined board, throughout (all surfaces of liner and medium), adding from 45% to 60% to the board weight. These coatings are visible on the surface of the board.

Wax helps the board to retain compression strength under short-term humid conditions. Board that has been curtain-coated retains more strength than dry wax impregnation alone. Cascaded wax provides the most water resistance of all.

But some wax-coated board is not recyclable because the wax is difficult to remove from the pulp. The wax formulations are blends of microcrystalline waxes and resins (typically polyethylene) with higher melting temperatures than paraffin. Extraction is an expensive process, typically not worth doing.

Although wax adds water resistance for short-duration uses, it does not necessarily improve long-term compression strength as much as does wet-strength containerboard and water-resistant adhesives. For longer-term wet storage, thermosetting resins can be added to the containerboard during manufacture of the paper and then cured. However, use of such wet-strength paper is limited due to embrittlement, cost, and recycling problems.

As discussed in Chapter 9, there have been advancements in water-resistant adhesives and treatments that do not cause problems in the recycling process.

There are also coatings to improve abrasion resistance (reduce scuffing), reduce slipperiness (increase coefficient of friction), or reduce or eliminate sticking of a tacky product to the interior surface (release agents). Coatings can increase the board's ability to dissipate static electricity to protect sensitive electronic products. Coating with volatile corrosion inhibitor, or vapor corrosion inhibitor (VCI), can be used to protect packaged metal objects from corrosion. Corrugated board can even be made flame retardant, raising the temperature at which it might catch fire and retarding the spread of flame.

14.2 CORRUGATED BOARD STRUCTURE

The structure of corrugated fiberboard is shown in Figure 14.2. Its two faces and corrugated medium use the physical properties of paper in an extremely effective way.

The structure is an adaption of the *engineering beam principle* of two flat load-bearing panels separated by a supporting structure, in this case by a rigid corrugated web. The structure resists bending and pressure from all directions. When the board bows in one direction (under compression in a stack, for ex-

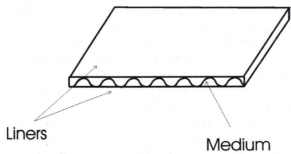

Liners   Medium

FIGURE 14.2. Corrugated fiberboard structure.

ample), the strength of the linerboard on the outside of the bow resists stretching and so resists the bend.

Box flutes normally run from the top to the bottom of a box, acting like columns to support the stack above. The flutes also act like arches resisting flat crush. They provide some cushioning, and they trap air, giving some insulation properties.

14.2.1 Corrugated Board Materials

Linerboard and medium, also called *containerboard*, is traditionally referred to by its basis weight (per 1,000 ft^2 or gsm) although independent corrugators may specify containerboard by its ring crush resistance.

14.2.1.1 Linerboard

Linerboard, the flat facing material, is usually made from unbleached kraft because of the need for strength. Kraft linerboard is made by the sulfate pulping of softwood, yielding long fibers. Although linerboard in North America is made from some recycled fiber, it usually has a percentage of virgin fiber to give it good tensile strength. In the United States it can still be classified as "virgin" with up to 20% recycled fibers (McKinney 1995). In 2011, 75% of the fiber for linerboard was kraft and 25% was recycled (FBA 2012). In Europe about 30% of linerboard is Kraft (FEFCO 2012).

Recycled fiber in linerboard is more common in other countries, especially those with few forest resources and high recycling rates. "Test" grade liners are widely used in Europe—the term "test" is used to indicate a liner made to a known quality of performance from sorted waste, though it has no strict definition. Mills often supply different grades of Test liner. Recycled liner is commonly dyed brown so that it looks as expected.

It is common to make linerboard on a *duplex fourdrinier* machine (or one with multiple headboxes), as described in Section 7.1.2. A dark, coarse high-yield kraft base sheet is first formed on the fourdrinier wire, and a lighter, cleaner, top ply sheet is formed on top of it to provide a better printing surface. The wire is vibrated during the forming operation to increase orientation of the fibers in the cross-machine direction as much as possible, in order to improve the compression strength of the ultimate combined board (flutes run in the cross-machine direction, and this is the most common stacking orientation for boxes).

When a white substrate is desired for printing higher quality graphics, *bleached kraft* is used. For the best quality, a clay-coated, calendered surface (*C1S bleached kraft*) is needed. In some food packaging applications that use the small E, F, and N flutes, a higher quality sanitary appearance is desired, and the board is made entirely of bleached kraft.

For most applications, however, the added expense cannot be justified. *White-top linerboard* (or the lower grade *mottled white*) is a lower cost alternative, a natural kraft sheet with a bleached top layer (or one made from recycled white grades of fiber).

For many years, the most common basis weight for linerboard in the United States was 42 lb (per 1,000 ft^2), or 205 g/m^2, typically about 16 points thick, but it is now available in a variety of basis weights, ranging from 26 lb to 90 lb (125–440 g/m^2). In Europe linerboard weights are typically lower. *Performance* or *test linerboard*, with a higher percentage of recycled content, more CD fiber orientation, and some strengthening additives, is increasingly used to provide greater strength at lower basis weights. Usually the two faces are the same basis weight, in order to avoid warping.

14.2.1.2 Medium

The corrugated *medium* is also unbleached paper made by the fourdrinier process. But it is made from a different kind of pulp. Medium is typically made from recycled corrugated board, or from virgin hardwood processed in neutral sulphite semi-chemical (NSSC) pulping. In the United States in 2011, 54% was NSSC with the balance being recycled stock (FBA 2012). In Europe most medium is recycled with NSSC being below 15% (FEFCO).

Paper made from this pulp would not satisfy the demands of most other kinds of packaging applications because of its low tear and tensile strength, but it offers several advantages in the corrugating operation. Hardwood and recycled pulp have shorter fibers so the undulations can be more easily formed in the machine direction without breaking the paper. The thicker walls of hardwood fibers give high stiffness and crush resistance. Furthermore, the semi-chemical pulping method gives a higher yield than the kraft process, so the cost per pound of pulp is lower.

Medium performs better with a higher percentage of recycled fibers than does linerboard. Recycling shortens fibers, which adds to the medium's formability (but reduces its tensile strength). In the United States, the most common basis weight for corrugated medium is 26 lb, typically about 9 points thick. The medium gives strength to the structure because of its curvature and columnar shape, not its thickness.

Table 14.1 gives some common basis weights (in the United States, per 1,000 ft²), and shows roughly equivalent common grammage grades in other parts of the world. The list is by no means exhaustive, and other BW/grammage containerboard is available.

14.2.1.3 Starch-Based Adhesive

Starch-based adhesive is most common because of its low cost, pumpability, and ability to create a bond at high speed between the liner and medium (because of heat gelatinization). A small portion of cooked starch acts as a carrier to keep the raw starch in suspension. Sodium hydroxide (caustic soda) and borax are used to improve adhesive properties.

The *gel point* is the temperature at which the raw starch granules begin to absorb water and form a gel, about 140–147°F (60–64°C). If the critical gel point is not reached, the bond will not form, causing blisters and delamination. The raw starch draws water from the cooked carrier starch and sticks to the paper fibers as it loses water. Once the starch gels, it stops penetrating into the paper and bonds to the fibers.

The water-soluble starch-based adhesive is inexpensive, forms a good bond, and can be easily recycled or biodegraded. The disadvantages of starch are that it makes boxes vulnerable to insect infestation and mold, and can allow the board to delaminate when it gets wet.

TABLE 14.1. Some Common BW and Grammage for Linerboard and Medium (adapted from Steadman (2002).

Linerboard		Medium	
U.S. BW (1,000 ft.²)	Metric g/m²	U.S. BW (1,000 ft.²)	Metric g/m²
26	125	26	125
33	150	28	140
38	175	30	150
42	200	36	160
47	225	40	200
69	330	42	200

14.2.2 Corrugated Board Recycling and Reuse

Corrugated board, over its history, has almost always had some recycled content. Early corrugated linerboard, called *jute board*, was made almost exclusively from wastepaper and old jute (burlap) bags on a cylinder machine.[26] The medium was made from straw recycled from grain harvesting. In 1940, 40% of linerboard and 32% of medium was made from recycled fiber (FBA 2012). During World War II, there was a large build-up of virgin kraft capacity in the United States, and for many years thereafter most linerboard and medium were made from 100% virgin kraft fibers, at much lower basis weights than comparable jute board (Gates 1954).

Beginning in the 1980s, an increasing amount of corrugated board (known to the recycling industry as OCC) began to be collected in U.S. recycling programs, and the corrugated fiberboard industry developed an infrastructure and technical capability to incorporate it into linerboard and medium. Today, corrugated board contains an increasing percentage of recycled fiber. In 2011, 24% of linerboard and 46% of medium in the United States was made from OCC and cuttings from box plants (FBA 2012).

In other countries, the fiber sources may include straw, mixed paper, or newspaper. Corrugated board made from these fibers is much less strong and predictable than kraft-liner board. World-wide, the corrugated fiberboard industry enjoys a reputation for environmental friendliness, and most corrugated boxes now proudly display a symbol advertising their recyclability and use of recycled content. Figure 14.3 shows two of the industry's recycled logos.

Corrugated board is the most highly recycled paper and packaging material today for reasons relating to volume, logistics, and technology. Since it represents almost 80% of the paperboard packaging produced, it is discarded in the highest volumes. Since corrugated boxes are used to ship goods to retail stores, it is easy and economical to collect empty boxes in large quantities from retailers and restaurants (who are left with a great deal of it) because they benefit by avoiding disposal costs. Most communities recycle corrugated board, although it is not always included in curbside collection because of its bulkiness. The U.S. corrugated board industry has developed a successful infrastructure for collecting it, coordinating through a network of board and box plants dispersed around the country, and more than 90% is recovered, compared to only 25% of other paperboard packages (EPA 2014).

Of the used corrugated board collected in the United States in 2005, 63% was used to make containerboard, 17% was used to make paperboard, and 17% was exported (FBA 2005). OCC from the United States has good value in for-

---

[26]The name *jute board* persisted long after jute was no longer used in it, to refer simply to linerboard made from recycled stock reinforced with some kraft fibers.

FIGURE 14.3. Corrugated recycles.

eign markets that lack their own fiber resources because of its high percentage of kraft content.

In Europe, where legislation has increased overall recycling rates, an actual shift in the European fiber balance has occurred. In the early 1990s the use of lower grades of recycled fiber from mixed wastepaper and old newsprint for medium increased, and the amount of virgin fiber for linerboard decreased, spurring research into ways to improve the board's properties, processing, and supply chain (Stefan 1993).

14.2.2.1 Effect of Recycled Content

Each time the kraft fibers in linerboard are recycled, the tensile, burst, folding strength, and density are reduced. This is caused by hornification, a reduction in inter-fiber bonding and shorter fibers. Hornification, as discussed in Section 6.5.3, is an irreversible closure of small pores in the fiber wall that makes the fibers resist swelling during rewetting, reduces their external surface area, and makes them more stiff, brittle, and susceptible to breakage. Strength is also lost because kraft fibers get shorter and some hardwood fibers are incorporated from NSSC medium.

Compared to pure NSSC, medium made with OCC has higher burst strength, due to the incorporation of kraft fibers, but it has more tendency to crack on the corrugator.

The first recycle causes the largest drop in combined board's burst, ECT, and FCT, and in the compression strength of boxes made from it (BCT). The strength properties can be somewhat improved by additional refining, but this further breaks fibers and creates more fines, causing the sheet to be less *free*, which slows down the dewatering process and requires a reduction in the speed of the papermaking machine. Compared to virgin NK/NSSC/NK board, recycled content has been found to cause a greater reduction in burst strength than in ECT (Koning and Godshall 1975).

Recycled fibers also increase the tendency of corrugated board to creep under load, particularly during cyclic conditions. This can reduce the working life of a box. But additional refining and pressing can somewhat mitigate this effect (Byrd and Koning 1978; Coffin 2005).

Subsequent recycles cause less strength reduction. After five times, most physical properties have stabilized (Nazhad 2005; Hubbe *et al.* 2007). The performance of recycled-content board depends on the characteristics of the recycled fiber as well as on its proportion. Since there is no way to know how many times a given fiber has been recycled, a constant stream of virgin fiber is required in the overall paper supply in order to maintain the quality of kraft containerboard.

Corrugated board can be made from 100% recycled fibers, as it is in many parts of the world where virgin fiber is scarce. The materials are *called recycled linerboard and medium*, although they have also been called *test liner* (when it complies with a performance specification), *non-test liner*, or *bogus liner and medium*. It should be noted that statements about the percent of recycled fiber contained in a certain grade of container board often refer to a monthly average, rather than being a specification for a particular sheet.

Paper scientists have investigated the relationship between recycled content and paper performance since the 1970s. The increase in recycling in the 1990s accelerated related research on its effect on box performance. *Box lifetime* research continues to add to our fundamental understanding of how corrugated boxes fail.

14.2.2.2 Reusable Corrugated Boxes

Nobody ever talks about it, but everyone knows that corrugated boxes can be reused because we reuse them for moving and storing our personal possessions. As the corrugated packaging industry faces increasing competition from plastic reusable packages, now is a good time to consider opportunities for reusable corrugated boxes.

Commercially, corrugated boxes are seldom reused. They are not designed for reuse, and there are few reverse logistics systems for returning boxes to use again. Since box dimensions and weight-bearing requirements vary by product, they are not interchangeable. Developing a corrugated box reuse program requires overcoming these obstacles.

Frito-Lay® is a good example of a supply chain that does reuse corrugated boxes. Their direct-to-store delivery system enables the driver, who also stocks the store shelves, to bring the empty boxes back to the factory in a closed loop for reuse. They developed machinery to form and close boxes with a "fan fold" closure that facilitates knocking down the boxes to a flat state for return. They have standard box sizes that can be used interchangeably for several different products, and all of their products are lightweight so the compression strength requirement is low. Although the number of times that they can reuse a box

(5–10 times) is less than for a stronger plastic reusable package, the cost is much lower.

In another example, Michigan State University on-campus students who are moving out of dormitories pack their belongings in corrugated boxes that have previously been used to ship food and supplies to campus. The campus housing service knocks down and collects boxes in anticipation of redistributing them at the end of the semester. MSU Stores, which purchases new boxes for campus departments, is working with MSU Recycling to explore more ways to reuse boxes by analyzing supply and demand.

A research agenda for corrugated box reuse would include seeking ways to make a box last through more uses, developing appropriate test methods, investigating supply chain logistics options, and, most importantly, solving the problem of matching supply and demand for various, but specific, box sizes.

14.2.3 Flute Configurations

The thickness of corrugated board depends on the size of the flutes and the number of walls. In common terminology, *faces* are the number of linerboards and *walls* are the number of corrugated medium layers.

Single-faced corrugated board is used as a wrapping material for items like light bulbs, and is shipped in rolls. *Double-wall* and *triple-wall* board are used for heavy-duty applications such as intermediate bulk containers for products like resin pellets and car parts. The most common type is *single-wall*, also known as *double-faced*. These are shown in Figure 14.4.

Flutes come in several different sizes. The most popular size is *C-flute* single-wall board. Eighty percent of the boxes in the United States today are made from it. In Europe, the most popular type is also C-flute (34%) followed by B-flute (13%) and double-wall BC-flute (12%) (FEFCO 2012).

Seven common flute sizes are produced in North America. Although the flutes are named alphabetically, the sizes are not in order of the alphabet but are rather a historical artifact based on the time of their adoption. The standard sizes, ordered from largest to smallest are K, A, C, B, E, F, N.

Historically, the first, named *A-flute*, has large flutes and for many years was the thickest board. The first use for A-flute single-face was as a cushioning wrap for bottles. Because of its thickness, it is used when cushioning is the primary concern. Its thickness also makes A-flute board the highest stiffness of any flute (except K), with high compression strength.

The second one to be produced was *B-flute*, with smaller flutes than A. It was introduced in response to the need for a board with more flutes per inch in order to give the board more flat crush resistance against the impact of printing. B-flute presents a better printing surface and makes better graphics possible. It

Single Face

Single Wall
(also known as Double Faced)

Double Wall

Triple Wall

FIGURE 14.4. Types of corrugated board.

461

also folds (scores) better, but it has lower top-to-bottom stacking strength than A-flute because a thinner board is less stiff.

C-flute is in between A and B flute (as in *The Three Bears*, it is "just right"). It compromises to get better stacking strength than B, and better printability than A. E (for "elite") flute came along next, and the letter D was skipped (although D-flute is produced in some other countries, but not in widespread use). E-flute is smaller than B for even better printing, and was designed to capture a share of the folding carton market.

Later, still smaller flutes such as *F-* and *N-flute* were introduced to further compete with paperboard for folding cartons. *F-flute* is an alternative to 30 point folding carton board, and *N-flute* competes with 25 point. They are generally used for retail and fast food packaging. They are stronger than paperboard and permit better quality printing than board with larger flutes because there is less washboard effect (hills where the flute contacts the liner and valleys in between). Flutes of size F and smaller are called *small flute* with trademarked names like *micro* and *mini* flute.

In addition to the standard sizes, new flute profiles—both larger and smaller—are being developed for specialized applications. An example is the newest *K-flute* (which is also called by some manufacturers *L-flute* and *S-flute*). It has the largest flutes and thickest combined board of all. In general, larger flute profiles give better stacking strength and cushioning. Board with smaller flutes is used for its graphics capability.

While there has been experimentation with square and triangle-shaped flutes, it is difficult to put sharp corners in the medium, and it tends to revert to a rounded shape. Also, some designs tended to roll over when pressure was applied, so the sinusoidal shape is generally used. Materials with more complex shapes, such as honeycombs, are sometimes used for blocking, bracing, or pallets.

Since the basic sine wave flute shape is similar for all standard sizes, boards with smaller flutes have more flutes per linear foot. The number of flutes per foot or the flute height have been recommended as better ways to identify flute size in TAPPI TIP 0302, "An alternative method for designating flute sizes," but the tradition is too strong for it to be overcome soon.

What is worse, there is no official listing of flute size and number of flutes per foot. There is variation in industry practice, and slightly different flute sizes and flutes per foot are cited as standard by different authorities. An individual manufacturer will typically have a set number of flute height and flutes per foot for each flute size produced, although some manufacturers produce a wide array of flute heights, flutes per foot, and even flute profiles, for a given flute designation. Table 14.2 illustrates the range of flute height and flutes per inch found listed by various authorities for the most common flute types.

**TABLE 14.2. Flute Characteristics
(Jonson 1999; Foster 2009; Dekker 2013).**

Flute Type	Flute Height (in)	Flute Height (mm)	Combined Board (in)	Combined Board (mm)	Flutes/ft	Flutes/m
A	0.157–0.193	3.99–4.9	3/16–1/4	4.8–6.35	33 ± 3	108 ± 10
B	0.087–0.118	2.2–3	1/8	3–3.2	47 ± 3	154 ± 10
C	0.137–0.145	3.5–3.7	5/32	4	39 ± 3	128 ± 10
E	0.039–0.071	1–1.8	1/16	1.6–1.7	90 ± 4	295 ± 13
F	0.029	0.7	1/32	0.8–1.2	125 (124–132)	
K	0.22–0.26		1/4	6.5	30	

An alternative to the number of flutes per foot is to specify the *flute pitch*—the distance between adjacent flute troughs (Figure 14.5). Of course, the flute pitch and the number of flutes per unit length are inversely proportional. This measure is used outside of the United States and primarily in machinery specifications. The *take-up factor* is the ratio of the length of the unfluted medium to the length of the combined board. Typical flute pitch and take-up factors for the common flute types are shown in Table 14.3.

Double- and triple-wall corrugated board may have different size flutes in each wall to combine the advantages of each. Double-wall boards are specified in order from outside in. For example, *BC-flute* double-wall board has B on the outside for its flat crush resistance and better printing surface, and C on the inside for cushioning.

Other variations include double-medium board and laminated linerboard, with two plies of either medium or linerboard. These both add further stiffness to the structure, but they slow down the corrugating process. In some

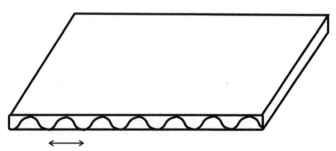

Flute pitch, distance between troughs,
= 1/3 in for A flute

FIGURE 14.5. *Flute pitch (distance between troughs).*

**TABLE 14.3. Flute Pitch and Take-up Factor
(FBA 2005; Jonson 1999; Dekker 2013).**

Flute Type	Flute Pitch (mm)	Take-up Factor
A	8–9.5	1.5–1.54
B	5.5–6.5	1.33–1.35
C	6.8–8	1.43–1.45
E	3–3.5	1.2–1.25
F	2.4	1.25
K	12	1.6

countries (for example, in the Middle East) an alternative to corrugated board is produced where a layer of expanded polystyrene foam is used in place of the medium. This can be more durable than corrugated in wet conditions and can be found used for wet and frozen seafood products.

14.3 CORRUGATED BOARD GRADES AND SPECIFICATION

Corrugated fiberboard can be specified by the basis weight (BW) of its containerboards and the flute type. Sometimes it is specified as a minimum combined weight of facings, which adds the BW of both, usually balanced, linerboards with the assumption that the medium is 26 lb BW. For example, a 42-26C-42 board (which was once an industry standard) would have a combined weight of facings of 84 lb/1,000 ft^2.

But the basis weight (or grammage) alone is not sufficient for establishing grades of corrugated board because properties vary based on the fiber content. While it may have been sufficient in the United Statesin the past when virgin kraft predominated, the increasing use of recycled fibers reduces strength for a given BW. How much depends on the source of the fibers. Therefore, the corrugated industries in several countries have established standards based also on strength or recycled content.

14.3.1 Corrugated Board Specification in the United States: ECT or Burst

The U.S. corrugated fiberboard industry has established grades of board based on the *edge crush test* (*ECT*) and the *Mullen burst test* which are described in Sections 8.4.6 and 8.4.7. A board's rating in the burst or edge crush test is stamped on the bottom of most U.S. corrugated fiberboard boxes.

Performance in these tests is related to performance in distribution, although each addresses a different protective need: the burst test judges puncture-resistance while the ECT is related to stacking strength. When additional properties like water resistance, surface smoothness, or high coefficient of friction are desired, these are also specified in terms of performance in tests.

It is important to note that there is no direct relationship between the ECT and the Mullen burst test, although both are related to the weight of the liners (increasing as basis weight increases). ECT is more related to the CD orientation of fibers and is less affected by the use of recycled fibers, whereas long virgin fibers are necessary for high burst strength results.

These tests predominate because of politics as much as their technical usefulness. Throughout the 1900s, the *American Association of Railroads and the American Trucking Association* collaborated with the corrugated fiberboard industry's trade association (which eventually became the *Fibre Box Association*) to standardize the properties of corrugated fiberboard. This alliance (described in Section 1.2.4) resulted in burst strength and (later) ECT becoming the standard way to specify corrugated fiberboard in the United States today.

In addition to Rule 41 and Item 222 (discussed below), the *Classifications* include a number of specifications for other package forms, for example crates (Item 245), fiber drums (Item 291), and multiwall bags (Item 200). There are also hundreds of package *exceptions* described in the *Classifications*; these are specific package designs approved over the years by packaging professionals employed by the UFC and NMFC (described next).

14.3.1.1 Item 222 and Rule 41 Burst Recommendations

Rule 41 is published in the National Railroad Freight Committee's *Uniform Freight Classification* (*UFC*), and the almost identical Item 222 appears in the National Motor Freight Traffic Association's *National Motor Freight Classification* (*NMFC*). Meeting the requirements of the freight classification earns a box certificate like those shown in Figure 14.6. This establishes the board grade that is used in specifications for purchasing.

The first "requirements" were based on the board's burst strength and combined weight of facings. The minimum requirements are established for a given box weight and size measured in *united inches* ($L + W + D$). The *bursting test and combined weight of facings standard* evolved into the standard that is still published and used today, shown in Table 14.4.[27]

[27]Today's burst test requirements are not much different from those published in 1922 when 200-lb. burst test board was used for up to 65 lb. and 65 united inches (Browder 1935), although the basis weight then was higher because the jute linerboard, made from wastepaper on a cylinder machine, was weaker (Gates 1954).

FIGURE 14.6. *Box certificates based on bursting test. Reprinted with permission of the National Motor Freight and Traffic Association.*

14.3.1.2 Item 222 and Rule 41 ECT Recommendations

The alternative, minimum ECT values, was adopted in the early 1990s as shown in Table 14.5. The ECT box certificate is shown in Figure 14.7.

When the carriers and the Fibre Box Association first started using the ECT values to replace Mullen requirements, the grades were equivalent, and Tables 14.4 and 14.5 above were combined (Fibre Box Association 1992). For example, a 42-26C-42 board with a minimum burst strength of 200 lb was equivalent to board with a minimum ECT of 32 lb

Since then, converters have found ways to improve ECT (like additives and CD fiber orientation), while at the same time reducing the basis weight of materials and increasing recycled content, creating so-called *high test* or *performance linerboard*. This has resulted in lighter-weight boards meeting the ECT requirement (for a given box weight), where heavier boards would have been required under the Mullen test. To reduce the implication of equivalency, the tables are now separated in publications. The following example shows how the tables are used.

Example question:

What burst strength and facing basis weight of single-wall board would meet the Item 222/Rule 41 requirement for a 25-lb package measuring 25″ × 20″ × 20″? What ECT would satisfy the requirement?

Burst strength answer:

The package weight is more than 20 lb and less than 35 lb, which would put it into the second row, but its size (25″ + 20″ + 20″ = 65 united inches) bumps it to the fourth row since the dimensions are more than 60 in but less than 75 in. Therefore, the minimum bursting strength requirement is 200 lb/in and the minimum facing basis weight is 42 lb/msf (combined weight of both facings is 84 lb, with an assumed standard basis weight for the medium of 26 lb). This is the good old standard 42/26C/42.

TABLE 14.4. Rule 41/Item 222 Burst Test Standard (NMFT and UFC).

Max Weight of Box Contents (lb)	Max Outside Dimensions (L + W + D) (in)	Min Combined Weight of Facings (lb/MSF)	Min Burst Test (psi)
Single-wall Corrugated Fiberboard Boxes			
20	40	52	125
35	50	66	150
50	60	75	175
65	75	84	200
80	85	111	250
95	95	138	275
120	105	180	350
Double-wall Corrugated Fiberboard Boxes			
80	85	92	200
100	95	110	275
120	105	126	350
140	110	180	400
160	115	222	500
180	120	270	600
Triple-wall Corrugated Fiberboard Boxes			Min Puncture
240	110	168	700
260	115	222	900
280	120	264	1100
300	125	360	1300

FIGURE 14.7. Box certificates based on edge crush test (ECT). Reprinted with permission of the National Motor Freight and Traffic Association.

ECT answer:

The minimum ECT requirement is met by board with 32 lb/in ECT, by the same interpretation of Table 14.5 above. It should be noted that 32 ECT board usually has a lower basis weight than the "good old" 42/26C/42, 200-lb burst board.

While there is no question that compliance with Rule 41 and Item 222 have prevented damage during their history, there is plenty of reason to question whether the recommended values are inflated, resulting in over-packaging in many supply chains. Adoption of the ECT alternative revealed how entrenched was the idea that transport carriers and the FBA had the right to require a certain grade of corrugated fiberboard. As discussed later in Chapter 15, the carriers no longer have the power to dictate the choice of corrugated board grade, and today's packaging professional has the advantage of a more holistic perspective on box performance. Performance testing of filled packages provides the ability to evaluate shipping container performance throughout a distribution environment.

But the freight classification "rules" have given the U.S. corrugated board industry a useful means to standardize grades, and have given the buyers an indication of its strength. Many small parcel shippers still specify burst test values, because the Mullen test better evaluates the board's resistance to the kinds of forces found in high speed sorting centers. Most shippers of palletized freight prefer to rely on the stacking strength assurance offered by ECT.

14.3.2 Corrugated Board Specification in Other Countries

Many countries do not have the forest resources to produce the same standard of corrugated board as found in North America. North and South America combined, produce a little more corrugated board than all of Europe, but Asia produces as much as both areas combined, almost half of all the corrugated board produced in the world (ICCA 2012).

In other parts of the world, there are a diversity of materials used to make corrugated board, including pulp made from wastepaper and fibers like straw. Many locations do not have the luxury of strong kraft fibers, and some coun-

TABLE 14.5. Rule 41/Item 222 Edge Crust Test (ECT) Standard (NMFT and UFC).

+Max Weight of Box Contents (lb)	Max Outside Dimensions (L+W+D) (in)	Min ECT (lb/in)
Single-wall Corrugated Fiberboard		
20	40	23
35	50	26
50	60	29
65	75	32
80	85	40
95	95	44
120	105	55
Double-wall Corrugated Fiberboard		
80	85	42
100	95	48
120	105	51
140	110	61
160	115	71
180	120	82
Triple-wall Corrugated Fiberboard		
240	110	67
260	115	80
280	120	90
300	125	112

tries even import OCC to "harvest" the kraft by repulping. Therefore, properties are less predictable for any given basis weight.

Other countries also do not have the historical collaboration between transportation carriers and their corrugated box industry.

14.3.2.1 European Corrugated Board Specification

Most mainstream corrugated packaging in Europe is traded on the basis of a specification agreed between the supplier and purchaser. Packer-fillers increasingly outsource packaging design and specification to their suppliers, in this case the corrugated board convertors. The market is sufficiently competitive that major suppliers provide high quality design services and focus on defining the lowest board grade to do the job in order to offer competitive quotes for a value added service. Many packaging design and innovation awards are given to innovative designs produced by corrugated convertors. These packaging suppliers use sophisticated performance models to convert customer requirements (pack size, product weight, target market, etc.) into a recommended grade which, wherever possible, is based on the convertor's higher volume materials.

The purchaser of the board (i.e., the packer-filler) is free to specify their preferred grade in material weight or in performance terms. Where purchasers specify their corrugated boxes, it is usually by material type and weight as this is the simpler of the two options. Specification is often based on sample trials. Containerboard is specified in terms of grammage (g/m^2); for example, UK liner grammage ranges from 125 to 410 g/m^2 and medium is 113 or 127 g/m^2 (Paine 1991).

Europe does not have region-wide transport bodies such as the NMFTA who are active in the specification of packaging performance. The risk for damage to goods or packaging in distribution is held by the consignor (the company shipping the goods) so carriers have no compelling interest in packaging performance or specification. Some major carriers do offer advisory services to encourage good product protection and to reduce complaints of damage. Some individual countries within Europe have specific guidelines for carriers, especially for rail transport, but these are not widely enforced or adopted.

The corrugated industry in Europe is represented by *FEFCO, The European Federation of Corrugated Board Manufacturers*. FEFCO organizes seminars, publishes industry statistics, product guides, technical information, newsletters, etc. It is well known for its international fibreboard case code which provides a method for defining case and fitment style; its illustrations are used in Chapter 15 and are available for free download on the FEFCO website. FEFCO also publishes specifications for recommended corrugated board grades, as do a number of member countries through either national trade associations or technical centers. FEFCO has a listing of European countries' corrugated standards, country by country (FEFCO 2014).

Recycled containerboard (material) grades are defined by CEPI (Confederation of European Paper Industries). The liners are categorized as kraft, testliner, and recycled liners made from recycled board. Kraft is made predominantly from virgin kraft pulp. Testliners are defined in three overall grades. Testliner 1 is the strongest on an index comparing SCT (its performance in the STFI test, which is similar to RCT, described in Section 8.4.6.1); Testliner 3 is weakest; and Testliner 2 is in between. There are several grades of white and brown recycled liners, based on coating and burst index. There are grades of medium (defined as flutings), including 2 grades of semi-chemical and three grades of recycled. A numbering system is given (CEPI 2007 and 2013).

Neither FEFCO nor CEPI provide a chart of flute sizes or flute definitions. Size charts are provided in most suppliers' technical data. Probably the most established formal definition of flute size is given in German standard DIN 55468-1 ("Packaging Materials–Corrugated Board–Requirements"). The definitions relate closely to those given in Table 14.3. It is possible to apply for a formal quality stamp to show conformance to DIN 55468-1 through the Wellpappe Corrugated Board Quality Assurance Scheme, though this is not widely seen.

In Europe, B-flute predominates with 35% of the market. C-flute is only 14% of the European consumption, with 13% B/C double-wall and 11.5% E-flute and microflute (FEFCO 2012).

14.4 CHOOSING AN ECT GRADE BASED ON ITS RELATIONSHIP TO BCT

ECT is significant because of its relationship to box compression strength, commonly known as BCT (test methods were described in Sections 8.4.6 and 8.4.7). BCT is one of the most useful properties of corrugated fiberboard boxes because it is an *indicator* of stacking strength. Boxes need to withstand stacking in factories, warehouses, and transport vehicles. They need to hold up in palletloads and sometimes support as many as 2–4 additional palletloads in a stack. This is the reason that boxes are usually designed so that the flutes run vertically, providing columns to support the load.

Stacking failures can cause damage and danger. Typical damage includes crushing, deformation, stress-cracking, and breakage. The danger comes from unstable stacks: stacking failure in the boxes on one side of a palletload can topple the whole stack, and material handling workers have been injured (and even killed) in such accidents.

The ECT (edge crush test) is significant for both the package converter and its customer because it is the basis for specification. Higher ECT is directly related to higher prices for corrugated board. It is a value that converters can control.

Converters use the relationship of the CD RCT of the three plies to ECT of the combined board (discussed in Section 8.4.6.2), aiming to increase RCT by orienting more fibers in the CD and by adding strengthening additives. ECT is also a good measure of flute formation and gluing.

The most common style of box, the *regular slotted container* (RSC), is a simple KD (Knocked down) tube shape with a *manufacturer's joint* along one edge and flaps on the top and bottom, as shown in Figure 14.8. The following two sections that deal with choosing an ECT grade are based on the common RSC, although the same theory applies to other styles that are discussed in Chapter 15.

For box buyers who choose an ECT grade based on its relationship to stacking strength, the place to begin is by determining the strength required. This is derived from either the anticipated height of the stack (H) or the anticipated number of boxes in the stack, the height of each box (h), and the weight (W) of each box.

If the anticipated height of the stack is known the number of boxes in the stack (n) is determined by dividing the stack height by the height of a box, and then rounding down (since you cannot put a partial box on the stack);

$$n = \frac{H}{h} \text{ rounded to next lowest integer}$$

Once n is determined, the required stacking strength of the box is calculated by multiplying the weight of one box by $n - 1$, the number of boxes stacked on the bottom box, since that box must support the heaviest load:

$$\text{Required stacking strength} = W(n-1)$$

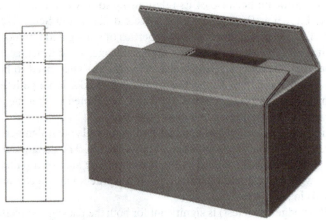

FIGURE 14.8. *RSC and RSC blank (FBA #0201). Reprinted with permission of Fibre Box Association, copyright 1992.*

However, as the following discussion shows, real stacking strength is a fraction of BCT. BCT must be discounted by variables that affect the board, box-making, filling, and the distribution environment. The next sections of this chapter explore the relationship between ECT, BCT, and real stacking strength to provide a perspective on how to choose among various ECT grades of board.

14.4.1 Predicting BCT from ECT: The McKee Formula

Compression strength of an empty box tested under TAPPI laboratory conditions (BCT) depends on the board's flexural (bending) stiffness, the ECT, and the size of the box perimeter. Stiffness comes from the fundamental plate physics of the panel itself. Stiffness increases with the thickness as described in Section 8.4.5.1, when the corrugated board is thicker (which is related to flute size) and the masses of the liners are farther from the center of the board.

The correlation between BCT, ECT, caliper, and perimeter was established in the well-known *McKee Formula* (McKee *et al.* 1963). It has been simplified to the following relationship:

$$BCT = 5.87 \times ECT \times \sqrt{tP}$$

BCT = compression strength of the box
  t = board thickness
  P = load-bearing perimeter

Be careful to use consistent units!

McKee model only applies to RSC with lengh (L) less than three (3) times the width (W) and the perimeter (p) less than seven (7) times the depth (D) as expressed in the relations below (McKee *et al.* 1961):

$$L < 3 \times W$$

$$P < 7 \times D$$

Example question:

Estimate the BCT of a 20″ × 15″ × 12″ box made from C-flute (5/32″ thick) with an ECT of 32 lb/in.

Answer:

$$BP = 2(20 \text{ in}) + 2(15 \text{ in}) = 70 \text{ in}$$

$$BCT = 5.87 \times 32 \ \frac{\text{lb}}{\text{in}} \times \sqrt{5/32 \text{ in} \times 70 \text{ in}} = 621 \text{ lb}$$

In metric units the same box has a perimeter of 1.8 m, the thickness is 0.004 m, and the ECT is 5.6 kN/m.

$$BCT = 5.87 \times 5.60\frac{kN}{m} \times \sqrt{0.00397 \times 1.78\ m} = 2.76\ kN$$

The results are the same, allowing for rounding errors.

BCT is defined as the predicted compression strength of an empty, perfectly-formed RSC in ideal conditions—a TAPPI T 804 laboratory compression test. The prediction does not take into account the depth of the box, although the box must not be too short or too tall, because depth does affect compression strength.[28]

The McKee formula takes into account the crushing of the box corners and edges and bending of the sides. Stress lines form from the corners. As a rule of thumb, a typical RSC carries two-thirds of the load in the four upright edges (and the area immediately adjacent to them) and one-third in the four side panels, as represented in Figure 14.9 (McKee *et al.* 1963). The *two-thirds corners, one-third sides* estimate can be used to estimate nonstandard situations like cut-outs and structures with more than four sides.

Example question:

For the box in the example above (20″ × 15″ × 12″, C-flute box, ECT = 32 lb/in, BCT = 621 lb), estimate the strength of each corner and side. How much weight can be carried by each edge (how much weight would one top corner withstand); how much is carried by each side panel?

Answer:

621 (2/3) = 414 lb carried by the four vertical edges = 103.5 lb/edge

621 (1/3) = 207 lb carried by the four side panels = 51.8 lb/side

Cut-outs added for handles or ventilation reduce the compression strength of that side panel. The most conservative estimate is to eliminate that panel's strength from the calculation, although the actual extent of the strength reduction depends on the size and placement of the hole(s). The greatest loss is when material is removed from areas of concentrated stress, usually extending diagonally from the corners of the box, and when material is removed too close to the scores (Peters and Kellicutt 1959).

---

[28]McKee and others have recognized that the parameters are linked to the dimensions of the box, and that the formula is less predictive of the behavior of a short box that fails inelastically. Recent research has used the analysis of several data sets to refine McKee's equation to account for elastic buckling versus inelastic buckling (Urbanik and Frank 2006).

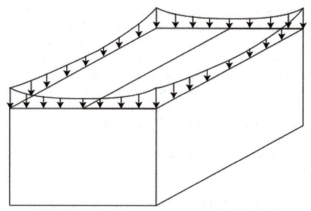

FIGURE 14.9. *Edges carry two-thirds of the load.*

Example question:

What is the most conservative estimate of BCT if the box above has ventilation and hand holes placed on two opposite sides, near to the scores or areas of concentrated stress?

Answer:

Eliminate the strength contribution of the two sides:

$$BCT = 621 \text{ lb} - (2 \times 51.8 \text{ lb}) = 517 \text{ lb}$$

The BCT of octagonal or hexagonal drums with more than four sides is greater than a rectangular box because of the additional upright edges. The BCT can be estimated by multiplying the McKee formula BCT result by 1.5, regardless of whether there are six or eight sides. More sides add more corners (which add strength) but also bigger angles (which detract from strength), as obtuse angles are weaker than 90-degree angles (Maltenfort 1988).

Example question:

What is the BCT of a C-flute hexagonal box with $P = 70''$, ECT = 32 lb/in?

Answer:

$$BCT = 1.5 \times 5.87 \times 32 \frac{\text{lb}}{\text{in}} \times \sqrt{\frac{5}{32} \text{in} \times 70 \text{ in}} = 932 \text{ lb}$$

The improvement in compression strength added by inserts and partitions can also be predicted, but this is more complex. Theoretically, the McKee formula can be applied to the perimeter measurement of the insert and added to the strength of the box. But since partitions do not have the same shape as a box, the prediction includes a factor that adjusts for weakening or reinforcing attributes, such as folds, cuts, and corners, which differ for each style of insert. In general, the more upright edges and supported panels, the higher the predicted BCT. Maltenfort (1988) presents a range of styles and associated prediction factors. The load-sharing ability depends on the height of the insert, which should be designed to fail at the same deflection as the box; failing together maximizes the overall box strength. Furthermore, attaching the inserts to the box sidewalls tends to increase BCT since this inhibits the ability of the side to bulge.

14.4.2 Factors that Reduce BCT

McKee provided the starting point for a rich vein of research by paper scientists into the factors that reduce a box's actual stacking strength from the predicted, ideal, BCT values. These include time, humidity, stress during conversion, stacking, and palletization.

Each study investigates a different set of factors, and generalizations are the subject of debate, especially when factors occur in combination. The subject is complex, and it is complicated by the fact that the ECT properties of board have changed much since most of the original research was performed in the 1960s and '70s. Periodic reviews can be found in Maltenfort (1988, 1989), Koning (1995), and Frank (2013).

14.4.2.1 Effect of Time and Humidity

The most significant factor that weakens boxes is humidity. Humidity affects not only the paper, but also the structure, since the board, joint, and closures are usually bonded with water-soluble adhesive.

High humidity storage conditions can severely degrade the strength of a stack of boxes. At 85% RH for three months, a box loses over 80% of its BCT. At 90% RH, the board's moisture content is about 20% of its total weight and it even feels soft and damp to the touch.

A related factor is time in the supply chain. Long-term storage, especially when coupled with high or cyclic humidity, can dramatically reduce stacking strength. The board weakens over time under load, and from the amount of transport and handling. A box under load loses about 40% of its strength in the first 90 days of storage at TAPPI conditions (discussed in Chapter 8). Longer storage reduces strength to about 50%, but when coupled with high humidity (90% RH for 3 months), the compression strength is reduced by nearly 90%, as shown in Table 14.6.

TABLE 14.6. Effect of Relative Humidity and Time on Load as Percentage of Initial BCT at TAPPI Conditions. Reprinted with permission of the Institute of Paper Science and Technology, from McKee and Whitsitt (1972).

	90 days	180 days	360 days
50% RH	60%	55%	51%
65% RH	43%	40%	37%
75% RH	32%	29%	27%
80% RH	23%	21%	20%
85% RH	16%	15%	14%
90% RH	12%	11%	11%

Although a constant high humidity condition is rare, there are many circumstances where the daily cycle produces high humidity for some time every day. This triggers the hysteresis effect (equilibrium MC differs depending on whether the humidity is going up or down) making paperboard and corrugated board subject to unexpected changes.

Cyclic humidity has been found to accelerate creep failure in a constant load test. Surprisingly, high/low humidity cycles reduce a box's time to failure even more than constant high temperature and humidity. Long cycles that last over a day are worse than short half-day cycles because the board absorbs, and then loses more moisture. As the box picks up and loses moisture, it expands and contracts, to a gradually decreasing height until it eventually collapses (Leake and Wojcik 1993).

The creep behavior of corrugated boxes is non-linear. Cycling the load gives more creep than the average load, and stress distributions are created because the heterogeneity of the material causes it to swell and shrink non-uniformly. Most of the creep occurs during the change from low to high humidity. At low humidity, the stress concentrates and the board stiffens. As the humidity rises, inter-fiber bonds weaken, the fibers swell, and dried-in strain is released. Accelerated creep is complex because of differences in fiber swelling, load nonlinearity, bonding, and dried-in strains, as well as factors related to the adhesive and the combination process of the board. An excellent review of the literature has been published by Coffin (2005).

14.4.2.2 Effect of Conversion

Box-making factors can reduce a box's compression strength. The box-making process is discussed in more detail in Chapter 15. The following conversion factors reduce BCT:

- Crushed board (even slightly crushed flutes from printing affect BCT)
- Flutes that are misaligned from scores
- Improperly formed scores
- Corners that are crushed or slotted beyond the score line
- Holes for handles or ventilation

The manufacturer's joint and closing method affect BCT too. The research into the effect of different types of manufacturer's joints finds that properly glued manufacturer's joints, with a sufficient bonding area, outperform taped or stitched joints. A box that is squared up performs better than if it is skewed. The box closing method affects BCT especially since a box that is taped closed has a tendency for the flaps to rotate inward, resulting in a higher BCT, but at a lower deflection (Frank 2014).

14.4.2.3 Effect of Stacking, Pallet Surfaces, and Unitization

The pallet and stacking pattern also affect stacking strength. Since the strength of a box is in its sides and upright edges, perfectly aligned column stacking best uses the boxes' compression strength. Alignment of the edges and corners creates a support beam structure within a palletload.

If boxes are misaligned or an interlocking pallet pattern is used, stacking strength will be lower. While column stacking retains 85% of a box's original compression strength, interlocking reduces it to about 50% (Kellicutt 1963; Koning and Moody 1966; Levans 1975). When boxes are interlocked, most of the supporting corners and edges rest on the weakest part of the box below it. However, an interlocking pattern is often preferred for the stability it lends to a load as it is handled and transported, especially if it is not stretch-wrapped or otherwise stabilized.

Pallet overhang and wide deckboard gaps also reduce stacking strength, since they affect the level surface of the bottom-most boxes in the stack, which are expected to carry the heaviest load. In worst cases the edges of the box coincide with spaces between the deckboards. A one-inch overhang can result in a 32% strength loss, and wide deckboard spacing can reduce strength by as much as 15% (Levans 1975). When palletloads are stacked, the bottom deckboards of the top load concentrate stress on the top boxes in the bottom load, further complicating stacking strength prediction.

14.4.2.4 Predicting Stacking Strength from BCT Using FBA Environmental Factors

The FBA recommends using the *environmental factors* shown in Table 14.7 to compensate for stacking and pallet pattern, and recommends multiplying the

TABLE 14.7. Time and Humidity Environmental Factors that Reduce BCT (FBA 2005, 3.13).

	BCT Loss	Multiplier	
Storage time under load	10 days—37% loss	0.63	
	30 days—40% loss	0.60	
	90 days—45% loss	0.55	
	180 days—50% loss	0.5	
RH under load (cyclical RH variation further increases compressive loss)	50% RH—0% loss	1	
	60% RH—10% loss	0.9	
	70% RH—20% loss	0.8	
	80% RH—32% loss	0.68	
	90% RH—52% loss	0.48	
	100%—85% loss	0.15	
Pallet Pattern		Best Case Multiplier	Worst Case Multiplier
Columnar, aligned	Up to 8% loss	1	0.92
Columnar, misaligned	10–15% loss	0.9	0.85
Interlocked	40–60%	0.6	0.4
Pallet overhang	24–40%	0.8	0.6
Pallet deck board gap	10–25%	0.9	0.75
Excessive handling	10–40%	0.9	0.6

relevant factors to give an *environmental multiplier* to be used in calculating the required BCT for a given situation. The FBA recommends applying these to BCT values generated by a box compression test. We draw an additional inference when we apply them to BCT values predicted from an ECT test.

For example, for a box stored in challenging conditions, 180 days at 90% RH in a poorly-interlocked palletload with a high degree of pallet overhang, the environmental factor is calculated as follows:

$$0.5 \times 0.48 \times 0.4 \times 0.6 = 0.0576$$

This means that the predicted stacking strength for the box with 621 lb of predicted BCT, from Section 14.4.1 above (dimensions = 20″ × 15″ × 12″), is only about 6% of its BCT: 621 × 0.0576 = 36 lb If the box's contents weigh 20 lb, boxes can only be stacked two-high (one box atop the bottom one).

The *environmental multiplier* is the inverse of the environmental factor. In our example above, the environmental multiplier is calculated as follows:

$$\text{environmental multiplier} = \frac{1}{\text{environmental factor}} = 1/0.0576 = 17.36$$

If we wanted to stack the boxes six high (five boxes on the bottom box), then the load that would need to be supported would be 5×20 lb $= 100$ lb The required BCT, then, would be:

$$\text{required } BCT = 100 \text{ lb} \times 17.36 = 1736 \text{ lb}$$

Following are some more sample questions, using FBA's environmental multipliers, based on the corrugated box predicted in Section 14.4.1 to have 621 lb of BCT (dimensions $= 20'' \times 15'' \times 12''$). *Note that these calculations assume that the BCT predicted from McKee's ECT formula is equal to the value obtained by testing the actual (empty) box in laboratory conditions.*

Question:

What is the predicted stacking strength of the 621-BCT box, adjusted for FBA's environmental factors, for intermediate conditions: 90 days of storage, 70% RH, good interlocking and only a little overhang?

Answer:

Multiply the "multipliers" by the BCT

Environmental Multiplier: $0.55 \times 0.8 \times 0.6 \times 0.8 = 0.21$

Predicted stacking strength $= 621$ lb $\times 0.21 = 130$ lb

In this case the 20-lb boxes can be stacked seven high. Note that while $130/20 = 6.5$, a box is indivisible; there are no half-boxes. Therefore, only six boxes can be stacked on top of the bottom box, for a total of seven boxes. Furthermore, since the box height is 12 in (1 ft), a stack height of 7 ft can be recommended (1 ft $\times$ 7).

Question:

What is the predicted stacking strength and acceptable stack height for this box in the best conditions: One-month dry storage in column stacks with no pallet overhang?

Answer:

Environmental Multiplier: 0.6 (the others are all 1)

Predicted stacking strength $= 621 \times 0.6 = 373$ lb

In this case the 20-lb boxes could be stacked 19-high, for a stack height of 19 ft

14.4.2.5 Predicting Stacking Strength from BCT Using ASTM D4169 Factors

ASTM D4169, Standard Practice for Performance Testing of Shipping Containers and Systems, proposes similar factors to convert BCT to stacking strength. It simplifies the effects of humidity, time, stacking and unitizing into three *assurance levels*, and then reduces BCT to an estimate of stacking strength with a single compensating factor.

The assurance level is chosen based on the "product value, the desired level of anticipated damage that can be tolerated, the number of units to be shipped, knowledge of the shipping environment, or other criteria. Assurance Level II is suggested unless conditions dictate otherwise." (ASTM D4169, 8.2) Assurance Level I is more severe, and Assurance Level III is less severe.

ASTM D4169 then assigns an *F-factor* from 3 to 10, as shown in Table 14.8. BCT is divided by the F-factor to estimate the maximum stacking load on the bottom box.

$$SL = \frac{BCT}{F}$$

where

SL = stacking load (load on bottom box in the stack)
BCT = laboratory compression strength
F = "F" factor

TABLE 14.6. ASTM D4169 Factors to Estimate Stacking Load from BCT.

	F-Factors and Assurance Levels					
	Warehouse Stacking			Vehicle Stacking		
Shipping Container Construction	I	II	III	I	II	III
1. Corrugated, fiberboard, or plastic container that may or may not have stress-bearing interior packaging using these materials and where the product does not support any of the load.	8	4.5	3	10	7	5
2. Corrugated, fiberboard, or plastic container that has stress-bearing interior packaging with rigid inserts such as wood.	4.5	3	2	6	4.5	3
3. Containers . . . where the product supports the load directly, for example, compression package.	3	2	1.5	4	3	2

This means that for the most severe stacking conditions in transit, and when the product does not help to support the box (shipping container construction 1 above), the BCT value should be divided by 10. Average vehicle conditions rate a factor of 7. The factor for warehouse stacking is lower, with a maximum of 8, and an average of 4.5.

Question:

What is the predicted stacking strength of a 620-lb BCT box, adjusted for ASTM D4169's F-factors for an average value, moderate damage tolerance, and moderate vehicle stacking shipping environment?

Answer:

Divide the BCT by the F-factor of 7 (for Shipping Container Construction 1, Vehicle Stacking Assurance Level II):

Predicted stacking strength = 620 lb / 7 = 89 lb

In this case the 20-lb boxes can be stacked only five-high, a stack height of 5 ft (since the boxes are 1 ft tall). Note that a stack this short would waste much of the space in a truck trailer, most of which average 8 ft of vertical clearance.

Clearly, these factors simplify a complex relationship. Like the FBA factors, ASTM D4169 applies them to laboratory-obtained BCT results, and so we draw an additional inference when we apply them to results that were predicted from ECT values.

Needless to say, the assumptions that are used for stacking strength prediction make a great deal of difference to the calculated result.

14.4.3 Fill the Box!

The final factor is the most significant one for a filled box: the load bearing capacity of the box's contents and how they fit. The contents of many boxes can be relied on (or designed to) to carry at least some portion of the load. The best way to find a filled package's compression strength is to test the filled package. Filled package testing is discussed in the next chapter, Section 15.8.1. But it can also be estimated from the individual product and box test results.

The decision about whether or not to let the product carry part of the load is based on the nature of the product. Metal cans, glass bottles, and appliances, for example, are durable and can add considerable strength to a tight *compression* package. Plastic bottles and folding cartons add less strength. And items that are flexible or irregularly shaped (like meat) may add none at all.

ASTM D4169 recommends applying a more lenient set of factors when the contents support the load directly in a warehouse, as shown for Shipping Container 3 in Table 14.8 above:

- 1.5 (for the low Assurance Level 3),
- 2 (for average Level 2) or
- 3 (for the highest Assurance Level 1)

These factors leave no doubt that BCT is different from real stacking strength. Even when the contents totally support the load, the box may need to have twice as much BCT as the load it will bear alone, to keep it square and secure in a stack.

Example question:

If the contents of the box in the previous examples (BCT = 621 lb) fully supports the load, what is the predicted warehouse stacking strength under average conditions? How many high can they be stacked in a warehouse or truck?

Answer:

In average warehouse conditions, $F = 2$ (Warehouse Assurance Level II)
621 lb/2 = 310 lb
310 lb/20 lb = 15.5; therefore 15 stacked on the bottom one (16 total)
Stack height is 16' (box is 12" tall)

In average truck conditions, $F = 3$ (Vehicle Assurance Level II)
621 lb/3 = 207 lb
207 lb/20 lb = 10.3; therefore 10 stacked on the bottom box (11 total)
Stack height is 11'
(Note, this is taller than the height of a trailer.)

In the more usual case where the contents provide a portion of the load bearing strength, the factor is arrived at by adding a factor for the percent of the load supported by the product to the factor for percent of the load supported by the box. Note however, that plastic containers and cartons are both susceptible to creep in compression, so they will have less effective strength in long term storage.

Most of the time, it is best to design a box with no headspace, to give the box and contents the best opportunity to support each other and to reduce cost. Likewise, inserts and other load-bearing components must be sized so that they work together with the box walls.

If the deflection at failure for the box is higher than that for the contents, the height of the box should be designed to be slightly taller in order for the critical deflection of both to be reached at the same time. For example, boxes for cereal cartons are often designed with a headspace equal to the difference between the deflection at failure for the case and the primary packages inside: if the case fails at 0.5× deflection and the cartons fail at 0.25″, the headspace should be 0.25″.

14.4.4 Applying the ASTM or FBA Factors to the McKee Formula to Choose an ECT Grade

The reason the previous sections that relate stacking strength to BCT are here in this chapter, rather than in the next one where box performance testing is discussed, is to lead back to the significance of ECT.

The FBA or ASTM D4169 factors, combined with the McKee Formula relationship, can be used to choose a minimum ECT grade of board, theoretically, to support a given stack height. The relationships are combined thus:

$$ECT = \left(\frac{H}{h} - 1\right)\left(\frac{WF}{5.87\sqrt{tP}}\right)$$

H = stack height
h = box height
t = board thickness
P = load-bearing perimeter
F = FBA environmental factor or ASTM D4169 F-factor
W = box weight

Note: In using this formula, it is important to round the first factor ($H/h - 1$) down to the integer, otherwise the required ECT will be over-estimated. In the examples presented here, the calculation results in an integer number of boxes, so no rounding is required.

Example question:

Using the McKee formula and ASTM D4169 F-factor for average conditions, what minimum ECT would theoretically enable a 20 lb C-flute RSC 20″ × 15″ × 12″ to be able to stacked to a height of 8 ft for transport, in the case where 100% of the load is carried by the contents, and the filled box weighs 20 lb?

Answer:

$$ECT = \left(\frac{96 \text{ in}}{12 \text{ in}} - 1\right)\left(\frac{20 \text{ lb} \times 3}{5.87\sqrt{\dfrac{5}{32}\text{in} \times 70\text{in}}}\right) = 21.6 \text{ lb/in}$$

Example question:

What minimum ECT would theoretically be needed to stack the same box at the same height under average conditions, in the case where none of the load is carried by the contents?

$$ECT = \left(\frac{96 \text{ in.}}{12 \text{ in.}} - 1\right)\left(\frac{20 \text{ lb.} \times 7}{5.87\sqrt{\frac{5}{32}} \text{ in.} \times 70 \text{ in.}}\right) = 50.5 \text{ lb./in.}$$

It is interesting to note the range, and the difference between theory and practice. The NMFC Classification (Table 14.5) would recommend 26 ECT for this box.

It must be emphasized that a number of simplifying assumptions have been made in these calculations. Assurance level II of D4169 probably gives a more cautious approach to reflect a wider range of potential conditions.

14.4.5 Why Not Specify BCT Directly?

This chapter has shown that ECT is the most common way to specify board in the United States, that grades vary in other countries, and that the relationship of the grades to box performance can be estimated from ECT, if a number of factors are considered.

This leads to the notion that in some cases it would be better to specify BCT directly. If the box is intended to support the load, specifying BCT would simplify the comparison of board from various sources. For example, firms sourcing boxes in many countries, where the board specification traditions vary, have found this to be a successful strategy.

In order to specify BCT, the stacking conditions and requirements need to be understood. Stacking strength requirements vary widely. For example, an increasing number of factories and warehouses use racks for palletload storage. Since palletloads in such a situation are unlikely to ever be stacked, there is less need for stacking strength to be built into individual boxes.

Knowing the maximum stack height can be used to optimize containers to reduce their cost. It makes sense for a supply chain to consider the economic tradeoff between investing in a fixed asset such as a racking system versus building high compression strength into every expendable container.

Although there are methods for testing and prediction, there is no guarantee of a corrugated fiberboard box's stacking strength, since there are many factors that affect it during the course of distribution. Furthermore, stacking strength is only one aspect of protection. Chapter 15, which shows corrugated box styles, specification, and cost estimation, ends with an introduction to true shipping container performance tests.

14.5 REFERENCES

ASTM (formerly American Society for Testing and Materials, now ASTM International). No date, annually updated. *Annual Book of ASTM Standards.* West Conshohocken, PA: ASTM International.

Byrd, V.L. and J.W. Koning. 1978. Edgewise Compression Creep in Cyclic Relative Humidity Environments. *TAPPI,* 61 (6), June: 35–37.

CEPI. 2013. European List of Standard Grades of Paper and Board for Recycling; Guidance on the Revised EN 643. Brussels: Confederation of European Paper Industries, CEPI. http://www.cepi.org/node/16884

CEPI. 2007. European List of Standard Grades of Recovered Paper and Board (EN 643). Brussels: Confederation of European Paper Industries, CEPI. http://www.cepi.org/topics/recycling/publications/EuropeanListofStandardGradesofRecovered-PaperandBoard

Coffin, D.W. 2005. The Creep Response of Paper. In *Advances in Paper Science and Technology: Transactions of the 13th Fundamental Research Symposium held in Cambridge: September 2005,* Vol. 2, edited by S. J. I'Anson, pp. 651–747. Bury, Lancashire: Pulp and Paper Fundamental Research Society.

Crowell, B. 1993, Increased Utilization of Recycled Fibers: Impact on OCC Performance. In *Recovery and Use of OCC: Analysis of Markets, Technologies, and Trends,* edited by Virginia Stefan, pp. 56–61. San Francisco: Miller Freeman.

Dekker, A. 2013. Corrugated Fibreboard Packaging. In *Handbook of Paper and Paperboard Packaging Technology,* 2nd ed., edited by Mark J. Kirwan, pp 313–339. Chichester: John Wiley & Sons.

Department of Agriculture Forest Service Report No. 2152, August. Reprinted in *Performance and Evaluation of Shipping Containers,* edited by George Maltenfort, 1989, pp. 274–279. Plainview, NY: Jelmar.

FBA (Fibre Box Association). 2005. *The Fibre Box Handbook.* Rolling Meadows, IL: Fiber Box Association.

FBA (Fibre Box Association). 2012. *Fibre Box Association Industry Annual Report, U.S. 2011.* Rolling Meadows, IL: Fibre Box Association.

FEFCO. 2012. FEFCO, *International Federation of Corrugated Board Manufacturers, Annual Statistics 2011.* Brussels: FEFCO. http://www.fefco.org/sites/default/files/documents/Fefco_AnnualEvaluation_2011.pdf

FEFCO. 2014. "World Standards: Comparison Testing Methods FEFCP/EN/ISO/TAPPI" and "National Standards Comparison Chart." Brussels: FEFCO. http://www.fefco.org/technical-documents/standards, accessed 4/1/14.

Foster, G.A. [CM26]Boxes, Corrugated. In *The Wiley Encyclopedia of Packaging Technology,* 3rd ed., edited by Kit L. Yam, pp. 162–170. New York: John Wiley & Sons.

Frank, B. 2014. Corrugated Box Compression—A Literature Survey. *Packaging. Technology and Science,* 27 (2): 105–128.

Gates, J.E. 1954. The Kraft-Jute Containerboard Controversy. *Fibre Containers and Paperboard Mills,* 39 (1) January. Reprinted in *Performance and Evaluation of Shipping Containers,* edited by George Maltenfort, 1989, pp. 13–22. Plainview, NY: Jelmar.

Goldberg, R.C., C.S. Hakkio, and L.N. Moses. 1986. Competition and Collusion, Side by Side: The Corrugated Container Antitrust Litigation. In Antitrust and Regulation, edited by Ronald E. Grieson, p. 105. Lexington, MA: Lexington Books.

Hubbe, M.A., R.A. Venditti, and O.J. Rojas. 2007. How Fibers Change in Use, Recycling, Bio-Resources, 2 (4): 739–788.

ICCA (International Corrugated Case Association). 2012. Annual Production Statistics. www. iccanet.org, accessed 520/12.

Impact Marketing Consultants. 2006. *The Marketing Guide to the U.S. Packaging Industry.* Manchester Center, VT: Impact Marketing Consultants.

Jonson, G. 1999. *Corrugated Board Packaging.* 2nd edition. Leatherhead, UK: Pira.

Kellicutt K.Q. 1963. Effect of Contents and Load Bearing Surface on the Compressive Strength and Stacking Life of Corrugated Containers. TAPPI, 46 (1): 151A–154A. Reprinted in *Performance and Evaluation of Shipping Containers*, edited by George Maltenfort, 1989, pp.141–143. Plainview, NY: Jelmar.

Koning, J. 1995. *Corrugated Crossroads: A Reference Guide for the Corrugated Containers Industry.* Atlanta: TAPPI Press.

Koning, J.W. and W.D. Godshall. 1975. Repeated Recycling of Corrugated Containers and its Effect on Strength Properties. *TAPPI*, 58 (9) September: 146–150.

Koning, J.W. and R.C. Moody. 1966. Slip Pads, Vertical Alignment Increase Stacking Strength 65%. Boxboard Containers, vo. 74, no. 4, pp. 56–59.

Kroeschell, W.O., Ed. 1993. *Corrugator Bonding: A TAPPI Press Anthology of Published Papers.* Atlanta: TAPPI.

Laufenberg, T.L. and C.H. Leake, Eds. 1992. Cyclic Humidity Effects on Paperboard Packaging: *Proceedings of a Symposium. Madison, WI: Forest Products Laboratory.*

Leake, C.H. and R. Wojcik. 1993. Humidity Cycling Rates: How They Influence Container Life Spans. *TAPPI*, 76 (10): 26–30.

Levans, U.I. 1975. The Effect of Warehouse Mishandling and Stacking Patterns on the Compression Strength of Corrugated Boxes. TAPPI 58 (8): 108–111. Reprinted in *Performance and Evaluation of Shipping Containers,* edited by George Maltenfort, 1989, pp.335–34. Plainview, NY: Jelmar.

Lincoln, W.B. 1945. Development of the "V" and "W" Board Weatherproof Corrugated Shipping Containers, *Fibre Containers and Paperboard Mills,* 30 (9) September. Reprinted in Performance and Evaluation of Shipping Containers. Reprinted in *Performance and Evaluation of Shipping Containers,* edited by George Maltenfort, 1989, pp. 8–12. Plainview, NY: Jelmar.

Maltenfort, G.G. 1988. *Corrugated Shipping Containers: An Engineering Approach.* Plainview, NY: Jelmar Publishing.

Maltenfort, G.G., Ed. 1989. *Performance and Evaluation of Shipping Containers.* Plainview, NY: Jelmar Publishing.

McKee, R.C., J.W. Gander, and J.R. Wachuta. 1961. Edgewise Compression Strength of Corrugated Board. *Paperboard Packaging* 46 (11): 70–76. Reprinted in *Performance and Evaluation of Shipping Containers,* edited by George Maltenfort, 1989, pp. 47–54. Plainview, NY: Jelmar.

McKee, R.C., J.W. Gander, and J.R. Wachuta. 1963. Compression Strength Formula for Corrugated Board, *Paperboard Packaging,* 48 (8) August: 149–159. Reprinted in *Performance and Evaluation of Shipping Containers,* edited by George Maltenfort, 1989, pp. 62–73. Plainview, NY: Jelmar.

McKee, R.C. and W.J. Whitsitt. 1972. *Effect of Relative Humidity and Temperature on Stacking Performance,* Institute of Paper Chemistry, Project 2695-9, November.

McKinney, R.W.J. 1995. Manufacture of Packaging Grades from Wastepaper. In *Technology of Paper Recycling,* ed. R.W.J. McKinney, 224–295. London: Blackie Academic and Professional Press.

Nazhad, M.M. 2005. Recycled Fiber Quality—A Review, *Journal of Industrial and Engineering Chemistry,* 11 (3): 314–329.

NMFTA (National Motor Freight Traffic Association). Annual. *National Motor Freight Classification (NMFC).* Alexandria, VA.: NMFTA.

NRFC (National Railroad Freight Committee). Annual. *Uniform Freight Classification (UFC)*. Chicago: Association of American Railroads.

Paine, F.A. 1991. *The Package User's Handbook.* Glasgow: Blackie (published in U.S. by AVI, New York).

Peters, C.C. and K.Q. Kellicutt. 1959. *Effect of Ventilating and Handholes on Compressive Strength of Fiberboard Boxes.* Forest Products Laboratory, U. S.

Perkins, S. and P. Schnell. 2000. *The Corrugated Containers Manufacturing Process.* Atlanta: TAPPI Press.

PPI Pulp & Paper Week. Published weekly by RISI. http://www.risiinfo.com/

Steadman, R. 2002. Corrugated Board, in *Handbook of Physical Testing of Paper,* 2nd ed, edited by Richard E. Mark, Charles C. Habeger, Jens Borch and M. Bruce Lyne, pp. 564–660. New York: Marcel Dekker.

Stefan, V., Ed. 1993. *Recovery and Use of OCC: Analysis of Markets, Technologies, and Trends.* San Francisco: Miller Freeman.

TAPPI (Technical Association for the Pulp and Paper Industries). 2000. 2000–2001 *TAPPI Test Methods.* Atlanta: TAPPI.

TAPPI (Technical Association for the Pulp and Paper Industries). 2000. *How Corrugated Boxes are Made.* Atlanta: TAPPI, CD Rom.

Urbanik, T.J. and B. Frank. 2006. Box Compression Analysis of World-wide Data Spanning 46 Years, *Journal of Wood and Fiber Science,* 38 (3), 399–416.

US EPA. 2014. *Municipal Solid Waste Generation, Recycling and Disposal in the United States, Tables and Figures for 2012.* U.S. Environmental Protection Agency. http://www.epa.gov/epa-waste/nonhaz/municipal/pubs/2012_msw_dat_tbls.pdf, accessed 6/27/14.

14.6 REVIEW QUESTIONS

1. What is the difference between an *integrated* and an *independent* corrugated board supply company?

2. What is the difference between a *sheet feeder plant* and a *sheet plant*, and what are the more descriptive names for these?

3. What kind of adhesive is generally used to combine corrugated board? What is the significance of the adhesive gel point?

4. Which side of the corrugated board is usually printed: the first or second face applied in the corrugator? Why?

5. What is the role of steam in the process used to make corrugated board? What is the role of heat?

6. What is the difference between wax impregnation, curtain coating, and cascading? Which provides the most water resistance?

7. Why is wax-coated board generally not recyclable?

8. What is meant by the engineering beam principle, and how does it apply to corrugated board?

9. Linerboard is made in what kind of pulping and papermaking process? What kind of process is used to make medium? Why?

10. Why is corrugated board the paper material with the highest recycling rate?

11. What is the effect of recycling on the performance of corrugated board?

12. How and why does recycled content affect the strength of corrugated board? What is meant by "irreversible hornification?"

13. Under what conditions are corrugated boxes commercially reused?

14. How many linerboards are there in triple-wall board?

15. List the flute letter designations in order of increasing flute size. List them in order of increasing number of flutes per inch.

16. Which is stiffer: A, B, or C flute? Why?

17. Why are ECT and burst strength the predominant basis for specifying corrugated board in the United States? What are the advantages of this, compared to other measures like basis weight or tensile strength?

18. Use the NMFC/UFC tables to determine which burst strength and facing basis weight of single-wall board would meet the Item 222/Rule 41 recommendation for a 20-lb package measuring $18'' \times 16'' \times 10''$. Which ECT would satisfy the recommendation?

19. What properties of containerboard and of converted boxes are related to ECT?

20. What factors in the papermaking process can be changed to improve ECT?

21. McKee Formula: Compression strength is related to the board's flexural stiffness and what other two factors? Upon what factor does the stiffness depend?

22. Using the McKee Formula, what is the expected BCT for a C-flute box with 26-lb ECT and dimensions of $18'' \times 16'' \times 10''$?

23. What proportion of a box's stacking strength is in its edges and corners versus side panels? What is the effect of adding cut-outs, inserts, or more corners (e.g., octagonal drum)?

24. Is box compression strength (BCT) the same as stacking strength? Why or why not?

25. What are the factors that reduce stacking strength from the predicted or measured BCT?

26. Which factor is most significant? Why?

27. List five ways that poor quality box manufacturing can reduce compression strength.

28. What three factors related to palletization affect stacking strength?

29. Use the FBA's table to estimate the stacking strength of a box with 535 lb BCT, if it is stored for 30 days at 60% RH in an aligned column stack with no pallet overhang or deck board gaps? If the boxes weigh 20 lb and are $10''$ tall, how high could they be stacked?

30. What if the box in the previous question has to survive worse conditions: 90% RH for 180 days with poorly interlocked load and excessive pallet overhang?

31. Use the ASTM D4169 table to estimate the stacking strength of the 535-lb BCT box under average warehouse stacking. What if is the estimated vehicle stacking strength under the same conditions? If the boxes weigh 20 lb, the product does not support the load, and boxes are 10" tall, how high are these stacks?

32. Answer the same question above as if the contents fully support the load.

33. Using the McKee Formula and ASTM D4169 F-factor for average conditions, what minimum ECT would theoretically enable a 10-lb C-flute RSC $18'' \times 16'' \times 10''$ to be able to stacked to a height of 8' for transport, in the case where 100% of the load is carried by the contents?

34. What minimum ECT would theoretically be needed to stack the same box at the same height under average conditions, in the case where none of the load is carried by the contents?

35. Under what conditions might it be preferable to specify BCT directly?

CHAPTER 15

Corrugated Fiberboard Shipping Containers

Corrugated fiberboard box designs have many similarities to folding cartons. The box blank is cut from the board, scored where it will be folded, and slots are cut for flaps. The most common traditional style, the *regular slotted container* (*RSC*), is a simple knocked down tube shape with a *manufacturer's joint* along one edge and flaps on the top and bottom, as shown in Figure 15.1.

As the use of corrugated board has become more diverse, the converting process has gained the capability to create ever more sophisticated packages and products. Today's corrugated board has gone beyond simple, straight-cut brown boxes to beautiful process color graphics and cleverly shaped displays, which will be discussed later in this chapter, along with the common RSC and some other standard box styles. But first, the process of making corrugated board into boxes is described.

15.1 BOX CONVERTING

The sheets of combined board from the corrugator plant are converted into *box blanks* in a separate operation. Sometimes this is done in a separate factory, as in the case of the *corrugator* (*sheet feeder*) *plants* that supply sheets to the *box plants* (*sheet plants*), or even by a separate independent company. A video tour of a box plant can be seen on the website of the International Corrugated Packaging Foundation (www.icpfbox.org).

A *box blank* can be straight-cut or die-cut. Shapes like RSCs with only straight lines are made in a straight line process by a *flexo folder-gluer*. More complex shapes are cut by a *die-cutter*, in a process similar to that used for cutting folding cartons, which may or may not be part of a flexo folder-gluer.

493

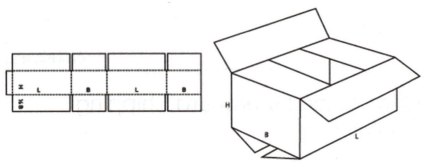

FIGURE 15.1. *RSC and RSC blank (International Fibreboard Case Code #0201). Reprinted with permission of FEFCO, copyright 2007.*

15.1.1 Flexo Folder-Gluer

The *flexo folder-gluer* is used primarily to make the least sophisticated (but highest volume and least expensive) box style: the RSC. It combines five operations that originally were done by separate machines: flexographic printing, scoring, slotting, folding, and gluing the manufacturer's joint.

Some box plants still use two separate machines: the *printer-slotter* and the *folder-gluer*. The traditional *printer-slotter* prints sheets and then cuts, creases, and slots them. Accumulated blanks are then fed into a separate machine, the *folder-gluer*, which folds the blank and glues or stitches the manufacturer's joint. The need to have two or three separate operations limits productivity on high volume jobs.

The flexo folder-gluer begins with a stack of sheets that have previously been slit to the width of the blank (which will be the height of the box), scored across the flutes (along the MD), and cut to the approximate length of the blank. As described in Chapter 14, this is typically done at the corrugator plant. These long flap scores will form the top and bottom edges of the RSC.

The stack of sheets is moved onto the *feed section*, which feeds them one at a time through a series of rollers that line them up for printing. The sheets are fed in the *short way*, in the direction parallel to the flutes.

As the name implies, the flexo folder-gluer employs a flexographic printing process, with polymer or rubber plates. The "rubber stamp" ability of flexo printing is a good match for the coarse and uneven surface of corrugated board. Many flexo folder-gluers have multiple print stations, although most printing on corrugated board is only one or two colors.

Next, the *scoring* section of the machine makes the RSC's four body score creases that run parallel to the flutes. It makes straight lines in only one direction, as each sheet passes through four pairs of mated heads on shafts that nip the board between them. These can be repositioned along the shaft to change box dimensions. Typically there are four since each box has four vertical edges. There may be two additional sets of heads to crush the board where the manufacturer's joint will be formed, so that the joint thickness will more closely match the thickness of the box walls.

The *slotter* section has similar shafts, but these have knife-like heads that cut the sheet to width (if necessary), cut slots to create the top and bottom flaps, and cut the manufacturer's joint. The *slotting heads* are lined up with the creasing heads, and are timed to put the slots at the lead and trail ends of the blank. Slots are most commonly three-eighths inch wide. *Slot strippers* pull away the slot material from between the blades.

The *folder-gluer* section next takes the blank and folds it along the second and fourth creases by using adjustable rods, belts, and folding rails. Folding brings together the first and fourth panels so that the joint can be made in the middle of the knocked-down box, in the same way that pre-glued tube-style folding cartons are formed. The adhesive is applied, using a wheel or extruder, to the manufacturer's joint. Water-based PVA adhesive is generally used and requires time to set. Hot melt adhesives are used when quicker setting is required.

In the last *delivery* section, the flexo folder-gluer counts the finished boxes, squares up and compresses them while the adhesive sets, stacks them in bundles of a predetermined number, and bands them together. It then transfers the bundle to a conveyor.

The simplicity of the RSC is key to its economy. The straight-line cuts and scores enabling a fast process that requires no tooling, aside from resetting the scoring and cutting heads, which is automatic on today's flexo folder-gluer. It is faster and less expensive than die-cutting. There is a certain elegance to the design and process.

The disadvantage is that the traditional configuration can only make simple shapes like the RSC and similar slotted styles. However, many flexo folder-gluers now have integral die cutting sections.

15.1.2 Die-Cutting

Some packages require shaped cuts that are more complex than simple slots. *Die-cutting* is used to make unique designs that require angular, circular, or other unusual cuts, slots, scores, vent, and hand holes and perforations. An increasing number of designs employ die-cut shapes.

Corrugated board die-cutting is very similar to the process used for die-cutting paperboard, described in Chapter 12, with *cutting and creasing rules* mounted on a wooden frame in either a rotary or platen machine. *Rotary die-cutters* use a circular, continuous motion, feeding the sheets between two cylinders: one has the cutting and creasing rules and the other acts as the anvil.

Rotary die-cutting can be done in-line or it can be done in a separate process after printing.

In-line rotary die-cutters can follow the print stations in a flexo folder-gluer machine and may be positioned either before or after the scorer-slotter section. This kind of configuration is commonly used to make special features like hand-holes in an RSC. For applications where the blanks are shipped as flat sheets, such as trays for canned goods, only printing and die-cutting are required, and so the creaser-slotter and folder-gluer are replaced by a *die-cut stacker*.

Platen (flat bed) die-cutters stamp out the blank from the sheet of corrugated board and a *stripping die* pulls away the waste sheet-by-sheet in a second press. Platen die-cutting is the most precise cutting method, and so it is good when tolerances are tight. But it is slower and more expensive than rotary die-cutting.

Die-cutting adds cost. If only straight cuts are needed, the printer-slotter operation is a better choice. The following three sections present more details about printing, slotting, and scoring, and the manufacturer's joint.

15.1.3 Printing

The liner that is applied first, in the singlefacer, is the one that goes inside the box because the glue lines are more pronounced. The smoother face is used outside the box and is printed. Flexography is used because the flexible printing plates best conform to the rough and bumpy surface.

It is difficult to get high-quality printing on corrugated board. The kraft linerboard has a coarse surface, and the flutes cause a *washboard* effect (the uneven surface has high and low points, like a washboard). The brown color does not reflect the light well enough to permit a full range of colors. Printing on corrugated board has traditionally been low tech, low quality, and one color (two at most). Furthermore, for many shipping container uses simple one- or two-color print is sufficient.

But printing on corrugated board is changing faster than any other printing industry. The surface of the linerboard has been refined by using less coarse fibers in the duplex fourdrinier process, and double-facing operations have been improved to make the surface more flat. But most importantly, the flexographic printing process has advanced.

Flexo plates are becoming thinner, anilox rolls are becoming finer, and higher-resolution process color is now possible. The quest for higher quality printing has stimulated improvements in inks and press skills over recent years. Although printing is still mostly one to two colors, because more colors add cost, many new flexo folder-gluers have four to six press stations, and it is possible to print four-color process-color halftones with good registration.

The flexo folder-gluer uses water-based inks. They dry quickly because they are easily absorbed into the porous board, and they are more environmentally friendly than solvent-based inks. The quick-drying inks make it possible to print, score, slot, fold, and glue all on the same machine.

In Japan and Europe, but to a lesser extent in North America, some microflute board is directly printed by offset lithography, similar to the way that paperboard is printed.

15.1.3.1 Preprint, Litho-Lamination, and Digital Printing

For even higher quality print, *pre-printed linerboard* is used in the double facer of the corrugator. Pre-print gives better accuracy in registration and permits the use of special inks and coatings. But preprint is only used for very high volume orders. Eldred (2007) estimates that the point at which preprinted liner is more economical than direct print is between three and six rolls, depending on the equipment.

There is a high cost for set up since the corrugator needs to be threaded with the preprint linerboard, and it runs slower than with ordinary kraft. Most preprint is printed with flexography, although gravure is used for longer runs with higher quality requirements (up to 10 colors). The printing can be on kraft, white (white-top or bleached), or a clay-coated paperboard.

Another high-quality printing option is called *litho-lamination*, or *litho-labeling*, where a lithographically-printed sheet is applied to regular double-faced board, before or after the blank has been cut and scored. The litho-lamination can range from spot labels to fully covering the box blank. They are usually printed by sheet-fed offset.

The most exciting development is *digital printing*. A new generation of six-color ink jet technology has been recently introduced for corrugated board. Digital printing is being used primarily to print point-of-purchase displays that are made in short runs, a market in which it competes with litho lamination. Although it is higher cost per imprint, digital printing eliminates the need for printing plates. The quality of print can be very good, and the image can vary from imprint to imprint. The current state of technology is still high-cost, slow, and uses expensive inks, but it is an active field of development that with a growing market.

Digital printing offers great opportunities for customized limited production runs. To pair it with cutting and scoring capability, either an automated sample table or a common die is used. The sample table adds cost and time for cutting, but is customizable. A common die, which can be used for many different printed images, is more economical, but is limited to the common shape. For example, a generic human shape could be used as the cut-out basis for various printed figures in a similar pose (imagine tailgate parties attended by cut-out images of coaches, players or fans).

15.1.3.2 Basic Elements Printed on Boxes

There are three basic elements printed on most corrugated boxes: the brand identity for the contents, the stock keeping unit (SKU) number, and the box maker's name or (in the United States) its certificate. Boxes for products shipped in less than truckload quantities (especially small parcel shipments) will also include the name, address, and customer number of the consignee, but this is generally attached as a label by the shipper.

The most important identification on a box is for the brand and the company that markets the product. Often it has two colors, black and a spot color to match a brand's trademark. It can be an elaborate and colorful promotion, but for many supply chains, a simple one- or two-color printed company name is sufficient and economical.

The *SKU* identification is the packer/filler's code for the product that is packed in the box. There is a different number for each brand, size, configuration, and flavor of product.

The SKU identification can be printed at the time the box is made in the high speed printing press, or it can be postponed until the box is filled and applied using an ink jet printer or a simultaneously printed label on the packing line. When coupled with a strategy wherein one size of shipping container is used for many SKUs, the postponement strategy can yield savings in purchasing cost due to volume discounts as well as reduced inventory.

The SKU code is used throughout supply chains to track inventory and match transactions to movement, so it needs to be easy to "read" by humans or machines employed by the supply chain. In the typical human-readable system, workers check the number on the package against documents like a bill-of-lading or order-picking list. He or she may later record transactions or movements in a computer system. Both the code-reading and data entry are prone to errors, which can be reduced by the use of a clear, easy-to-read SKU code.

Automatic identification—bar codes and radio frequency identification (RFID)—can improve the quality and timeliness of logistical information, but only when the whole supply chain is using a common coding system. A bar code can be printed either at the box plant or at the filling location, with the added complication that the quality that is required to make a bar code readable is sometimes difficult to achieve on corrugated board, especially using on-demand ink jet printing. An RFID tag can be applied by the box maker and reprogrammed by the packer/filler to reflect SKU information. There is potential for printing an RFID antennae and applying the chip directly to packages, and active research in the subject, but commercialized systems are not yet available.

The box maker's certificate is usually placed in full view on an outer flap on the bottom of a box. In the United States, it gives the box maker's name and location and lists either the minimum combined weight of facings and the minimum burst test or the edge crush test value for the combined board, as discussed in Chapter 14.

Other printing that may appear on corrugated boxes includes information about recyclability (and/or recycled content) and specific handling instructions, like "This End Up" or "Keep Dry." A standard set of instructional symbols can be found in the National Motor Freight Classification (NMFTA), the same book where the motor carriers' other packaging recommendations are published. International symbols are standardized by the International Standards Organization (ISO). Some examples are shown in Figure 15.2. If the product is a hazardous material, the government (DOT) requires special markings to identify the product, the shipper, and the package type and manufacturer (Title 49, U.S. Code of Federal Regulations).

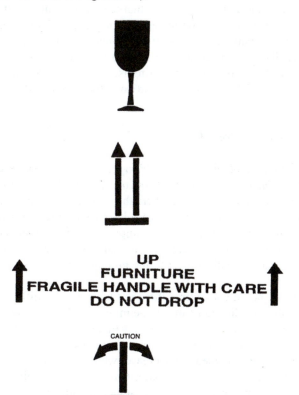

UP
FURNITURE
FRAGILE HANDLE WITH CARE
DO NOT DROP

CAUTION

TOP HEAVY

FIGURE 15.2. Special instructional symbols. Reprinted with permission of the National Motor Freight and Traffic Association.

15.1.3.3 Print for Marketing and Advertising

Most corrugated boxes have minimal printing because it adds to the cost. But box-makers increasingly find opportunity in adding the value of higher quality printing because it adds to their profitability. They are investing in better presses and better trained crews.

Improved printing capabilities have given corrugated board a new role in merchandising *point of purchase* (POP) displays and cases with bold graphics. A POP display lets the manufacturer control the retail frame for a product. POP displays offer the opportunity to advertise and capture a shopper's attention at the moment when he/she is most likely to purchase a product.

High quality printing can be a powerful advertising medium in a retail store or wholesale market. For example, the shipping containers for fresh produce have a tradition of proudly identifying the grower, to give a competitive advantage in wholesale and retail produce markets. A colorful end-of-aisle standalone structure sets the stage for an eye-catching array of consumer packages. Corrugated board manufacturers have found many creative ways to structure and decorate displays. When the box is also the package that the consumer buys, as for electronic products, power tools, housewares, and toys, its graphics must have all of the information and advertising required for self-service shopping.

15.1.4 Scores and Score Allowances

In order to make a blank that will fold to form a box, it is necessary to score and slot it. *Scoring* or *creasing* wheels or rules are similar to cutting blades.

Unlike the folding carton score, where the raised ridge of the score is on the inside, there is no similar ridge (or bead) because corrugated board contains void space. The scoring blade simply compresses a groove and collapses the corrugated structure. The box walls are folded towards the groove (so this is the opposite direction with respect to the groove, compared to a folding carton).

In corrugated board converting, there is such a distinct difference between the scores in the two directions that some in the industry have even attempted to give them different names. One reason is that they are usually made on different equipment, sometimes by different companies.

Scoring is properly defined as the creases which cross the flutes, and is generally done on the end of the corrugator. *Creasing* runs parallel to the flutes, and is commonly done on the flexo folder-gluer. In industry practice, however, the two terms are often used interchangeably.

Slotting (removing the narrow slice of board from between the flaps) is necessary to permit the flaps to fold. The slot must be deep enough to permit folding. But too deep, beyond the score line, significantly reduces the box's compression strength, because so much of the top-load is carried by the corners. The high speed printer-slotters or flexo folder-gluers are not as precise as die-cutting, and their dimensional tolerance is about one-sixteenth inch.

Just as with folding cartons, a score allowance must be added to box dimensions, but unlike folding cartons, the score allowance is not specified by the user, but is added separately by the box maker. *The buyers of boxes only specify the inside dimensions.* The box maker is expected to make the box fit, and adds the allowances.

The three-dimensionality of corrugated board must be skillfully accounted for in score allowances, because the score occupies space on the blank, and the fold subtracts space from inside the box. (This is easy to forget, and the first box made by almost everyone is too small.) Furthermore, the inner and outer flaps are all folded on the same scoreline, but do not occupy the same space when the box is closed. Each box manufacturer has its own scoring allowance standards based on the board and the type of scoring wheels used; they are not uniform from one company to the next.

As a general rule of thumb, the score allowance is equal to the thickness of the board. The reasoning behind this is that if a piece of corrugated board is scored or creased along two parallel lines, the distance between the facing panels bent up at 90° from the main panel is approximately equal to the original distance between the lines less the thickness of the corrugated board, as shown in Figure 15.3.

For C flute with a thickness of five-thirty-seconds inch, the score allowance for side-to-side dimensions is usually between three-sixteenth inch and one-quarter inch. In general, the thickness is rounded up to the closest one-sixteenths inch increment greater than the actual width to insure having an adequate inside dimension. The allowance is added to each of three of the panels in the length and width dimensions.

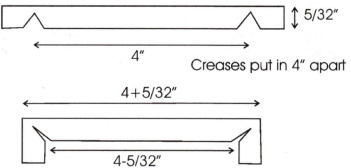

FIGURE 15.3. *Rationale for score allowance equal to board thickness.*

The panel that is adjacent to the manufacturer's joint has slightly more or less added allowance depending on whether the joint is lapped inside or outside the box. If the joint is inside, a smaller allowance is added in order to make the resulting box's flaps close squarely. If the joint is outside, a larger allowance is added to this panel to give enough board to go around the outside of the corner.

The allowance for the top-to-bottom dimension adds about twice the thickness of the board to the body panels, since two flaps will need to fit into the space on the top and bottom. When the board is especially thick, the inner flaps are scored on a different line from the outer flaps.

Perforations or slits through one face can be used to replace or augment scores when a fold of greater than 90° is needed or when board is especially heavy and stiff. Perforations are also used to make easy opening features. They are common in point of purchase display packages where the retailer can simply tear off one panel to reveal the display. However, poorly made perforations reduce board strength and can easily tear open too early during distribution.

15.1.5 Manufacturer's Joint

The *manufacturer's joint* fastens together the sides of the blank along one edge. It is usually a minimum of 1-3/8" wide with a glue width of 1-1/4".

It can be inside or outside of the box. Inside is more common because it gives a square outside dimension. On the outside, it has the advantage of not interfering with automatic case packing equipment. However, a joint on the outside can affect the alignment of boxes on a pallet, as shown in Figure 15.4, and can also result in snagging on conveyors and shelves.

The most common way to make a manufacturer's joint is with water-based PVA adhesive (also called *cold glue*) or, less commonly, with hotmelt adhesive in the high-speed flexo folder-gluer machine. A PVA-glued joint is strong and inexpensive. It can fail when it gets wet, but the starch-based adhesive used to combine the board plies will usually fail first. Hot-melt adhesive makes a joint that is strong dry or wet, and cures quickly.

Although it is now rare, manufacturers' joints can also be joined off-line with a *stitcher* (which uses staples), a *taper* (which uses tape), or a gluer. An

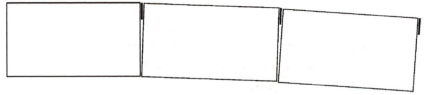

FIGURE 15.4. *Outside manufacturer's joint can affect alignment of adjacent boxes.*

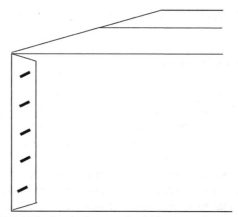

FIGURE 15.5. *Stitched joints should have staples at an angle.*

unlapped, taped joint is the least strong, but since there is no double-thick joint, the knocked-down boxes lay flat with no bulge on the palletload.

A stitched (stapled) joint is often used for large and heavy boxes with flowable non-food contents, since a glue bond is limited by the internal bonding strength of the board. To maximize the strength of the stitching, the staples should be at an angle to the box edge, flutes, and the machine direction of the paper, as shown in Figure 15.5. Stitching is expensive because it is usually done off-line and is slow (although modules can be added to flexo foldergluers; this is more popular outside North America). Stitching is not used when the contents could be damaged by the staples (for example, food or a liquid-filled bag-in-box). It is also avoided as a method for closing boxes when the staples could pose a danger for the person opening the box.

15.2 COMMON CORRUGATED BOX STYLES

There are a number of standard box styles. They are designated by name (such as *regular slotted container*), acronym (like *RSC*), and by an International Fibreboard Case Code (e.g., #0201). The Case Code has been developed by the FEFCO (the European Federation of Corrugated Board Manufacturers (also known as the Fédération Européenne des Fabricants de Carton Ondule) and EBSCO (European Solid Board Organization). It has also been adopted by ICCA (the International Corrugated Case Association). The American FBA (Fibre Box Association) publishes the codes in its Fibre Box Handbook (FBA 2005). This chapter refers to the International Case Code numbers and their common names in the United States.

There are five basic forms: slotted containers, telescope boxes, folders, rigid (Bliss) boxes, and self-erecting boxes. Corrugated board can also be used to make display stands, pallets, and slipsheets. Some common forms of these are described next.

5.2.1 Regular Slotted Container

The RSC is the most popular and basic box style, shown in Figure 15.1. It is inexpensive and strong. The RSC is the most economical box style; there is very little manufacturing scrap because all flaps are the same length, one-half of the container's width. The straight-line scoring, slotting, and folding is a less expensive process than die cutting.

Since most boxes have a shorter width than length, this means that the flaps on the long sides meet at the center when closed, but the flaps on the short sides do not. The flaps that meet are always closed second, on the exterior where they can be securely sealed. The International Case Code for RSCs is #0201.

The most economical RSC in terms of material usage has proportions $L{:}W{:}D = 2{:}1{:}2$. This uses the least square footage of board for a given cubic volume (the greatest volume for a given square footage of board), and minimizes the size of the flaps. These dimensions result in a square front face and boxes that interlock efficiently.

Example:

What are the dimensions of an RSC containing 4 cubic feet that uses the least amount of C-flute?

$$L{:}W{:}D = 2{:}1{:}2$$
$$\text{Let, } W = \text{unknown}$$
$$\text{Then, } L = 2W$$
$$D = 2W$$
$$\text{Vol} = LWD = 2W(W)(2W) = 4W^3 = 4$$
$$W = 1, L = 2, D = 2$$

Of course, there are many considerations besides just minimizing the amount of board, especially since the contents have dimensions of their own based on merchandising considerations and packing geometry. For example, cylindrical products sold by the dozen are usually packed in a 3×4 configuration. Other dimensional considerations include the width of the board web (sometimes a small change in dimensions can reduce scrap) and optimizing pallet patterns.

15.2.2 Other Slotted Container Styles: HSC, CSSC, OSC, and FOL

The *half slotted container* (*HSC*) is the same as the RSC, but it has an open top with no flaps, as shown in Figure 15.6.

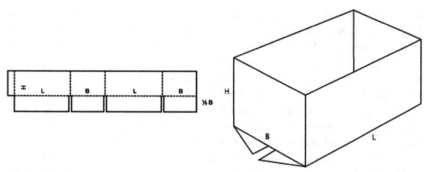

FIGURE 15.6. *Half slotted container (HSC) #0200. Reprinted with permission of FE-FCO, copyright 2007.*

The *center special slotted container* (*CSSC*) is like the RSC except all of its flaps meet at the center, as shown in Figure 15.7. Since most boxes have a shorter width than length, this means that the flaps on the short sides must be longer than those on the long sides in order to meet in the center when closed. Therefore, this style uses more material and creates more manufacturing waste than the RSC, due to the blank having longer flaps on two panels and wasting a strip of scrap on the other two.

The chief advantage of the CSSC style is its flat top and bottom, since both the top and bottom have a full double thickness of corrugated board, providing a level base for the products inside. It is also a more squared-up style, because making the flaps all meet reduces the possibility of parallelogramming when closed. Side-to-side compression strength increases substantially compared to an RSC because all flaps meet.

There are several variations on the CSSC style. In the *side special slotted*

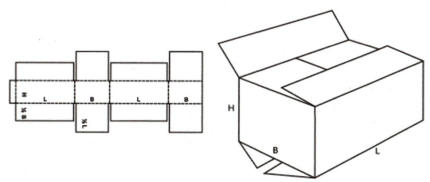

FIGURE 15.7. *Center special slotted container (CSSC) #0204. Reprinted with permission of FEFCO, copyright 2007.*

container (SSS) all pairs of flaps meet, but not at the center of the box. The *center special overlap slotted container (CSO)* and *center special full overlap slotted container (SFF)* have differential length outer flaps of varying degrees of overlap, with inner flaps that meet in the center.

The *overlap slotted container (OSC)* has flaps all of one length (like the RSC) but they are long enough for the outer flaps to overlap, as shown in Figure 15.8. The overlapping flaps are often closed with staples. This is a style that is good for long products where there is a long gap between the inner flaps, because the sealed overlap prevents the outer flaps from pulling apart.

The *full overlap slotted container (FOL)* has all flaps the same length as the width of the box, with the outer flaps coming within one inch of complete overlap. It is similar to the OSC except with more overlap, as shown in Figure 15.9.

The FOL is a particularly strong box that stands up to rough handling. The extra layers of board add cushioning. It also has good strength when stacked on its side, since the full overlap flaps can provide compression strength to make up for the loss of compression strength when flutes are horizontal. It is also an expensive style that uses much more corrugated fiberboard than does a common RSC.

The slotted container that uses the least board doesn't even have a Case Code number. Called a *gap-flap* box, none of the flaps meet, leaving a gap in the center. This is used for products or consumer packages, like cereal cartons, that help to provide compression strength and have a dimension such that they won't fall through the gap. The gap-flap style is even more economical than an RSC. It is easy to open without a box-cutter because the gap gives a place to grip the flap and pull it open. It does require a special casing machine capable of gluing the partial flaps the right amount in the right place.

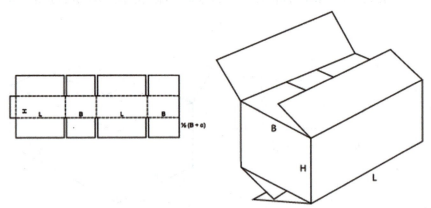

FIGURE 15.8. *Overlap slotted container (OSC) #0202. Reprinted with permission of FEFCO, copyright 2007.*

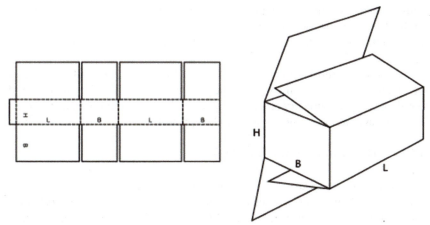

FIGURE 15.9. *Full overlap slotted container (FOL) #0203. Reprinted with permission of FEFCO, copyright 2007.*

15.2.3 Telescope Styles

Telescope style boxes have a separate top that fits over the bottom. The truck and rail classifications call these *telescope boxes* if the cover extends over two-thirds of the way down the sides and *boxes with covers* if the cover is shorter.

Full telescope boxes with the cover the same depth as the bottom are used for produce, like bananas, where they can provide very good compression strength and bulge resistance due to the double layer of sidewalls. Full telescope boxes are favored for reuse because they are strong and easy to repeatedly open and reclose.

Partial telescope boxes and boxes with covers may depend on the sidewalls of the bottom box for compression strength. On the other hand, as in boxes used for reams of office paper, the box can be filled higher than the level of its sidewalls with the cover resting on the contents rather than the sidewalls. Only sturdy and square contents like office paper will support such a topload.

It takes two blanks to make a telescope style box. The bottom one must have a slightly smaller footprint than the top to ensure a good fit. There are two styles for the blanks. Two trays are used to make the *full telescope design* (*FTD*) and *design style with cover* (*DSC*). The second, the *full telescope half slotted* (*FTHS*) style, is made from two half slotted containers (HSCs). These styles are shown in Figure 15.10.

The FTHS style is key to the marketing of bananas. Most banana boxes today are full telescope style, but with short gap-flaps. The telescope style is needed to give compression strength, since the banana shape does not help to support the load. The flap gap helps to facilitate air flow, which is crucial because bananas are ripened by the application of ethylene gas. They are gassed in the box, in transit or storage chambers, to trigger ripening at the right time to be sold. The box also makes it easy to check for ripeness because it can be reclosed.

Furthermore, most banana boxes continue to bravely serve after their tour of banana duty. Most of them are reused, primarily as boxes for salvaged and unsalable goods in grocery reverse logistics operations, where they are favored for their large numbers, uniformity, and reclosability.

15.2.4 Folders and Wrap-Around Cases

Folders and *wrap-around cases* are formed from a single blank that is scored to fold up around a product, as shown in Figure 15.11. The most common types are variations on a *five panel folder*. The fifth panel completely covers one side. In addition, the closed box can have overlapped layers of flaps on each end, which gives good stacking strength. Five-panel folders are used

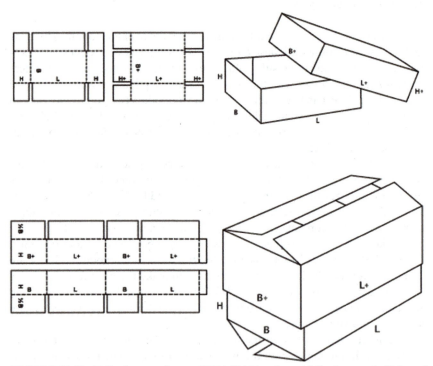

FIGURE 15.10. *Full telescope design (FTD) #0301 style and full telescope half slotted (FTHS) #0320 styles. Reprinted with permission of FEFCO, copyright 2007.*

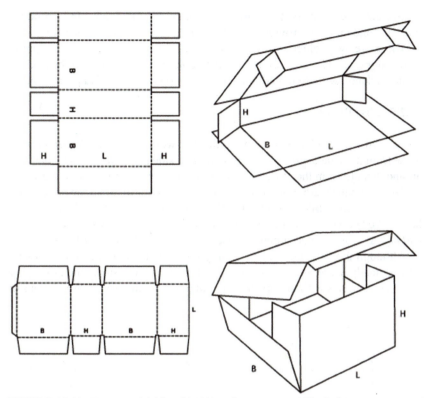

FIGURE 15.11. *Five-panel folder (#0410) and wrap around blank (#0406). Reprinted with permission of FEFCO, copyright 2007.*

mostly for packages in which the depth is the shortest dimension, especially for long products like vinyl siding.

A *wrap-around blank* is similar, but in place of the fifth panel it has a shorter manufacturer's joint. Wrap-around cases are formed in a machine that wraps the blank around the product and glues the end flaps and joint. Hot melt adhesive is used since it must be set before the box can be handled or stacked.

Wrap-around cases are typically less expensive than RSCs because they use less board for the same volume of product. This is because the flaps are on the smaller ends, rather than the top.

A wrap-around case provides a tighter pack than an RSC, because the RSC needs to be slightly larger in order to give room for filling. The wrap-around case's tightness is a virtue for products like cans and jars because it reduces the void space that allows them to bang together inside a case that is dropped. Wrap-around cases provide less compression strength than RSCs because some flutes are horizontal, but they are commonly used for sturdy products like cans and jars that provide enough strength on their own for stacking.

There are two folder styles designed for books in small parcel shipping, shown in Figure 15.12. These rely on the book to provide much of the strength in the package, with #0403 adding a cell on two ends for added protection.

15.2.5 Bliss Boxes

Blessed with the happiest name in the box world, the *rigid*, or *Bliss box*, is made from three pieces of solid fiberboard: two side panels and a body that folds around them to form the other sides, bottom, and top, as shown in Figure 15.13.

The side panels are glued using special equipment to make the six or more joints at the packer/filler's plant. Once the joints have been sealed, the box cannot be knocked down, hence the name *rigid box.* They are also known as *recessed end boxes*. This style is sometimes used to make solid fiberboard beer cases.

Although the Bliss box was originally made from solid fiberboard, they are now typically made from corrugated fiberboard and are used in the grocery industry, sometimes without tops for point of purchase displays. With no flaps on

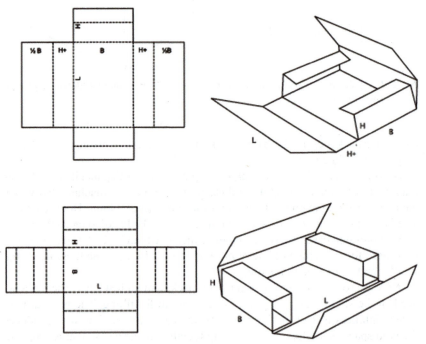

FIGURE 15.12. *One piece folder (OPF) #0401 and OPF with cell end buffers #0403.*

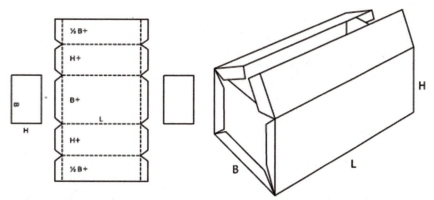

FIGURE 15.13. *Bliss box also known as rigid box (#0606). Reprinted with permission of FEFCO, copyright 2007.*

the bottom, this style uses less board than an RSC and has good compression strength along the joints. There are several variations of the style, including one without a top.

15.2.6 Self-Erecting and Snap-Bottom Boxes

Self-erecting and *snap-bottom boxes* have die-cut bottoms that cleverly fit together and/or are pre-glued in such a way that the packer-filler can manually erect the bottom of the box without using any adhesive or equipment. Such designs are especially useful for small scale operations that cannot justify the investment in equipment for automatically erecting boxes and find that it is too labor-intensive to manually erect and secure the bottom of RSCs.

Self-erecting boxes have a pre-glued automatic bottom that simply folds into place. Snap bottom (also known as *1-2-3*) boxes have a bottom that interlocks in a 3-part process: the largest flap is folded in first, then the two side flaps, and the final flap locks everything into place. Figure 15.14 shows these styles. The tops of these boxes can be RSC-style or tucked.

15.2.7 Fresh Produce Trays

The *fresh produce tray* is an open-top shipping container and display tray for fresh fruits and vegetables. The tray is formed in the field in tray-forming equipment that erects the tray and glues together the corners. There are some unglued, self-locking designs, but they lack the stability of glued corners. A typical produce tray is shown in Figure 15.15.

Sometimes called the *common corrugated footprint tray* (*CCF*), most have

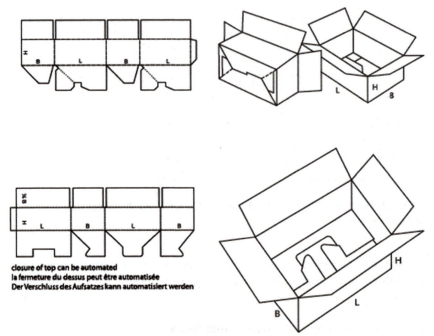

FIGURE 15.14. *Self-erecting (#0711) and snap bottom (#0216) boxes. Reprinted with permission of FEFCO, copyright 2007.*

a standard footprint of 60 cm x 40 cm (23-1/2″ × 15-11/16″). The standard outside $L \times W$ dimensions are supported by the International Standards Organization (ISO 3394 2012), the U.S. Fibre Box Association (FBA 2013), and the European Federation of Corrugated Board Manufacturers (FEFCO 2008).

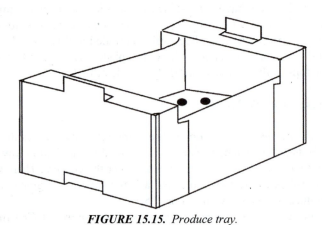

FIGURE 15.15. *Produce tray.*

This common footprint fits "five down" on a standard 48″ × 40″ U.S. pallet or a standard 120 × 100 cm international pallet, or four down on a standard 120 × 80 cm Europallet. The depth can be varied to suit the product depending on the dimensions, weight, and demand for the product, to optimize shipping density and protection.

The standards also include a half-size, 40 × 30 cm (15-11/16″ × 11-11/16″) (and a quarter-size that is rarely used). Products most appropriate for the 40 × 30 cm tray are generally smaller, with a lower level of demand, like mangos, avocados, papayas, strawberries, passion fruit, mushrooms, okra, hot peppers, figs, and limes (ITC 2002).

The footprint dimensions were chosen to match the international standard size for reusable plastic produce containers (RPCs), with which they compete. As a result, the two types of containers, and many varieties of produce sourced from anywhere in the world, can fit together in a modular system. They make tidy and standard mixed palletloads and displays, and can be automatically sorted in distribution centers. They facilitate international trade.

The trays have features to facilitate stacking and display. Some styles have a narrow support shelf along two top edges, which can be colorfully printed to add brand identity in a store display. They have standard-positioned interlocking tabs on top and matching slots on the bottom, 2-1/2″ (64 mm) wide × 5/8″ (16 mm), to better secure a column stack and prevent sliding. While most are open-top, there are also styles with flap or full telescope covers. Above all, they are designed to prevent "nesting," which in this case means that the geometry and features need to provide enough compression strength so that the containers do not "nest" into the container immediately below forcing the produce itself to carry the weight of the stack.

Retailers rely on the standard footprint for building modular displays. An empty tray can be easily swapped for a full one, which improves the retailer's stocking productivity, especially on a busy shopping day. The ability of the CCF trays to carry colorful graphics is a distinct advantage over the equivalent reusable plastic containers, giving a means for branding and advertising.

Produce trays also need to facilitate cooling and maintain their strength under very humid conditions. They may be filled in the field where it is wet, the produce transfers its own moisture to the board, and condensation is common during transfer across a loading dock. Ventilation cut-outs are necessary for quick chilling, but they weaken the board. Wet strength paper, wax coating, and water-resistant adhesive are often used because of the high moisture content of fresh produce.

FEFCO and FBA have designated a special voluntary mark for trays that meet their standard criteria. The criteria can be found in the technical specifications on their websites (FBA 2000, FEFCO 2008). The CCF criteria, however, are very general and the specific design, combination of board, and features can vary.

The Spanish corrugated board industry (the Asociación Española de Fabricantes de Cartón Ondulado) has developed some of the most comprehensive designs for produce trays. Their *Plaform* proprietary licensed system has recommended constructions and test methods including vibration, compression, and bottom-sag. The standards result in tray specifications tailored to the characteristics of the type of produce. For example, a tray for heavy tropical pineapples would have a stronger board combination than would a tray for lightweight raspberries in plastic clamshells. Trays for wetter produce have more waxed plies than do those for dry products (AEFCO 2009).

15.2.8 Point of Purchase Displays

Point of purchase (*POP*) *displays* are gaining popularity. They can be a powerful advertising medium, attracting shoppers at the moment when they are most ready to buy. Manufacturers increasingly use highly-decorated corrugated board to provide a frame to showcase and advertise their products in the store. This trend has stimulated the development of higher-value four- and six-color printing on corrugated board.

Display trays are made from a solid bottom piece of corrugated board with its edges folded up to form sidewalls and its corners joined. The top is usually open, although a cover can be attached. There are many variations. The sidewalls can be short, as in the short one inch- high trays that are used to provide a solid bottom for a shrink-wrapped case of cans and bottles. Or they can be tall, as in the trays used in telescoping boxes.

One of the most common types of display trays used in the grocery industry is shown in Figure 15.16. It has high sidewalls on the two long sides and corners to provide support to the products within, and it has low walls on the ends to make the products easy to see.

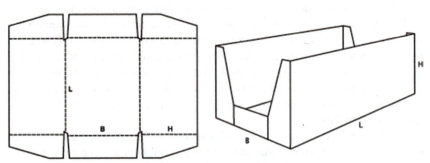

FIGURE 15.16. *Display tray (#0460). Reprinted with permission of FEFCO, copyright 2007.*

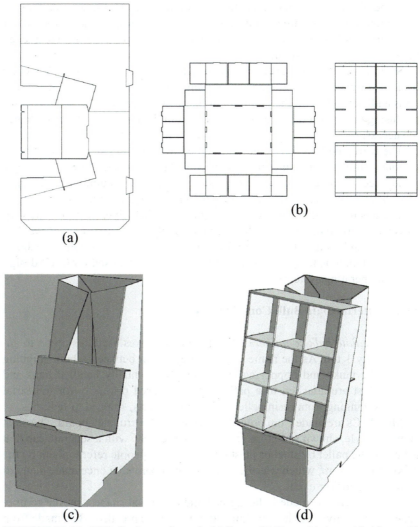

FIGURE 15.17. *POP display and stand: (a) Base; (b) Tray; (c) Assembled base; (d) Assembled display. Reprinted by permission from Dennis Young, 2014.*

Display trays are used in the store to form end-of-aisle displays and are common in high volume discount stores. They cut the cost of building displays because there is not a top to remove and the product can be displayed right in the case. In many cases, full or partial palletload displays are assembled by the supplier, and the retailer has only to set it on the floor for a pre-built display.

Stand-alone POP displays, like that shown in Figure 15.17, are used for special promotions. They are used in stores for a few weeks, and the stock may or may not be replenished as it is sold. They present the product at the ideal level, between eye and waist-high.

Stand-alone POP displays must be easy to set up since they may be set up by a retail worker in the store, who is unfamiliar with how to do it. The instructions and method are in the grand tradition of "Tab A in Slot B," which can be confusing in the best of circumstances, so a simple design has the best chance of success.

The corrugated display stand is designed to hold trays that are refilled, or to hold a shipping container that has been suitably opened and mounted. The stand needs to be easy to maintain and sturdy to resist humidity and abuse. Most have bases and headers that have the highest quality printing.

There is no standard style display stand which has given rise to some very clever designs. Examples can be found in publications like *Out of the Box* (Lianshun and Lang 2008) and *Package and P.O.P. Structures* (Dhairya 2007), which also include templates on a CD-ROM that can be used for CAD design on most operating systems.

15.2.9 Intermediate Bulk Containers (IBCs)

Intermediate bulk containers (*IBCs*) are large boxes used to ship up to approximately 1 ton. As the name implies, they are in between a bulk shipping mode (like tanks and hopper cars) and smaller manually handled bags and boxes. Most are attached to a pallet. IBCs are used primarily for industrial products, ranging from resin pellets to assembly parts, shipped to factories.

Most IBCs are made from double- or triple-wall corrugated board. The most common style, shown in Figure 15.18, is a large HSC with a separate cap that is held to the pallet by steel or plastic bands. Some people refer to them by the nickname *Gaylord*, which was the name of a company that once made many of them but no longer exists.

IBCs can also be made in hexagonal and octagonal shapes, similar to fiber drums, which gives greater stacking strength. There are also IBCs made from steel and plastic, as well as flexible IBCs (*FIBCs*) made from woven fabric (primarily polypropylene).

A style closely related to the IBC and to the telescoping style is the tube-and-cap design shown in Figure 15.19, used for appliances, furniture, and assembly parts. It is made from three corrugated parts—a base, a tube, and a cap—and a pallet. The corrugated tray is placed on the pallet, the product is placed in the tray, the tube of corrugated board is pulled down around the product and fitted into the tray, and the cap, which is also in the form of a tray, is put on top and strapped down. There is a similar package used for large ap-

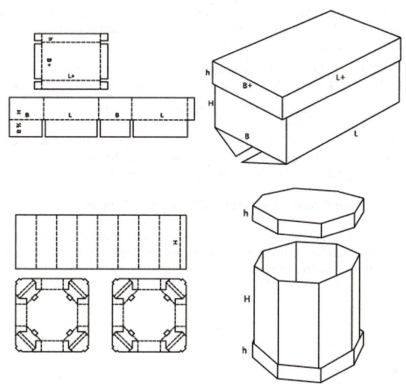

FIGURE 15.18. *Intermediate bulk containers (#0312 and #0351). Reprinted with permission of FEFCO, copyright 2007.*

pliances, called a *baseloid*, in which the top cap is more securely attached and used as a lifting assist.

15.2.10 Corrugated Boxes for Extreme Environments

When corrugated board is expected to perform in extreme situations, especially when water and weather resistance is needed, a higher standard of specification is required. The U.S. government (General Services Administration and Department of Defense) during World War II developed a special series of weatherproof and waterproof containers (Lincoln 1945). Although most of the constructions would not be acceptable today from an environmental point of view (because of asphalt impregnation and rubber adhesives), they have left a legacy of test methods and specifications.

Today, more acceptable wet strength and water resistant treatments and adhesives have been developed, and the government standards have been now transferred to ASTM. The only things that have not changed are the names; they go by the V (weatherproof, but also for "victory"), and W (water resistant) series, with names like V3c and W5c. The standards are:

- ASTM D4727/D4727M: Corrugated and Solid Fiberboard Sheet Stock
- ASTM D5118/D5118M: Fabrication of Fiberboard Shipping Containers

These two standards also reference the test standards for boxes used in extreme situations, like water and fire resistance.

V- and W-board are primarily used for military applications. The organization to which the military packaging professionals belong is the National Institute of Packaging, Handling, and Logistics Engineers (www.niphle.org).

15.2.11 Other Creative Corrugated Fiberboard Containers

There are many other container designs possible using corrugated fiberboard. The standard styles are popular primarily because they are easy to make and fill using standard equipment.

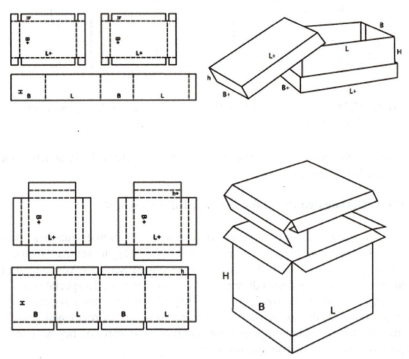

FIGURE 15.19. *Tube-and-cap and baseloid designs (#0310 and #0325). Reprinted with permission of FEFCO, copyright 2007.*

Like paperboard, corrugated fiberboard can be used to make many non-standard container designs, limited only by what can be folded into shape with the help of cuts and scores from a thick three-dimensional board. Corrugated board is a very versatile material, capable of making many kinds of cleverly shaped folders, wrappers, bins, and bikini boxes (so called for their minimal coverage). Resources like *The Packaging Designer's Book of Patterns* (Roth and Wybenga 2013), *Packaging Templates* (Hai 2009), and competitions sponsored by the corrugated industry are the source for many good ideas.

For a new design to be useful, however, it must be capable of being made and filled. And to be useful on a large scale, these processes must be mechanized. A new shipping container design can involve a great deal of development work.

15.3 SOLID FIBERBOARD BOXES AND SLIPSHEETS

Solid fiberboard was developed in the early 1900s, at the same time as corrugated board. It is similar but without the corrugation, made by gluing two or more plies of containerboard together. The containerboard can be made from recycled paper or virgin kraft. The caliper can range from 35 to135 points (0.89–3.43 mm), but most is 70–80 points (1.8–2 mm).

For most uses, corrugated board is preferred, because of its high strength to weight ratio and its low cost. But solid fiberboard can be made to be stronger, tougher, and more water-resistant, and so it is preferred for applications like slipsheets and reusable boxes, described in the following sections.

15.3.1 Solid Fiberboard Boxes

Solid fiberboard is heavier and more expensive than corrugated board, and IS used primarily for reusable boxes or applications where water-resistance is needed. Boxes can be made in most of the same styles as corrugated boxes. Solid fiberboard can be cut and scored using rotary or flatbed processes, although the material is harder to bend and cut compared to corrugated board.

The standard grades resist moisture better than corrugated board, and this can be significantly improved by adding internal sizing, kraft facings, or an extrusion-coating of PE. There are also coatings to add resistance to grease and oil.

Solid fiberboard is more popular in Europe, where the largest market is for poultry boxes, accounting for about 60% of solid fiberboard box use there. They are used for vacuum-packed meat and poultry. Fresh fish is packed wet and then frozen in boxes made from PE-coated board. Likewise fruits, vegetables, and flowers may be packed wet in solid fiberboard boxes (Kirwan 2013).

Because of its durability, puncture-resistance, and high tensile strength, solid fiberboard is also used to make boxes for reusable beverage bottles, games, and shoes.

15.3.2 Slipsheets

The primary use for solid fiberboard in the United States is for *slipsheets*, which can substitute for pallets to serve as a base for a unit load. A slipsheet is a heavy high tensile strength sheet of corrugated or solid fiberboard. They can also be made from plastic, but fiberboard is generally less expensive and easier to recycle.

A slipsheet fits the footprint of the load (usually 48″ × 40″ in the U.S. grocery industry and 1,200 × 1,000 mm internationally), and has tabs (usually 3″–4″, 7.5–10 cm wide) extending along one or more sides, as shown in Figure 15.20. The number of tabs depends on from how many sides it will be handled, from one to four, similar to two-way versus four-way pallet designs.

The disadvantage of slipsheets is that they require a systematic change to special equipment throughout the logistics network, every place that loads are handled. A slipsheet alone is not stiff enough to support a unit load during material handling. They need to be handled with a lift truck that differs from a fork lift because it has a solid platform plate instead of forks. It has a gripper that extends to grab the tab on the front side of the slipsheet. Then the gripper retracts, pulling the slipsheet onto the solid plate. The lift truck can then drive the load to its next position.

Slipsheets are most common in the grocery industry where wholesale and retail distribution centers commonly have the specialized equipment necessary to receive them. Often, slipsheeted loads are placed onto pallets when they are received at a distribution center to facilitate handling and storage in the facility.

The advantages offered by slipsheets are that they are less expensive than pallets and save cubic space in transport. The space (or cube) advantage increases in importance with the transport distance, since it directly affects transport cost. Slipsheets are most often used for lightweight products like cereal and paper products that do not reach the weight limit for truck transport; the use of slipsheets can enable the shipper to put more cargo in a trailer, thus reducing transport cost.

The cube advantage is being used by some electronics manufacturers who ship globally. They find that by using slipsheets they can utilize all of the space

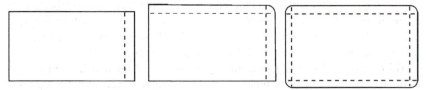

FIGURE 15.20. *Slipsheets: one-way, two-way, and four-way tabs.*

in a sea container as efficiently as if it were hand-loaded, with the added advantage of being able to mechanize and improve the productivity of loading. Such shippers usually find that they can easily justify the purchase of the special lift trucks for their distribution centers at both ends of the transport route because of the savings in transport and handling costs.

Solid fiberboard slipsheets are made from a lamination of plies of kraft linerboard and cylinder board; the thickness is expressed in points. Corrugated fiberboard slipsheets are described in terms of flute size and the Mullen burst test value.

The most important physical properties of slipsheets to specify are high tensile strength (especially across the score line between the tab and the platform), high stiffness, and high coefficient of friction. High tensile strength is necessary to be able to grab and pull the tab without breaking it. Stiffness prevents sagging. The coefficient of friction must be sufficient to prevent the packages in the unit load from sliding off of the slipsheet.

15.4 OTHER CORRUGATED FIBERBOARD PACKAGING APPLICATIONS

Corrugated fiberboard is a versatile material that has been used beyond the packaging realm, to make furniture and other structures.

Four applications are its use for insulation, cushioning, dunnage, and pallets. Although corrugated board is not the usual choice for cushioning or insulation, its structure provides both. The properties of corrugated board are not as predictable as those of plastic foams and the weight is considerably greater, resulting in higher freight costs.

15.4.1 Insulation

The insulating properties of corrugated fiberboard are about the same as for an equivalent thickness of plastic foam, due to the air trapped between the faces (Ramaker 1974). This provides a benefit for shipping refrigerated products, because the box alone can protect for a short period. On the other hand, it is a problem for shippers of freshly harvested fruits and vegetables that need to be quickly chilled, and is the reason that most produce boxes have ventilation holes.

Corrugated insulating materials have been developed with polyethylene or foil laminations, to trap and/or reflect heat.

ASTM D3103—Thermal insulation quality of packages—provides a method for testing insulated packaging, with sensors inside and outside of a package stored in a controlled environment.

15.4.2 Cushioning and/or Dunnage

Corrugated fiberboard's collapsible flutes are not as resilient as plastic foams, but they are capable of absorbing shocks and of amplifying vibration. The walls of a shipping container made from corrugated board provide a thin cushion and give surface protection.

Forms and blocks made from built-up layers of corrugated board are used as interior packaging, usually intended as a blocking or dunnage material to separate the sidewalls of the box from the product. This is not to be confused with a cushion designed according to predictable cushion curves (which trade off cushion density and thickness against weight bearing area). The cushioning characteristics depend on whether the material is used in an edge crush or flat crush orientation (Paine and Gordon 1973).

Problems in performance as a cushion include fatigue sensitivity due to gradual compaction of the flutes, RH sensitivity, and tendency to create dust. Also, consumers tend to perceive corrugated as a cheap, although environmentally favorable, material. Corrugated cushioning is most often used for light weight relatively non-fragile products.

ASTM D1596—Shock absorbing characteristics of package cushioning materials—provides a means for testing.

Dunnage that braces products inside containers, such as edge protectors, corner blocks, and cushioning materials, can be made from built-up corrugated board structures or honeycomb board.

15.4.3 Honeycomb Fiberboard

Honeycomb fiberboard has two kraft faces separated by a series of honeycomb-like cells (Figure 15.21). It is thicker than corrugated board and is used to protect corners of furniture and to provide clearance between the package sidewall and the product. Similar honeycomb structures made from thin paper are used as void-fill cushioning materials. Honeycomb board is also sometimes used in corrugated pallets.

15.4.4 Corrugated Fiberboard Pallets

Although most pallets are made from wood, corrugated fiberboard can also be used to make pallets. Corrugated pallet styles are similar to wooden ones. They can have a single or double deck, stringer or block design, and two- or four-way entry. The decks are made from double- or triple-wall corrugated board, with blocks or stringers made from built-up blocks of corrugated board glued together, usually vertically, or other structures separating the decks.

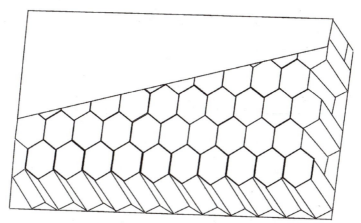

FIGURE 15.21. *Honeycomb board.*

Corrugated pallets are light weight and are used for light weight loads. They are inexpensive and disposable, used only once and then recycled along with other corrugated board. They are less durable than wood, and more vulnerable to water. They have little beam strength and cannot be placed in racks without added support.

15.5 BOX SPECIFICATIONS

Corrugated box specifications generally include the material properties of flute and combined weight of facings and burst strength or ECT, as described previously in Chapter 14. Some more progressive users simply specify box compression strength, since box compression strength depends on the box construction as well as ECT.

Specifications also include the description of the box style, such as RSC, a drawing of the box blank and its dimensions. The blank layout notes whether it shows the inside or outside of the case. It may show the direction of the flutes, although, if not specified, it is assumed that the flutes are vertical to add to compression strength. The manufacturer's joint can be shown on either side. Figure 15.22 shows a typical RSC specification.

Dimensions specified by the buyer always refer to inside measurements, unless otherwise noted. This is different from how paperboard cartons are measured from the centers of their scores. The box-maker adds score allowances to each panel and the manufacturer's joint. Most dimensions, including allowances, are rounded up to the nearest one-sixteenth inch.

As with cartons and bags, the conventional way to express dimensions is in order of $L \times W \times D$. Shipping container dimensions are based on the ge-

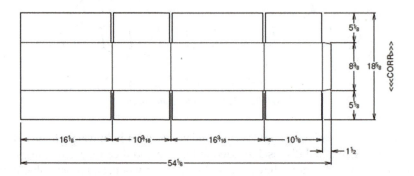

QR-014A REV. 2 10/8/2012

FIGURE 15.22. *RSC specification. Reprinted by permission of Landaal Packaging Systems, 2014.*

ometry of the contents and the desired pallet pattern. Computer programs like CAPE™ and TOPS™ can be used to optimize the dimensions of the primary consumer package, shipping container, pallet pattern, and trailer loading plan. It is important to remember that the inside dimensions are different from the outside dimensions because of the corrugated board thickness.

A tight fit for the contents is generally best for safe transport, because it makes best use of the strength of the box and the contents. A looser fit may be desirable to facilitate automated case packing equipment or when the contents cannot provide any stacking support.

When dimensioning a box, one should always physically place the contents into the prototype to verify calculated dimensions. As simple as this may sound, there have been many cases where dimensions based on sound geometry calculations proved not to fit the actual product, due to unexpected variations in product dimensions that add up to a big difference in the case size needed.

Spot colors on corrugated boxes are commonly specified according to the Glass Packaging Institute's FLEXO color guide, generally referred to as the GCMI ink standards. The samples are printed on actual kraft and white-top linerboard to show exactly how the color will look on a corrugated box (GPI 2009).

15.6 BOX COST ESTIMATION

The supply of corrugated boxes is very competitive, and prices fluctuate with market demand. It is useful for a buyer of boxes to be able to judge whether a price quote is fair in the current market.

Although there are many factors such as volume discounts that affect the price of boxes, there is an easy way to estimate the cost. It is similar to the technique used for cartons and bags, based on the size of the blank, the basis weight, and published *transacted boxboard prices*.

PPI Pulp & Paper Week gives a reference standard for current prices of 42-lb BW kraft linerboard and 26-lb BW semi-chemical medium. It gives up-charges for other basis weights, recycled, and *performance* containerboard. Prices given are per ton (2,000 lb).

There are four easy steps to estimate the cost of a standard RSC:

1. Determine the area of the blank from the inside dimensions, adding flaps and 2.5″ to $2(L + W)$ to account for the score allowances and manufacturer's joint. This is the amount of board used to make the blank.

2. From the blank size and basis weight, determine the amount of material in the two liners and the medium. The medium has a take-up factor that varies by flute (1.43 for C-flute, shown in Table 13.2).

3. Find the market cost of the containerboard materials and multiply it by the number of pounds in the blank.

4. Double the cost of material, using this as an estimator for the conversion cost of containerboard to box. Research has shown that the cost of a finished RSC, including one- to two-color flexographic printing, is generally double the cost of the material.

Example:

If a 42-26C-42 RSC measures 20″ × 15″ × 12″, what is its estimated cost if the transacted price (from *PPI Pulp & Paper Week*) is $800/ton for 42-lb linerboard and $700/ton for 26-lb medium?

1. The blank will occupy $[(2 \times 20″) + (2 \times 15″) + 2.5″] \times [(2 \times 7.5″) + 12″]$ = 1,957.5 in^2, which equals 13.6 ft^2.

2. Linerboards: 2×13.6 ft$^2 \times 42$ lb/1,000 ft$^2 = 1.142$ lb
   Medium: 1.43×13.6 ft$^2 \times 26$ lb/1,000 ft$^2 = 0.506$ lb

3. $(1.142 \times \$800/2,000$ lb$) + (0.506 \times \$700/2,000$ lb$) = \$0.634$ estimated material cost.

4. $\$0.634 \times 2 = \$1.27$ estimated cost of box

An estimation method like this can be used to judge the effect of changing carton size or basis weight. For example, if the linerboard in the example above was reduced from 42-lb to 33-lb basis weight and assuming equal cost/ton, the estimated cost would be $1.07, a reduction of 20¢/box (which multiplied over 50,000 boxes would be a significant savings of $10,000).

An estimation method like this can be used to judge whether a supplier's price quote is fair. As with the other cost estimation examples in this book, it should be noted that prices also depend on volume and the degree of decoration or design complexity.

15.7 CASE PACKING AND SEALING

Shipping containers can be packed and sealed either mechanically or manually. High-volume case-packing of ordinary product shapes is usually mechanical. Typically, a machine erects the box and seals the bottom flaps, assembles the contents in an array, drops them in, and then seals the flaps. As with cartoning, machinability is an issue in case packing. Properties of the corrugated board like stiffness, coefficient of friction, and score bend force affect its machinability.

A great many shipping containers are packed and sealed manually. Some examples are large goods (like furniture and appliances), highly variable case contents (like catalog/internet sales), and products that are manufactured on too small a scale to justify mechanization.

There are aspects of shipping container design that can make manual operations more productive, like standardizing package sizes to minimize variations and board usage. There is even new technology for making boxes on demand, from fan-folded corrugated board in a fixed width (box height) and variable footprint, which minimizes board use and empty space in boxes for internet sales fulfillment centers.

Box flaps can be glued with cold glue (PVA), hot melt adhesive, or tape. Glued flaps add more rigidity to the box than does tape. PVA glue is used for high-volume long runs. It takes longer to set than does hot melt adhesive, but the cold glue has plenty of time to set if the flaps are on the top and bottom and the box is palletized.

Tape can be plastic and pressure sensitive (PS) or made from paper with a water-activated adhesive. PS tape is the simplest closure method, especially if the operation is not automated. For tape to be effective, there must be a strong bond, and so firmly pressing it down onto the board is essential. Rough handling tests show that gluing makes the strongest closure, followed by reinforced paper tapes, and then by pressure-sensitive plastic tapes.

15.8 BEYOND MATERIAL PROPERTIES: DISTRIBUTION DYNAMICS PERFORMANCE TESTING

As packagers become increasingly accountable for the range of functions of packaging, the need to prevent damage is still an important focus. And for this, a new generation of performance test standards has been developed. The most common performance tests judge a package's ability to prevent damage from the following four forces:

- Impacts, because packages are often dropped during handling
- Vibration generated by vehicle dynamics
- Mishandling by forklift, clamptruck, or slipsheet handling
- Warehouse stacking compression

Package performance in impact, vibration, and compression tests depends not only on the properties of the materials, but also on how the package is closed and sealed, and the nature of the contents. For corrugated boxes, the performance also depends on the relative humidity, which can be thought of as a fifth force.

A significant and growing number of companies (the more sophisticated and savvy ones) even specify packages using performance test requirements, like BCT or impact resistance. Others simply use the tests to evaluate and compare alternative package constructions.

While it is beyond the scope of this book to describe and discuss these tests, Table 15.1 provides a partial list of standard test methods published by ASTM and TAPPI. ISTA standards, like Item 180, combine impact, vibration, and compression tests. An excellent review of the subject has been published by Goodwin and Young (2011).

TABLE 15.1. Shipping Container Performance Test Standards.

Property	ASTM	TAPPI	IS0	ISTA
Compression	D642	T804	12048	
Impact	D5276 (free fall) D5487 (shock machine)	T802	2248	
Shock Fragility and Cushion Testing	D3332 (fragility) D1596 (cushion curves)			
Horizontal Impact	D4003	T801		
Vibration	D999 (resonance and fixed low frequency) D4728 (random)		2247 (fixed low frequency) 8318 (variable frequency) 13355 (random)	
Impact, Vibration, and Compression Test Cycle	D4169 D7386 (single parcels)		ISTA tests Item 180	

15.8.1 Test Cycle Performance Tests

The 1990s was a period when packaging professionals began to develop test cycles that reproduce a series of potentially damage-causing forces, including impact, vibration, and compression. As shown in table 15.1, ASTM D4169 and ISTA test cycles were developed, with test intensity based on the product weight, size, and value.

Under the slogan "Do a 180," leaders in IoPP, ASTM, and ISTA (International Safe Transit Association) proposed and prevailed in adding the alternative *Item 180* to the *National Motor Freight Classification*. Unlike Item 222, Item 180 has no necessary relationship to corrugated fiberboard's material properties. In fact, a distribution package does not need to be made from corrugated board to pass the tests. It is a true package performance specification, with tests for compression, vibration, and impact of filled packages. The test methods represent development over many years and incorporate elements from ASTM and ISTA tests.

Item 180 offers a choice of stacked vibration using a top load equal to the usual superimposed load or a separate vibration and static compression test. The weight of the load is based on the number of packages that can be stacked 108" high in a trailer with an average freight density of 10 pounds/cubic feet (although the static test adds extra weight as a design factor). The package is vibration tested for 1 hour.

The impact test requirement depends on whether the package is attached to a pallet. For unpalletized packages, the drop height ranges from 10" to 24", depending on the weight; for packages under 40 lb (which is most common), the drop height is 24". They are dropped on the top, bottom, two sides, a bottom edge, and a bottom corner. For palletized loads, raised edge, incline impact, and forklift handling tests are used.

Use of package performance testing represents a great advance in the requirements of carriers. It is a much more robust evaluation method than the transportation carriers' early focus on test results from small pieces of corrugated board. It better protects carriers from damage claims. The Item 180 box stamp, shown in Figure 15.23, certifies that the package meets the test requirements.

15.8.1.1 Transport Liability and the Status of Transport "Requirements" Today

For over 100 years, U.S. transport carriers have been involved in setting shipping container standards. When corrugated board was invented in the early 1900s, transport was highly regulated. The railroads had the legal designation of *common carriers*, meaning that they were available for hire without

NAME — CITY — STATE

FIGURE 15.23. *Box certificates based on Item 180 performance test. Reprinted with permission of the National Motor Freight and Traffic Association.*

discrimination. Rates were regulated by the government and based on the *classification of freight*; shippers of high value finished goods paid a higher rate than shippers of low value raw materials.

Common carriers were liable for loss and damage at the goods' full value. Strict liability was the justification for the carriers' packaging requirements; carriers wanted to ensure that packages they carries were adequately strong. The presence and compliance with the box certificate ensured, in theory, that if the goods were damaged in transit, the common carrier would assume liability for the damage to the extent stipulated in contracts and tariffs.

The technical reasons for the change to performance oriented packaging parallel the political and supply chain changes that occurred in the late 1900s. In 1980 transportation was deregulated and in 1991 the Interstate Commerce Commission (ICC), the government authority under which the carriers could collectively act, was itself abolished.

Transport deregulation made it clear that the firms in the supply chain that owns the goods bear more responsibility for packaging sufficiency than do carriers. It was a political by-product of the trend towards increasingly well managed supply chains. Transportation, even though it is still the largest logistical cost, is merely a *third party* service function. Transport carriers never own the goods (although they do have liability for damage during transit). They only serve the supply chains that do make the investments. By deregulating it, the government forced the transport industry to be more competitive.

Transport liability law requires only that the shipper not be negligent in its packaging, not that it meets a board specification requirement. The shipper can prove this with the results of performance testing or documentation of successful past shipments.

The balance of power has shifted to manufacturers, retailers, and their supply chains. The supply chain is accountable for packaging from beginning to end, with considerations that go far beyond a simple material property or transport carrier interests. The supply chain benefits from packaging protection, utility, and communication, but wants to do so at the lowest cost. For example, the current trend to more display-ready packages encourages perforations and open tops, clearly weakening containers, but not below adequacy for the purpose, in the service of a marketing strategy.

Packaging decision makers today look beyond simple material properties and approach packaging problems from a system-wide perspective. A new coalition of package users, often through their memberships in ASTM, ISTA, IoPP, and NIPHLE, have developed test standards that evaluate the whole package and can be tailored for specific supply chains.

Where does that leave the carriers' packaging "requirements" for corrugated fiberboard burst or ECT properties? There are many cases where the corrugated board grade recommended has more material, and is higher cost, than is required to prevent damage. Since the carriers' only concern is damage prevention, particularly when the shipper bears the packaging expense, it stands to reason that carriers would prefer maximum protection, rather than weighing marginal protection improvements against cost. Furthermore, excessive packaging provides an excuse for rough handling. Some small parcel carriers go so far as to recommend doubling the Item 222 burst factors to compensate for the brutality of their handling methods.

There is no doubt that the original burst and ECT requirements were useful, with a long history of success. And there can be no mistaking the significant role that they have played in standardizing corrugated fiberboard shipping containers in the United States. The corrugated fiberboard industry successfully defined itself as tied to transport, and the carriers' rules have been a strong barrier to competition and innovation.

15.8.1.2 Hazardous Materials Shipping Container Regulations

The only truly regulated shipping containers are those used for hazardous materials. The regulations are performance-based, and are under the jurisdiction of the U.S. Department of Transportation because the whole transportation infrastructure is at risk from leaks and accidents (US CFR 49). The U.S. standards are harmonized for the most part with United Nations' recommendations, and are similar to those in other countries.

The regulations classify the materials' hazards, and require a more severe cycle of tests for packages for materials that are more hazardous. The tests include impact, vibration, compression, and some tests specific to a package failure mode, like external water spray or internal pressure. Retesting is periodically required, as are special markings designating the manufacturer and hazard class.

15.9 SUBSTITUTES FOR CORRUGATED FIBERBOARD BOXES

Corrugated boxes are the primary type of shipping containers today for five very good reasons. They are lightweight, strong, inexpensive, easy to recycle, and can be colorfully printed, compared to most alternatives like wood or plastic boxes. But there are alternatives for some products in some supply chains. Three examples are shrink and stretch-wrap bundles, reusable plastic totes, and blanket-wrap.

Shrink-wrap and stretch-wrap bundling works well for products like cans and jars that do not require the compression strength provided by a corrugated box. The LLDPE film simply keeps off the dust and holds the case together. Most shrink- or stretch-wrapped bundles also incorporate corrugated members like trays, bases, caps, or edge protection.

The choice of shrink or stretch wrap depends on the relative cost and the nature of the product. For case sized bundles, like those used for cans, jars, and roofing shingles, shrink bundling is most common. For large packages like those used for appliances and furniture, stretch wrapping is more economical because it uses less energy. In either case, the wrapping is easy to mechanize.

The advantages of shrink or stretch wrapping, compared to using corrugated boxes, are that it is usually less expensive and more adaptable (one roll fits all). The product visibility helps to reduce shipping errors and damage.

Reusable plastic containers (*RPCs*) are most successfully used in well-managed supply chains that are short in time and distance. The reason is that they are from 5 to 10 times more expensive than an expendable corrugated box. The supply chain needs to be well enough organized to either negotiate and apportion the cost for purchase, tracking, and returning a fleet of containers, or to involve third parties from whom containers can be rented. The supply chain needs to be short in time in order to keep the containers moving, and short in distance to minimize return costs. Two notable examples of supply chains that successfully use reusable shipping containers are the U.S. automobile assembly industry and the fresh produce industry supplying some grocery chains.

The advantage of RPCs is that, in select applications and when the logistics are well managed, a reusable shipping container system can reduce costs. They can be an environmentally favorable asset when the environmental burden is low (energy, materials and water required for manufacturing, transport, cleaning, and recycling). The disadvantages are that logistical costs and supply chain variations can easily increase costs beyond that of the expendable (recyclable) corrugated fiberboard box.

Blanket wrap is a form of packaging that uses moving van blankets, decking, and strapping to protect unboxed furniture when it is shipped in full truckloads. The truck is professionally loaded at the factory and unloaded by the shipper's contractors at destination, so protection is maintained throughout. However, this is not suitable for less-than-truckload shipping nor for small products.

15.10 DISTRIBUTION PACKAGING FUNCTIONS BEYOND PROTECTION

Although a primary function of shipping containers is to protect the goods, there are other shipping container requirements that solve problems and reduce cost of logistics. While details of these are beyond the scope of this book, this section provides a short introduction to the subject also known as *Packaging Logistics*. A review of the subject has been published by Twede (2009).

Shipping containers affect the cost of every logistical activity and have a significant impact on the productivity of supply chains. Transport and storage costs are directly related to the size and density of packages. Handling cost depends on unit-loading techniques. Inventory control depends on the accuracy and timeliness of manual or automatic identification systems. Customer service depends on the cost to unpack and discard shipping containers. An integrated logistics approach to packaging can yield dramatic savings.

Some examples of how packaging can reduce supply chain costs include:

- Unitization increases the number of vehicles that can be loaded/hour.
- Cube and weight minimization reduces transport costs (more items/trailer).
- Case count in order quantities improves order-picking efficiency (items picked/hour).
- Modular case dimensions and standard identification improves cross-dock efficiency (items sorted/hour).
- Automatic identification improves the accuracy and timeliness of supply chain information.

Packaging decisions often trade off packing costs against supply chain productivity. Some examples include:

- E-commerce fulfillment businesses want to use a minimum number of box sizes for hundreds of different items, but the optimum number of sizes depend on the items' dimensions as well as minimizing cube, packaging materials, and packing cost.
- Reusable shipping containers reduce packaging material cost but increases logistics cost and complexity.

• Postponing packaging in a supply chain can reduce transport and inventory costs, but increases packaging cost.

Shipping container decisions require a holistic understanding of protection, productivity, information management, and supply chain strategy, as well as understanding the packaging technology.

15.11 REFERENCES

AEFCO. 2009. Normas de Fabricación Plaform®. Madrid: Asociación Española de Fabricantes de Envases y Embalajes de Cartón Ondulado. http://www.afco.es/pdf/Folleto_Normas_de_Fabricacion_Plaform_2009.pdf, accessed 6/10/13.

ASTM (formerly American Society for Testing and Materials, now ASTM International). No date, annually updated. *Annual Book of ASTM Standards.* West Conshohocken, PA: ASTM International.

Dhairya. 2007. *Package and P.O.P Structures*, book and CD-ROM. Germany: Index Books.

Eldred, N.R. 2007. *Package Printing*. Plainview, NY, Jelmar.

FBA. 2005. *The Fibre Box Handbook*. Rolling Meadows, IL: Fibre Box Association.

FBA. 2000. The Corrugated Common Footprint, http://www.fibrebox.org/Info/CCF.aspx, accessed 6/5/13.

FEFCO. 2007. International Fibreboard Case Code, 11th ed. Belgium: FEFCO, http://www.fefco.org/sites/default/files/documents/FEFCO_ESBO_code_of_designs.pdf, accessed 5/25/13.

FEFCO. 2008. Common Footprint Standard. Belguim: FEFCO, http://www.fefco.org/technical-documents/common-footprint-standard , accessed 6/10/13.

Goodwin, D. and D. Young. 2011. *Protective Packaging for Distribution: Design and Development.* Lancaster, PA: DEStech.

GPI. 2009 FLEXO Color Guide, 10th ed. Washington, DC: Glass Packaging Institute.

Hai, J., Ed. 2009. *Packaging Templates,* book and CD-ROM. Berkeley: Gingko Press.

ICPF, International Corrugated Packaging Foundation. http://www.icpfbox.org

International Safe Transit Association. Updated periodically. ISTA Test Protocols. East Lansing, MI: ISTA. http://www.ista.org/pages/procedures/ista-protocols.php

ISO 3394. 2012. Complete, filled transport packages and unit loads - Dimensions of rigid rectangular packages, 3rd ed. Geneva: International Standards Organization

ITC. 2002. Packit Product Module: Fresh Fruit and Vegetables. Geneva: International Trade Centre.

Jonson, G.. 1999. *Corrugated Board Packaging.* 2nd edition. Leatherhead, UK: Pira.

Kirwan, M.J. 2013. Solid Board Packaging, in *Handbook of Paper and Paperboard Packaging Technology*, 2nd ed., edited by M.J. Kirwan, pp. 353-384. Chichester: John Wiley & Sons.

Maltenfort, G.G. 1988. *Corrugated Shipping Containers: An Engineering Approach.* Plainview, NY: Jelmar Publishing.

—, ed. 1989. *Performance and Evaluation of Shipping Containers.* Plainview, NY: Jelmar Publishing.

McCaughey, D.G. 1995. *Graphic Design for Corrugated Packaging.* Plainview, NY: Jelmar Publishing.

McKinlay, A.H. 1998. *Transport Packaging.* Herndon, VA: Institute of Packaging Professionals.

National Motor Freight Traffic Association. Annual. *National Motor Freight Classification* (*NMFC*). Alexandria, VA.: NMFTA.

National Railroad Freight Committee. Annual. *Uniform Freight Classification* (*UFC*). Chicago: Association of American Railroads.

Paine, F.A. and G.A. Gordon. 1973. Cushioning Properties of Corrugated Board, *International Paper Board Industry*, 16 (8):10–15.

PPI Pulp & Paper Week. Published weekly by RISI: http://www.risiinfo.com/

Ramaker, T.J. 1974. Thermal Resistance of Corrugated Fiberboard. *TAPPI*, 57 (6): 69–72.

Roth, L. and G.L. Wybenga. 2013. *The Packaging Designer's Book of Patterns*, 4th ed. New York: John Wiley and Sons.

Shao, L. and L. Hu. 2008. *Out of the Box: Ready to Use POP Packaging*, book and CD-ROM. Singapore: Page One.

Shulman, J.J. 1986. *Introduction to Flexo Folder-Gluers*. Plainview, NY: Jelmar Publishing.

TAPPI. 2000b. *How Corrugated Boxes are Made*. Atlanta: TAPPI, CD Rom.

Twede, D. 2009. Logistical/Distribution Packaging. In *The Wiley Encyclopedia of Packaging Technology*, 3rd ed., edited by Kit L. Yam, pp. 677–684. New York: John Wiley & Sons.

US CFR 49. U.S. Code of Federal Regulations, Hazardous Materials, Title 49 CFR, Parts 172-179.

15.12 REVIEW QUESTIONS

1. What style of corrugated fiberboard box is most common? Why?

2. What five operations are performed by the flexo folder-gluer?

3. Which has a higher production rate and lower cost: flat or rotary die cutting?

4. What are the limitations in printing natural kraft corrugated board?

5. What kind of printing process is most common for corrugated board? How many colors are generally printed on corrugated boxes?

6. Describe what is meant by litho-labeling and preprint.

7. Imagine a potential application for digital printing on corrugated board, and discuss why digital printing would be a good choice.

8. What is meant by an SKU code, and how are they used?

9. What information can be found on a box maker's certificate?

10. Why is there a trend to printing corrugated board with high-quality process color graphics?

11. How are the dimensions of a corrugated fiberboard box specified by the buyer? How does this differ from how a paperboard carton's dimensions are specified?

12. Are corrugated scores folded toward the groove or toward the ridge?

13. Approximately how much is added to an RSC's L and W panel dimensions for a score allowance? Why might this vary by panel? As a general rule, how much is added to the depth?

14. What circumstances make it desirable to have the manufacturer's joint on the inside versus the outside of the box?

15. What are the four ways that a manufacturer's joint can be attached? Which method is the least expensive?

16. What is the most common style of corrugated fiberboard shipping container? What are its virtues?

17. What is the advantage of die-cut boxes compared to slotted containers?

18. What are the most economical proportions of an RSC? What are the most economical dimensions for an RSC with a volume of 2 ft^3?

19. What differentiates the RSC, CSSC, OSC, and FOL corrugated box styles? Provide a translation of the acronyms and draw the blanks.

20. What are the advantages of the CSSC style? What is the chief disadvantage?

21. What is meant by a telescoping box style, and why is it favored for reuse?

22. What is the difference between a 5-panel folder and a wrap-around box? Draw the blanks.

23. What is the two chief advantages of a wrap-around box over an RSC? For what kind of products is a wrap-around box appropriate?

24. How does a Bliss Box differ from an RSC? Draw the blank.

25. What are some good applications for the self-erecting and snap-bottom styles?

26. What is a POP display? What are some design considerations?

27. What are the advantages of produce trays that conform to a 60 cm × 40 cm footprint? Where can you find standards for produce trays?

28. What is a "Gaylord?"

29. How can corrugated board be made more water and weather resistant? Where can standards for water and weather resistant board with names like V3c and W5c be found?

30. What is the organization to which military packaging professionals belong?

31. What is the primary use for solid fiberboard in the United States?

32. 32. What are slipsheets? What are their advantages over pallets? What systemic change do they require? What properties should be specified?

33. What advantages does a corrugated pallet offer, compared to a wooden one? What are the advantages of a wooden pallet compared to a corrugated one?

34. Discuss the insulation and cushioning properties of corrugated board, compared to plastic foam.

35. Draw the blank for an RSC with dimensions $15'' \times 12'' \times 10''$, labeling all dimensions as for a specification, including score allowances. Similarly, draw a blank for a CSSC with the same dimensions.

36. Estimate the cost of the boxes in the previous question, made from C-flute (tu = 1.43), 42/26/42 BW containerboards, assuming that the market price for liners is $800/ton and medium is $700/ton. Estimate the cost if the liners are a lower basis weight, 33 lb (assuming the same price/ton for containerboard).

37. What are the three most common methods for sealing the flaps on a corrugated box? Which is simpler for manual operations? Which ones give the box more rigidity?

38. What are the four primary forces and sources of damage that shipping containers aim to prevent?

39. What are the main categories of shipping container performance tests?

40. How did transport deregulation weaken the power of Rule 41 and Item 222?

41. Why do manufacturers and their supply chains have more at stake in packaging decisions than do carriers?

42. What are the tests required by Item 180? How are they different from a material test requirement?

43. What kind of shipping containers are regulated by the government? What aspects are regulated?

44. What are the most common substitutes for corrugated fiberboard shipping containers?

45. What are some functions of shipping containers that go beyond the properties of corrugated fiberboard?

46. How can shipping containers improve supply chain productivity and information system accuracy?

CHAPTER 16

The Future of Wood, Paper, and Paperboard Packaging

This book shows the evolution of wood and paper based packaging. It describes the most common forms and how they have changed from the early twentieth century to today. As the markets for consumer and industrial products have grown, so have paper- and wood-based packaging. We are heirs to a craft that continues to change.

The wood and paper-based packaging industries in North America are mature, and growth has slowed considerably due to competition from plastics. Plastics have replaced paper- and wood-based packaging for many uses. Corrugated containers have been able to hold their position mainly because most products are still shipped in corrugated boxes. They have good strength at a low cost, and their high recycling rate makes them environmentally acceptable. But the use of plastic reusable shipping containers and shrink bundles is growing. Paperboard cartons have been losing market to plastic pouches, and producers have responded by making stronger, lighter weight board, and by emphasizing its graphic versatility. Wooden pallets are still popular, but plastic pallet applications are growing.

In the worldwide growth of packaging, there are still many applications for which wood- and paper-based packaging add value. The squared-off world of distribution, filled with flat shelves and walled trailers, will continue to favor rectangular-shaped fiberboard boxes on stiff wooden pallets. The colorful billboard-like retail displays will continue to exploit the high quality printability of paper labels and paperboard cartons. For many uses, wood, paper, and paperboard will continue to be the low cost alternative. The environment will continue to benefit from the renewable source and their favorable carbon footprint compared to synthetic polymers. And consumers will continue to benefit from economical and attractive packages that have functional benefits in storage and use.

539

But a strong message from the historical pageant in Chapter 1 is that packaging forms come and go. Some have a longer lifespan than others, and some traditional forms simply morph into new ones. While this final chapter does not presume to predict the future, it does review some of the competitive forces that will shape it.

16.1 STRENGTHS

The wood and paper industries have a plentiful source of supply, environmental benefits, and technical advantages of rigidity and printability. Paper, paperboard, and corrugated board have the greatest strength-to-cost ratio of any packaging material.

Trees in North America and Scandinavia are plentiful and sustainable. In these regions, wood, paper, paperboard, and corrugated board remain low-cost packaging materials. They can also be valuable exports; North America has exported wood and fiber products since colonial times. The resource base is renewable, and with proper forest management should remain so for the foreseeable future. North America has a well-developed industry and infrastructure, so supply and the ability to export is assured.

Wood and paper are recyclable and biodegradable. They have an environmentally friendly reputation. Paper has the highest recycling rate of all packaging materials, and so the more we make, the more we are able to recycle. Printing paper is recycled into packaging too, and paperboard is a good market for old newspaper. Used paper and wood packaging can also provide good waste-to-energy fuel, or (in the long run) they biodegrade and some can even be composted, which makes them an ecological packaging material choice. This benefit is particularly strong in Europe where packaging regulations focus on increasing recycling, including waste-to-energy.

Paper and paperboard are excellent substrates for printing. They create vibrant in-store billboards to carry point of purchase communications and consumer information. After all, printing was the reason that paper and the machinery to make it were invented. From ancient Asian calligraphy to Gutenburg's press to today's folding cartons, paper has been valued for the information and graphics it can carry.

Paper and paperboard are strong, stiff, and they hold a fold. They run fast on conversion equipment. Rigid paperboard forms fit well on store and home shelves, and the flat smooth surface is a merchandising asset for branding. They can be combined with plastic films or closures to make many innovative and convenient packages that add significant value.

Corrugated boxes provide good stacking strength, which is especially required in distribution when the contents are not rigid. Corrugated board gives some cushioning and temperature protection. Corrugated box forms are adaptable in moderate-sized lots to different sizes and styles. The printability is unequaled by any other form of shipping container, a factor growing in importance for point-of-purchase display cases. On the other hand, many corrugated boxes are overly strong for a single use.

16.2 WEAKNESSES

The industry's weakness primarily stems from its commodity orientation and physical property limitations. There are also some environmental costs.

The paper and paperboard industry is less agile than the plastics industry, its main competitor. The large scale of the integrated producers makes it difficult to enter or exit the industry. The main package forms—cartons, crates, and corrugated board—are made by a mature industry.

Much of the paper-based packaging industry suffers from *marketing myopia* (Levitt 1960), in a short-sighted focus on selling paperboard, rather than developing solutions for its customers' packaging problems. It is product-oriented rather than customer-oriented.

The large scale of manufacturing, especially in containerboard, means that the supply is relatively inelastic. The industry competes on price rather than market differentiation. When demand falls, in the short run producers have to keep their mills running, and so prices can even fall below production costs. The containerboard industry has a history of suffering from a focus on mill production and commodity orientation, although there are some indications that this is changing.

The paper and paperboard industry's integration, concentration, and culture reduce the likelihood that paper is combined with other materials. The industry is mature, and there is a low level of technology changes to take advantage of emerging packaging opportunities and threats. The industry focuses on the materials, rather than on packaging systems and markets, though most of the major players have operations throughout the supply chain.

Wood, paper, and paperboard have a number of physical property limitations compared to plastics. They are hygroscopic and can weaken significantly in humid conditions without a water-resistant coating. They require glue, nails, or adhesive coatings to form a seal. Paperboard is less durable than plastic materials, and tends to buckle and delaminate under impact. Wooden pallets are vulnerable to infestation, which poses international trade problems, and wooden packages are heavy, which increases shipping costs. The biggest limitations are for food packaging, because uncoated paper is not a barrier to water vapor, oxygen, or pathogens that deteriorate food; but when it is combined with plastic, paper can even contain liquids.

Besides the environmental benefits, there are also environmental costs. Harvesting wood and making paper are energy-intensive, cause air and water pollution, and create by-products with little value. There are environmental problems with large scale deforestation, especially when old-growth forests are destroyed, and with greenhouse gas emissions. The life cycle may contribute to global warming, but less than comparable glass, metal, or synthetic plastic packaging. The impact of paper manufacture varies with region and process; countries with plentiful hydroelectric power fare better, as do integrated chemical pulp mills where power is generated from timber and pulp byproducts (though yields are lower than for energy-intensive mechanical pulp mills). Most packaging materials are produced from regenerated forestation although these "farm" forests are large and lack biodiversity.

16.3 THREATS

There are two types of threats to the wood, paper, and paperboard packaging industries: loss of traditional markets and substitution by alternative materials, especially plastics.

Consumption patterns have reduced the demand for some paper-based packaging forms. For example in the United States, successful anti-smoking efforts have reduced demand for cigarette packaging, traditionally the top use for paper-based packaging.[29] Soaps, traditionally sold in the form of powder packed in cartons and bars in paper wraps, have been largely supplanted by liquid detergent and body wash, which are sold in plastic bottles.

Furthermore, the packaging industry is dynamic and highly competitive between materials because there are substitutes. Competition will continue, and even accelerate as innovations grow in other materials.

An increasing number of packaging professionals understand how to compare substitutable materials on an objective basis. They have no loyalty to any specific material and are more amenable to alternatives than in the past when packaging suppliers had more control over their customers' packaging. The packaging user information network is developed in University programs and in professional organizations which are not dedicated to a single material form, like the *Institute of Packaging Professionals* (*IoPP*), the *Institute of Food Technologists'* (*IST*) *Food Packaging Division, ASTM*, the *International Association of Packaging Research Institutes* (*IAPRI*), the *National Institute of Packaging and Handling Engineers* (*NIPHLE*), and the *International Safe Transit Association* (*ISTA*).

Most chapters in this book describe substitutes for wood and paper-based packaging, and most of the substitutes are plastic. The use of plastic bags, pouches, shipping containers, and pallets is growing. Even books and currency can be printed on plastic sheets. Such threats from competing materials will

---

[29]A little like digging your own grave, the cartons are also forced to carry warnings against use.

continue as plastics become cheaper and easier to recycle, their properties become easier to tailor, and their print quality continues to improve to even give a more brilliant high-gloss surface than paperboard. New bio-based plastics like polylactic acid (PLA) even compete with paper on the basis of sustainability.

Newly-industrialized countries that are now developing their packaging supply industries are investing in plastic packaging. Most countries do not have the necessary forest resources for a large scale paper industry. Plastic fabrication technology is more versatile and can be operated on a smaller scale.

Exacerbating the threat from plastics is the threat from within the wood and paper-based packaging industries themselves. The focus on cost, mill production, and traditional forms has led to a lack of vision concerning ultimate consumers and the value that can be added by a paper-based package. The cost of the materials is well understood, but sometimes the value is not.

16.4 OPPORTUNITIES FOR THE FUTURE

The relatively low cost of North American forest resources will continue to make them useful for packaging. The greatest opportunities are to improve the value of wood, paper, paperboard, and corrugated fiberboard. The industry has the opportunity to move from a commodity, price-competition focus to a marketing segmentation approach. It can segment its markets to provide functional solutions that add real value to products.

A focus on value begins with consumers. This section outlines consumer, supply chain, and global value, as well as packaging technology and sustainability opportunities.

16.4.1 Consumer Value

There are opportunities to create new benefits for consumers, with value that can be measured. These include adding properties that extend product freshness, developing package features that save time, and improving the experience of consumption.

There are opportunities in understanding the needs and emotional response of consumers. Interviews and ethnography are recommended research techniques that provide detailed feedback. In ethnographic research, consumers are observed shopping, reading labels, and using packages in a natural setting. They are asked about how they interact with each package—what they expect, what they need, what they wish for, and how the packaging works in the environment where it is used.

Figure 16.1 shows some key dimensions of four emotional drivers: safety, wellness, gratification, and convenience (Peters *et al.* 2013). Packaging that appropriately enhances these dimensions can add value to a product. The fol-

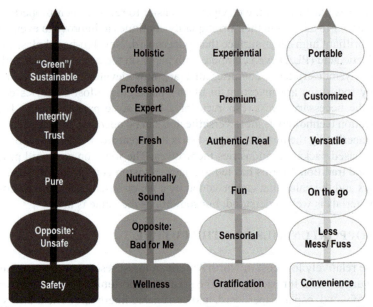

FIGURE 16.1. *Array of emotional drivers and key dimensions of packaging. Reprinted by permission of HAVI Global Solutions, from Peters, Higgins and Richmond (2013) Creating Value through Packaging, p. 105.*

lowing sections provide some examples of how the tactile and graphic capabilities of paper and paperboard provide good opportunities for emotional connection.

16.4.1.1 Gratification: Premiumization

Paper and paperboard have the opportunity to convey a premium image. Fancy gift wrap and decorated set-up boxes have long been used for this purpose. An otherwise ordinary product, dressed up in paper-based packaging, can aim for a higher priced target market. This has been such a successful strategy for the glass bottle industry in competition with plastic bottles, that beverage marketers have coined a new term for it: *premiumization.*

Now, as the use of "flimsy" plastic pouches is growing, stiff paper wraps and paperboard cartons have the opportunity to differentiate (premiumize) higher-priced product lines. For example, as the price of cigarettes has risen, soft-packs have been largely supplanted by flip-top cartons that convey a higher value.

In another example, Apple® has a strategy to give consumers a delightful experience when they open the box. The stiffness of paperboard and corrugated board are used to emphasize the products' premium quality in creative designs that reflect an almost obsessive attention to detail. The strategy aims at emotion, and its value is consumer loyalty (Lashinsky 2012).

Paperboard's high-quality printing has always been a strategic opportunity. For example, Kleenex® has a strategy for its variously-printed cartons to fit home decors and/or moods ranging from dainty to rugged. Some cookies and cereal cartons are printed with scratch-and-sniff spots on their cartons to provide a sensory experience while shopping, and others are printed with fun games to entertain during consumption.

Texture, embossing/debossing, gloss or gold, color and graphics, stiffness (or softness), thickness and weight—all of these factors can contribute to a premium image. These same factors can be manipulated to convey a variety of other tactile and optical illusions. Depending on the target market, they can convey an image that is rustic, high-tech, old-fashioned, sexy, or juvenile. The most powerful image is that of naturalness and sustainability, a natural competitive advantage compared to plastic alternatives.

Premium packaging convertors have invested in high quality printing and coating presses with special effects like metal finishing techniques, 12-color printing, "soft touch" inks, and matte and gloss varnishing. While print quality on plastics continues to improve, this illustrates the opportunity for further differentiation for premium brand markets, albeit at a high investment cost.

Another opportunity is to design e-commerce packages that deliver gratification along with the product. A shipping container that can deliver a shiver of anticipation, present the product in an attractive way, and provide a means for easy disposal or return can increase the emotional value of an internet purchase.

The superior printing ability of paper also provides an opportunity for brand-owners to better communicate with consumers through their mobile devices. The ability to "read" bar and QR codes with a smart phone enables the package to be a portal to a vast amount of information that would never fit on the package. They offer an opportunity to enhance consumption and consumer involvement with the brand.

16.4.1.2 Convenience

Opportunities to provide convenience seem to have been overlooked in paperboard packaging. There are few hassle-free easy-open features, and tuck-tabs never did work well for reclosing. But plastic packaging is worse, so much worse that consumers have coined the phrase *wrap rage* to describe "heightened levels of anger and frustration resulting from the inability to open hard-to-open packaging, particularly some heat-sealed plastic blister packs and clamshells," (Wikipedia 2014).

There are opportunities to use paperboard's properties, like tear-ability and delamination-ability, to develop packages that are easier to open. There are opportunities to develop new systems that are easier to dispense and reclose for dry products like pet food, cereal and snacks, as well as liquids.

There are opportunities to add plastic closures to paper-based packaging. For example the plastic screw-top closure adds significant value to gable-top cartons, and a plastic zipper can add value to a multiwall pet food bag.

The liner bags in cartons for dry foods are notoriously difficult to open and reclose, and so there is an opportunity to develop *linerless cartons*. This would require cooperative development by suppliers of paperboard, plastic, and machinery.

16.4.1.3 Safety and Wellness: Food Packaging

There is a demand for food packaging that contributes to healthy living, convenience, and cost-effective shopping. There is a demand for food packages that stimulate appetites, promote health, promote trust, convey variety, and save time (Smithers Pira 2013). In combination with plastic and other coatings, paper and paperboard can deliver these advantages.

A good example is the strategy of Tetrapak®, to combine strong, stiff SBS with PE in a sterile filling process to pack minimally-processed, shelf-stable milk. This packaging system has revolutionized milk distribution in countries that lack refrigeration. The secret to its success is the integration of material, machine, and process.

There are opportunities to develop new paper-based packaging systems to extend food shelf-life. Packaged food needs protection from physical, biochemical, and microbiological deterioration, as well as protection against organoleptic changes. Surface treatments to increase oxygen and water vapor barrier, water resistance, and grease resistance, as well as reducing the migration of undesirable components into the food, are an active area of research (Andersson 2008).

There are opportunities to develop paper-based packaging for minimally-processed food. In the past, the folding carton industry has dominated for dry processed food and frozen meals. Cartons' colorful graphics even stimulate appetites, which plays a key role in the shopping experience and the decision to consume a ready-meal.

The new wave of minimally-processed and "fresh prepared" food presents an opportunity for new forms of paperboard packaging to extend shelf-life, indicate freshness, cook in, and attractively present a meal. For example there is great demand for microwaveable packaging that produces a "home-cooked" quality of food, including browning and differential heating, as well as a stiff plate or bowl from which to eat it. There is also a demand for such packaging to minimize waste and/or be recyclable.

16.4.1.4 Safety and Wellness: Health and Beauty Packaging

The demand for health and beauty products, ranging from medicine to anti-wrinkle cream, is growing, along with the aging population. There are opportunities to develop package solutions that promise benefits, reassure, or comfort consumers, or remind them how and when to use a product. Paperboard, itself, has a kind of reassuring, natural image that tends to invoke trust.

An example is *compliance packaging* for a multi-dose drug regime that combines a plastic blister and a paperboard folder that indicates which pill to take at what time on each day. Another example is the proliferation of anti-aging skin-care products packed in jars or tubes that are, in turn packed in cartons to convey health and beauty messaging.

Cartons and labels can also add authentication features to packages for drugs, cosmetics, and other product categories that are vulnerable to counterfeiting. The risks of counterfeit drugs can be illness and death of consumers, as well as economic vulnerability in the supply chain. Anti-counterfeit packaging incorporates special printing effects and/or serial numbering.

16.4.2 Supply Chain Value

Future opportunities for shipping containers go far beyond simply providing stacking strength and protecting the contents. Shipping containers affect the cost of every logistical activity. Their value lies in their ability to reduce logistics costs.

As supply chains grow increasingly integrated, there is an increasing need for packages to reduce cost and seamlessly interface with logistics information, transportation, and material handling systems. There is also an opportunity to segment markets, to supply targeted solutions for specific needs.

High-speed, high-value material handling and logistics require a total supply chain visibility that is made possible by automatic identification and standardization. There is increasing demand for automated systems that are flexible and scalable, like robotic handling and high-speed truck loading/unloading. There is an opportunity to develop packaging systems that optimize throughput, for example in automated order-picking systems. There are opportunities for sensors, standard data formats, and identification protocols to enable real-time cloud-based tracking by GPS (Gue 2014).

There are opportunities for intermodal containerization at levels smaller than the current 20- or 40-foot containers but larger than pallets:

> *"Global-standard, smart, modular, designed-for-logistics containers (replacing current cartons, boxes and pallets) are needed that can accommodate raw materials and finished goods. These include unit-load, carton*

and transportation containers. These containers should be reusable and reconfigurable, providing the ability to take them apart and combine components to support a variety of sizes and shapes while supporting efficient automated handling," (Gue 2014).

There are opportunities to improve corrugated boxes for fresh fruits and vegetables. Functional opportunities include adding wet-strength to the board as well as developing systems to control product ripening or degradation. Possible examples include adding a time-release material to retard bacteria, or structures that control relative humidity.

Emerging urban logistics options include shared-use distribution facilities and dynamic routing, both of which will favor standardization of dimensions and identification. New small-package delivery systems will require lightweight, secure package options to enable unattended deliveries to unsecure locations—or even delivery by drones (Gue 2014).

In the brave new world of the *physical internet,* products are encapsulated in "physical packets" designed for efficient logistics and seamless interfaces. There will be opportunities for containers that are standardized and automated, with benefits in throughput capacity, storage density, and reduced product damage. Containers will be equipped with smart tags, unique identification, sensors, standard data formats, and identification protocols to enable real-time cloud-based tracking by GPS (Gue 2014). Increases in e-commerce small parcel distribution have been a big opportunity for shipping container manufacturers, and this creates new opportunities for new shipping container solutions. On one hand, this is a threat because reusable plastic containers are often the best choice for automated and modular systems. But the corrugated box and wooden pallet industries have an opportunity to work with system developers to develop compatible packaging systems that take advantage of the strength, low cost, and sustainability of wood and fiber-based packaging.

And at the end of the supply chain, there are opportunities for shipping containers to add value in retail stores. High-quality printing that is now possible for corrugated board offers point-of-purchase advertising opportunities. Since most shipping containers have two large sides, these have the opportunity to serve as billboards during distribution, for the brand and its colors. Since shipping containers are often seen in displays and during handling operations, this is just one more opportunity to reinforce the brand identity to consumers and material handling workers alike (Wallentin 2011).

There are opportunities throughout a supply chain for packaging improvement, but the greatest opportunity is to integrate logistics operations and marketing needs with packaging solutions that are sustainable, safe, and ergonomic. An excellent literature review has been prepared by Aziz *et al.* (2012).

16.4.2.1 Power of Cube[3]

Most of the trends described above favor standard-dimension packages that fit unit loads or automated systems. On the other hand, there are also opportunities to optimize the trade-off between standardization and cube minimization, opportunities that take advantage of the easily-variable dimensions of a corrugated box, compared with the fixed dimensions of reusable plastic shipping containers.

The greatest opportunity can be found in products that are packed in various dimensional arrays. In such cases, *a standard package is always too big for what is inside it.* The opportunity is to find cost-effective, maybe even flexible, ways to pack tightly with a minimum inventory of empty boxes. An example is fresh produce. Because fruits and vegetables vary in dimensions, standard-sized produce packages always waste space.

Another example is *fulfillment* packaging used for internet and other home-delivery sales. Typically, an internet order-fulfillment company will choose a limited number of sizes to reduce box purchasing cost in terms of volume discounts and carrying cost, as well as to run in their automated systems. But this results in wasted space and requires extra dunnage material to fill it. Furthermore, some small parcel carriers charge a *dimensional weight* premium for partly-empty boxes that waste space.

Space is a "final frontier" in shipping, because empty space wastes transportation cost (at least for products that cube out a trailer). There is an opportunity for shipping container and pallet industries to minimize cubic volume. Some companies excel at this; and for global marketers like IKEA®, it is a source of competitive advantage. IKEA's boxes and pallet patterns are designed be packed tightly with no voids, to minimize space, and thereby reduce transportation cost.

One company that found an opportunity for corrugated boxes to reduce space is Packsize®, which has developed an on-demand system for dimensioning corrugated boxes, based on a limited number of board widths. The system determines the cube of an array of products, and then makes a box to fit them. This innovative system adopted for online order fulfillment by Staples® was honored in 2013 by the Council of Supply Chain Management Professionals' annual Innovation Award for saving corrugated board (15%), avoiding 60% of void-fill, and reducing cube by 20% (PR Newswire 2013).

16.4.2.2 Reusable Shipping Containers and Pallets

Transportation deregulation and the increasing cost of disposal have caused many supply chains to pursue opportunities for reusable containers. The trends favor standardized plastic reusable packages (RPCs), which are more durable than corrugated boxes. But RPCs have a high investment cost, high reverse logistics cost, and are prone to loss (theft as well as casual repurposing).

There are opportunities for multi-use fiberboard shipping containers that are much less expensive than RPCs. These may survive fewer uses than plastic, but would be much less costly to purchase and replace.

The challenges for corrugated box reuse include designing a box to perform through more uses, developing appropriate test methods to predict the number of uses, investigating supply chain logistics options, and, most importantly, solving the problem of supply and demand for various, but specific, box sizes and graphics.

16.4.3 Global Value

The largest production of paperboard and corrugated board will continue to be in the regions with the greatest wood resources: North America and Scandinavia.

There are opportunities to export to fiber-poor countries, helping those countries' packaging industries to create markets for paper-based packaging. Paper, paperboard, and containerboard could become more valuable exports. Global marketing favors global suppliers, and multinational integration and partnerships in the paper-based packaging industry is likely to continue.

16.4.4 Packaging Technology Opportunities

There are a set of opportunities address specific technologies. These include combining paper with traditional or bio-based plastics, tailoring properties for packaging, digital printing technology, and new package shapes.

16.4.4.1 Combine with Other Materials

Rather than concentrating on vertical integration of forests and mills, there are opportunities to horizontally integrate with plastics converters and equipment manufacturers to develop packaging solutions beyond traditional materials and forms. These are opportunities to take a more *packaging system approach* rather than the industry's traditional production orientation.

There are opportunities to develop barrier, sealing, and forming technologies to use the best qualities of wood, paper, and board in combination with the best qualities of other materials. Combinations of paper, plastic film, foil, and metallization can deliver a broad range of protective properties and sealability at a minimal weight, cost, and use of resources.

In such structures, paper's contribution is its printability, strength, machinability, and stiffness. Furthermore, the porosity of paper can be used to advantage for products that benefit from some air (like crusty bread and crisp french fries), dry products in pouches that require air evacuation, and medical products that are sterilized through the package using gas.

There are opportunities to develop hybrid packages made from paper with plastic components. Spouts, zippers, and snap-caps widen the range of products that can be packed in a bag, carton, or canister. Shrink-wrapping can be stiffened with paperboard or corrugated board.

Although combination materials can reduce waste, it is not usually practical to recycle them in single-material systems. But all paper and paper/plastic laminations now offer the opportunity to be usefully recycled as waste-to-energy fuel, which could be a significant energy source in a world where half of the wood harvested is still used for fuel.

New developments in bio-based plastics like PLA may offer opportunities for plastic-paper combinations that are biodegradable or recyclable in bio-refining systems. New systems for producing nanocellulose (from cellulosic fibers) offer opportunities to strengthen paper or coat a barrier against oxygen, water vapor, or grease/oil for food packaging (Innventia 2014).

There are also opportunities for new kinds of shipping containers made at least in part of fiberboard. Freeing corrugated boxes from carriers' classification schemes has removed a major barrier to innovation. Although the RSC still predominates, there is a significant trend to alternatives like shrink-wrap (often with corrugated fiberboard trays) and styles like the Bliss Box for more efficient, low-cost structures. The strength of fiberboard, the right amount in the right place for processing and distribution, is key to these structures. There is an opportunity to work with machinery manufacturers to develop automated case-packing solutions that take strategic advantage of the strength and stiffness of fiberboard (PMMI 2008).

This is a period when many food packaging researchers are experimenting with antimicrobial additives to plastic film (e.g., Joerger 2007), or other active packaging that interacts with the product (e.g., Singh 2011), films that "breathe" (e.g., Allan-Wojtas *et al.* 2008), and compostable plastics (e.g., Ammala *et al.* 2011). But little of this research has addressed fiber-based packaging. There is an opportunity to create and test paper-based structures for some of these applications.

16.4.4.2 Think Outside of the Tree

Wood-based packaging is popular and inexpensive in countries like the United States with plenty of forest resources, where trees grow on plantations.

But paper can be made from other vegetable materials, and these can also play a major role in the production of organic products needed as fuels and feed-stock for plastics.

Examples are the residues from alcohol derived from sugar cane or beets and all sorts of oilseeds like caster, rape, sunflower, corn, soya, etc. In many instances it is possible to get two products from a single crop, for example obtaining oil or alcohol, and leaving the fibrous residue as a feedstock for paper or board.

There are opportunities to increase the use of paper feedstocks that may be more efficient and environmentally sustainable than trees, through an increase in waste paper recycling and the use of non-tree fiber resources. Packaging is one of the most significant uses for waste paper, which is a growing resource as recycling programs increase. Other forms of cellulosic fiber, such as straw, bagasse, and other agricultural residues, hold great potential, especially for making thick types of board.

16.4.4.3 Smart Packaging and Paper Electronics

There are opportunities to leverage the printability of paper to enable new identification methods, product tracking, and freshness sensors. If such enhancements can be simply printed in line onto paper or paperboard with other converting operations, they can become affordable for a wide range of products.

The emerging field of printed *paper electronics* may offer many opportunities to electrify packaging. Ink-jet and flexography can print metal oxides, organic, or hybrid materials that use paper as a substrate, dielectric, or semiconductor (Martins *et al.* 2013).

Paper has shown promise as a substrate for sensors, biological assays, RF antennas, electronic circuits, and transistors. The capillary property of fibers in paper offers potential in microfluidic devices. Researchers are even building light-emitting devices on paper and paperboard to make luminous displays, and experimenting with how to print a power source for them.

> *"Remarkable advances in materials science and simpler fabrication methods are setting the stage for a whole new breed of cheap, bendable, disposable, and perhaps even recyclable electronics. And some of the most exciting work in this field is happening with paper. . . . You'll see it first in markets where low cost—not high performance or small area—is the main consideration. . . . Paper has the potential to extend the reach of electronics into areas we might never have considered before," (Steckl 2013).*

Imagine the potential for electrified packaging: breakfast cereal cartons that wake you up, medicine bottles that transmit messages from your doctor, or a food package surface that recommends serving options when it is heated!

The low cost and high strength of paper-based packaging recommend it as an electronic substrate. Electronic paper and board offer the opportunity to further blur the distinction between product and package, to develop disposable electronic products. The paperboard industry, as the rate of gadget obsolescence increases, could be well-positioned to serve the demand for inexpensive, recyclable electronic devices.

16.4.4.4 Small-scale Demand

There is an opportunity for smaller-scale production of specialty papers, boards and conversion for the growing supply of small-scale artisan products. There can be high value in hand-made products, like soaps and foods, which is enhanced by creative packaging. But most U.S. suppliers currently operate on too large a scale for such small orders. Sources can be found in Asia, where packaging is aggressively marketed to entrepreneurs in the United States via the internet, but long-distance supply is inconvenient and risky. Most entrepreneurs would perfer to buy locally.

The paper bag and rigid box forms will continue to offer value for a small business's products that benefit from a traditional package appearance. The growing production of boutique foods and hand-crafted items creates a demand for the natural texture of wood, paper, and paperboard. Paper is easy for craftspeople to use since it is adaptable to many printing and converting methods on various scales

There are growing opportunities for digital printing, as technology is steadily improving in terms of quality and throughput. Digital printing is the fastest-growing segment of packaging, expected to more than double between 2014 and 2018, as the market for printing trends towards shorter runs (Smithers Pira 2014).

The customizable scale of digital printing opens the opportunity for small-scale conversion, especially when combined with computer-aided package manufacturing. The first packaging use of digital printing was for labels. First introduced as a sheet-fed paper-printing process, now web-fed printers, as well as machinery for printing of paperboard and corrugated board, have been developed. This technology enables new marketing strategies like *mass customization* postponement, distributed printing and personalization.

16.4.4.5 Think Outside of the Box

The carton's display panel, strength, cupboard-friendliness, and reassuring familiarity are unequaled by flexible or plastic packages. But the box form has not reached a state of total perfection and grace, and there are opportunities to improve it. For example, there are opportunities to develop cartons with better closures and barrier properties.

Furthermore, paper's ability to hold a score and fold offers opportunities to develop creative origami-like rigid shapes that go beyond the box to differentiate a product. While creative shapes are relatively easy to design, they are harder to fill and seal, and so there is an opportunity to develop automated solutions.

There are opportunities to develop new paper-based structures such as three- dimensional molded packages or thermoformable paper, as alternatives to plastic containers and cushioning materials (Innventia 2014). There is opportunity to develop semi-flexible envelope-like solutions for internet sales.

16.4.4.6 Improve Paper and Board Properties

Past paper research was mainly focused on paper for printing, but in this electronic communication age, paper and board manufacturers are turning to packaging for new markets. We need more research on properties for packaging like barrier, strength, recyclability, and compostability.

There is an opportunity to continue to do what the fiberboard industries have always done best: develop paper-based structures with a higher strength-to-weight ratio, or surfaces with better graphic capability.

The biggest technical opportunity is to overcome paper's hygroscopicity. As shown throughout this book, the history of paper based packaging has included many inventions to make paper more water resistant, from wet-strength pulp additives to water-resistant adhesives to plastic coating and laminating. Of course, this needs to be done without sacrificing repulpability for recycling.

There are opportunities to reduce corrugated box basis weight. Specifying on the basis of ECT and burst in too many cases result in overpackaging. The corrugated industry has developed a robust understanding of moisture interactions, and can reasonably predict box performance. But there is still much to learn, and so we finish this section with a quote from a pre-eminent researcher in the industry:

> *Even with our best models, our ability to predict the performance of our materials is 10 times or more behind material models for most other packaging. This arises in part from the inherent variation in using a natural product and the inhomogeneity of our product compared with metals or polymeric materials, but it also comes from the complexities and variations of the structures the industry produces. Our best models deal primarily with typical 'single-wall' RSC boxes, and there is still much to be done to improve those models let alone expand them to multiwall and small flute fiberboard, including its inherent compression strength. Significant work remains to define the influence of scoring profile and flaps and to incorporate those effects into our models to provide better guidance to the box maker.*

> *Those of us working with corrugated packaging also need to expand our understanding and modeling more robustly beyond the RSC, to more complex designs incorporating dividers, multiple components, or significant corner structure, and to better account for what happens when parts of these boxes are absent or removed. A model that helped illuminate how the different structural members on a complex design truly interact to provide load support would be invaluable, both from a predictive standpoint and from the opportunity it would provide to refine and optimize box designs. . . . Research still has to take us beyond our first-order understanding to relating the performance to a single box to boxes in a unit under various conditions (Frank 2014).*

16.4.5 Sustainable Value

One of the most important assets of wood and paper-based packaging materials is the perception that they bring additional value by being sustainable. Trees, with proper management, are a renewable resource. They take carbon dioxide out of the air, converting it to biomass (and generating oxygen for animals, including humans, to breathe).

Consumers understand that fiber-based packaging is easy to recycle, biodegradable, and strong for its light weight. It has a good image for naturalness; the natural source is easy to understand and the materials can return to nature. There is an opportunity to promote the following aspects of its ecological footprint (Wallentin 2011):

1. It burns easily in incinerators. "Every ton of paper or board that travels down the supply chain is pure bioenergy that can replace fossil oil. After recycling, the board represents energy equal to 420 litres of oil and retrieving that energy can avoid the emission of 1200 kg of fossil carbon dioxide."

2. Paper mills increasingly use their own waste/residues (bark and lignin) to generate energy, so fuel use is falling.

3. The forest industry plants at least two trees for every one they cut down.

4. Trees bind more CO_2 than they give away. During tree growth, for every 6 moles of CO_2 and H_2O captured from the environment to make one mole of glucose, 6 moles of oxygen are released in the environment. Trees will release CO_2 only during the biodegradation process ($6CO_2 + 6H_2O \rightarrow C_6H_{12}O_6 + 6O_2$).

5. Trees absorb CO_2 as they grow and then partly bind it in the ground via roots and stumps; young trees harvested for paper absorb more CO_2 than old ones.

6. The fiber that we use for packaging in the United States comes from trees in the Northern hemisphere, not rain forests.
7. Fiber can be used four to seven times before it is too short to be recycled.

There are many opportunities to continually improve the paper industry's ecological footprint. These include reducing the energy consumption, water/air/soil pollution, and waste generated during the manufacture processes, as well as increasing recycling rates. Increasing use of renewable energy (hydroelectricity, biomass combustion, solar, wind) and decreasing use of fossil fuels for recycling as well as production of virgin materials will further improve the already "green" profile of wood and paper packaging. The paper industry can help support the emergence of composting systems for food and contaminated paper, as a viable way to achieve managed biodegradation of paper packaging. The industry has already made significant strides in use of recycled fiber; these can be continued and expanded.

But there are tradeoffs. Often the best way to reduce package size and mass is to combine materials, such as the many paper/plastic combinations that have been discussed throughout this book. But, combinations of materials generally reduce end-of-life options. It is inherently more difficult to recycle packages that contain more than one material, compared to the single-material options. Many of the paper coatings that improve performance, especially by improving resistance to moisture and barrier properties, greatly decrease the biodegradability (compostability) of the resulting material. Similarly, increasing recycled content generally decreases environmental impacts, but this may not be true if the performance of the material is affected so much that the package has to be made heavier. Optimizing performance in one stage of a package's life without considering the impacts on other stages is not likely to optimize the system as a whole.

There are some useful guidelines. While certainly there are exceptions, generally using less material reduces environmental impacts (as long as the overall performance remains in the acceptable area). Often this is true even if the light-weighted material is not as recyclable as the heavier material. Using recycled materials usually reduces impacts. When designing packages that are intended to be recycled, it is imperative to avoid doing things that cause problems for the prevailing recycling processes. Similarly, if a package is intended to fit into the (slowly) growing operations for composting of municipal organic wastes, it must not contain components that would cause problems in these operations (e.g., coatings that significantly slow down the biodegradation process).

Life cycle analysis (LCA) can be a useful guide in considering the sustainability and overall environmental profile of wood and paper-based packaging systems. Often a large part of LCA's value is in identifying the parts of the process that are the major contributors to undesired impacts. These are the places where system changes are likely to have the greatest benefit in reducing impacts—the places where investment in process or product changes can have the greatest effect.

16.5 AT LAST

Paper- and wood-based packages are an economical, environmental, and functional choice for many uses. Compared to plastics, they offer the opportunity for production and consumption to be more closely connected to the ecosystem. They are produced from a plentiful renewable source. They give second life to trees that sequester carbon. And when they are finished, they know how to return to the earth. Paper-based packaging can help contribute to a sustainable economy in the twenty-first century.

What will the future hold for wood- and paper-based packaging? You, the packagers of the future, will be the ones who will decide!

16.6 REFERENCES

Abramovitz, J.N. and A.T. Mattoon. 1999. *Paper Cuts: Recovering the Paper Landscape.* Washington, DC: Worldwatch Institute, Paper #39.

Allan-Wojtas, P., C.F. Forney, L. Moyls, and D.L. Moreau. 2008. Structure and Gas Transmission Characteristics of Microperforations in Plastic Films. *Packaging Technology and Science,* 21 (4): 217–229.

Ammala, A., S. Bateman, K. Dean, E. Petinakis, P. Sangwan, S. Wong, *et al.*. 2011. *An Overview of Degradable and Biodegradable Polyolefins.* 36 (8): 1015–1049.

Andersson, C. 2008. New Ways to Enhance the Functionality of Paperboard by Surface Treatment—A Review. *Packaging Technology and Science,* 21 (6): 339–373.

Arzoumanian, M. 2003. Industry Faces New Realities, *Paperboard Packaging.* Dec., 30–31.

Azzi, A., D. Battini, A. Persona, and F. Sgarbossa. 2012. Packaging Design: General Framework and Research Agenda. *Packaging Technology and Science,* 25 (8): 435–456

Frank, B. 2014. Corrugated Box Compression—A Literature Survey. *Packaging. Technology and Science,* 27 (2): 105–128.

Gue, K., Ed. 2014. *Material Handling and Logistics U.S. Roadmap.* Jointly prepared and published by 14 industry associations. http://www.mhlroadmap.org/

Innventia. 2014. *Papermaking Towards the Future.* Stockholm: Innventia, http://www.innventia.com/en/Projects/Ongoing-projects/Papermaking-towards-the-future/

International Institute for Environment and Development. 1996. *Towards a Sustainable Paper Cycle.* Geneva: World Business Council for Sustainable Development.

Joerger, R.D. 2007. Antimicrobial Films for Food Applications: A Quantitative Analysis of their Effectiveness. *Packaging Technology and Science,* 20 (4): 231–273.

Lashinsky, A. 2012. *Inside Apple: how America's Most Admired—and Secretive—Company Really Works.* New York: Business Plus.

Levitt, T.. 1960. Marketing Myopia, *Harvard Business Review,* 38 (July-August): 24–47.

Martins, R., L. Pereira, and E. Fortunato. 2013. 29.4: Invited Paper: Paper Electronics: A Challenge for the Future. *SID Symposium Digest of Technical Papers,* 44 (1): 365–367.

McDonough, W and M. Brangart. 2002. *Cradle to Cradle; Remaking the Way We Make Things.* New York: North Point Press.

Milogrom, J. and A. Brody. 1974. *Packaging in Perspective.* Cambridge, MA: Arthur D. Little.

Peters, J., B. Higgins, and M. Richmond. 2013. *Creating Value through Packaging: Unlocking a New Business and Management Strategy.* Lancaster, PA: DEStech.

PMMI. 2008. *Secondary Packaging Market Research Study; Findings of a Market Research Study that Assesses the Current Use and Trends of Secondary/Transport Packaging.* Reston, VA: Packaging Machinery Manufacturers Institute. http://www.pmmi.org/Research/content.cfm?ItemNumber=907

PR Newswire. 2013. Staples and Packsize Win 2013 Supply Chain Innovation Award. PR Newswire, Nov 07. http://www.prnewswire.com/news-releases/staples-and-packsize-win-2013-supply-chain-innovation-award-231016181.htm

Selke, S.E.M. 1990. *Packaging and the Environment.* Lancaster, PA: Technomic.

Singh, P., A.W. Ali, and S. Saengerlaub. 2011. Active Packaging of Food Products: Recent Trends. *Nutrition and Food Science,* 41 (4): 249–260.

Smithers Pira. 2013. *The Brand Owner Trends Report: Consumer Packaging to 2018.* Leatherhead, UK: Smithers Pira.

Smithers Pira. 2014. *The Future of Digital Printing to 2018.* Leatherhead, UK: Smithers Pira.

Smithers-Pira. 2014. *The Future of Corrugated Board Packaging to 2019.* Leatherhead, UK: Smithers-Pira

Steckl, A.J. 2013. Electronics on Paper: Paper Electronics Could Pave the Way to a New Generation of Cheap, Flexible Gadgets. IIEE Spectrum, January 29. http://spectrum.ieee.org/semiconductors/materials/electronics-on-paper

Wallentin, L.G. 2011. *Packaging Sense.* Göteborg, Sweden: Lars Wallentin, http://packaging-sense.com/

Wikipedia. 2014. "Wrap Rage." http://en.wikipedia.org/wiki/Wrap_rage, accessed 4/15/14.

16.7 QUESTIONS FOR THE FUTURE

1. Assume a scenario in which the paper industry does not change its packaging products or production methods over the next 50 years. Present an argument supporting which segments of the market might first fall victim to competition from plastic, and which might remain strong. How could the paper industry defend against these threats?

2. Imagine a paper-based packaging form that has not been discussed in this book. What would need to happen to implement this creative package?

3. What are some ways that wood and paper based packaging materials and their production methods might change in the future?

4. Imagine 10 specific features that could be added to a paperboard carton and/or paper bag to add value for a consumer. How much more do you think that someone would be willing to pay for each?

5. From your current viewpoint, imagine and describe a job that you might enjoy, a job which deals with paper-based packaging. What will be the most useful things from this text that you will need to know for that particular job?